# Horsedrawn Plows
## and plowing

# Horsedrawn Plows

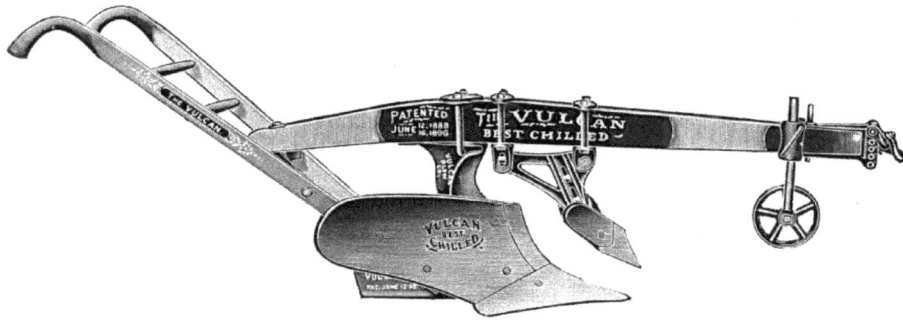

# &

# plowing

## L. R. Miller

**Horsedrawn Plows and plowing / Revised Edition**
Lynn R. Miller
Copyright © 2018 Lynn R. Miller

Publisher      Davila Art & Books
                PO Box 1627
                215 N Cedar St.
                Sisters, Oregon 97759
                (541) 549-2064

Distributed through www.smallfarmersjournal.com
or www.davilabooks.com
Author information at www.lynnrmiller.com

First printing First Edition 2000
First release Revised Edition 2018

ISBN 978-1-885210-24-5

Also by Lynn R. Miller:

*The Work Horse Handbook*
*Art of Working Horses*
*Training Workhorses / Training Teamsters*
*Haying with Horses*
*Horsedrawn Tillage Tools*
*Horsedrawn Mower Book*
*Starting Your Farm*
*Old Man Farming* (essays)
*Farmer Pirates & Dancing Cows* (essays)
*Why Farm: Selected Essays and Editorials*
*Ten Acres Enough: A Farm for Free* (pending)
*Thought Small: Poems, Prayers, Drawings & Postings*
*The Glass Horse* (novel)
*Talking Man* (novel)
*Elastic Signature: Notes on Painting*
*Horses at Work* (out-of-print)
*The Complete Barn Book* (out-of-print)

*On the front cover: Top, Bucher & Gibbs Hillside rollover walking plow circa 1900. Center, photo by Kristi Gilman-Miller of L.R. Miller resting with Cali, Lana and Barney hooked to a Pioneer 16 inch walking plow. Bottom picture, Moline Good Enough No. 3 Sulky Plow in factory colors.*

*Back cover: All images in factory colors. Top left, Moline Best Ever Sulky Plow. Top Right, Moline's Little Dutch Sulky plow. Next row, C Series Moline Walking Plow. Next Row. Le Roy Wood Beam Chilled Walking Plow. Bottom Right, Moline's Rotary Dutchman Disc plow.*

Dedication:
For my mentor,
farming partner
and dear departed friend

Ray Drongesen

*Ray, after you finish reading this let's go for lunch at Aggie Lou's and then hook a colt or two to my Oliver gang. I miss your cigar smell and your Nebraska Dane smile. What a lasting and wonderful beginning you gave me. I miss you.*

# Horsedrawn Plows & Plowing

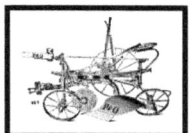

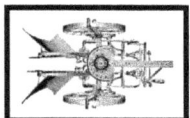

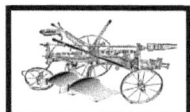

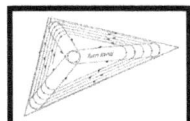

## Preface

That primal image of plowing - a sweet and vital contradiction - for me it has always been the key, the cardinal essence. I've never felt the call to pay much attention to views of tractors tearing and flipping the earth. Instead I've often felt yearning for that first picture - a gentle vista of two or three horses, a walking plow and a man moving with poetic purpose along the edge of a land of perfect, steaming, crumbling earth. And the contradic-

tion came in the unlikely, yet balanced combination of powerful, careful animals wed to an oddly shaped *tuning fork* or *short bent butter-knife* of a tool; and that followed by an old man seemingly fueled by an invisible yet casual vigor. Witness the magic. Pure magic. How else to explain it? The horses. No matter the training, they had to be so close to unpredictable, feral, mannerless, explosiveness. Didn't they? After all they are tethered animals borrowed away from their preferences. The plow. Wood and/or steel does not explain away the mirage of frailty of that deep and sorely tested crooked throat. And the man? A crippled slave to memories and labor-accelerated aging, yet from somewhere emanates the balanced effortless strength in his gentle, following touch to plow and team. Magic. To the arrogant eye a magic which seemed just maybe 'no big deal'. The heart raced by the logic to spit 'I can do that'. The heart raced and the gut said, 'I HAVE to do this.' Deep bit by the bug of that primal plowing.

Thirty years later when I first presumed with pained vanity to put together this book and share a love affair with a process, I could say, with experience, that it was and is magic. Now beyond, at forty-five years later, and with this new reissue, I know it remains the magic of a certain balance married to a specific process. As magic it is still illusive. Yet, for those who would strain to follow their heart, it can be learned.

Many slight of hand tricks lose their magic (or attraction) for us when we learn the secrets. Not so with plowing, least ways not for me. The work I have done with *Small Farmer's Journal* often had me traveling to work horse events where I regularly witnessed horse plowing of the highest caliber. And I am still drawn to the image. Now in old age and with a measure of plowing time, I find I am equally drawn to the master and the novice. The master because I never tire of reveling in the uncountable subtleties each different human and team combination brings to this poetic dynamic form. The novice because I want to cautiously reach out and adjust the setup, relax the teamster, lengthen the lines.

*A view of an old English plowman*

I do not come to this writing assignment as a master plowman. I come as a meddler, and an entire cheering section. I come as one who loves the practice, strives to be better at it, and wishes to share simple mechanical mysteries and the joy of the process.

I have no idea whether others of you will feel the same attraction to plowing with horses or mules. But that does not stop me. The only way you might stop me is to walk away and even then I'll continue mumbling about adjustable beauties.

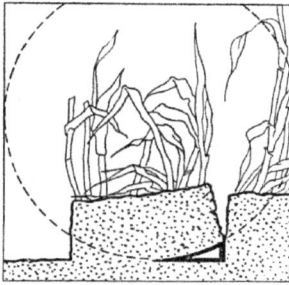

*the 'cone zone' of the approaching moldboard*

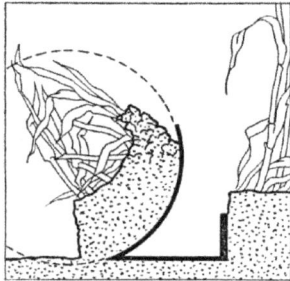

*the lift and turn...*

When I first started plowing with horses regularly it was because I had to. It was to get a field ready to plant vegetables and another ready for grain and I had no tractor. My first using plow was a loaner from my dear friend and mentor, Ray Drongesen. The plow was an Oliver 23B two-way riding plow which had one beam removed altogether. This modification was done for interesting competitive reasons. The Oliver 2-way has two wheels directly in line rather than the customary three offset wheels found on the most popular foot-lift framed sulkies. Ray was an ardent competitor at draft horse plowing matches in riding plow divisions. He had discovered that for the opening crown (or back) furrows, and the clean-up dead furrow, his modified plow was absolutely superior. It functioned essentially as a walking plow with following, non-steering, wheels and a convenient seat.

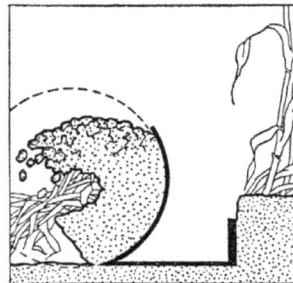

*the fold of the trash and the crumbling...*

With my horses, Goldie and Queenie and Bud and Dick, I plowed. Day after day after day, I plowed. And each day Ray or Charlie Jensen or someone else would step in and adjust something, change something. I was dangerously arrogant, in truth I knew little or nothing of the craft. I began to learn best in moments when I was alone and something wasn't working right. Why? I would ask. Why, did it work okay yesterday and now it doesn't? What needs changing?

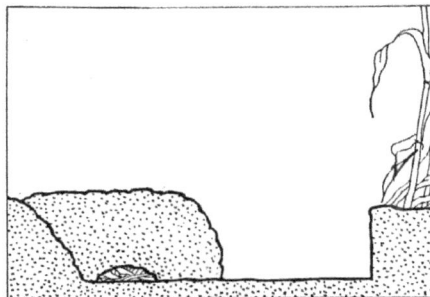

*the finished furrow.*

Now, we jump ahead to today, over the top of many miles of furrows.

This is a book which needed to be done. Not because anyone felt it had a market, least of all this author. This book needed to be done because in the doing a vast scattered wealth of knowledge and craftsmanship would, once and hopefully for all, reside in an accessible lump. We refer to this book as a lump because it is not the result of a traditional creative process. This book is a gathering of all the information I could find at the time. There will be a few who will be disappointed to discover that their prize plow is not represented here. We are so exhausted and pleased to have completed our end of this massive task that we feel no need to apologize for any such omission. For we feel we have accomplished our goal. Our goal was to preserve - through collection, catalog and publication - the mysterious and wonderful craft of horsedrawn plows and plowing.

By presenting the measurements, the formulae, the undergarments of this wonderful procedure we have, we know, given many people access and an opportunity to learn the craft. This presentation we see as prelude to a planting process which is central to preservation. It has become critical in this dawning of the age of synthetics and artificiality to protect and preserve that which is parent. To narrow the analogy, we now know it is important to preserve parent plant varieties. We do this, in part but no less directly, by saving the seeds of old varieties. But, to jar-up, tag, and shelve old seeds, though important, is in and of itself not enough. "The seeds need to be successfully planted", the plants they create grown and their seeds gathered if we are to save the variety. The same is true with valuable 'relic' technologies and methods such as plowing with horses.

Imagine a time in the far future that there is a need to prepare areas to plant crops because people are hungry. Imagine the clever young

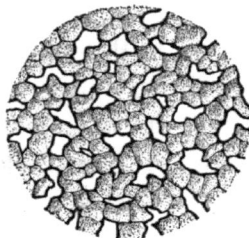

*A microscopic view of what a fine seedbed's soil particulate should look like.*

person standing with draft animals he or she has succeeded in befriending and harnessing. Imagine, because petroleum fuels are available only to despots and corporate heads, that they must reinvent the process of turning and stirring the soil by hand and muscle. They'll do it because they have to. It is my contention that something as simple as this lump of a book will help and it will also go a long ways towards historical and cultural continuity. It is our hope that this book hollers out, "hook your horses and go plow!" This book is a tagging, packaging, and shelving process - just like that seed saving analogy. But it must go further. It must work to assist the "planting" process because if plowing with horses is to be preserved as a craft people will need to actually do it. To this end we feel we may have succeeded because this big old lump of a book does seem to cause an itch in anyone attracted to moist, soft ,waiting seedbeds.

We do hope that masters of the craft of plowing with horses will welcome this book. It is natural for those who enjoy a capacity with specific craft-based mysteries to want to "keep the secrets" lest their cultural power and position fall away. We are certain that "experts" will rush to "correct" and "criticize" this volume. And we're happy in that certainty because it will be additional proof that we've succeeded in helping to keep the subject alive. Stripping away some of the "mystery" of plowing with horses will only deepen the deserved respect we have of those "experts".

For the uninitiated who happen upon this volume, we wrap up this preface to this second edition by noting we are well into the twenty-first century, and **yes** we are advocating animal-powered agricultural enterprise. We know we are not alone in our steadfast belief that for some, maybe many, animal-powered family-scaled farming is a viable and important option.

Now, let's go plowing! LRM

## Chapter One
## Introduction
# Basic Principles of Plowing - Function

The plow is an age old soil preparation tool. By design it is meant to be drawn down into and against the soil a short depth and to lay that soil over or stand it up disrupted.

There have been many actual and so called advances in most farming implements and machinery but the common plow has not been essentially improved or modified in any important particular, except as regards to mechanical construction, since the days of the early Greeks and Romans.

In what might seem like a minor contradiction to what has just been stated, it may be observed that modern plows are well designed to cause increased pulverizing of the soil during the turning action. The total modern plowing action ideally aerates the soil and incorporates the surface vegetation and assorted organic matter into a deeper layer.*

The first plows were nothing more than sharp sticks pointed and pushed through the soil disturbing only to a depth of two or three inches.

The common plow of the last two centuries is essentially a wedge-shaped instrument which is forced through the soil to loosen it. The topsoil is forced aside thrown up or turned over. This action separates the soil particles in a loosening action. The loose soil occupies more space than the compact soil allowing passage of water. Water will pass down into the loosened soil, however, there is less "capillary" action to draw water back up. Rendering plowed ground to a finer particle size, by further field tillage, allows for finer passageways between particles and hence greater capillary action.

If you wish to turn and stir your soil with depth and vigor the plow is the tool you need. For the power required, it is the most efficient of all tillage implements.

**Essential parts of the plow:** The plow, as it is drawn down into and through, takes up a ribbon of soil, lifts and more or less inverts that ribbon with a twisting motion. This is accomplished with the three wedges of which it is composed and of which two have flat surfaces. These are the bottom of the plow and the landside. The third wedge has a curved surface which twists the ribbon of soil or

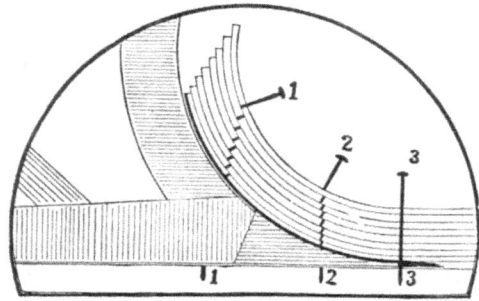

Fig. 3. Illustrating the pulverizing action of the moldboard

furrow so that it is partially or wholly inverted. This is made up of the moldboard and of the cutting edge attached to it, the share.

The critical thing about the plow is the way the angles of these various wedge faces work together. Do they lift and turn the furrow acutely, or at a flat angle?

Figure 1. With horses unhooked mid-furrow we are able to illustrate, dramatically, what is meant when we say plowing is the slicing of **ribbons** of top soil. This photo demonstrates how the twisting action of the moldboards pulverize the soil and lay over the sod or crop residues. The look of the surface of these furrows suggests that this ground should be harrowed the same day it is plowed to avoid a hardening off.

Does the moldboard have a short, complete twist or a long, partial twist? Is the furrow of earth cut entirely free from the soil below or only partially so that it must be broken off in turning? Do the bearing or friction surfaces polish or scour easily, wear well and avoid clogging? Are those parts which are subject to the most wear easily replaced? How is the weight and thrust of the plow carried? Does that weight and thrust unduly pack the soil? How are all these things related to the draft and handling of the plow?

**Action of the plow on the soil.** As the plow advances through the soil the furrow that is taken up

Figure 2. Long furrows of perfectly 'loosened' soil.

*(If properly used which includes questions of timing, relationships of the history of a field's plowed depths, intelligent address of the materials on the top of the soil before plowing, the best use of attachments and timely followup tillage.)

*Figures 4 & 5. Six Belgian horses draw a Pioneer two bottom plow during the 1998 Horse Progress Days in Mt. Hope, Ohio. Front view and rear view. Note the excellent loose furrows.*

# Descendants of the Modern Plow

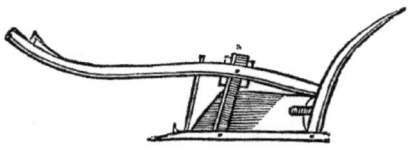

*Fig. 7. The Dray Plough of 1708.*

*Fig. 11. The Herefordshire Wheel Plough (circa 1700).*

*Fig. 13. Four Coulter (knife) Plough (1750).*

*Fig. 8. The Sussex Plough of 1700.*

*Fig. 12. Common two wheeled Plough of Jethro Tull's time (1750).*

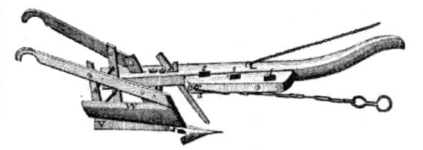

*Fig. 14. Opposite view of Four Coulter Plough.*

*Fig. 9. The Lincolnshire Plough (1700).*

Thousands of variations in basic plow (or 'plough', as it's said in England) design were seen all over the world. The British isles were a hotbed of innovation during the eighteenth century with many unique farming theories being tested, the most important of which were the brainchild of Jethro Tull.

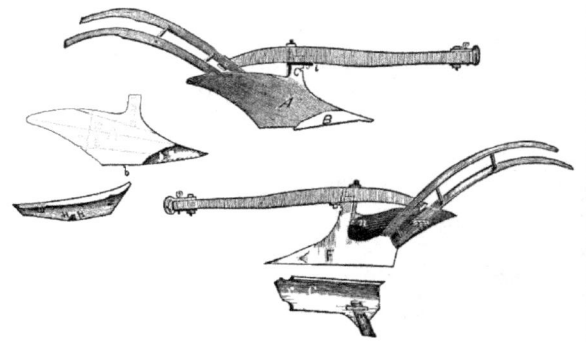

*Fig. 15. Felt by some to be the first modern cast iron plow, designed in New York, 1819, by Jethro Wood.*

*Fig. 10. The Cambridgeshire Plough (1700).*

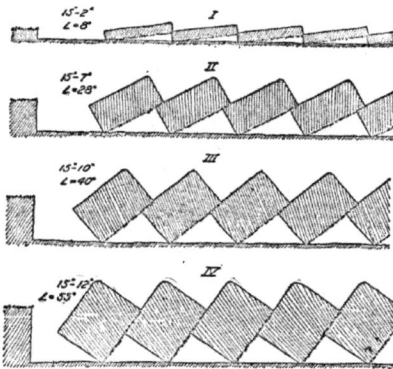

Fig. 6. The relation of width to height of furrow can have an important bearing on the effectiveness of plowing. Of those illustrated II and III are frequently thought the most satisfactory. However, the inventive and intelligent farmer will find excellent uses for the 'skim' plowing illustrated in I and the deep plowing of IV.

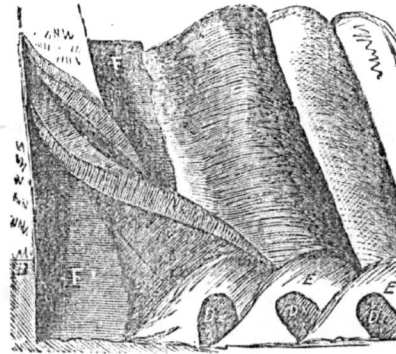

Fig. 17. With the use of a 'jointer' or 'skimmer' (a small plow mounted ahead of the primary plow) sod and/or a top dressing of manure may be curled into a filling at the center of the furrow (D). There are mixed opinions on the value of this practise. It used to be held that this allowed a fine working of the top layer of soil (E) without dragging up clods, roots and debris. And it was felt that the decaying sod or manure was at a good location for slow release plant food. Some modern thought holds that this curl of buried material heats, forms gas and is too slow to decay.

is cut free from both the land and the bottom, and slides up the curved moldboard, which twists the furrow, resulting in the very complete fracture of the soil and, in that way, pulverizing or powdering the soil. The soil tends to be divided into a series of layers in both the horizontal and the vertical directions that slide over each other like the leaves of a book when they are bent (see fig. 3). This is an imperfect example because the curvature is not the same in different parts of the moldboard and consequently the pulverizing action is not the same at different positions on the plow surface. At any rate, if the soil is in proper moisture condition for tillage - that mellow, moist condition when the soil breaks freely but shows no shining surfaces when it slides on the metal - it is thoroughly broken into fine crumbs that fall from the moldboard in a loose, mellow mass. If the ground is in sod the soil is pulverized just the same but the crumbs or granules are held together by the rootlets and the furrow to some extent retains its form. If the soil is too wet it puddles*; if too dry, it breaks up in large lumps.

Again using the example of the leaves of a book, it may be seen that the deeper the furrow is turned, just as the larger the number of leaves bent, the greater is the sliding of the soil granules. That's to say, deep plowing pulverizes the soil more than shallow plowing. In the same way a wide furrow is pulverized more than a narrow one. Consequently, the efficiency of plowing - other things being equal - is in favor of a deep wide furrow, or large furrows.

**Types of plows in relation to the soil**.
Although the modern moldboard plow has a design that is, for the most part, universal in its suitability for

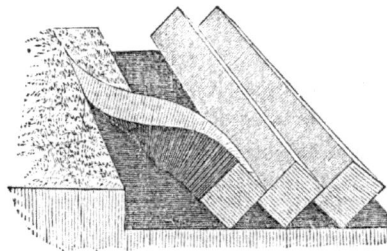

Fig. 16.  A simplified diagram of how the soil in plowing is ideally cut in a ribbon (furrow), which is twisted, and then lapped upon the former furrow or ribbon.

different soil conditions (given a range of appropriate attachments i.e. coulters, jointers, knives etc.) not so long ago specialized moldboards were designed for certain conditions. At the turn of the century it was felt that different soil conditions required different types of plows, the difference being in the shape of the cutting and twisting surfaces - the share and the moldboard. For sandy and rather dry soil a short point and a rapid rise on the share, together with a short and rather abrupt twist to the moldboard, were considered desirable. On clay soil, also a little dry, much the same form was recommended, but with a longer and more slender point and a little flatter rise on the share. Both these forms give the vigorous pulverization that was sought, with a minimum of draft.

On the other hand, on land bearing a heavy sod, a still longer point and a longer and more gradual turn of the moldboard were felt to be required. This form cuts the roots cleanly as they are drawn taut by the advanced point and the easy turn of the moldboard lays the furrow over in a continuous line without breaking it.

The exact type of older style plow that is best for each soil cannot always be predicted in advance because of peculiarities not fully understood. It must be found by trial. In the dark, red limestone soils of Cuba, for example the modern moldboard plows have not been made to work successfully. They refuse to scour and fill up with earth so that they act merely as cultivators and do not invert the soil but only stir it.

If a soil is plowed when too wet it is puddled* by the same action that would otherwise pulverize it, and dries down into a bricklike mass. If plowing at just that time is necessary, the later fitting (harrowing, discing, etc.) should be done as soon as the soil has dried down to the point where puddling will

* Puddling: Very fine-grained soils (i.e. clay) have a tendency, because of the close contact of particles, to move in upon themselves at all times. When they become very wet this tendency is increased. If stirred at the proper stage, as the excess of moisture is leaving, this tendency is easily overcome. If, however, this cultivation is done before a sufficient amount of water has been removed, all parts of the soil subjected to the pressure of the horse's feet and the tools used will be compressed still more, excluding the air and reducing the capillary capacity of the soil. These portions, when dry, form hard clods, and this cementing action in soils is referred to as "puddling". Take any clay soil in hand when wet and work it, compressing out the air, and the result will be 'brick'. Puddling or 'bricking' can be disastrous to the soil for several years.

Figure 18. Two teams, two walking plows, two men. Some might see inefficiency where others of us would see craftsmanship. Some might see drudgery where others of us would see soul-stirring hard work filled with accomplishment and a possibility for commraderie.

not occur and before thorough drying had occurred.

**Depth of Plowing.** There is no exact depth for all plowing. The depth varies with soils, crops and conditions. Light open soils that do not need to be loosened are better plowed very shallow unless some other factor, such as the working in of organic matter, enters into the scene. Any compact soil, such as clay, needs to be plowed as deeply as possible to get the benefit of being loosened up and better ventilated. Well-drained soils can safely be plowed deeper than poorly drained soils. (Applicable to bare land as well.)

Most crops, including corn, potatoes, beans and many vegetables, prefer a mellow but rather compact seedbed. For these, deep plowing is best. Wheat and oats on the other hand do best in a very compact seedbed; plowing, especially deep plowing shortly before seeding, frequently reduces their yield.

Still another consideration is the use of organic matter and the general maintenance of the productive capacity of the soil. The old-time standard recommendation (call it the conventional wisdom of our fathers) is that organic matter - manure, stubble, sod, weeds, and green manure crops - be plowed under relatively deep. Since the use of these materials **should be** practised on most farms it would appear this is the same as recommending that deep plowing be practised under nearly all conditions. *(However the good farmer is ever mindful of the incredible diversity of conditions, limitations and opportunities which surround that which we call "soil condition" and the good farmer is open to experimentation with all practises most particularly those so simply adjusted for as in the*

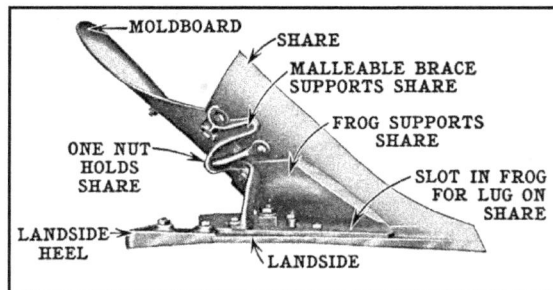

Fig. 19. Plow bottom (seen from underneath) with the parts named.

*case of plowing depth.)*

**What is deep plowing?** It is common practise to plow from 4 to 6 inches deep but this may be considered shallow plowing. For the sake of our discussion and certainly recognized in many farming circles, six to eight inches is a medium depth and anything more than 8 inches is deep plowing. The limit for depth of plowing is indefinite and determined largely by the power available to pull that plow, the cost of the operation and the supply of organic matter. If these are adequate the plowing could even be several feet deep as was the case in Hawaiian sugar-cane plantations (where 2 to 3 feet was not uncommon.)

All that said we need to state some cases against deep plowing.

1. Most subsoils are relatively barren of organic matter and corresponding acids. That is they will not at once grow crops if exposed at the surface. They must be weathered and incorported with organic matter. Therefore, if the plowing of any upland soil has been shallow, the operation of deeper plowing should be practised gradually by increasing the depth only an inch or so in any one season.

2. The subsoil is generally lacking in organic matter. A balance must therefore be maintained between the depth plowed and the amount of the organic matter turned under. As the supply of organic matter is increased, the depth of plowing may be increased.

3. The depth of plowing must be adjusted to the width of the furrow. Usually the furrow slice should not be turned over flat but should be set somewhat on edge. It should rest at an angle of 30 to 40 degrees with the

# Modern Plow Bottom Styles

Fig. 20. General purpose bottom. A gentle slow-turning moldboard. Highly adaptable to diversified farms. Good for clover, timothy, and alfalfa sod. Adaptable to heavy loam and gumbo soils if not too wet. Not suited for sticky, waxy or overly wet soils.

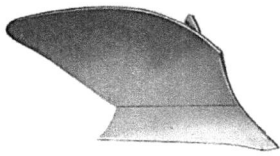

Fig. 21. Stubble Bottom. Shorter and more abruptly curved. Turns soil quickly, covers trash well. Use only on old grain lands, or ground farmed with regularity. On old established sod it doesn't work well. If you plow fast with this bottom it will throw chunks and furrows will roll back on you.

Fig. 22, Blackland Bottom. For gummy, stiff soils. Designed to be pulled faster than usual and behind tractors. Less surface, steep. Specialized.

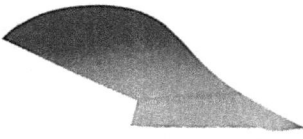

Fig. 23. High Speed General Purpose. Narrow waisted, longer, sloping moldboard. Designed for high speed tractor plowing in most soil types. Not recommended for high- quality plowing.

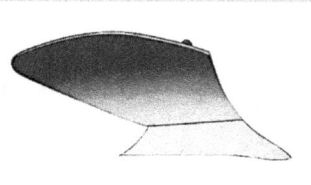

Fig. 24. Ash Bottom. Specially designed for plowing soils which tend to fall back and not turn over. Not for sod. Works well in heavy clay soils.

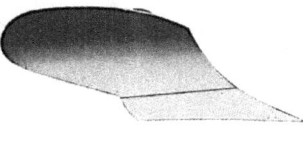

Fig. 25. Slat Bottom. For sticky soils that do not hold together sufficiently to apply enough pressure against the moldboard. Design reduces friction. Use on heavy gumbo, clay and black waxy soils. If scouring is a problem this may be the bottom to try.

Fig. 26. Chilled bottom. For sandy, gravelly and gritty soils. Flint rock textures and clay soils can be handled by this bottom. The chill of the metal is important for the extra resistance to wear required of abrasive soils. Affords excellent aeration for sandy soils.

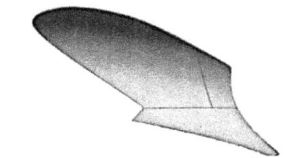

surface of the land. By this arrangement the organic matter is better distributed through the soil and there is a good layer of free soil at the surface to be formed into a seed bed; moisture can move into the soil along the face of the furrows and return upward through the edge of the furrow; and roots can penetrate more easily than where the soil is completely inverted. To get this position of the furrow it should be approximately twice as wide as deep. If too narrow, the furrow will be set too steep; if too wide, it will be set too flat. A furrow 10 inches deep should be 16 to 20 inches wide. To increase the width of the furrow in proportion to its depth requires an increase in the size of the plow and of the draft power.

**Draft of the plow.** (By 'draft' we mean the resistance offered when pulled forward.)  In one older farm text the draft of the moldboard plow was calculated as follows:

*(1) The draft per square inch cross-section of furrow cut increases as the width increases.*

*(2) The draft per square inch cross-section of the furrow cut decreases as the depth increases.*

*(3) The sharper the curve of the moldboard, the greater the draft of the plow.*

*(4) When the soil is either too wet or too dry, the draft is increased.*

*The draft of the plow is ordinarily 5 to 9 pounds per square inch of cross-section for the sharp-curved or fallow-ground plow and from 4 to 8 pounds for the long-curved or sod plow. For a furrow 6 inches deep and 14 inches wide, this would be about 600 pounds in the first case and 500 in the second. For a furrow 9 inches deep and 16 inches wide, it would be about 700 to 900 pounds respectively.*

In Lyberty Hyde Bailey's turn of the century <u>'Cyclopedia of American Agriculture'</u> he gives this information on draft of plows.

*The draft of plows under ordinary field conditions ranges from 400 to 500 pounds. Sometimes in preparing oat stubble for wheat it will run as high as 600 pounds. The conditions of soil, both as to texture and moisture content, affect the draft materially, but under general conditions it may be divided as follows: Twelve per cent of the draft is due to the work of turning and pulverizing the furrow-slice; about 33 per cent is due to the friction of the pull on the sole and landside; and 55 per cent is consumed by cutting and lifting the furrow-slice. The sulky plow embodies a feature that reduces the draft somewhat in proportion to the amount and kind of work done. The presence of wheels, and especially of the third wheel, which works in the furrow, makes it possible to reduce the length and width of the landside without reducing the ease of working the plow. This, in fact, is one of the important features in the adaptability of the sulky plow.*

*The jointer, or skim-plow, as an attachment, is one of the most important improvements that has been made on the plow. Its intelligent use renders a much broader utility of the plow than would be possible without it. Tenacious sod can be worked satisfactorily if the jointer is set deep enough to break the rigidity of that part of the furrow-slice that remains uppermost. On the other hand, stubble and cultivated lands are rendered more* friable by its judicious use. When properly adjusted, it lessens draft and prevent s the furrow-slice from being turned over too flat. Thus, soil warms up more rapidly in the spring, has a greater moisture-holding capacity, and is left in such condition that subsequent tillage by harrow or drag is facilitated. The jointer also aids materially in turning under manure or rubbish that may be on the surface of the soil.

The disc-coulter, while serviceable in clean sod lands, is not adapted to lands covered with stones or any material that cannot readily be cut. In tenacious sods the disc-coulter lessens the draft somewhat by cutting away the furrow-slice from the land, thus reducing the amount of work required in lifting and turning the furrow-slice.

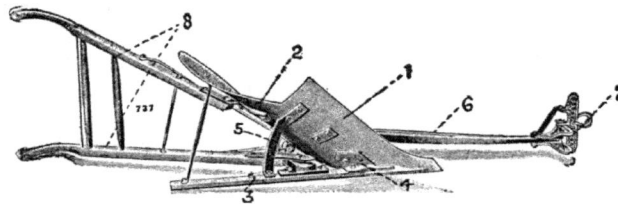

Fig. 27. Bottom view of a walking plow. 1. share; 2, moldboard; 3, landside; 4 frog; 5, brace; 6, beam; 7, clevis; 8, handle.

Other features of the adaptation of the plow are found in the length and set of the handles. This effect is manifested more by the ease of handling the plow than on the work which it does. In the best plows the handles are about the same length as the beam and are set with the left handle slightly over the land. With this proportion and set, the handles will be at an angle of about 30 degrees to the direction of the beam. A very slight variance from what is correct in this position will be readily detected by the plowman after a day's work. The straight line drawn from the hames of the harness and passing through the point of attachment at the bridle of the beam, should pass also through the center of draft of the plow, which is located behind the moldboard and about 2-1/2 to 3 inches from the side and bottom of the furrow. If this straight line is deflected downwards at the bridle of the plow, perhaps a slightly better effect is produced, as this does not make it necessary to use a pitch point on the plow. The pitch point (also referred to as the 'suction') increases the draft out of proportion to the work done and is solely for the purpose of holding the plow in a tenacious, stony soil or in correcting poor application of draft. The line of hitch or pull should also be about in line with the direction of the landside; that is, a plow cutting more or less than its standard width does so at a disadvantage because undue friction is brought on the landside.

Within the Bailey quote are references to principles, adjustments and measurements which will be addressed as we proceed with this book. At the risk of confusing the reader by jumping ahead, it is important to understand that with plowing, most all features and aspects have an affect on the quality and relative ease of the work.

**The plow sole.** Where soil is continually plowed to the same depth, the weight of the plow and the trampling of the furrow horse and farmer pack the soil slightly, just below the furrow line, so that it becomes somewhat of a hardpan, called a plow sole.

(Important to note, however, that relative to the packing dynamic of tractor plows, with the rolling pin action of several tons of vibrating tractor pressing down onto rubber or steel furrow wheels, horse and man have negligible effect.)

This compaction can be avoided with horsedrawn plows (not so much with tractor plowing) by varying the plow depth from year to year. A system or plan of deep plowing to avoid the hardpan formation and keep organic matter near the surface consists of plowing shallow to moderately deep when the sod or manure is being turned over, say for corn. At the next plowing the soil is turned as deeply as practical. Instead of leaving the organic matter on the surface or deep in the soil, this brings it back fairly near the surface where it continues to be rather rapidly available. In other words, plan the plowing depth variation to accomplish two goals, reduction of hardpan and the stirring and lifting of organic matter to the level best suited for the scheduled crop.

**Horsedrawn versus Tractor Plows.** Simple facts: Horses, mules and oxen provide the motive power today (2000 AD) for more plowing, worldwide, than do tractors and internal combustion engines. Tractor pulled plows can cover far more acreage, much faster, than animal-powered plowing. However, the horsefarmer is closer to his work and therefore able to be far more discriminating in the practise of his craft. Because of the low speed of animal power the twisting action of the moldboard is given full opportunity to maximize the pulverization of the soil. It is far too easy for the impatient tractor operator to accelerate and 'throw' chunks of soil as he plows. As was mentioned before, horsedrawn plows have only a slight soil compaction factor. In other words, all things being as equal as possible, horsedrawn plows are capable of far superior work. But to acheive this goal requires a firm grasp of all the elements involved including soil type and condition, seasonal concerns, the animals utilized,

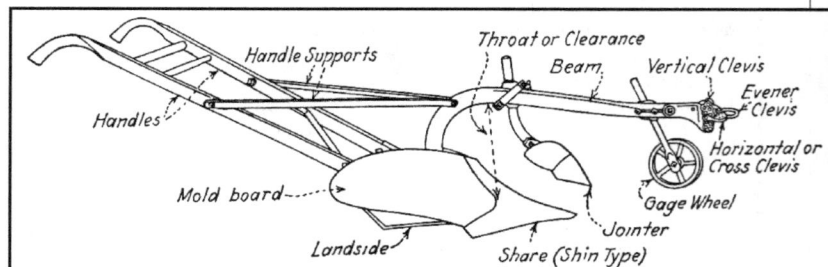

Fig. 28. Side view of a walking plow with parts named.

the implements and the hitching variables. Successful plowing with horses is all about craft and craftsmanship.

There are those, and this author is one, who believe that the twenty-first century will serve up major adjustments to the sociological, technological and scientific arrogance of the twentieth century. The human stomach and the human spirit must both be fed. Computers and genetic manipulation will not feed the stomachs nor the spirit of the mass of humanity. It is a telling spiritual paradox of the first order that a craftsmanship-based farming can and does feed both the soul of man and his carbohydrate needs.

Animal-powered plowing is considered by many to be an artifact of the past, but the protection of the technology and method is important because all formerly predominate relic ways of working have a growing power, relevance and feasibility every day that society moves deeper into the abstractions of electronics, chemistry, and gene splicing. Knowing how to do it all by hand, having that capability, is security.

**The Time to Plow**. Some will tell you that the proper time to plow should be determined not so much by the time of year or the season as by the *condition* of the soil. Others will argue that the soil *type* and the use you have for the tilled field should have more bearing on when to plow. It is the opinion of this author that the thinking observant farmer can often make exceptions to these rules with outstanding success. It is also the opinion of this author that all farming should be viewed and planned for in multitudes of seasons (over several years) if fertility is the goal. Rotations of crops, procedures, exposure, and rest all may contribute to better soils. And it is no coincidence that field work, plowing most particularly, becomes easier as the fertility, and the corresponding friability, of the soil improves. Here are some _suggested_ guidelines for when to plow:

Cultivation should not begin on any soil that is either too wet or too dry. (Horseplowing can be started on soils which are too wet for tractors but caution should be taken to avoid 'bricking'.) The sandy soils may be plowed when containing more moisture than clay soils, but even these should be firm, and elastic when ready for plowing. Same way, no soil should be plowed when too dry because this simply breaks off and turns up large, irregular-shaped chunks of earth which are in no condition to cultivate. Also, dryland plowing requires additional power (read animals) and more time to properly finish tillage. The best time to plow the soil is when the plow leaves an inverted furrow in a mellow, elastic and crumbling condition. There should be moisture sufficient enough

to make the plow clean (or scour) well, but the mold-board should not leave a "shiny" surface on the soil. *If the furrow is shiny and the moldboard is muddy*, it is too wet. *If the moldboard is shiny and the furrow*

Fig. 29. This gentleman from 1912 is doing a superb job of plowing with his team and walking plow. The quality of the furrows is the result of understanding the equipment, the animals, the process and a marriage of all those things to the type of soil and its condition on the day of plowing. When it all works together it is a masterpiece of craftsmanship.

*crumbly*, the soil is just right. If the soil crumbles readily in the fist it is generally speaking in good condition for plowing.

**Seasons: Fall and Spring Plowing.**

In most regions of the North Temperate Zone, where winter is severe and springs are late, fall plowing is preferable to spring plowing for a majority of the farm crops. This is due to the time saved, the great labor convenience and to the beneficial effect of freezing and exposure on many soils. The frost crumbles the moist soil into a mellow condition and settles it somewhat. Small grains especially do best on such a seed-bed. Green crops and stubble turned under in the fall have a longer period to decay. It is possible to get on the soil with fitting tools (harrows, discs, etc.) earlier in the spring than would otherwise be possible. Where

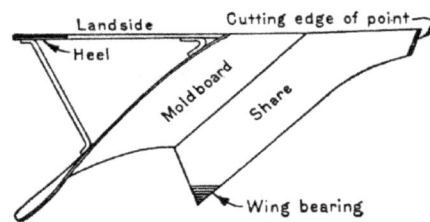

Fig. 30. A view from above of a walking plow demonstrating the bearing and cutting points. The point and wing of the share take on the majority of the wear along with the heel of the landside.

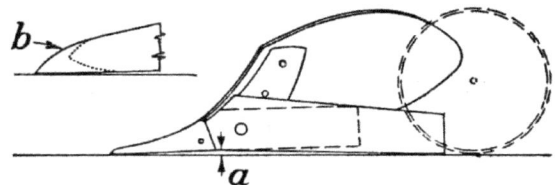

Fig. 31. A sideview demonstrating the share point, hardened and the suction of the share point. The gap at 'a' creates a suction which, when needed, helps to hold the plow in the ground. This can be important if the desired or required plow depth coincides with a hardpan layer which might cause the plow point to pull up and out of the ground. Care should be taken not to confuse a maladjusted plow hitch with the effect of hardpan.

there is dry weather in late spring or summer, the saving in moisture is an important factor.

Light sandy soil is *usually* best plowed in fall but where it is exposed to heavy wind during the winter this is not advisable because it is likely to be blown away. If it is fall plowed it should be thrown up in as large ridges as possible as these break the force of the wind and largely prevent erosion. With the fall plow weeds can be destroyed to better advantage and the furrow-slice is given time to settle down against the subsoil and to establish good capillary connections for moisture. In semi-arid regions, wherever the soil is sufficiently moist, fall plowing tends to conserve snow moisture during the winter and early spring, because plowed land will absorb moisture more readily than hard, unplowed land. Fall plowing also relieves much of the labor rush that occurs in the planting season (spring), and makes it possible for the farm manager to give more careful attention to the pulverizing of the seed bed and to the work of planting. In regions where the growing season is comparatively short, and where the spring season is also short, spring plowing delays seeding, and may cause injury to the crops from a late harvest. Fall plowing is of greater advantage on sandy soils than on clay soils, because moisture is harder to conserve for crops, and a fall plowed seed bed has better capillary connections with the subsoil than one that is spring plowed.

The judicious (let's say intelligent) use of a combination of spring and fall planting allows the horsefarmer to spread his labors over the year alleviating much of the concern about the powersource speed. Paradoxically, because the work is customized in such a way, and divided between seasons, the soil benefits and superior yields may be realized.

With many clay soils, in regions where spring rains are plentiful, spring plowing is regarded preferable for certain crops, such as corn and barley. The seeds of these crops germinate best and early growth is most rapid, if comparatively warm temperatures prevail in the soil. Spring plowing for heavy soils will usually provide somewhat warmer temperatures than fall plowing, and, when practiced in a region of abundant spring rainfall, the conservation of winter moisture is not important. Spring plowing for root crops, such as potatoes, sugar beets, and mangels, is considered best on types of soil that are inclined to become compact during the winter months, if fall plowed. Deep spring plowing provides a mellower area for roots to develop in than fall plowing, if the soil is heavy and the spring season moist.

If spring plowing is practiced because of hard, dry soil conditions in the fall or of insufficient time in the fall, rapid evaporation of moisture from freshly plowed land can be easily checked by the harrow and the sub-surface (or

Fig. 33. Twelve head of Belgian horses were hitched to this 4 bottom tractor trail plow at the 1996 Topeka Indiana Horse Progress Days.

roller-) packer. A soil condition in spring plowing similar to that in fall plowing at the spring season can be created with the packer and the harrow. If spring plowing is packed the same day it is plowed, all air spaces will be eliminated from the furrow-slice and the plowing will be crushed down against the sub-soil with good capillary connections for moisture. If the packing is followed immediately by surface harrowing whatever moisture is in the plowing will be securely locked up. The work of packing can be done well with a roller-packer or a common disk harrow with disks set straight ahead. Packing and harrowing spring plowed land in the semi-arid regions, the same day land is plowed, is essential to a good seed bed. It is not so essential to spring plowed land in humid regions, but is, nevertheless, desirable in preparing land for small grains. If clods are thrown up by fall plowing, frost and water are given time to crumble the clods and close up the air spaces; but with spring plowing the case is different, and a spring plowed seed bed is likely to be cloddy and full of air spaces, unless it is packed and harrowed the same day as plowed.

### Plowing Practice
### All Soils Cannot Be
**Plowed Alike.** Soils vary so in texture and character of subsoil that the best results are not secured by uniform methods of plowing. Local experience is often essential to an understanding regarding the best plowing practice for a certain soil. As a rule, clay and clay

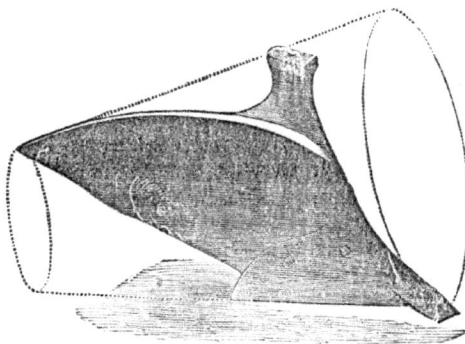

Fig. 32. In 1867 a Mr. Solomon Mead came up with the geometric notion that the marriage of the plow's moldboard configuration to a cone would result in superior efficiency of application. This was a radical concept for the time and it appears his theory was applied to most of the better plows from that time forward.

loam soils should be plowed deeper than sandy or sandy loam soils. The sandy soil is naturally porous and too much loosening of the soil is undesirable, as it may destroy good capillary connections in the seed bed. The clay soil, on the other hand, is naturally retentive of moisture and deep plowing will usually benefit aeration and warmth in the soil.

Relatively deep plowing, to eight inches, has become the standard among most farmers in the United States. Sod breaking is now commonly done to a depth of five inches, whereas 80 years ago three to four inches was formerly thought to be the correct depth. When land is plowed six to eight inches deep, a much better seed bed is provided for young plants than with shallower plowing. Deep plowing prevents an excessive run-off of rain water (and resultant erosion), and also provides a comparatively large soil area in which the roots of young plants may quickly penetrate to absorb moisture and plant food. The movement of air throughout the seed bed is also helped when deep plowing is practiced, and air and warmth are as essential to seed germination and plant growth as moisture. In times of drought a deep, mellow seed bed is less likely to bake and dry out as a shallow seed bed especially if fall plowing has been practiced or spring plowed land packed with the roller-packer. The liberation of available plant food from inert forms in the soil is facilitated when deep plowing is practiced, because more favorable temperature, moisture and aeration are provided for the presence of the soil bacteria that assist chemical changes.

Experience has shown that when deep plowing is considered on ordinary prairie or timber clay loam soils in humid regions, on which shallow plowing has been previously practiced, it is advisable to increase the

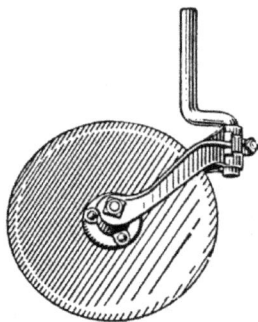

Fig. 34. A clear demonstration, from 1917, that two horses on a sulky plow can do an enviable job of deep plowing.

depth of the plowing gradually rather than to increase a bunch in one year. It is not unusual for very poor crops to follow a radical change in the depth of plowing. This is almost always the case if the subsoil is different in character from the surface soil or if tests show the subsoil to be more acid than the surface soil. In semi-arid regions, or on any soil area where the subsoil contains more lime than the surface soil, and where the subsoil is of the same character as the surface

Fig. 36. Rolling coulter with swivel bracket.

soil, a quick change in the depth of plowing will not usually cause poor crops. If it is known that the subsoil is more acid that the surface soil, a quick change in the depth of plowing can be effected without much danger by plowing under green manure crops and also liming the soil to correct acidity. A good top-dressing of manure, together with lime, on freshly turned, deep plowed land, would also overcome the problems arising out of a quick change in the depth of plowing.

There is little evidence to show (except in the most unique circumstances) that deep tillage is sufficiently profitable to supersede ordinary six to eight inch plowing on the vast majority of North American farms.

**Walking Plows.** For small tracts, plots or fields the walking plow is a superior implement. It does, by its nature, require somewhat more finesse on the part of the operator. Within the twentieth century there have been hundreds upon hundreds of design and application variables many of which are covered later in this book. For the small diversified family farm utilitzing

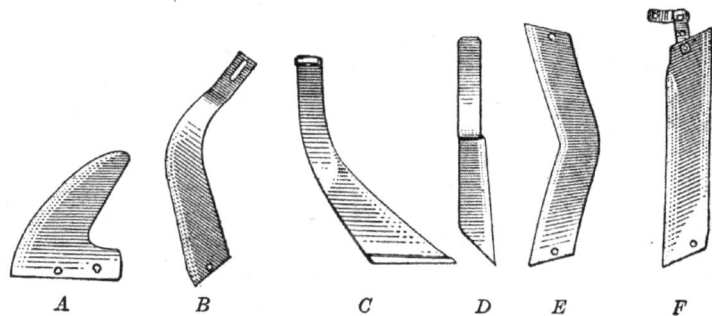

Fig. 35. Plow knives or knife coulters as they are some times referred to. A, fin; B, knee cutter; c, standing cutter; D, hanging cutter; E quincy cutter; F, reversible cutter.

animal-power, this author recommends at least a good walking plow or a combination of a walking and wheeled plow. More on this subject in the walking plow chapter.

**Wheeled plows.** The wheel plows of the sulky type are sometimes preferred to walking plows. The sulky plow doesn't work on rolling ground for contour plowing because the furrow is not easily turned up the hill on one-half of the return pass. This difficulty, and the disadvantage of the dead-furrow are both eliminated by the use of the double sulky or two-way plow. This implement carries two plows, one a right-hand and the other a left-hand plow, with which it is possible to plow continuously on one side of the land. The furrow can thus be always turned down the hill; dead-furrows are avoided and the surface is kept smooth.

The side-hill walking plow is an earlier model

than the double-sulky plow, designed to avoid the necessity of traveling around the land. The moldboard is attached by a hinge joint and can be turned from side to side so that one may continually plow on one side of the land. The moldboard cannot be given the proper curvature for difficult soils.

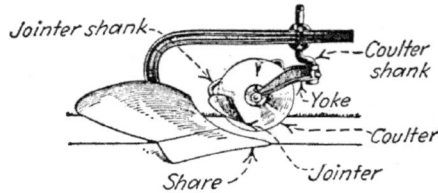

Fig. 37. Combination rolling coulter and jointer showing how the hub of the coulter is set over the point of the share.

**Disc plows.** A single disc performs all the functions of point, share and moldboard. The landside thrust is taken by a wheel since these plows must always be of the wheel or sulky type.

Disc plows are most effective on hard clay land and where weeds, stubble and rubbish need to be turned under. They have a tremendous pulverizing action but they are not suited to sod land. The curvature is such that the sod is broken up and not properly turned. For stony and for especially hard soil, the cut-out disc is better than the solid disc.

**Plow attachments**. These play an important part in the efficient operation of the moldboard plow. Where land is in sod or bears a considerable growth of vegetation, it is hard to keep the shin sharp enough to cut it easily. Therefore, an additional cutting edge is often advisable. The devices for this purpose are the coulters of various types and the jointer. A coulter is a blade or knife attached so as to cut the furrow free from the landside before the share and moldboard begin to turn it. A small triangular blade attached to the landside just back of the point is known as a fin coulter. It may be drawn out in a blade form so the tip reaches above the surface of the soil. This is one of the best types for stony soil as the stones are lifted out of the way. Blade coulters are attached to the beam and usually extend downward and forward. Their action is very much like that of the fin coulter but some power is saved since the entire face of the furrow is not cut free.

For land having a tall growth, especially of vines, the blade coulters are not suited because the vegetation collects on them. There and on sod land the disc or rolling coulter is better. It also is attached to the beam and is set to cut about an inch outside the landside so that the furrow has a sloping rather than a vertical face. This makes it easier to turn over at the next round. It is usually set down so as to cut about half the depth of the furrow but may be set deeper. This disc coulter is not well suited to stony land as even a small stone on the surface striking the coulter is likely to throw the plow out of the ground.

The jointer serves a double purpose. It is a miniature plow attached to the beam in the position of the coulter, where its sharp forward edge performs first the functions of the coulter to cut the furrow free. The curved surface then turns over the edge of the furrow so that any growth such as grass or a green-manure crop not more than a foot in height is turned under and does not stick out when the furrow is laid under. Any rubbish or stubble is also thrown over where it will fall in the bottom of the furrow. (see Fig. 17) The jointer is a very handy attachment adapted to a wide range of conditions, except where there are vines. A disc or

rolling coulter can be fitted with a jointer attached just back of the edge on the furrow side. In this way the advantages of the rolling coulter are combined with those of the jointer.

The position of the coulter or jointer with reference to the plow point is important. In nearly all cases it should be a short distance back of the point. The coulter should be in such position that the knife edge meets the roots in soil after it has been raised and the roots somewhat stretched by the advance of the point. This makes the cutting easier. (More on setup a little later.)

On muck soil first plowed and on rooty land the coulter is sometimes slanted backward instead of forward and extends from the beam to a position considerably back of the point. Thus roots are better gripped and may be cut. At the same time the plow is better held in the ground by the point.

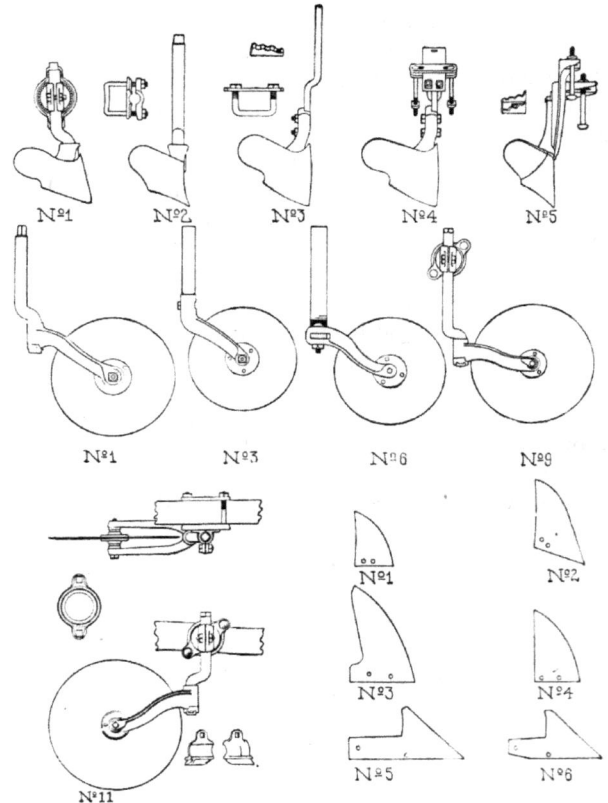

Fig. 38. A selection of Parlin & Orendorff plow attachments. Top row, Jointers. Second row, rolling coulters. Bottom right, fin knives.

This chapter has been a broad overview as introduction to the subject of animal powered plows. In the following chapters each aspect of the subject will be covered in greater depth.

Chapter Two

# Horse Plows - Design Basics

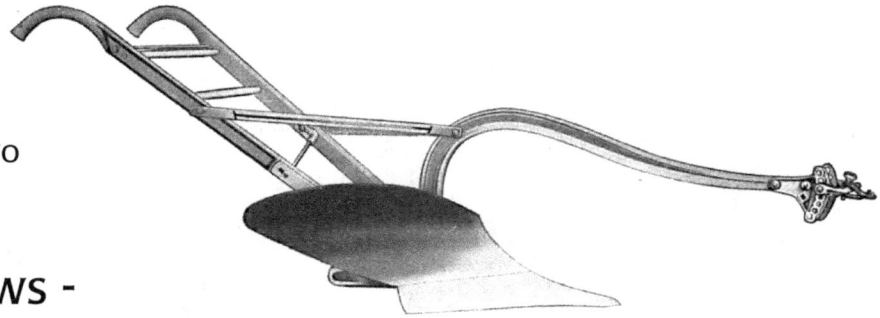

Fig. 39. John Deere Walking Plow

In the first chapter we offered a broad overview of plowing and animal powered plows. In this chapter we will discuss those basic elements of plow design which are somewhat universal to all the makes and models. The chapters which follow this one will go into detail on each category from walking to riding and so on. There is a giant chapter in this book which offers engineer cuts of as many makes and models as we could locate at press time. Any make or model may have some design or operational aspect that is unique to that specific plow. We have uncovered evidence of close to ten thousand variables in animal-drawn plow design and have long ago given up on our goal of including every one in this book.

This author is concerned that the reader be able to quickly and easily find the information he or she seeks. That said we have the challenge of not repeating basic information too often. Therefore we've organized the book so that you MAY have to go over this chapter, then the chapter specific to the type of plow, and perhaps a couple of other points in this text. It is hoped that the reader's enthusiasm for the subject, coupled with this book's hoped-for usefulness, will cause a complete reading.

**The Horse Plow**. There are many different types of plows, but all serve the same purpose. The listing plow, the middlebreaker, the disc (or disk) plow, and

regular moldboard plow are all used in stirring and pulverizing the soil in preparation for planting. And all of these classic plow types have been designed and manufactured specifically for animal power in riding and walking configurations. Now, in the dawn of the twenty-first century, readers may be surprised to learn that animal-drawn plows are still being manufactured around the world. And in North America one will find the very best of modern horsedrawn plows being

Fig. 41. A three abreast of Fjord Draft Ponies hitched to a new Pioneer Sulky plow at the '98 Horse Progress Days in Ohio.

redesigned, improved and manufactured new by several companies.

It is impossible to reconstruct the numbers with real accuracy but every indication from historical records suggests that from 1890 to 1938 over a hundred million horsedrawn plows were manufactured in the U.S. and Canada alone. And that production, and somewhat discriminating market, resulted in the research and development of the most advanced and efficient animal-drawn plow designs to be found any where in the world.

We differentiate in this case between animal drawn and tractor drawn plows for interesting reasons. The brute excessive force of wheeled tractors, coupled with remote hydraulic hitches, made it possible to build mammoth many-bottomed plows with little or no concern for optimum plowing and efficiency of draft. This compared to the abiding concern in animal drawn plows for maximum efficiency with restricted draft and

Fig. 40. From 1920 a photo of four mid-sized farm chunks hitched abreast to a John Deere foot-lift sulky plow. This is an awkward hitch with the off, or right, horse having to walk on plowed ground.

Fig. 43. Again from the 1920's, this time a huge steam tractor with a 14 bottom plow leaving the sod strip evidence of long air tunnels. A typical view of how speed and power will cover great tracts of land but produce decidedly inferior work to what a good farmer can do with one or two bottoms and good horses, mules or oxen.

Fig. 45. Below. A new exceptional Pioneer-built two-bottom riding plow. These plows, and many other animal-drawn implements are made and field-tested in Dalton, Ohio.

Fig. 42. An old photo from the 1920's showing good work being done by a six-up of farm chunks pulling a two bottom foot-lift gang plow

Fig. 44. A furrow scene at the '98 Ohio Horse Progress Days. One six-up of Belgians follows another, each pulling two bottom gang plows of new design and manufacture (see one to the right). Notice the outstanding job of plowing, that soil will only require a gentle timely harrow pass to be ready for planting.

Fig. 46. A remote hydraulic activated two-way plow built by White Horse Machine of Gap, PA. With this implement all furrows fall in the same direction and dead furrows or gullies are avoided.

Fig. 47 White Horse's new hydraulic-lift two bottom gang.

# Plow Bottom Parts

Fig. 48

Design Basics

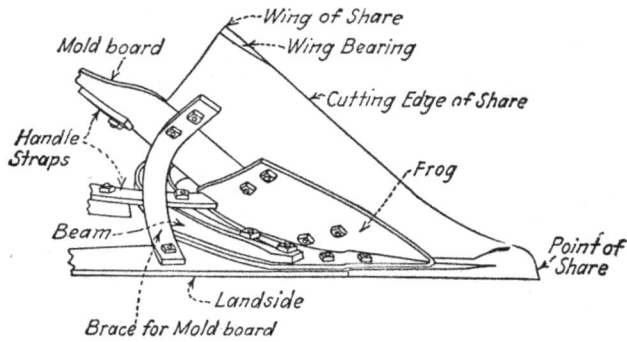

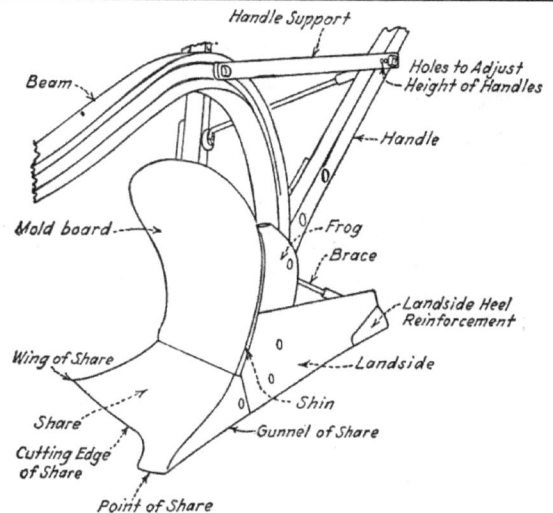

slower speeds of operation. While it is true that animals may successfully pull plow bottoms specifically designed for tractors (all other tractive concerns being equal) the opposite is not necessarily true.  Hook a big tractor on to a horse plow and plow for a half day at four miles per hour and that plow will likely be destroyed. This does not mean that the animal drawn plow is inherently weaker. It means that the tractor power and indifferent operator can be brutish in effect. *(It may be like comparing a fly rod to a deep sea fishing pole in the hands of a blindfolded wrestler or a fine small carving knife to a meat cleaver in the hands of a political campaign manager.)*

## Plow Bottoms

Leaving aside, for the moment, the specialty plows of unusual design, the key component of all plows - or the base of the plow, be they riding or walking styles, is the plow bottom.

A plow is no better than its bottom. No matter how well the frame may be built, how strong its beam, how modern its

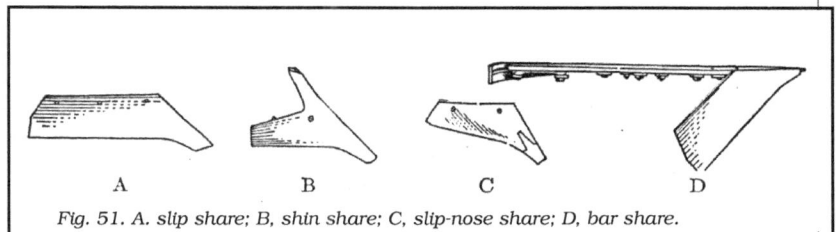

design, the plow will only perform as good as its bottom. If the bottom fails to scour (or clean) and turn the soil properly, the seed bed will be uneven and lumpy, making proper seedbed preparation difficult. If the bottom turns an even furrow, covers trash well and pulverizes the furrow slice as desired, a uniform seed bed will be easily within reach.

Costly delays at plowing time are often caused by plow bottoms that refuse to scour. The problem may be in a poor match of bottom design to soil type. It may be in the adjustment of the hitch, or it may be due to a dull or improperly-set share. It may be the result of a pitted, corroded and/or rusty share. Regular attention needs to be paid to the conditon of the share.

Fig. 51. A. slip share; B, shin share; C, slip-nose share; D, bar share.

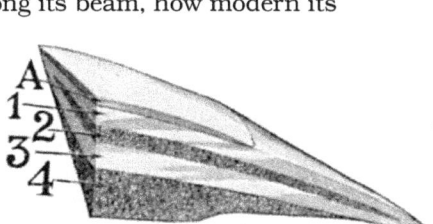

Fig. 49. JD soft-center steel share point. A = patches of hard tool steel. 1 & 3 = Hard steel. 2 = Soft Steel. 4 = Steel landside, lap welded. Note relative thickness of layers.

Fig. 52. Heavy lines show proper shape of sharp share points for good penetration. Dotted lines show how worn points look before sharpening.

Fig. 50. A JD illustration of genuine soft-center steel. 1 and 3 Hard high-carbon steel , 2 soft, tough steel with hard steel fused to it.

**Parts of the Bottom**. The plow bottom consists of share, landside, moldboard and frog. (Fig. 48) The share and landside act as a wedge in the soil, cutting the furrow loose from the subsoil much as a wedge splits a log. The curved surface of the upper part of the share and the moldboard inverts the furrow slice. In passing over this curved surface, the furrow is twisted and broken, and the soil is pulverized, mixed and aerated.

**Types of Bottoms**. Different types of soil require different shapes of bottoms to accomplish the results desired in plowing. That said we need to differentiate between the "bottom" and those individual parts tied directly to best performance in certain soil types.

The *moldboard* is that part of the plow just back of the share. (see figure 48). It receives the furrow slice from the share and turns it, in a twisting and lifting

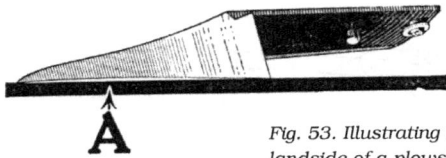

Fig. 53. Illustrating underpoint suction in landside of a plowshare.

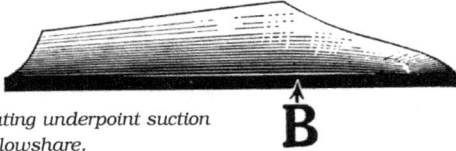

Fig. 54. Illustrating underpoint suction in throat of a plowshare.

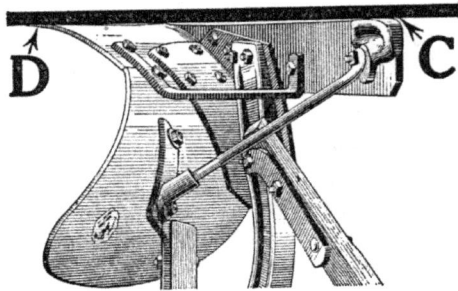

Fig. 55. Wing bearing, point D, is necessary to smooth running of walking plows. To measure the wing bearing, place the straightedge across the heel of landside at C and wing of share at D. The wing bearing is the amount in contact with the straightedge at D.

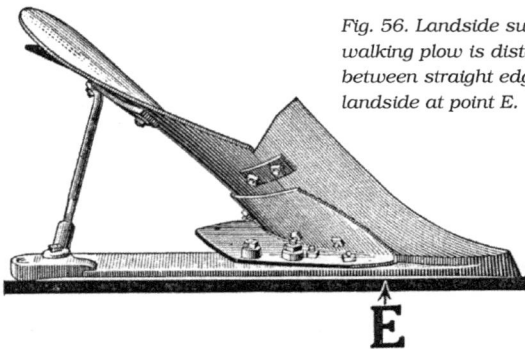

Fig. 56. Landside suction on a walking plow is distance between straight edge and landside at point E.

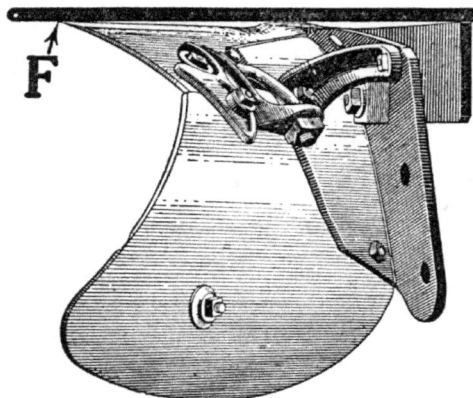

Fig. 57. Riding and tractor plows do not require wing bearing at F.

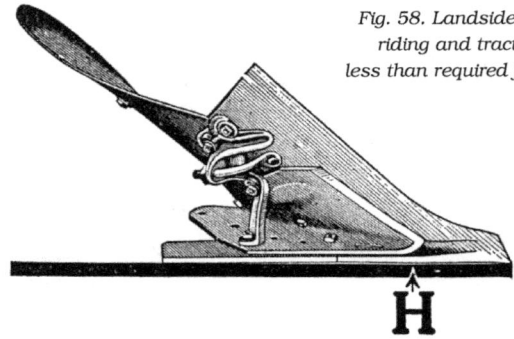

Fig. 58. Landside suction for riding and tractor plows is less than required for walking plows.

action, into a ribbon of pulverized flipping earth (see figures 1 & 2 in chapter one). It is upon the moldboard that the furrow slice is broken crushed and pulverized. Different soil types and soil moisture conditions call for different moldboard shapes.

The texture of the soil and the amount of moisture it contains determine whether it should be thoroughly pulverized or merely turned over, to be pulverized with other implements. A mellow loam soil and soils of similar texture should be plowed with a bottom that will pulverize well, while a sticky, wet clay soil should be plowed with a bottom that will break it as little as possible, leaving the pulverizing to be done with other machines.

The pulverizing effect of a plow depends upon the shape of its bottom. A bottom with a long, gradual curve in the moldboard turns the furrow slice gently and disturbs its composition but little. The other extreme is the short, abruptly curved moldboard that twists and shears the soil as it passes over, making a mellow, well-pulverized furrow. *(The pulverizing effect produced by the curved surface of the moldboard is illustrated in Fig. 2 in Chapter One.)*

Between these two extremes are many types of bottoms designed to meet many different soil conditions, but for general use, bottoms may be classified as breaker, stubble, general purpose, slat moldboard and blackland, (Figure 63).

The *breaker** is used in tough sod where complete turning of the furrow slice without materially disturbing its texture is desired. This type of moldboard is constructed so as to have a long, gradual, auger-like twist, the object being to turn the slice completely upside down covering all vegetable matter thoroughly.

*Stubble*** bottoms are especially adapted to plowing in old ground where good pulverizing of the soil is desired. This moldboard is one that is broader and bent more abruptly along the top edge causing the furrow slice to be thrown over quickly, pulverizing it much better than the other types of moldboards.

*General purpose* bottoms meet the demand for bottoms that will do good work in stubble, tame sod,

---

*  breaker: takes its name from the 'unbroken' prairie land sod, a land which had never seen the plow. Such lands had dense tough wiry sod which required a specially designed bottom named for the fact that the 'breaker' plow 'broke' the unbroken prairie sod.
** stubble: this term, in the context in which it is being used, comes from 'stubble ground' or 'old ground' meaning soil that has been cultivated from year to year and on which the stubble of plants from the previous crop still stand.

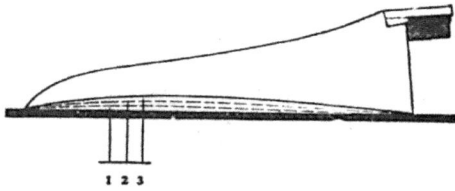

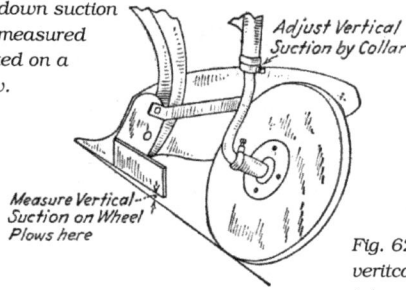

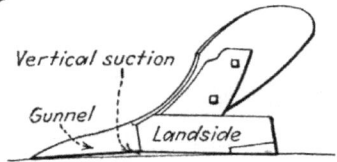

Fig. 61. (Right) How the vertical or down suction should be measured and adjusted on a wheel plow.

Adjust Vertical Suction by Collar

Measure Vertical Suction on Wheel Plows here

Vertical suction

Gunnel     Landside

Fig. 62. On a walking plow the amount of veritcal suction should be measured at the intersection of share and landside.

Fig. 59. (Above) Share suction: 1, regular -3/16 inch for light soil easy penetration; 2, deep -5/16 inch for ordinary soil that is dry and hard; 3 double-deep - 3/8 inch for stiff clay soils, gravel land, and other soils where penetration is difficult.

1 2 3

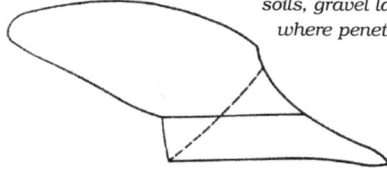

Fig. 60. (Left) An outline of a plow bottom showing area, below dotted line, which receives 75 percent of the draft when plowing, illustrating the necessity of keeping shares sharp for light draft of plows.

# PLOW BOTTOMS

Fig. 63

At the turn of the century the Moline Plow Co. advertised a manufacture of over 2,000 different plow bottom styles, shapes and sizes. These drawings were taken from the Moline literature to illustrate some of the bottoms most common to all makes .

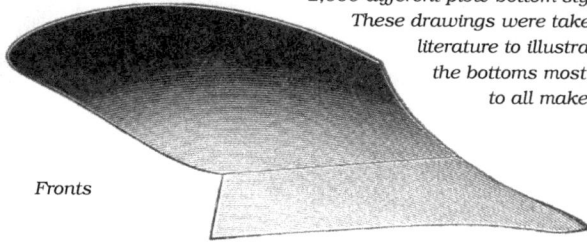

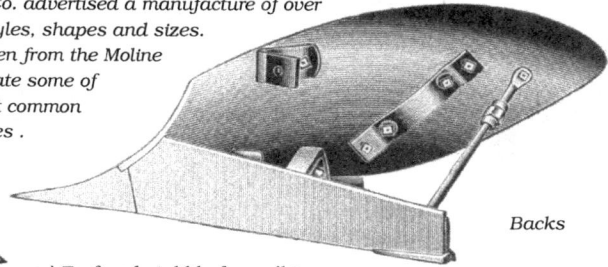

Fronts

Backs

a.) Turf and stubble for walking plows.

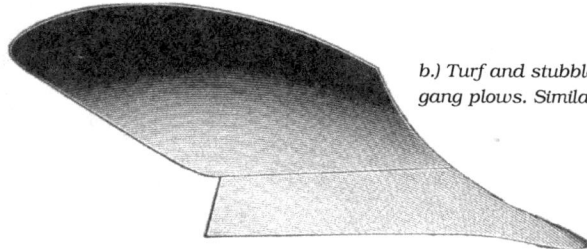

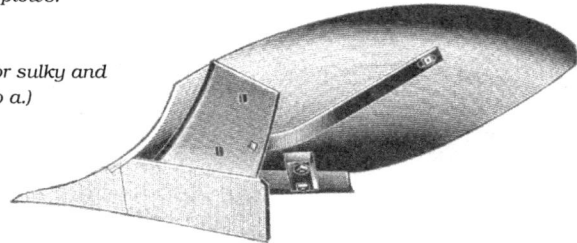

b.) Turf and stubble for sulky and gang plows. Similar to a.)

The Turf and stubble bottoms are considered general purpose, for some sod, old ground, clover, timothy or blue grass. Good for heavy sticky soils, desirable for deep plowing.

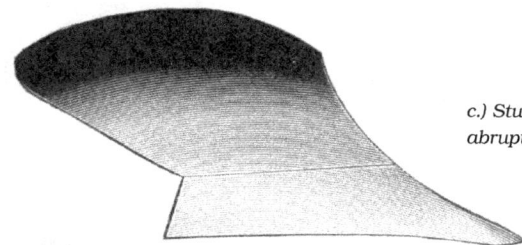

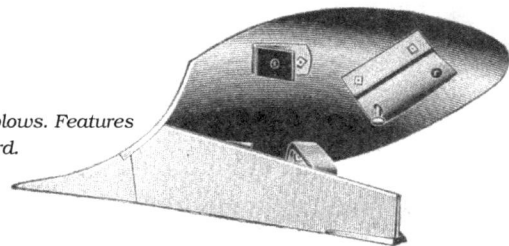

c.) Stubble for walking plows. Features abrupt upright moldboard.

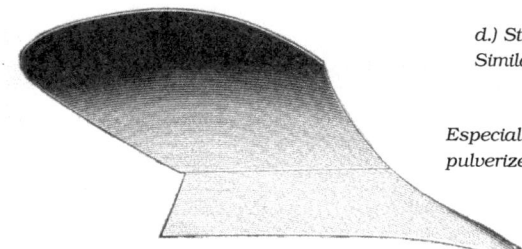

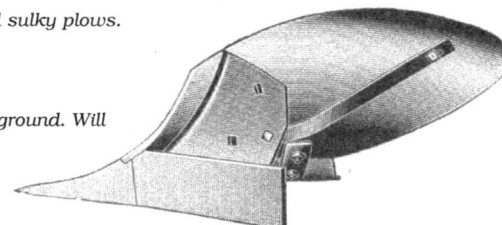

d.) Stubble for gang and sulky plows. Similar to c.)

Especially adapted for old ground. Will pulverize more thoroughly.

# Plow Bottoms

*Fig. 63 continued*

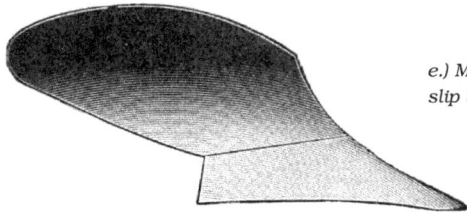

e.) Mixed Land Bottom with a
slip heel on the landside.

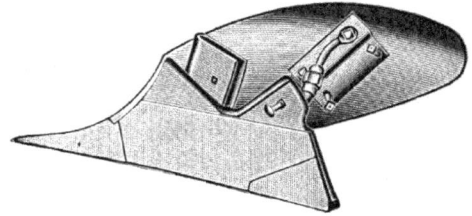

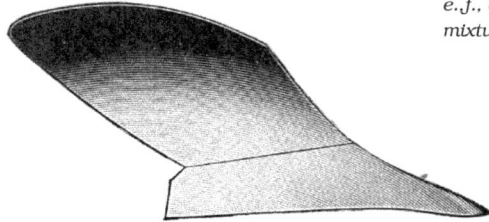

e.,f., & g. are designed for soils containing a
mixture of sand and clay or light loam.

f.) Mixed land bottom
designed for a walking
plow

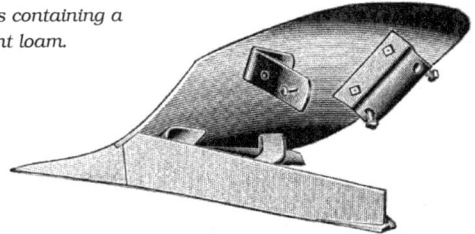

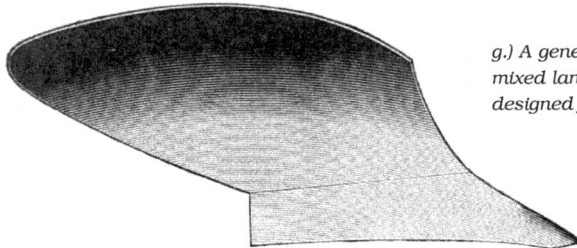

g.) A general purpose
mixed land bottom
designed for riding plows.

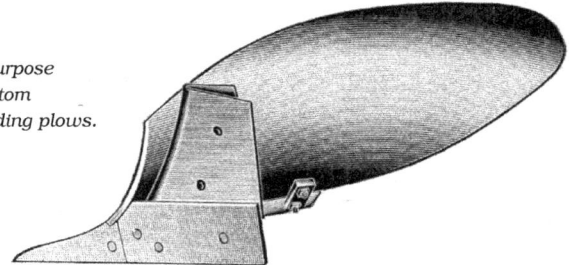

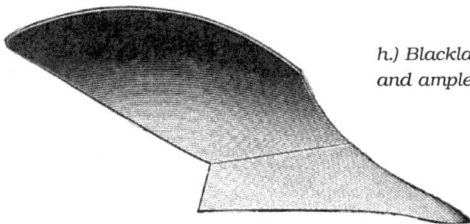

h.) Blackland walking plow bottom with long share
and ample suction.

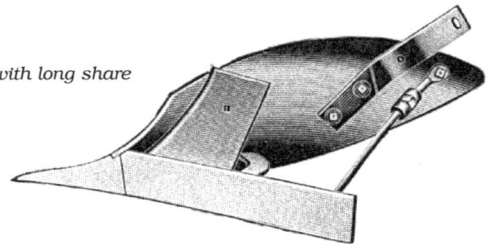

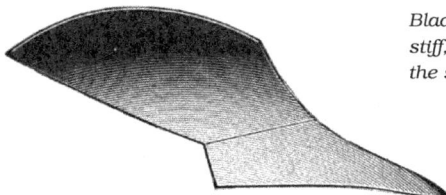

Blackland bottoms are designed for the
stiff, black, difficult, waxy lands found in
the southern U.S.

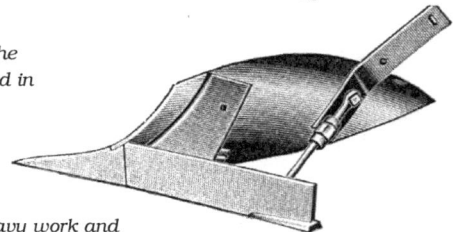

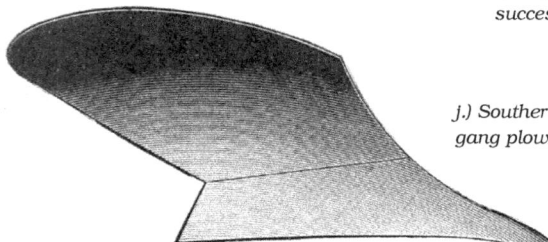

i.) Sugar land bottom for heavy work and
successful in mixed lands.

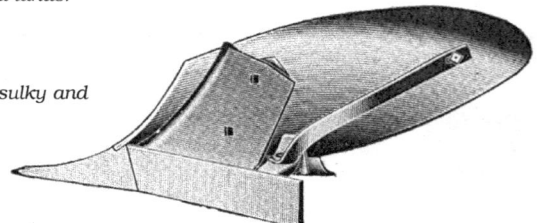

j.) Southern blackland sulky and
gang plow bottom.

*Fig. 63 continued*

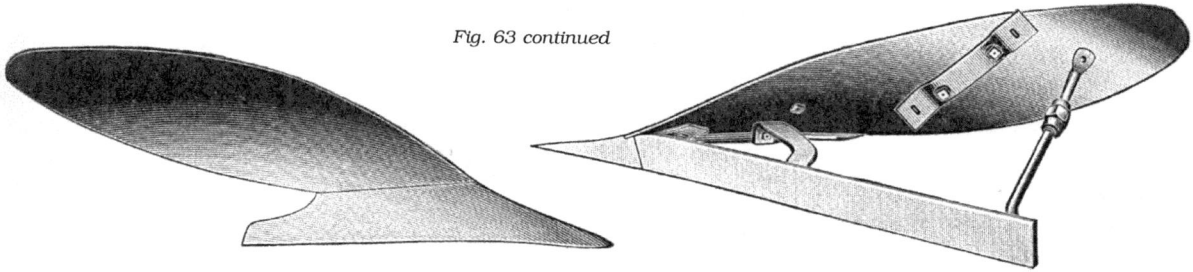

*k.) Western-style breaker bottom for walking plows.*

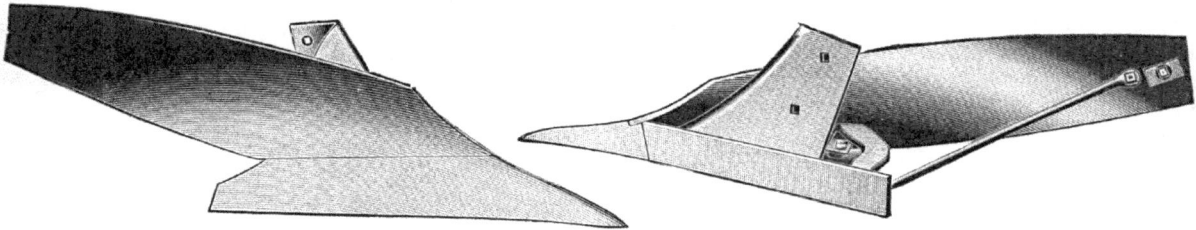

*l.) The breaker bottom designed for sulky or gang plows. Can be equipped with moldboard extenders*

# Plow Bottoms

*m.) Deep furrow bottom.*

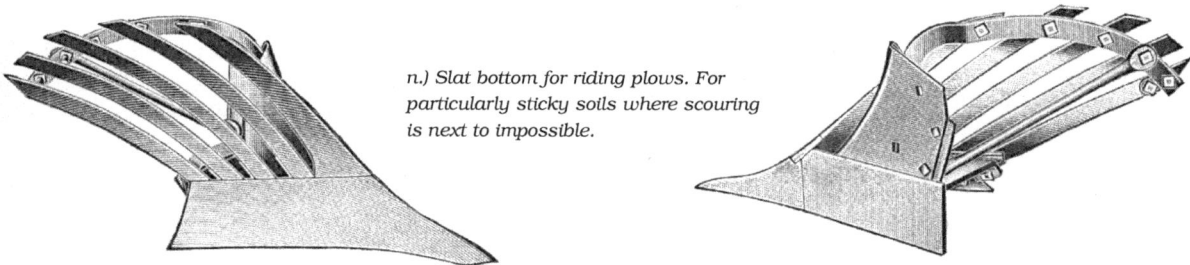

*n.) Slat bottom for riding plows. For particularly sticky soils where scouring is next to impossible.*

old ground, and a variety of similar conditions. The general purpose is designed to do satisfactory work in the varying conditions found on the average farm. It has less curvature than the stubble.

The *slat* moldboard bottom is used in loose, sticky soils, and the *blackland* bottom in gumbo and "buckshot" soils. In both types of soil, scouring is a serious problem.

There are a number of variations of these general bottom shapes built to meet a wide variety of soil conditions.

**Materials Used in Bottoms.** There are three classes of moldboards according to the construction; namely the solid, slat and rod.

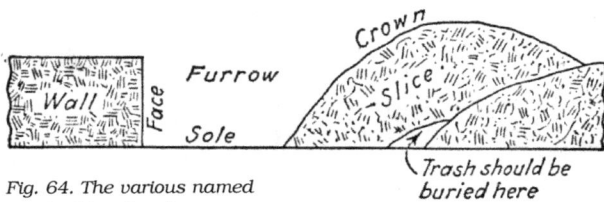
Fig. 64. The various named parts of the plow furrow.

The solid-type has no perforation or sections cut out to break the smooth surface of the plow.

Slat moldboards have exposed long strips leaving only half of the surface to come in contact with the furrow slice. This is important in soils which will not scour (or clean off the moldboard).

The rod type of moldboard looks like the slat style and consists of round rods attached to the plow in such a manner as to form a surface upon which the furrow slice will be turned. There will be little if any pulverizing of the soil with this type of board. It is found to be practical and useful in some soils of the prairie type that are sticky and will not shed as they should from solid moldboards.

Classified according to materials used in manufacture, there are two kinds of plow bottoms - steel and chilled cast-iron. (On some very cheap plows straight cast iron is used.) Steel bottoms may be either solid steel or hardened soft-centered steel. The latter bottoms are in more general use. In some sections, the soil conditions are such that a combination of chilled-iron shares and soft-center steel moldboards is used with excellent results. Both steel and chilled bottoms are used by some farmers who have varying soil conditions on their farms.

**Soft-Center Steel Bottoms**. A very highly-polished fine-textured steel moldboard is necessary to good scouring in sticky, fine-grained soils. Soft-center steel has the necessary hardness and thickness for good scouring and long wear on the outer surfaces, and strength enough in the inner layer to withstand shock and heavy loads in difficult soils. The outside layers are very high in carbon, extremely hard steel. Between these two hard layers in a layer of soft, tough steel, the hard steel having been fused to it (Fig. 50). In genuine soft-center steel all three layers are uniformly thick. There is no outcropping of soft spots, no thin places in the outer layers to wear through rapidly. Fig. 49 shows a cross section of genuine soft-center steel share point, illustrating how it is reinforced by lap-welding.

**Solid Steel Bottoms**. Solid steel bottoms are used in soils where scouring as a rule is not difficult. They are made of solid steel and are not tempered. They should not be used in sandy or gravelly soil, as they would tend to wear too rapidly. Solid steel shares are sometimes used with soft-center

steel moldboards where soil conditions do not require the more costly soft center steel shares for scouring.

**Chilled-Iron Bottoms**. Plow bottoms made of chilled-iron are used mostly in sandy or gravelly soils where the share and moldboard must withstand the scratching and hard wear of a soil of this type. The material used in these bottoms is extremely hard. Its good wearing qualities are due to a process called "chilling".

In casting chilled shares, a piece of metal called a "chill" is placed into the mold along the cutting edge and point where the finished share is to be chilled. When the hot metal comes into contact with the "chill", the sudden cooling leaves the grain of the metal at right angles to the surface. Thus, the dirt rubs the ends of the grain in the metal when passing over the share. A smooth and long-wearing surface results. Chilled shares may be sharpened by grinding, but, because of their low cost, it is usually more satisfactory to replace worn shares with new ones.

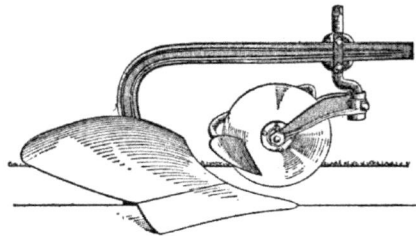
Fig. 65. Proper adjustment of combination rolling coulter and jointer for ordinary plowing

Fig. 66. When used alone, the coulter should be set 1/2 to 5/8 inch to the land, and the hub should be about 3 inches to the rear of share point.

**The Frog**. The steel frog, the foundation of any plow, holds the bottom parts and beam together. It is an irregularly shaped piece of metal to which the share, landside, and moldboard are attached. If you remove the frog all other parts are useless because they cannot be held in the proper positions. In most plows the beam is also attached to the frog. The landside and moldboard are bolted solidly to the frog. Frogs are generally made of steel, malleable iron, or cast iron. Steel is used more extensively than any other material for making frogs. It is a light, strong, durable, and easily shaped material. The malleable iron frog is used and comes into play on plows that do not require the frog to be made thin and small. If the frog is not very large, it is often necessary to reinforce it to prevent bending or breaking. Once the plow's frog is bent, it's impossible to return to the original shape in a useful form. The cast iron frog is used only on the cheaper plows of the single horse variety, and of those, commonly it is the ones with a wooden beam. In these economy plows the frog and standard are made in one piece.

**The Share.** The share provides the cutting edge of the plow. The main parts of the share are the point, the

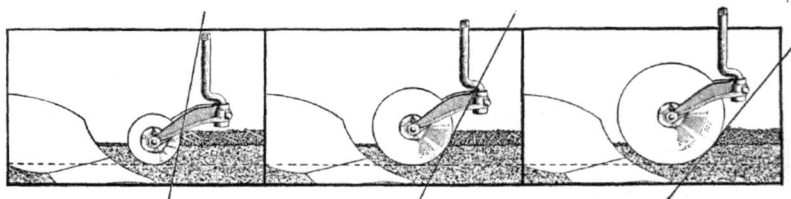
Fig. 67. Large rolling coulters mount trash more readily than small ones thereby insuring a clean-cut furrow with less draft.

*Fig. 68. These lovely drawings are from a circa 1914 Parlin &
Orendorff (P & O) equipment catalog. They illustrate their Ohio
Clipper Plows made primarily for clay and sandy soils. To quote the
catalog, "Our claim on these plows is that a farmer can buy what is*

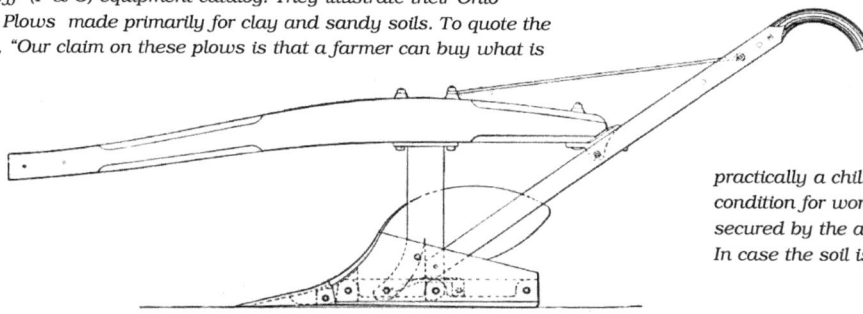

*practically a chilled plow, and then when the soil is in
condition for work with a steel plow, one can be
secured by the addition of a steel share and landside.
In case the soil is of varied nature, a suitable plow for*

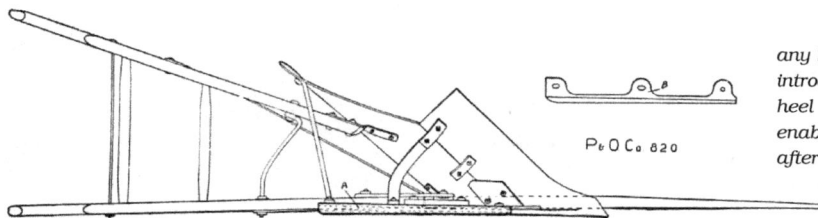

*(To the right is the wood beam model.
Below is the steel beam model.)*

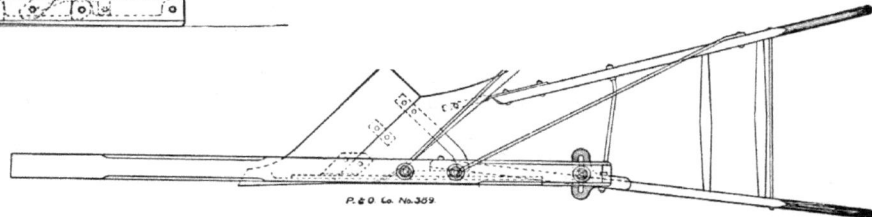

*any kind of ground is secured. A new feature which is being
introduced on these plows is the long and heavy shoe on the
heel of the landside. This shoe is adjustable, and will
enable the operator to get the desired suction of the plow
after the point of share becomes worn or rounded."*

wing, and the cutting edge or throat. The point is the
part which first enters the soil. The wing is the outside
corner of the cutting edge. The cutting edge extends
from the point to the wing. This edge is usually curved
and what is sometimes called the throat of the share.

The four main types of shares are illustrated in
figure 51. The slip share has no extension to form the
landside as does the bar share. The shin share has an
extension to form the cutting edge or shin for the
moldboard. When a share is replaced with a new edge,
both the cutting edge and the shin are then new. The
slip-nose share is one where the point is detachable.
The materials used in making shares are plain crucible
steel, soft-center steel, chilled cast iron, and cast iron.

**Detachable shares**. Most riding plows and tractor
plows of the moldboard type now have quick-detachable shares. To remove or replace the share, it is
necessary to loosen but one nut (Fig. 199 on page 76).
This quick-detachable feature saves time when shares
are removed for sharpening.

Certain types of chilled-iron plows have a detachable shin piece that serves as a cutting edge for the
moldboard (Fig. 51b).

This shin-piece gets the hardest wear of any part
of the moldboard and may be easily removed when
worn and replaced by a new part at small expense.

**Sharpening Plow Shares.** The share is the most
vital part of the bottom. It is the "business end" - the
premiere part in all of the work that a plow does. Draft,
penetration, steady-running and good work all depend
upon the share. It is also the part which usually wears
out first. Note Fig. 60, which shows the area of the
plow bottom that is responsible for 75 per cent of the
draft when the plow is at work. This illustration clearly
shows the importance of keeping the share in good
cutting condition at all times. Note, also, Fig. 52 . A

dull share may cause poor penetration and may greatly
increase the draft of a plow. A sharp, correctly-set
share adds to the efficiency and good work of the plow
bottom.

Many farmers have shop equipment for sharpening their plow shares, while the great majority depend
upon local blacksmiths or mechanics for this service.
In either case, it is well to know how shares should be
sharpened.

When sharpening soft-center or solid steel shares,
the point of the share should be heated to a low cherry
red (not too hot, 1470 degrees). The shares should be
placed in the forge flat (along the cutting edge or throat)
and not vertically. If placed vertically, heating cannot
be confined to the edge. Heat only a small portion at
one time and begin at the share point and work back
toward the wing. Hammer only on the top side with the
lower side flat on the anvil. This is important because
the thicker layer of hard can be drawn out over the soft
steel in the center and the thin layer of hard steel on
the under side. If the hammering is done from the
underside, the soft center will be left exposed and very
likely the top layer will be loosened and parts flake off.
While sharpening care should be taken not to destroy
the suction. Hammering should be done at a cherry red
only. Working the share at a high heat destroys the
quality of the steel. Only as much as can be hammered
should be heated at one time. The body of the share
should not be heated while sharpening, but should
remain cool to prevent warping and disturbing of the
fitted edges.

Should the share get out of shape or the fitted
edges become warped during the sharpening process,
put the blade in proper shape before hardening. This
can be done best at a black heat.

Soft-center steel shares should be hardened after

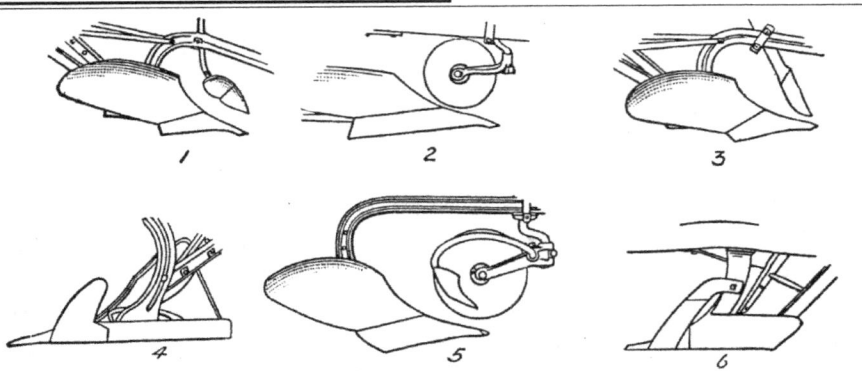

Fig. 69. Plow attachments: 1. jointer, 2. rolling coulter, 3. hanging cutter or plow knife, 4. fin knife, 5. combination coulter/jointer, 6. knee cutter.

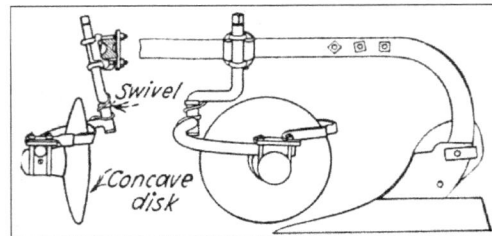

Fig. 70. Swivel disc jointer.

sharpening. To do a thorough job of hardening, it is necessary to prepare the fire to heat the entire share uniformly to a cherry red. Care should be used in getting the heat uniform. The share is taken from the fire and dipped into a tub of clean, cold water with the cutting edge down. Care should be taken to keep the blade perpendicular during this process.

Solid steel shares should not be hardened.

**Sharpening chilled and cast-iron shares.** This material cannot be heated and hammered to draw out the edge because of the brittleness of the material. Instead of hammered they must be ground and this should be done on the upper side. A safe rule to follow in sharpening any share, whether it be soft-center or crucible steel, chilled or cast iron, is to work from the upper side.

**Repointing shares**. Shares that are badly worn or have been sharpened a number of times should be repointed by welding a six inch piece of steel bent U-shaped to both the lower and upper side of the point.

**Hardening Shares**. The usable life of shares might be extended by applying a hard metal to the cutting edge with a torch. Care should be taken to get the hard metal well distributed on the bottom side of the cutting edge.

**Suction of shares**. As the share, and indeed the entire plow bottom, moves through the soil, cutting a ribbon, it is possible to affect the ease and regularity with which a certain depth is maintained by creating a suction factor. Imagine a perfectly flat sharp knife cutting through soft dirt. That blade might tend to wander on its own or as the result of any minor or major resistance (i.e. compaction, gravel, root balls, etc.). If the knife is designed with a slight and appropriate amount of 'cup' to one side of the blade, it's movement through the dirt would create a moving air pocket which would result in a suction against the middle of that blade. This suction could be designed to help hold the blade true to a desired path. This phenomenon is planned for, to varying degrees, in many plow shares. Some modern plows, such as the Pioneer models, which employ Oliver Raydex bottoms can be used with a short nosed share with no suction (if the work is being done in regularly farmed soil of good friability). Understanding share suction and the relative need for share suction can make the difference between a miserable time of plowing

and something close to heaven. On horsedrawn plows there are other points of the bottom where suction may be employed to advantage.

There are three points of bearing on an ordinary walking plow, namely, the point of the share, the wing of the share, and the

Fig. 71. Six Ohio belgians pull a two-bottom Pioneer riding plow at the '98 Horse Progress Days. Note the exceptional quality of the plowed ground and the clean straight furrow wall.

heel of the landside. These three points are the only points that actually come in to contact with the furrow sole, as can readily be seen as the plow rests on the floor. The curvature from the point of the share to the heel of the landside makes the 'vertical suction'.

The amount of bearing at the wing will greatly influence the operation of large walking plows. Plows mounted on wheels do not require wing bearing, as they are controlled by the lead of the furrow wheels.

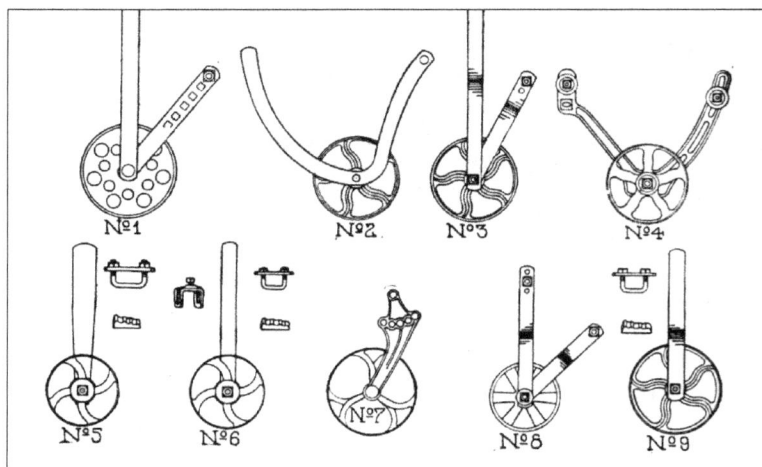

Fig. 72. An array of old walking plow depth wheel attachments from the Oliver Plow Company.

Design Basics

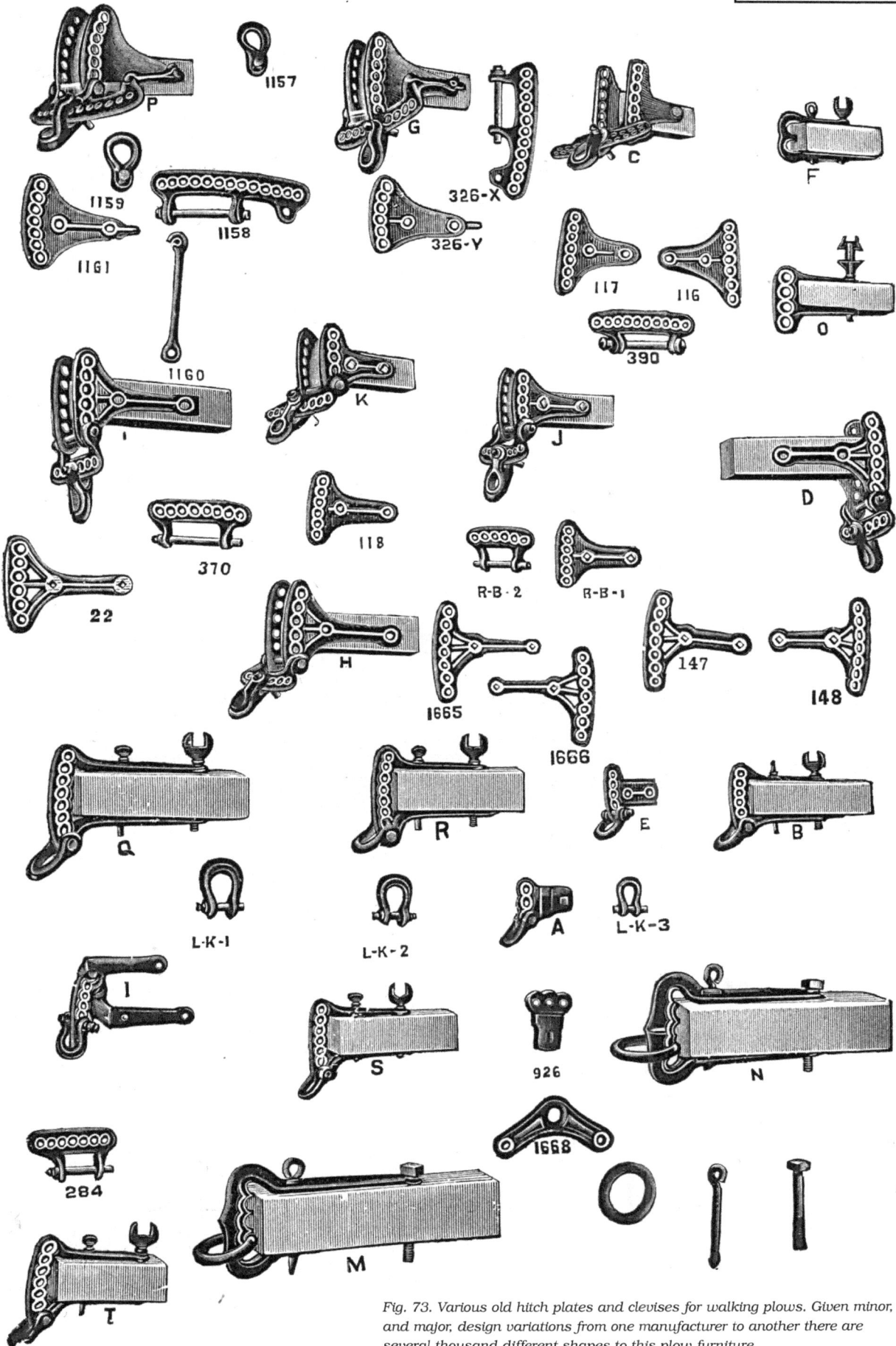

Fig. 73. Various old hitch plates and clevises for walking plows. Given minor, and major, design variations from one manufacturer to another there are several thousand different shapes to this plow furniture.

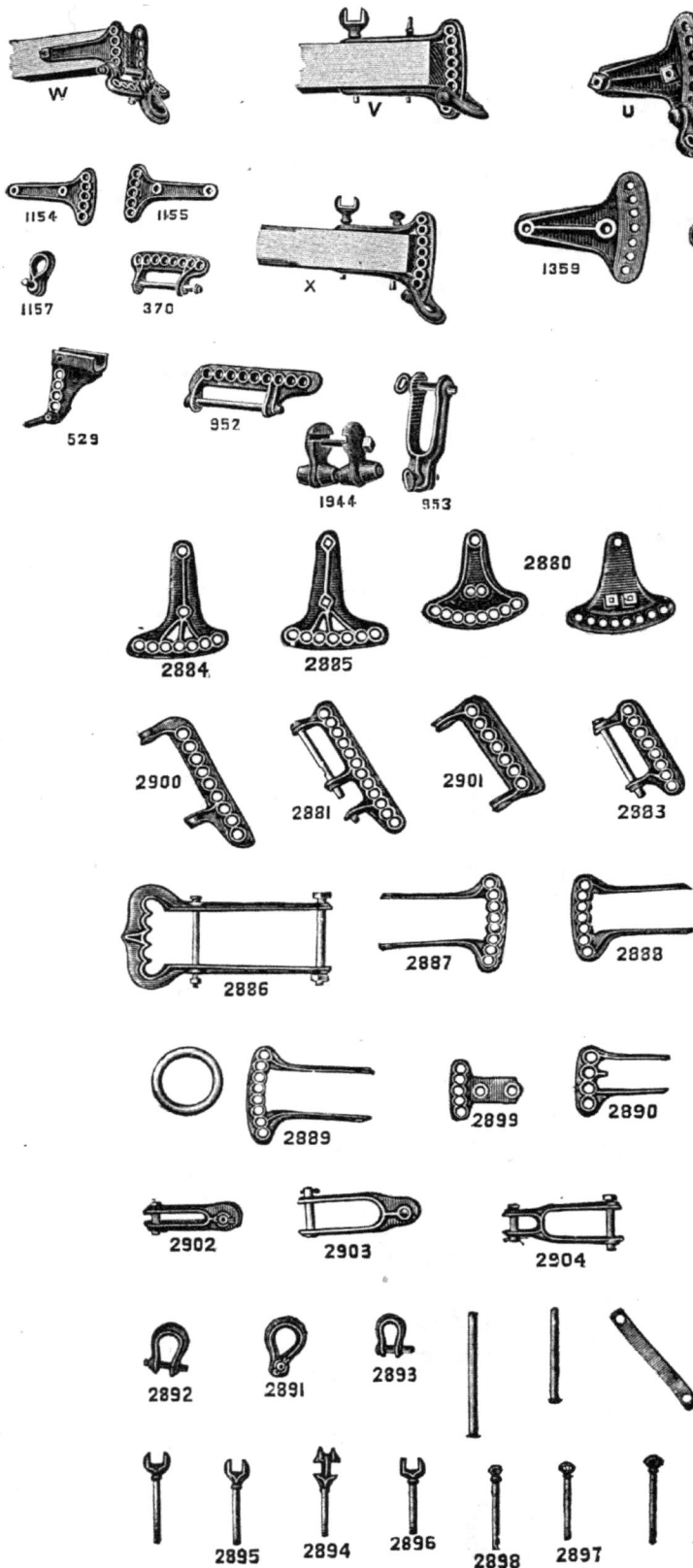

Fig. 74. Additional walking plow hitch plate designs plus clevises and hitch pins including some with wrench heads.

The amount will vary from almost nothing to 1 $^{1/2}$ inches according to the size of the plow and soil conditions. The wing bearing for a 12 inch walking plow is about $^{3/4}$ inch, 14 inch plow 1 $^{1/4}$ inches and for a 16 inch plow 1 $^{1/2}$ inches. The amount of wing bearing is measured as shown in Fig. 55.

**Vertical or Down Suction.** In this case we are talking about both the actual cup or downturn of the share and (in the case of wheeled plows) the relative operational angle of the plow. So this can be the actual bend downward of the point of the share to make the plow penetrate the soil to the proper depth when the plow is pulled forward. In which case the amount of suction will vary from $^{1/8}$ to $^{3/16}$ inch depending on the style of the plow and the soil it was made to work in. This suction can be measured on a walking plow by placing a straightedge on the bottom of the plow extending from the heel of the landside to the point of the share, then measuring vertically the greatest distance from the straightedge to the plow bottom (fig. 62).Or in the case of wheeled plows this can be the operating angle of the plow bottom (with more or less share suction). On all moldboard plows mounted on wheels, it will be noticed that the heel of the landside does not touch the floor when properly set; the vertical suction in this case will be the amount the heel of the landside is elevated above the floor (fig. 61). Ordinarily this is about $^{1/2}$ inch with average length landsides.

**Horizontal of Land Suction.** This is the amount the point of the share is bent out of line with the landside (fig. 56). The object of this suction is to make the plow take the proper amount of furrow width. Horizontal suction is measured by placing a straightedge on the side of the plow extending from the heel of the landside to the point of the share, then measuring horizontally the greatest distance from the straightedge to the plow bottom. The amount is usually about $^{3/16}$ inch.

**The Landside.** This is that part of the plow bottom which slides along the face of the furrow wall. It helps to counteract the side pressure exerted by the furrow slice on the moldboard. It also helps to steady the plow while being operated. Landsides are divided into three kinds, high, medium, and low. The size is determined by the style of the plow and the depth of the furrow it will cut. Landsides are considered low up to 4 inches; medium from 4 to 6 inches; and high from 6 to 8 inches in height. There are many types of landsides varying according to the shape. Some are made rectangular, others are flanged in various ways. Some are flanged on both sides at the bottom forming an inverted letter *T*. Those that are flanged only on one side form the letter *L*. They may be inclined inward under the plow to prevent the plow from sinking down at the rear. Since the heel of the landside forms one of the points of bearing, it is sometimes necessary to replace it, due to the wearing and the loss of suction in the plow bottom. Some landsides are built with detach-

Design Basics

**ONE
12" PLOW**

**12" GANG PLOW**
Total Cut 24"

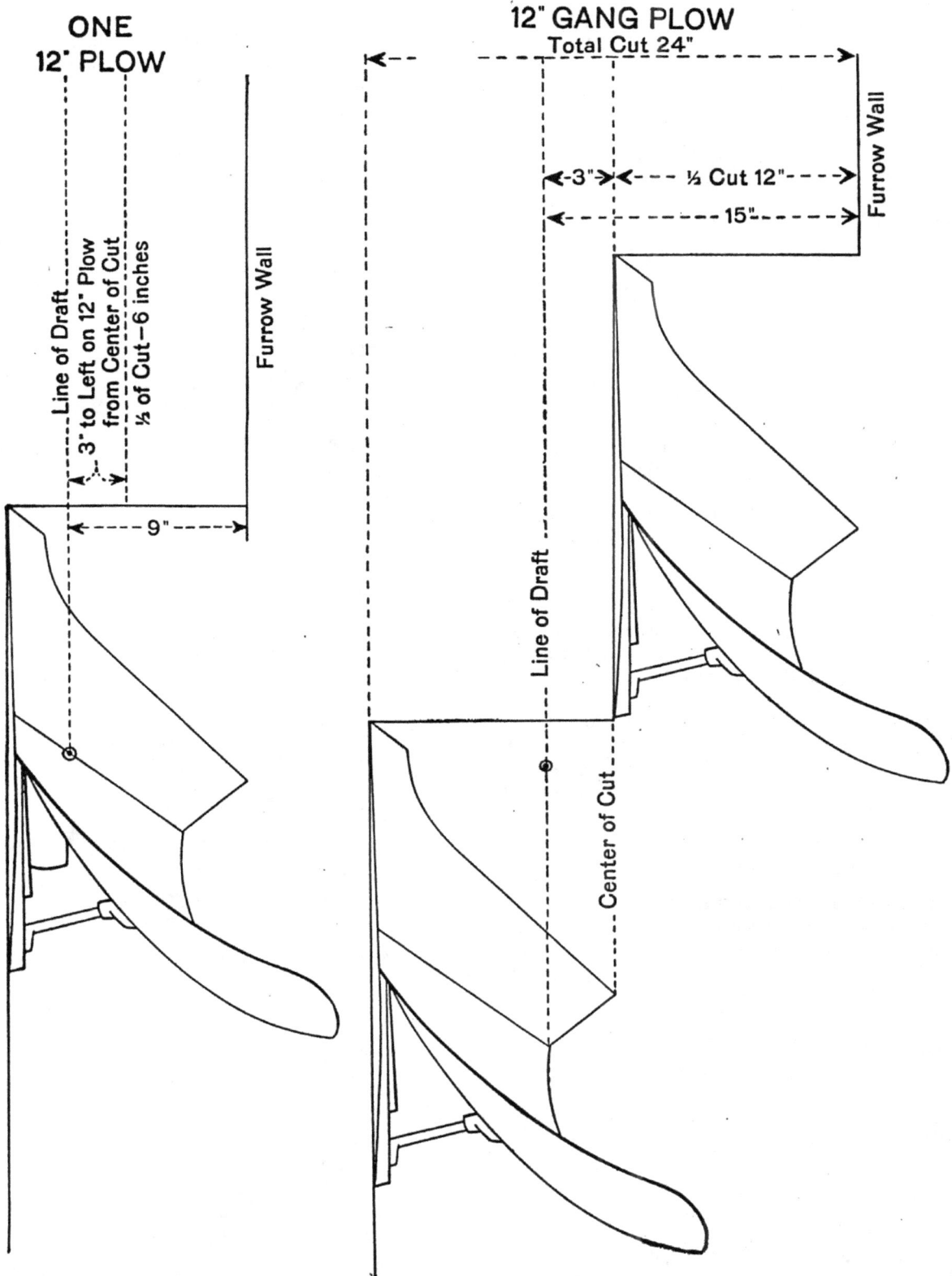

Fig 75. One of the most important illustrations in this book. You might want to flag this page for quick reference. Note the line of draft in each case. And note the point of draft. It is critically important that these two features be understood in order to set up the proper horizontal hitch point and the vertical point of contact with the plow beam. The line of draft is, in both cases, 3" left of center of cut. This important subject is covered (repeated) in each category of plows along with the chapter on Hitches.

Fig. 76. This photo of three Percherons pulling a Case foot lift riding plow in the heavy wet clay spring soil of Western Oregon demonstrates the resulting challenges. That ground will need to be harrowed the same day in order to avoid bricking.

able heels (Fig. 48) that may be renewed when necessary. The materials used in making landsides are: cast iron, solid steel, and soft-center steel.

**Setting Shares for Suction**. The plow bottom may be held to its work by the under-point suction of the share. Such suction is produced by turning the point of the share down slightly below the level of the under-side of the share. (See Fig. 53) The amount of suction necessary depends upon the type of plow and existing soil conditions. Stiff clay soils are harder to penetrate than light loam soils and require more suction in the share point.

Landside suction (see Fig. 58) in a plow share holds the bottom to its full-width cut. It is produced by turning the land suction, as well as the down suction.

The importance of having the correct amount of suck in the share cannot be emphasized too strongly. Too little under-point suction will cause the plow to "ride out" of the ground and cut a furrow of uneven depth. Too much will cause "bobbing" and heavy draft. In both cases, the plow is difficult to handle. If the landside suction is too great, the bottom tends to cut a wider furrow than can be handled properly, and the reverse is true when the landside suction is not sufficient.

The setting of walking plow shares and of riding and tractor plow shares is discussed separately, as the shares are different.

<u>Directions for setting walking plow shares</u> are as follows: Set the point of the share down so there is $1/16$ - to $1/8$ - inch suction, or clearance, under landside at point "A" (Fig. 53). The clearance, or under-point suction in the throat of the share should be $1/16$- to $1/8$-inch at point "B" (Fig. 54). All 12-, 14-, and 16-inch walking plow shares should have a wing bearing. The correct wing bearing (point "D", Fig. 55) is as follows: 16-inch plow, $1$-$1/2$ inches; 14-inch plow, $1$-$1/4$ inches; 12-inch plow, $3/4$ inch. A straight edge placed at rear of the landside (point "C") and extending to

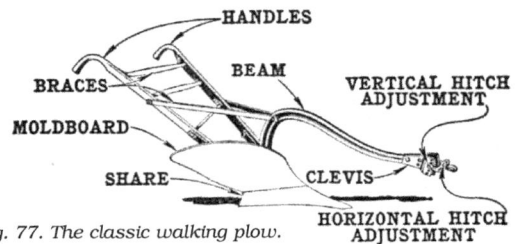

Fig. 77. The classic walking plow.

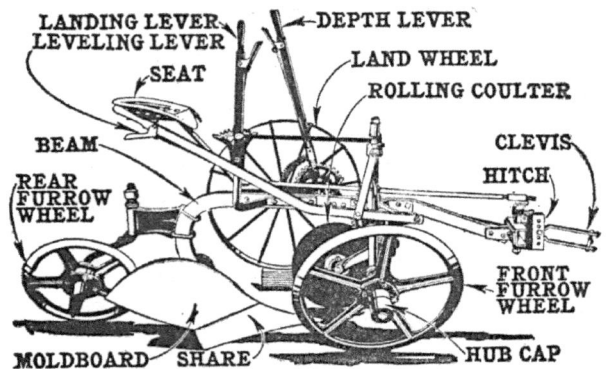

Fig. 78. The classic frameless style of riding plow with wheels mounted direct to the beam.

Fig. 79. The classic foot-lift frame sulky plow where the beam hangs in suspension.

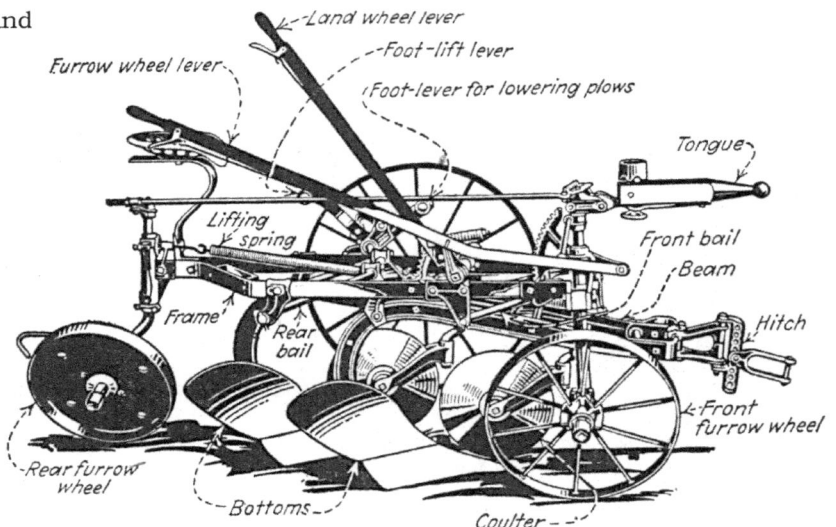

Fig. 80. The classic foot-lift frame gang plow.

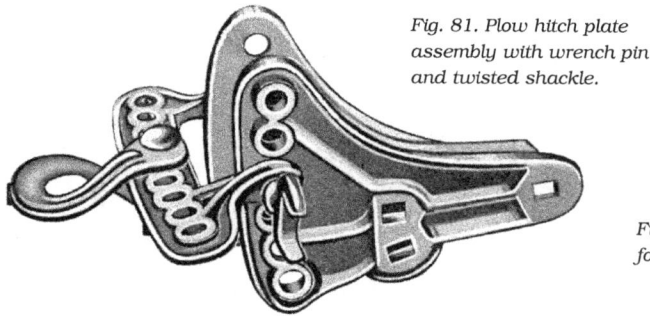

Fig. 81. Plow hitch plate
assembly with wrench pin
and twisted shackle.

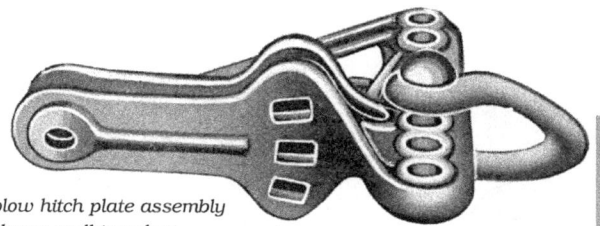

Fig. 82. A plow hitch plate assembly
for the one-horse walking plow.

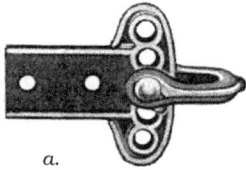

*a.*

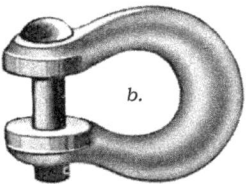

*b.*

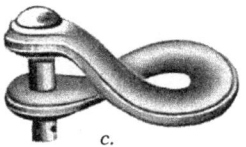

*c.*

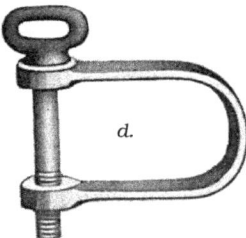

*d.*

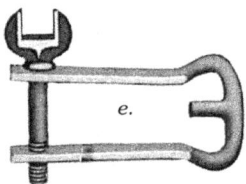

*e.*

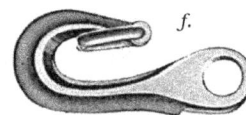

*f.*

Fig. 83. a. Simple small plow hitch
plate. b. single horse (direct)
shackle. c. twisted plow evener
shackle. d. flat threaded evener
shackle. e. two position evener
shackle with wrench pin. f. evener
hook with lock ring (in up
position).

wing of share should touch back of edge (point "D", Fig 55). When sharpening the share, care must be taken not to turn point to one side or the other. When fitted to the plow, there should be about 1/4-inch clearance or landside suction at "E" (Fig. 56).

For riding and tractor plows, set the point of the share down until there is 1/8- to 3/16-inch suction under the landside at point "A" (Fig. 53). See that clearance in throat of share at "B" (Fig. 54) is at least 1/8-inch. Set edge of share at wing point "F" without wing bearing (Fig. 57). For landside, set should be about 3/16-inch clearance at "H" (Fig.58).

**Care of the Bottom.** The plow bottom will give the best satisfaction when given the best care. If kept in good condition, it will give little scouring trouble. If permitted to rust, it may cause any amount of hard work and lost time.

*One of the first rules a plowman should learn is to polish the bright surfaces of his plow bottoms and apply a light coating of oil whenever the plow is not in use.* Strict observance of this rule will save many hours of difficulty in getting a rusted surface repolished. *A heavy coating of a good hard oil or grease should be applied to the bottoms when storing the plow from season to season.*

Plow manufacturers paint or varnish the surfaces of new plow bottoms to protect them from moisture from factory to user. This protective coating must be removed before the plow is taken into the field. This can be accomplished best by

## Plow Designer's Reach

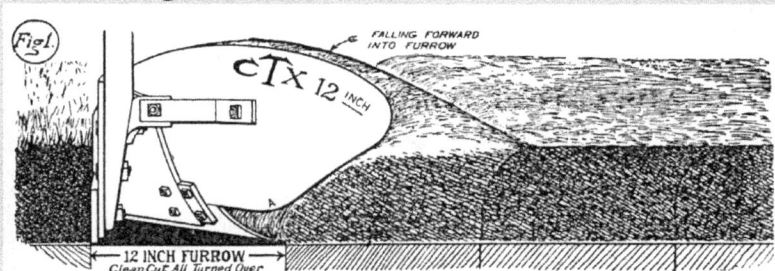

No Air Spaces in "CTX" Plowing

Note Air Spaces at B, C and D

Fig. 84. Beyond the basics of plow design many manufacturers offered a variety of special features with claims they improved the performance whether specific to a soil condition or in general. This 1922 cut, above, demonstrates how one company tried to sell an extended moldboard, share size remaining same, as a general improvement to plowing. The picture looks good but the claim of complete even pulverization is hard to accept.

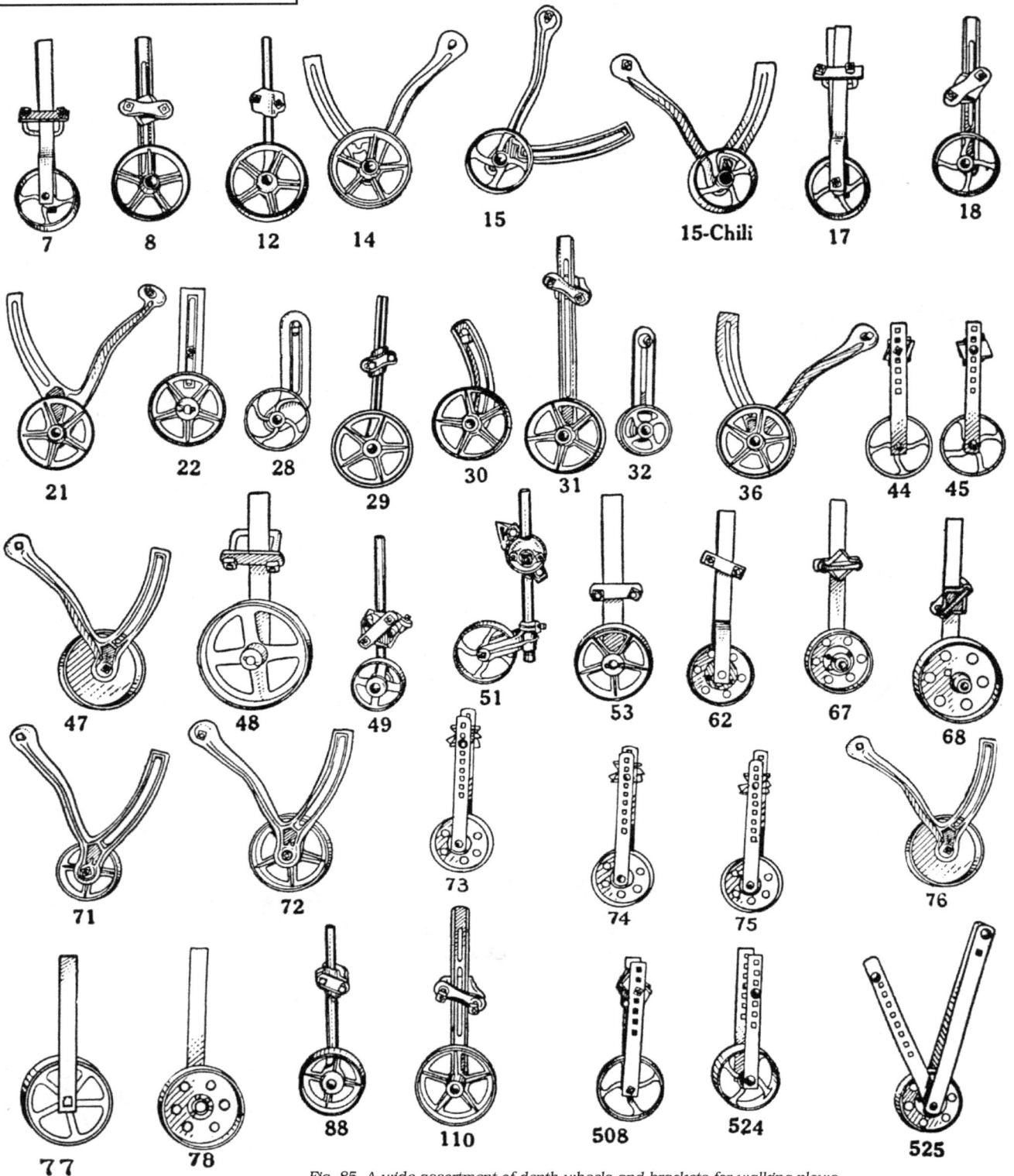

Fig. 85. A wide assortment of depth wheels and brackets for walking plows.

means of a paint and varnish remover which is obtainable at most paint stores. **Be careful not to scratch the polished moldboard and share surface**.

If the new plow is not to be used immediately after the protective covering has been removed, the bottoms should be oiled or greased, as the metal rusts readily if exposed to the air after paint removal.

In case a plow bottom becomes badly rusted, working it in a coarse sandy or gravelly soil will aid in restoring a land polish.

## Plow Attachments

**Rolling Coulter and Jointer**. One of the most important duties of the plow is to cover the stubble, stalks or other trash usually found on the surface of a field. Thorough covering of such matter hurries its decomposition and makes cultivation of future crops less difficult than when trash is left on top of the seed

Design Basics

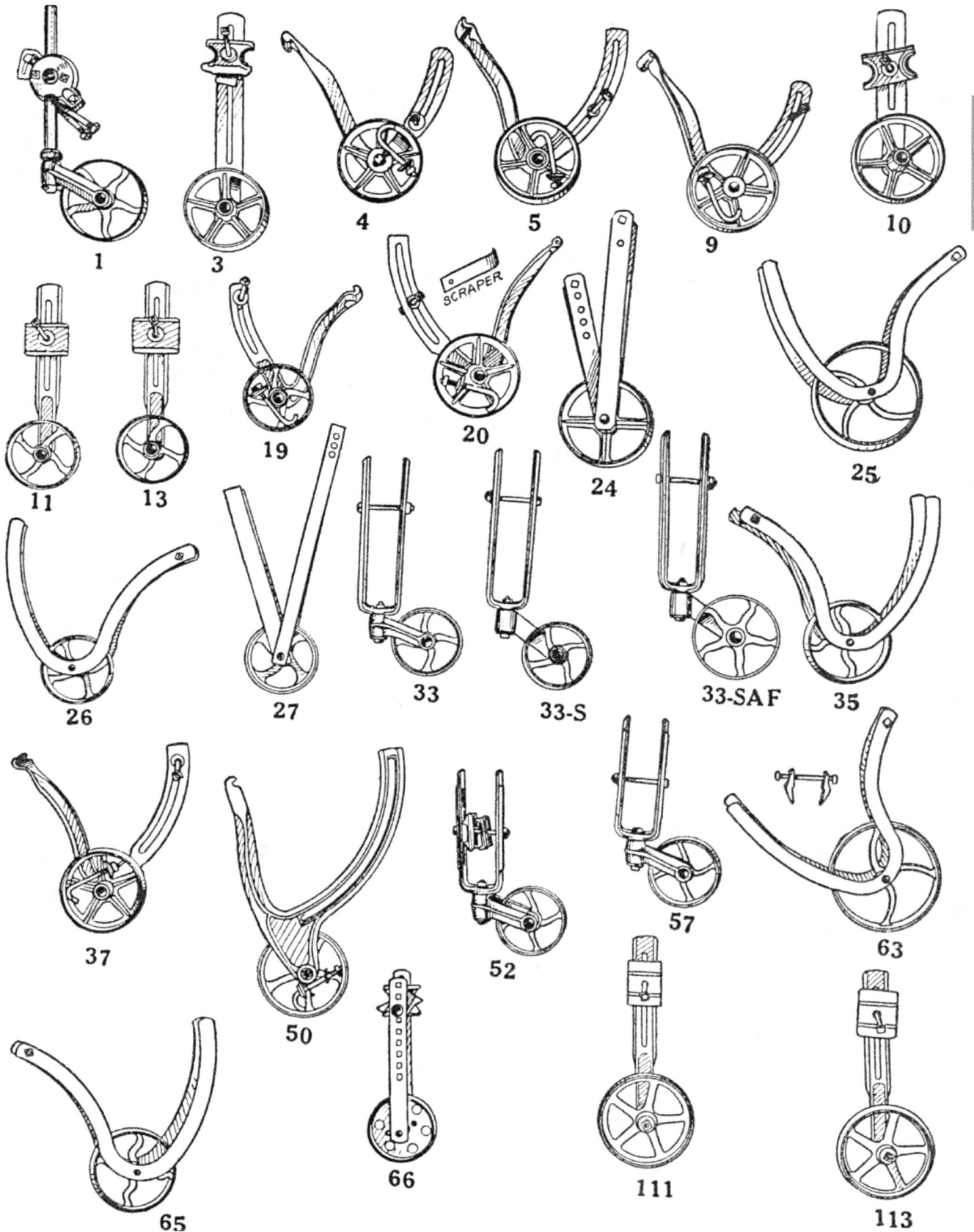

Fig. 86. Another set of assorted walking plow depth wheels and brackets. The major plow manufacturers of the turn of the century offered many hundreds of variations on each attachment.

Fig. 87. Jointers for both walking and riding plows.

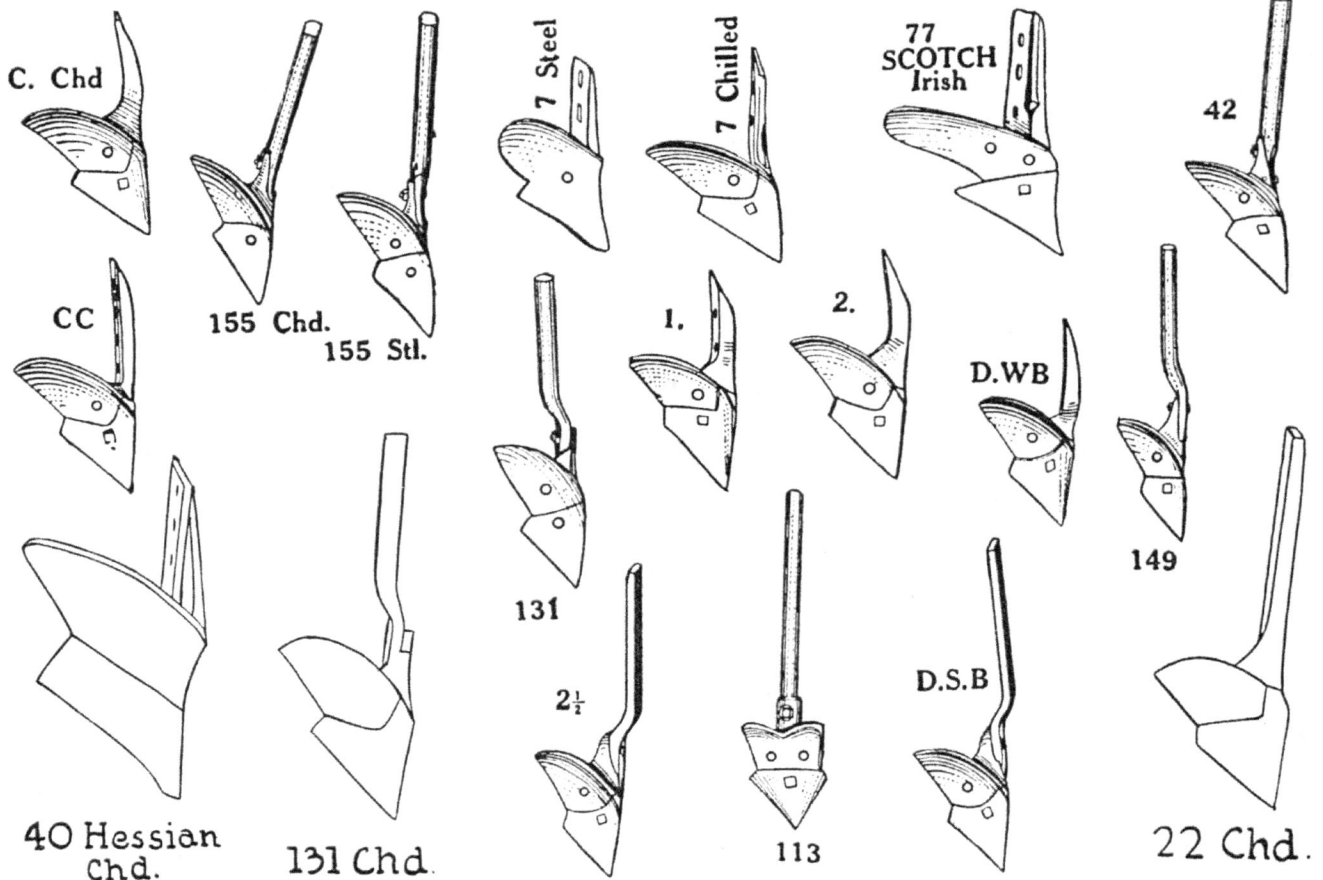

C. Chd    7 Steel    7 Chilled    77 SCOTCH Irish    42

CC    155 Chd.    155 Stl.    1.    2.    D.WB

131    149

2½    D.S.B

40 Hessian Chd.    131 Chd.    113    22 Chd.

bed to clog cultivating machines.

**The Jointer** is the most important single attachment found on plows, whether walking plows, sulky plows, or gang plows. The jointer is in reality a miniature plow, the purpose of which is to cut a small furrow off the main furrow slice, throwing it toward the furrow in such a manner that all stubble and trash are buried in the bottom of the furrow. A mistaken impression is held by some that the rolling coulter takes the place of the jointer. The functions of the two are quite different. The jointer should be set that its point is above or just back of the point of the main share. The jointer should run 1 1/2 inches to 2 inches deep. It should also be set to run 1/4 to 1/2 inch to the land side of the shin of the plow.

**The Coulter.** In some quarters 'coulters' are any and all plow attachments which cut ahead of the moldboard. This is why they are also called 'cutters'. In modern plow lexicon, in the western U.S., coulters are usually of the rolling type as in Fig. 67. The knife blade 'coulters' are commonly now referred to as knives.

The purpose of coulters and knives is to cut the furrow slice from the land, thus leaving a clean-cut furrow bank. If no coulter is provided, the shin of the moldboard must do the work of a coulter; but since in the absense of a coulter the face of the share begins to lift the furrow slice before the shin has had an opportunity to cut it off, a ragged job is sure to result.

The rolling coulter is not well adapted to the walking plow though it is sometimes used. The fin or

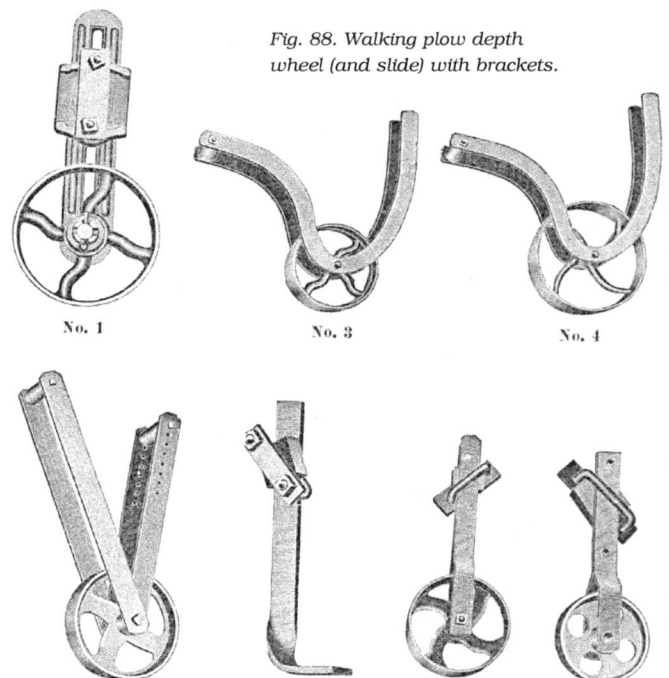

Fig. 88. Walking plow depth wheel (and slide) with brackets.

No. 1    No. 3    No. 4

the hanging cutter when used on walking plows improve the quality of the work and materially decrease the draft.

When the rolling coulter is used alone, it should

Design Basics

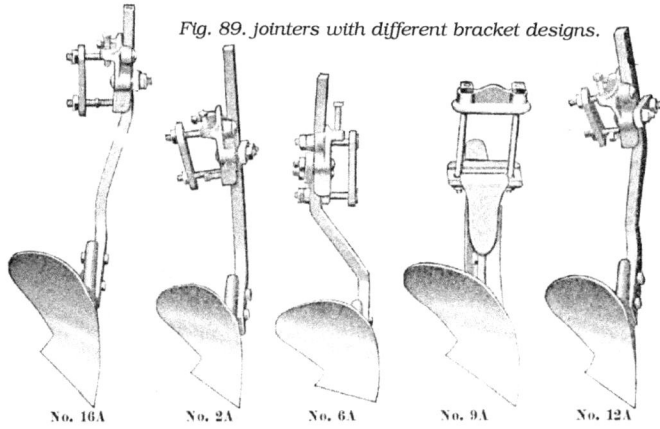
Fig. 89. jointers with different bracket designs.

No. 16A    No. 2A    No. 6A    No. 9A    No. 12A

aids to clean plowing surface trash and aids in securing a clean furrow wall, reducing the draft on the cutting edge of the plow bottom.

To get the best results with the combination rolling coulter and jointer, shown in Fig. 65, the hub of the coulter should be set about one inch back of the share point and just deep enough to cut the trash, about three to four inches in ordinary conditions. The jointer should cut about two inches deep. There should be about 1/8-inch space between the jointer and the coulter blade.

Keeping the rolling coulter sharp and well oiled and the jointer sharp will add greatly to the efficiency of their work.

**Depth wheels** (sometimes also known as *gauge* or *beam* wheels). The effect of the depth wheel is to cause the plow to run more steadily and thus to make the work easier for both man and team. The hitch should be adjusted so as to cause the wheel to press firmly but not too heavily on the ground. It should be set near the end of the beam and on a line with the landside. The use of a depth wheel always results in a more uniform

be set about 1/2- to 5/8-inch to the land (Fig. 66). The hub of the coulter should be about three inches behind the point of the share. In soil that does not scour well, more pressure on the moldboard can be secured by setting the coulter farther to land. If there is considerable trash on the field, the coulter should be set just deep enough to cut it - if set too deep, it pushes instead of cuts trash. The larger rolling coulters prove more effective in trashy conditions, as they mount trash more effectively than smaller coulters. (See Fig. 67) When plowing sod, the coulter must be deep enough to cut the roots below the surface, usually about one inch shallower than the share is cutting.

All other coulters or cutters (read knives) should, when it is possible, be set back of the point of the share, at a depth equal to about one half the depth of the furrow, and from one fourth to one half an inch outside the line of the shin.

The rolling coulter and jointer attachments for riding plows and tractor plows have proved to be big

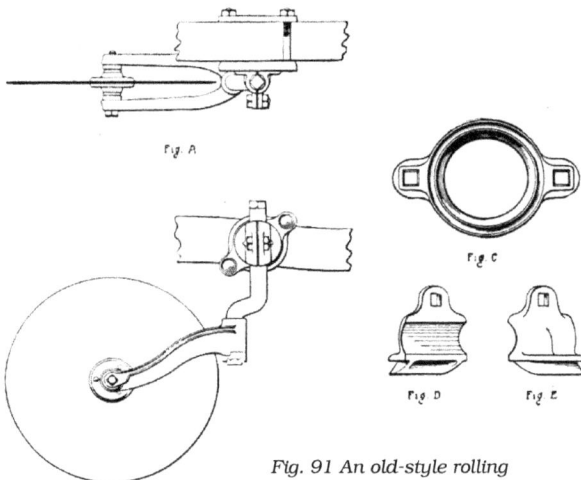
Fig. 90. Jointer and rotating bracket.

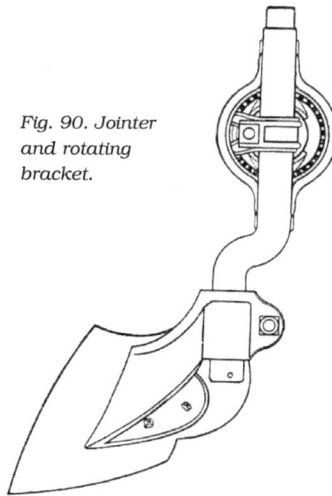
Fig. 93. A heavy-duty Jointer and adjustable bracket for a wood beam walking plow.

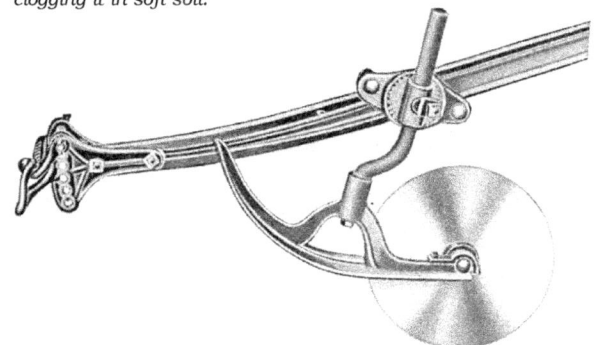
Fig. 91 An old-style rolling coulter and bracket.

Fig. A    Fig C    Fig D    Fig E

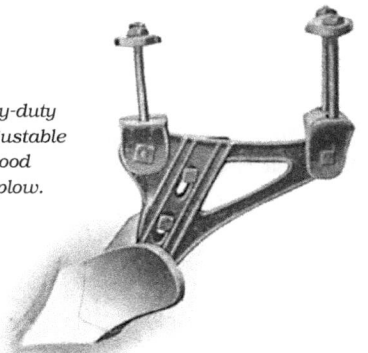
Fig. 92. A highly unusual 'moon bracket coulter'. The Avery Plow company came out with this special application walking plow coulter at the turn of the century. The patented 'shoe' prevents choking and clogging of plow or coulter when turning growths of mammoth clover, pea, strawberry, or other vines, stalks, weeds, or Johnson grass; or matted and tangled growths of any kind, no matter how high or dense. The shoe presses down and holds firmly all weeds and trash and prevents tough grass from doubling over the blade edge and clogging it in soft soil.

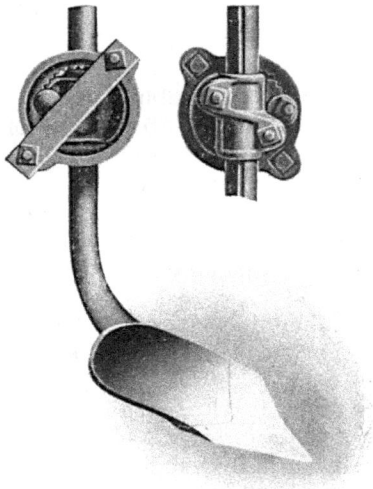

Fig. 98. Sharpening a rolling coulter. The blade is allowed to turn as it is sharpened. Extreme care should be observed when doing this.

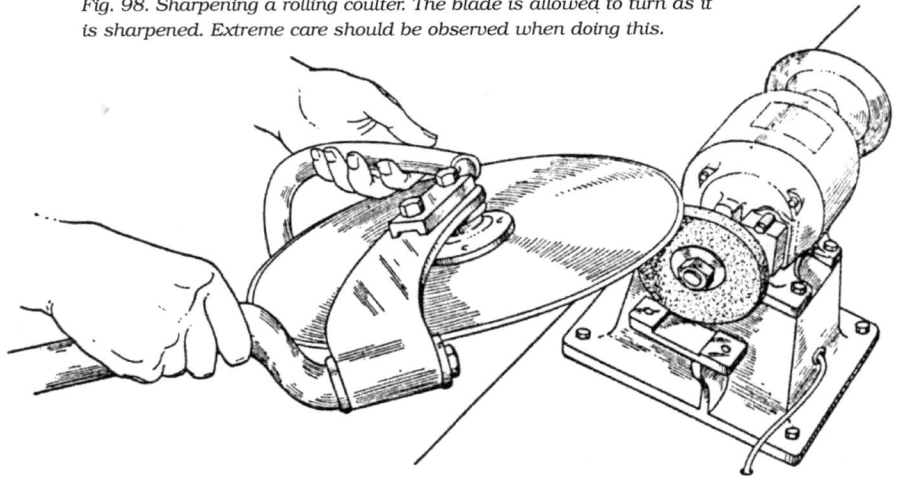

Fig. 94 A steel beam jointer assembly.

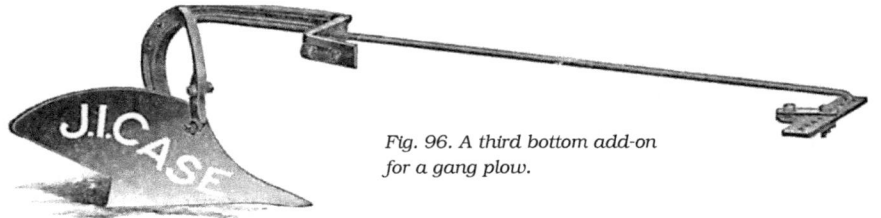

Fig. 96. A third bottom add-on for a gang plow.

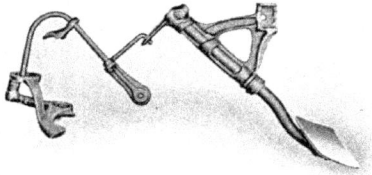

Fig. 95. A roll-over jointer set.up for hillside rollover walking  plows.

Fig. 97. An add-on seat assembly.

Fig. 99. This photograph shows a combination rolling coulter and jointer well set and working properly. Additional information on coulters and jointers appears in Chapter Four.

furrow and, in most cases, in a material reduction in draft. Two wheeled trucks are used to a limited extent, with favorable results, but their weight is against their common use.

**Miscellaneous Attachments**. There are many different lesser attachments for plowing including *weed hooks*, *grass rods*, and *trash chains*. These all are designed to help turn over or lay down tall vegetion while plowing (see figure 100). Illustrations of these options can be seen throughout the next several chapters.

Scattered throughout this book are unusual plow attachments as far flung as add-on bottoms, wheel assemblies, seats, lifting mechanisms, furrow harrows, subsoilers, umbrellas, etc. In most cases these attach-

ments increase the draft and thereby the power required to efficiently pull the plow. In a few cases these more obscure attachments might reduce draft.

*As a side note: one of the men who taught this author to plow, Charlie Jensen, rigged his favorite riding plow with a cup holder and a bracket to hold his transistor radio. Neither of those 'attachments' are available from plow manufacturers of today or of old. Which is a shame because we believe they might actually reduce the draft of some plows.*

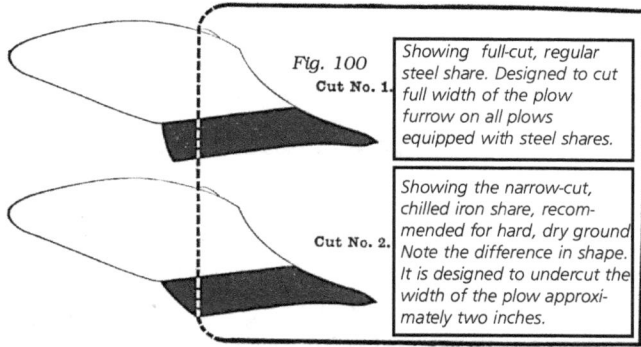

Fig. 100
Cut No. 1. *Showing full-cut, regular steel share. Designed to cut full width of the plow furrow on all plows equipped with steel shares.*

Cut No. 2. *Showing the narrow-cut, chilled iron share, recommended for hard, dry ground. Note the difference in shape. It is designed to undercut the width of the plow approximately two inches.*

Fig. 101. A John Deere gang plow doing good work in hard ground with narrow-cut cast shares.

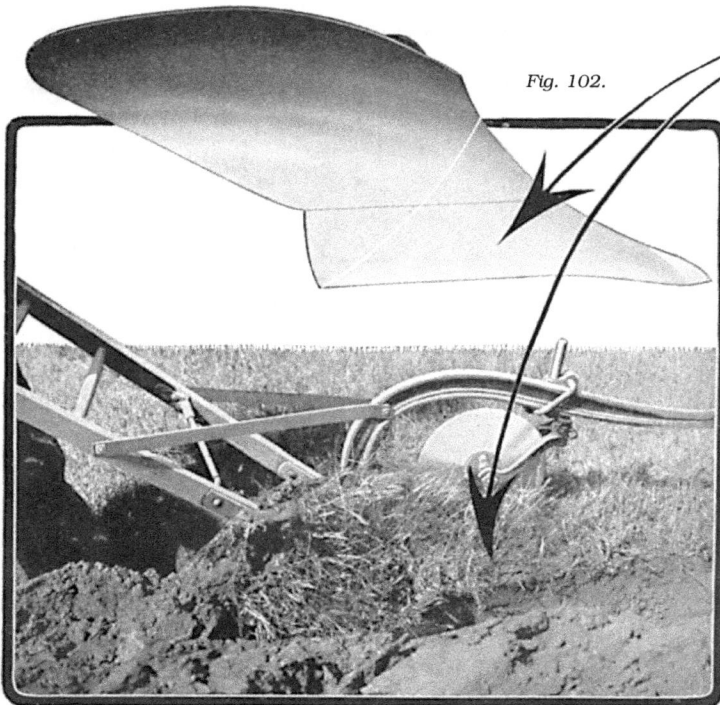

Fig. 102.

(A) Seventy-five per cent of the draft is created by the work of that part of the bottom below the white line drawn across the face.

thinking the ground will just be too difficult to get through.

Even where horses are not used, and the plowing is done with the tractor, the hot dry dusty field makes it harder on both the tractor and the operator; consequently, anything that can be done to lighten the draft and make plowing easier ought to be appreciated.

The use of the narrow-cut cast-iron shares for plowing when the ground is dry and hard is a good option. But it is important that the soil is not of a wet, heavy or sticky nature, necessitating steel shares. Fig 100, on this page, illustrates quite clearly the difference in the general shape of the narrow-cut cast share, compared with the full-cut steel share, as used on a 14 inch plow. The difference on the other sizes of plows is proportionately the same. In addition to being narrower at the wing, the share has greater clearance back of the point. This is what is commonly called "suction". This greater clearance permits the share to be used longer before the point becomes worn back enough to destroy this clearance or suction, necessitating a new share.

The narrow-cut cast share will penetrate much easier than it will when the full-cut steel share is used and will hold to the ground much better; it will pull lighter, and, due to the fact that the entire furrow slice is not cut off, the action of the moldboard in breaking the rest of the furrow over this, will have a pulverizing effect that will result in a better job of plowing.

This should not be taken as a recommendation for a cast share for year round plowing, because in the late fall plowing after the rains have come, and in spring plowing, the cast share would not scour, and would cause a bunch of trouble. These are recommended then for just the conditions described; hard soil, in summer and early fall plowing.

**Sharp Shares.** By keeping your shares sharp, and replacing them when necessary, your plows will run lighter.

Note the plow bottom with a line drawn across it in fig. 102. This shows clearly where the greatest draft occurs on a plow bottom. It has been estimated by plow engineers that 75 per cent of the total draft of the plow bottom in cutting and turning the furrow slice occurs in front of the line shown. Note that the share takes up the major portion of the area in front of this line, and if the share is dull or improperly shaped, it will materially increase the draft at this point. A sharp share will penetrate the ground much easier than a dull share. A sharp knife or a sharp ax is much easier to use than a dull one and does better work. Same way with the plow. The plow must cut its way through the soil; it pulls much easier when its share

### Shares for Summer and Fall Plowing

In the summer and early fall when the average weather is quite warm and the soil is frequently dry and sometimes very hard, plow draft is always a problem. Some plowmen will actually opt not to plow

*Additional information on coulters and jointers appears in Chapter Four.*

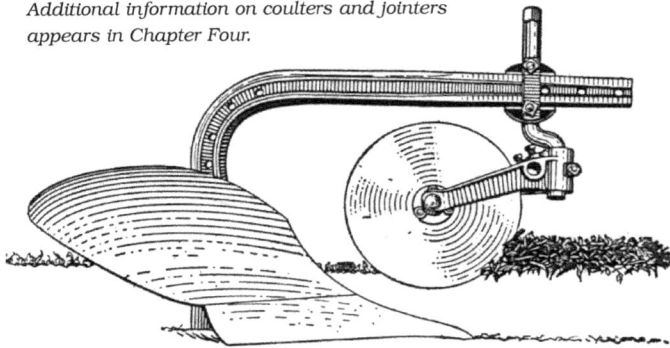

*Fig. 104. Point of jointer should be set slightly ahead of point of plow.*

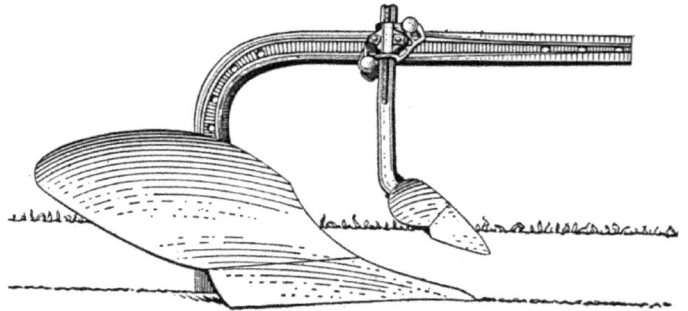

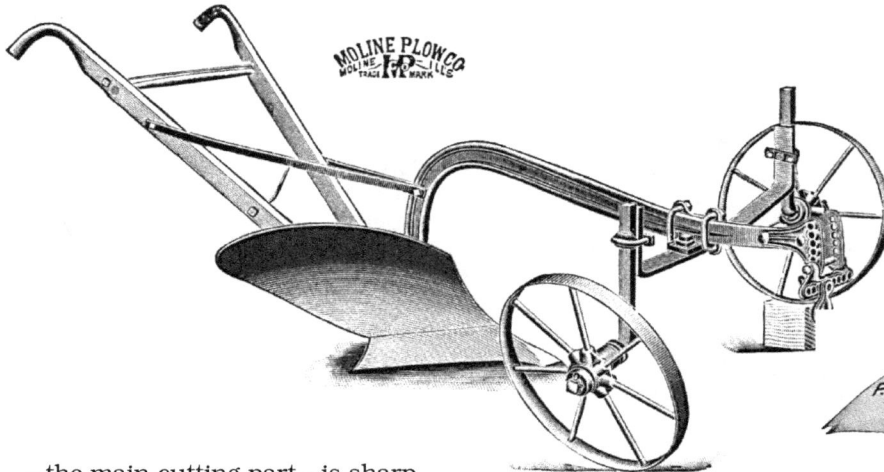

*Fig. 103. Coulter properly adjusted to cut trash*

*Fig. 105. The Moline Plow Company offered, as did other plow companies, a set of 'truck wheels' to fasten on to the end of the walking plow beam. These worked in the same way that depth wheels did, helped to reduce draft and hold the plow at a certain adjusted for depth. However, this particular option created a challenge for the plowman at the end of the furrow. 'Deadhead' travel with this plow require that the handles be held up and this was, is, hard work.*

- the main cutting part - is sharp.

There is still another factor that enters into the draft of the plow, and some other tools where the cutting edge becomes blunt. When the cutting edge is sharp and properly shaped, some other parts of the plow share and plow bottom are almost entirely free from friction, but when the cutting edge is allowed to wear blunt, these parts are exposed to wear and friction, which not only increases the draft of the implement, but rapidly wears away these parts.

Often the plow landside and the frog which supports the share become almost worn away, because the share had not been properly sharpened and shaped, and often the sleeve and block on cultivator shovels get in the same condition. Just imagine the power that is wasted in pulling these blunt implements through the field, and, in addition, worn-out parts of the implement that should never have become worn.

Refer back to page 33 of this chapter and to Chapter Twelve for sharpening and care information.

*Fig. 106. P & O Plow company offered this tag-on umbrella for its riding plows.*

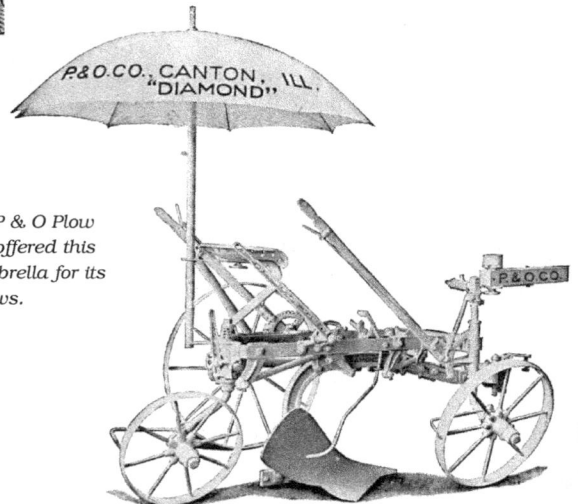

*Fig. 107. With this truck wheel assembly P & O was able to allow the plowman to lift the front end up enough that the plow ran, when out of the ground, on its landside heel.*

## Heavy-duty Fundamentals

Much of the information which immediately follows will be of questionable value to the person just interested in doing a good job of plowing. Pass it by if you choose. Some of the information has already been covered and/or will be in the next chapters. It's offered here, in this form, for the academics and scientists hungry for formulaic discussions of weight. To steer the anxious farmer from this section we offer it without illustration.

This material is taken from the 1937 volume of Farm Machinery & Equipment by Harris Pearson Smith, A.E., originally published by McGraw-Hill.

**Center of Resistance of a Plow Bottom**. - A point where all three of these forces meet is considered to be the center of resistance of load. It cannot always be determined just exactly where this point will be on a plow bottom, but it will usually come within the range of the following dimensions for a 14-inch bottom. Vertical forces will be in equilibrium 2 to 2-1/2 inches up from the floor; the horizontal forces 2 to 3 inches to the right of the shin; the longitudinal forces 12 to 15 inches back from the point of the share. Briefly, we can say that the center of resistance of any moldboard plow bottom will be on top about where the share and moldboard intersect and to the right of the shin. If two or more bottoms are used, the center of resistance will be the average of all the centers. For a two-bottom plow it would be half way between the center of resistance for the two bottoms. On a three-plow outfit it would be at the center of resistance of the middle bottom. Of course, the style of bottom as to shape, type of share and moldboard will influence the point where all the various forces acting on the bottom will be in equilibrium.

**Influence of Friction on Design**. - After taking all the above principles into consideration, they will resolve themselves into one general principle of plow design that must be considered in every type of plow, no matter whether it be stubble, general purpose, or sod. That principle is that friction will be the greatest at the point of the share and gradually decrease backward to the end of the moldboard. This can be seen readily on any plow bottom after considerable use. The greatest amount of wear is shown to be at the point and gradually decreases backward to the tip of the wing of the moldboard. This is why the stubble moldboard, which has a greater amount of curvature, gives better pulverization to the furrow slice. It is also seen that this type of moldboard will pick up the soil quicker and turn it over harder than any other type. That makes this type of plow more adaptable to plowing the loams and the sandy loam soils. The general purpose moldboard has a less amount of curvature than that of the stubble and it is in this class that the blackland type of plow will fall, because the curvature is not so pronounced as that of the stubble moldboard.

**Influence of Speed on Design**. - In the last few years there has been much agitation regarding the designing of plows for high speeds. It is not so difficult to design a plow for high speeds as it is to obtain pulverization. The bottom designed for high speeds must have gradual curves, which approach closely those of the sod type of plow. It can be seen readily that it is not necessary to have the moldboard as wide in this case in order to lift and invert the furrow slice. The higher velocity will carry the soil up over the moldboard, throwing it farther to the side. Much difficulty is likely to result from plows for high speed which must incorporate a

plow bottom of long slopes. They may scour well while going at a high rate of speed but when the speed drops to 2 or 3 miles per hour, the question is, will they continue to scour at this speed? Will they do the same type of work as at the higher speed?

## Draft of Plows

There is no doubt that plowing is the job on the farm that takes the most power. It is important that every effort should be made to reduce this power used to the minimum in keeping with good practice. The following factors must be considered in determining the actual draft of the plow:

Depth of plowing, width of plow, character of soil, moisture, previous treatment of soil, smoothness of surface, shape of moldboard, sharpness of share, rigidity of plow, and speed.

**Draft as affected by Depth of Plowing**. - It is almost impossible to make a plow run at a constant uniform depth no matter how well it may be adjusted. Naturally, the deeper the plow penetrates the soil, the more draft there will be. Tests indicate that a 14-inch plow will increase in draft an average of 92 pounds for each inch increase in depth. Taking into consideration the whole United States and the various conditions encountered in the various parts of the country, and the different types of plows used, whether walking, gang, or tractor plow, the average depth of plowing for all these conditions will be around 5 inches. The draft of any plow can be determined by an instrument called a dynamometer, which registers the pull or draft of the plow over a measured distance. Then, knowing the speed of the team and the time it took to travel this distance, the horsepower, as well as the average draft per unit of the cross-section of the furrow slice, can be determined.

Horsepower = $\frac{\text{force x distance traveled in feet per minute}}{33,000}$

**Plow Draft as Affected by the Width of the Furrow**. As in the case of the depth of the plow affecting the draft of the plow, the width of the furrow will also affect it. Naturally, the wider the plow, the more soil will be turned over and the more draft there will be. The width of furrow that a plow cuts cannot be controlled absolutely. This is especially true in the walking plow. With the gang plow, however, the bottoms are spaced equally and if the plow is properly adjusted to cut its proper width of furrow, and all other factors are working perfectly, each bottom will cut a constant uniform width of furrow. It does not follow that each furrow slice will give the same resistance. There are some natural influences that will affect the resistance of furrows. Usually, the draft of the plow is given in the number of pounds pulled per square inch of the cross-section of a furrow slice. To determine this, the depth and width of furrow must be considered. The number of square inches in a cross-section of a furrow slice can be determined by multiplying the depth by the width of the furrow. That is, a plow going 6 inches deep and cutting the furrow slice 14 inches wide will give a cross-sectional area of 84 square inches. Then, if the total draft for the whole plow is 500 pounds, the draft per square inch would be 500 divided by 84, or 5.95 pounds per square inch.

**Character of Soil**. - The character of the soil, whether it be sandy, clay, loam, or blackland will have a great deal to do with the number of pounds pulled or draft of the plow. A less number of pounds is required to pull a plow in sandy soil than in a stiff gumbo or clay soil. The draft of the plow is affected by soil conditions as well as the type of soil, and will range from 2 to 3 pounds up to 20 pounds per square inch.

**Moisture**. - The amount of moisture in the soil will also affect the draft or total pounds required to move the plow. The amount of moisture frequently will determine the time when the plowing should be done, whether it should be in the spring or in the fall. When there is a good season, or plenty of moisture in the soil, the draft is not so great as when plowing is attempted when the ground is hard and dry. When the ground is very hard, the plow will not penetrate the soil easily. This, in itself, will indicate that the draft will be increased, owing to lack of moisture.

**Previous Treatment of the Soil**. - The draft of the plow will be influenced to a considerable extent by the previous treatment to which the soil has been subjected; that is, whether it has been properly plowed and cultivated, whether the crop planted on the soil before was cultivated, whether it was harrowed by a disk harrow before being plowed, or allowed to go untreated. The amount of straw and organic matter that may have been covered by a previous plowing will also affect the draft because organic matter will cause the soil to become mellow and break up easier.

**Smoothness of Surface**. - If the surface of the soil is uneven, naturally, the water will collect in the low places and leave the high places without the required amount of moisture. Then, of course, when the plow comes to the moist places it will plow easier than where there is a lack of moisture, so that the draft is affected directly from the unevenness of the ground. While, if the plow is a gang plow of the unit type and the surface is not even, some of the plows will go much deeper than others, causing an overload. Up and down hill causes heavy draft one way and light the other.

**Shape of Moldboard**. - As has already been indicated, the draft of the plow will be affected by the shape of the moldboard, whether stubble, general purpose, blackland, or sod. Tests indicate, to some extent, the difference of draft as affected by the shape of the moldboard. The results of these tests show that the stubble moldboard gives a greater draft than that of the sod moldboard, and the general purpose come in between these. The conclusions were that the more abrupt the curve, the greater the draft. The less curvature there is to the moldboard, the less the pulverizing action upon the furrow slice and, naturally, the less pressure will be exerted upon the surface of the moldboard, resulting in less draft. The tests at Ames, Iowa, show that the type of bottom did not materially influence the draft; that an increase in speed produced about the same increase in draft with any type of bottom. Upon analyzing the results, it is shown that a sod bottom has a long section of furrow slice which is carried on the share and moldboard and it must be pushed off. The greater area in contact results in a corresponding increase in frictional resistance and draft.

**Sharpness of Share**. - The share must cut the furrow slice loose from the ground and a large percentage of the draft of the plow results from cutting the furrow slice loose. As a result of some tests to determine the draft necessary for cutting and turning the furrow slice, and the draft of the plow alone, the conclusion is:

*The draft of the plow on the ground, 18 per cent; draft due to turning furrow slice, 34 per cent; draft due to cutting slice, 48 per cent.*

Thus, it is seen that practically 50 per cent of the total draft of the plow is used in cutting the furrow slice. A test was run to determine the effect of dull shares and sharp shares upon the draft of the plow. In a test on sandy loam soil the difference in draft of a sharp share was almost negligible. In a field of bluegrass sod there was a difference of 14 per cent in favor of the sharp share. In soil that is soft and mellow the sharpness of the share will not matter so much, but if there are many roots or the soil is comparatively hard or lacks moisture, a sharp share is to be advocated

**Hitch**. - The angle of hitch will also affect the draft of the plow. If the angle is short and sharp and the implement hitched close to the point of power, there will be a tendency to lift the plow which will take some of the weight of the plow off the ground and slightly decrease the draft. The reverse will be true if the hitch is farther away.

**Rigidity of Plows**. - Some plows may not be constructed rigidly enough to secure a uniform depth of penetration and a uniform width of furrow. It is important that they should be, because of the effect that the depth and width of the furrow will have upon the draft of the plow.

**Speed**. - Some tests have been made to determine the effect of speed on the draft of plows. All these tests have shown conclusively that there is an increase in draft as the speed increases. The results of the tests were as follows:

*In clay loam speed 1 mile per hour - draft, 100 per cent. Speed 2 miles per hour - draft 100 to 114 per cent. Speed 3 miles per hour - draft 128 per cent. Speed 4 miles per hour - draft 142 per cent.*

*Tests in Iowa black-loam soil gave the following results:*
*Speed 1 mile per hour - draft 100 per cent. Speed 2 miles per hour - draft 117 per cent. Speed 4 miles per hour - draft 126 per cent.*

The conclusions were: that an increase of the field speed of a plow with a general purpose moldboard, from 2 to 3 miles per hour, resulted in an increase of draft from 8 to 12 per cent, varying with the soil. Doubling the speed will result in an increase of draft from 16 to 25 per cent. The amount of work accomplished is increased from 50 to 100 per cent, respectively. It is to be remembered that practically 50 per cent of this task of plowing is consumed in cutting the furrow slice. The conclusions reached by Collins in his tests in Iowa, in 1920, were that the increase in draft, due to speed, is applied to that part of the total which is required for turning and pulverizing. This varies with the speed form less than one-third to about one-half the total draft of the plow within a range of 2 to 4 miles per hour.

Studies made in Ohio indicated that the average increase in draft, due to increased speeds, with two bottoms

## Rate of Travel

| Miles per hour | Feet per minute | Miles per hour | Feet per minute |
|---|---|---|---|
| 1 | 88 | 4 | 352 |
| 1¼ | 110 | 4¼ | 374 |
| 1½ | 132 | 4½ | 396 |
| 1¾ | 154 | 4¾ | 418 |
| 2 | 176 | 5 | 440 |
| 2¼ | 198 | 5¼ | 462 |
| 2½ | 220 | 5½ | 484 |
| 2¾ | 242 | 5¾ | 508 |
| 3 | 264 | 6 | 528 |
| 3¼ | 286 | | |
| 3½ | 308 | | |
| 3¾ | 330 | | |

was 1.17 pounds per square inch of furrow slice for each mile per hour increase in speed.

**Chart showing acres covered per hour with different widths of implements at various speeds.**

| Ft.-In. | Acres per Hour | Miles per Hour |
|---|---|---|
| (1) - 12 | 1/8 | 1 |
| 14 | 5/32 | |
| 16 | 3/16 | 1 1/4 |
| 18 | 7/32, 1/4 | |
| 20 | 9/32, 5/16 | 1 1/2 |
| 22 | 11/32, 3/8, 13/32 | 1 3/4 |
| (2) - 24 | 7/16, 15/32, 1/2 | 2 |
| 26 | 9/16 | |
| 28 | 5/8, 11/16 | 2 1/4 |
| 30 | 3/4 | 2 1/2 |
| 32 | 7/8 | 2 3/4 |
| 34 | | |
| (3) - 36 | 1 1/8 | 3 |
| 38 | 1 1/4, 1 3/8 | 3 1/4 |
| 40 | | |
| 42 | 1 1/2, 1 5/8 | 3 1/2 |
| 44 | 1 3/4, 1 7/8 | 3 3/4 |
| 46 | | |
| (4) - 48 | 2, 2 1/8 | 4 |
| 51 | 2 1/4, 2 3/8 | 4 1/4 |
| 54 | 2 1/2, 2 5/8 | 4 1/2 |
| 57 | 2 3/4, 2 7/8 | 4 3/4 |
| (5) - 60 | 3, 3 1/8 | 5 |
| 63 | 3 1/4 | 5 1/4 |
| 66 | 3 1/2, 3 3/4 | 5 1/2, 5 3/4 |
| 69 | 4 | 6 |
| (6) - 72 | 4 1/4 | 6 1/4, 6 1/2 |
| 75 | 4 1/2, 4 3/4 | |
| 78 | 5 | 6 3/4, 7 |
| (7) - 84 | 6, 5 1/2 | 7 1/4, 7 1/2 |
| 90 | 6 1/2 | 7 3/4, 8 |
| 96 | 7, 7 1/2 | 8 1/2 |
| (8) - 96 | 8, 8 1/2 | 8 3/8, 9 |
| 102 | 9, 9 1/2 | 9 1/4, 9 1/2 |
| (9) - 108 | 10, 10 1/2 | 9 3/4, 10 |
| 114 | 11 | |
| (10) - 120 | 12, 11 1/2 | |

Usual Range of Horse Speeds

**Rate of Plowing**. - One 14-inch plow bottom pulled at the rate of 2-1/2 miles per hour will plow approximately 11/32 or 0.3 acre in 1 hour. Some time, however, must be allowed for turning which will depend on the shape and size of the field and how it is laid out. For example, with a two-plow outfit in a field 80 rods long, where lands of average width are struck out and the turning is done on headlands, about 6 per cent of the time is spent in turning at the ends. The chart above shows the acreage plowed with different width plows when drawn at different speeds.

A four-plow outfit, of course, will accomplish about twice as much as the two, if both are run at the same speed; and a six-plow outfit twice as much as the three-plow outfit. One acre contains 43,560 square feet, or 160 square rods. A 14-inch furrow 1 mile long equals 6,160 square feet.

**Draft as Affected by Attachments**. - A study found that two 10-foot wires increase the draft about 2 per cent, an ordinary jointer absorbs about 7 per cent of the power when used with a coulter, and the jointer alone requires less power than a combination coulter and jointer. They found that the covering wires, coulter, and jointer together absorb between 10 and 15 per cent of the total power required to pull the plow with attachments.

**Effect of Grades**. - When on a grade, the effective drawbar pull of a tractor is lessened 1 per cent for each per cent of grade. For example, the weight of the tractor ready for work with an operator and a three-bottom plow is approximately 7,600 pounds. To negotiate a 10 per cent grade with this outfit would require an additional power equivalent to a pull at the drawbar of 760 pounds.

**Other Factors Affecting the Draft of Plows.** - Scouring of plows will influence the draft. If there is a smooth, polished surface for the soil to slip over, it is obvious that there will be less friction and draft.

The weight of the implement cannot be overlooked. A heavy bulky machine will pull heavier than one that is light.

**Draft of Disk Plows**. - Practically all draft tests on plows have been made with the moldboard type. The few made with the disk plow seem to indicate that it is slightly lighter in draft than the moldboard, when plowing under similar conditions and turning the same volume of soil. The type of soil is the greatest external factor to consider in the draft of any plow. In very hard ground, it is often necessary to add weight to the wheels to force the plow into the soil. Of course, the added weight will create more draft.

Factors incorporated in the plow are very important. The bearings of the disk-plow blade affect the draft. According to tests conducted, a plain cone bearing will pull 23 per cent heavier than a ball or roller bearing.

The type of scraper used to clean the disk will also affect the draft. Tests indicate that the revolving type gave slightly less draft than the spade type.

Fig. 108. A tall crop of timothy is plowed under for fertilizer value. A trash or drag chain is hooked to the end of the double tree and back to the plow beam with an adjusted arch set to fold over the grass for full cover by the plowed ground. See Fig. 150 for another view of drag chain set up.

# Chapter Three

# Walking Plows

*In this chapter we'll discuss what a walking plow is, how some of the models vary, and how to use the plows with a single horse, a team of two, and a three abreast. Actual field layout is covered in Chapter Eight. Additional information on hitch variables and eveners can be found in Chapter Ten. Augmenting what you will find in this chapter, those new to the work of plowing with horses will find additional infomation in chapter nine.*

**The Ordinary Walking Plow**. - To dispense with what should be obvious. What makes a walking plow a walking plow is not that the plow walks but rather that the operator walks.

The walking plow was the first truly successful type of plow developed. Many men worked for centuries upon

the development of the walking plow but the steel plow was not developed until near the middle of the eighteenth century. Some people believe that the first successful walking steel plow was invented by John Lane, Sr., in 1833.

Fig. 109. A steel beam P & O stubble plow for use in 'old ground' regularly farmed.

Fig. 110. A bottom view of a P & O wood beam stubble plow.

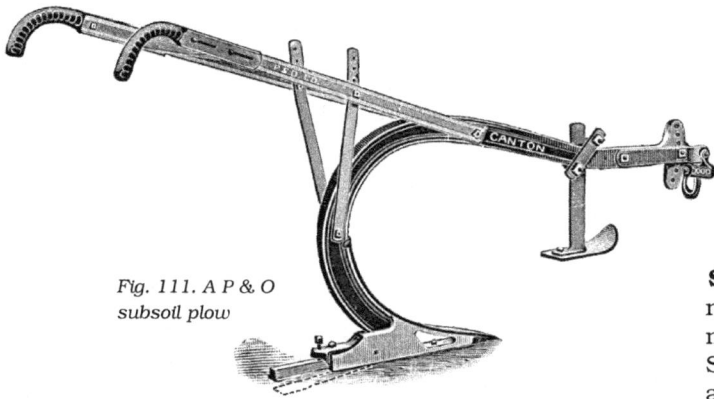

Fig. 111. A P & O
subsoil plow

Walking plows are referred to according to the material in the bottom as steel, chilled and, in a few cases, cast iron. The various parts composing the walking plow, such as the share, moldboard, landside, frog, beam, handles, and clevis, already have been discussed and/or illustrated. There may be right-handed or left-handed plows according to the direction in which they throw the furrow slice. Most north american plows throw right, although amongst some Amish communities left is favored. There appears to be no measureable or practical reason for right versus left. It appears to be a case of cultural preference.

## Special Walking Plows

**The New-ground Plow**. - (See Fig. 155) This plow is especially designed for ground that has been cleared of brush, leaving the soil filled with troublesome roots. It is built with the idea of simply breaking the surface of the soil and at the same time, cutting all roots. It's bottom is constructed with the share and moldboard in one piece. Wheels are sometimes used to take the place of the landside. It also has a hanging coulter of the knife type to aid in cutting the roots. The beam may be made of either wood or steel. The share size varies from 7 to 10 inches.

**Reversible Hillside Plows**. - Hillside plows consist of walking plows where the moldboard and share are hinged at the bottom and can be reversed either to the right or to the left. The plowman is able to make a right-handed plow into a left-handed plow by swinging the bottom underneath to the left. They are used in fields where all the furrow slices are to be thrown in the same direction, as on hillsides, from which they get their name. They are good plows for experimental

plots and irrigated fields. They are also good for plowing out irregular shaped fields and in corners. No dead furrow is left when this plow is used. One disadvantage of this plow is that the perfectly symmetrical arrow-shape required of the moldboard is not generally suited for all soil conditions. This means that specifically difficult lands will challenge this plow.

**Subsoil Plows**. - In some parts of the country it is necessary to break the subsoil to aid in the retaining of moisture and to give a larger root zone for the plants. Such a plow is called a subsoil plow. Instead of having a share and moldboard as in the ordinary walking plow, (See fig. 111.) these parts are almost entirely done away with. Extending downward from the beam is what is called the standard, which is made of steel. The front edge of this standard is sharpened, making a heavy knife. The shoe is attached to the bottom of the standard. This shoe is constructed on the order of a small share which has considerable vertical suction. The walking type of subsoil plow is used in the bottom of the furrows behind the ordinary walking plow. This allows the subsoil plow to penetrate to a greater depth,

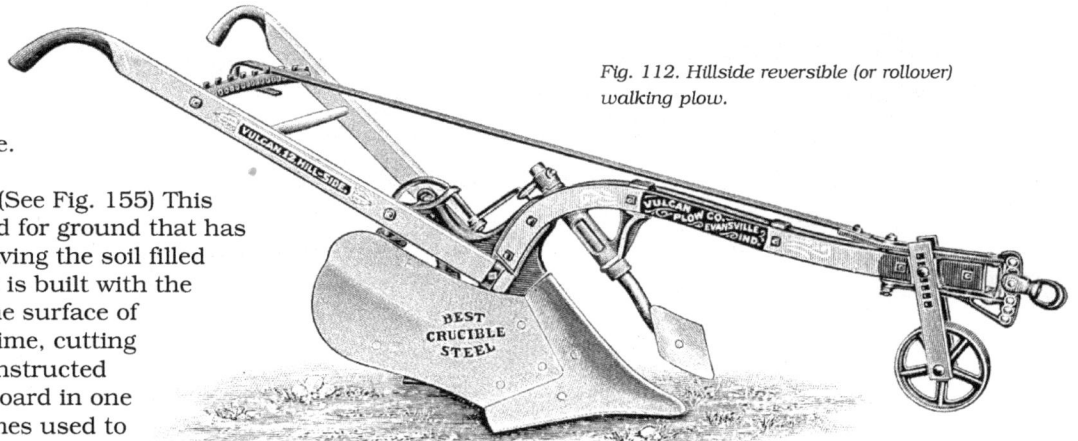

Fig. 112. Hillside reversible (or rollover) walking plow.

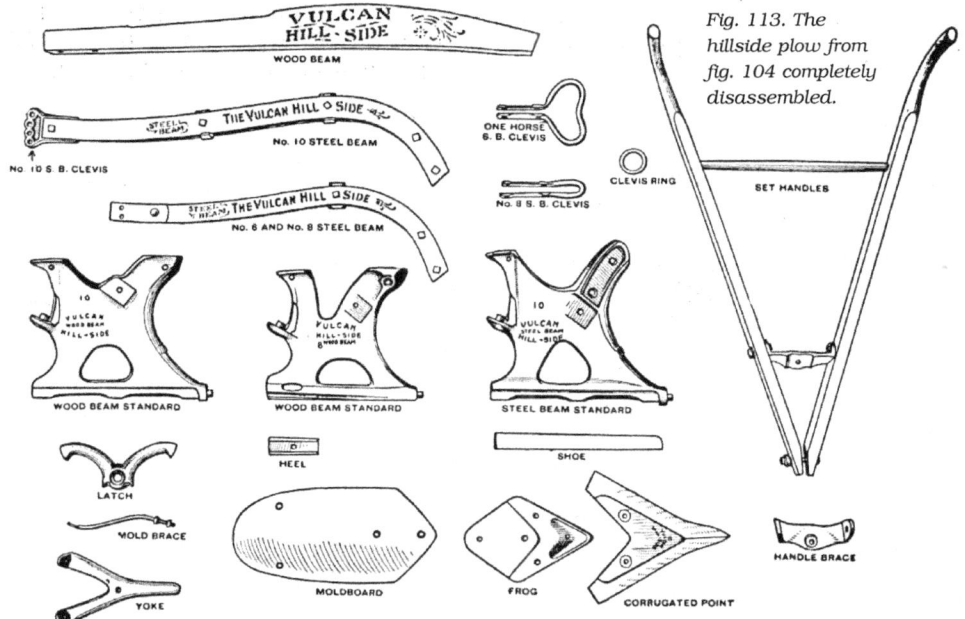

Fig. 113. The hillside plow from fig. 104 completely disassembled.

up ditches. Here it is called a ditcher. However, it is more commonly known as a middlebreaker. It is constructed with two moldboards, one for turning the soil to the right, the other for turning it to the left. The share is a doublewing affair to take care of both the right and left boards. This plow, instead of having a landside, has what is called a rudder; it acts in about the same way as a landside on an ordinary walking plow. There is a knife or rudder blade attached to the bottom of the rudder which cuts down into the soil and prevents it from dodging to the side.

**The Georgia Stock**. - The Georgia Stock may be classed as a walking plow. This stock consists of a beam, handle, and a shank where the plow shape or shovel is attached. It is a one-horse, one-man outfit. A poor job of plowing is done with it and it is a very hard

*Fig. 114. Jim Claffey with Cap and Bud at a recent New Jersey plow match. Note the line passing over one shoulder and under the other, the safest system.*

*Fig. 115. Two new Pioneer walking plows employing Oliver raydex bottoms. Note the drag bar on the right handle. This permits the plowman to lay the plow over on the moldboard without having the wood handle drag in the dirt. The author plows with one of these beauties and can attest to the superior work they are capable of.*

*Fig. 116. Halsey Genung with Blaze (furrow) and Frank on the land. Note Halsey's relaxed manner. It is finesse and not brute force that makes the plowman.*

loosening the subsoil beneath the furrow slice.

**Middlebreaker**. - This is a special type of walking plow which gets its name from the work it is required to do. In the South, where the middles in between the rows are burst out, it is called a middlebreaker. In the semi-arid sections of the country, where the crops are planted in the bottom of the furrow, it is called a lister. This same tool may be used in an irrigated country for opening

*Fig. 117. Another New Jersey plowman, John Allen with three fine Belgian cross mules, Dolly, Judy, & Polly. Note the use of the jocky stick on the outside, off, mule to keep her head away from the center mule.*

Fig. 118. Roger Burger does a championship job of plowing with two fine draft horses, Smokey and Max.

Fig. 119. James Brown plows deep with Dick and Charlie at the Howell Living History Farm Plowing Match for 1999

tool to adjust to do good work under the most favorable conditions.

**Vineyard Plow**. - The vineyard plow is a special built plow for working in vineyards, where it is necessary to plow close to the vines, yet, at the same time, prevent the handles from injuring the fruit and foliage. The handles on this plow are adjustable to the side to allow such work to be done.

The above plows are a smattering of the different styles that were manufactured. A walk through Chapter Eleven will quickly show the hundreds of variables that were once available.

**Distribution and Use**. Once upon a time every farm had at least one walking plow. Today maybe half the walking plows in north america are yard ornaments. Which is unfortunate because, for this

Fig. 120. Wayne Bird and Edd Feller muscle walking plows out to the contest yard at Howell

author, no other farm implement better represents or symbolizes the wonderful potential of a growing craft-based small farm population. An argument might be made that every person studying agriculture should be required to master the walking plow. The reason is that this tool at work teaches so much about tilth and soil health. So, for today, we say every farm with draft animals, large or small, needs a walking plow for plowing gardens and small plots.

A staggering variety of steel and chilled cast-iron walking plows were built, up until World War II, to meet a wide variety of soil conditions. One excellent new model is currently being manufactured by Pioneer Equipment of Dalton, Ohio.

Fig. 121. Rick Balzano, behind the plow, and David Reath on the lines with a four-up span of oxen on a walking plow in New Jersey. The oxen are Bright and Star in the lead and Bo & Luke at the 'wheel'.

The discussion of plow bottoms in the previous chapter paints a detailed picture of this most important part of the plow. Plan on referring to those pages for the particulars of set, suction, etc.

Fig. 28 illustrates and names the parts of a general-purpose walking plow. Other types of walking plows are similarly constructed.

## Now, or very soon, to plowing...

**Suggestion:** Read through the remainder of this chapter, then return to the details of Chapter Two, _BEFORE_ you attempt to actually plow with the walking plow. And this applies, in our humble opinion, even to those with some real plowing experience. Ninety-nine percent of the problems folks have with learning to use the walking plow, making their plow go true and easy, and getting horses to work properly can be traced back directly to one or more inaccurate settings, adjustments (including 'sharpness'), dimensions, and hardware. **If the plow is not moving true you do not have the strength to physically correct it by pushing and/or pulling on the handles.** If care is taken to understand and implement all the particular applicable information in Chapter Two, and what immediately follows, you will find the walking plow to be easy to master and, very likely, a process full of a sense of accomplishment and exhilaration.

Since plowing conditions vary (and sometimes dramatically*) the plow must be readjusted frequently to suit these changes. Conditions vary not only from one field to another but in the same field from day to day according to the amount of moisture in the ground.

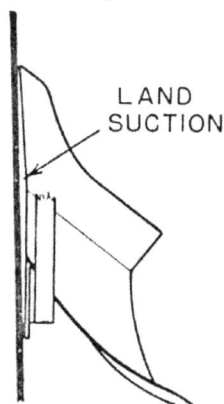

A correctly adjusted plow does better, easier work. And when you do a good job of plowing you save time later when it comes to fitting the ground for seed.

**The Right Size Plow.** Before reviewing key adjustments we need to touch on plow size and design relative to **How Many Horses**. Horse drawn plows, both walking and two-way sulky plows (see Chapter Five) are designed for either two horses or for three. (An important exception are those walking plows designed specifically for one horse.) It is difficult to

LAND
SUCTION

Fig. 122. Make sure all the suction aspects of your plow bottom are set. See Chapter Two.

correctly adjust a two-horse plow when using three horses. Here are some helpful things to look for when selecting plows. **For a two-horse plow, the beam should be parallel to the landside.** The difference between a two and three horse walking plow is in the 'landing of the beam'. (Fig. 123) On a two-horse plow, the beam is parallel to the landside of the plow and the center of the beam is from 1 ¹/² to 2 inches in from the landside.

**For a three horse walking plow the beam should be bent toward the land** or 'landed'. It is not parallel to the landside, but bent toward the land. The center of the beam will be about ¹/² to 1 inch in from the landside. The landside edge of the beam will be directly over the landside edge of the plow point. (See Fig. 123)

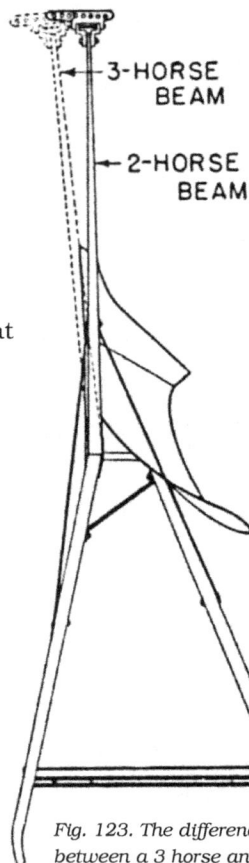

← 3-HORSE BEAM

← 2-HORSE BEAM

Fig. 123. The difference between a 3 horse and a 2 horse walking plow.

The classic two-horse walking plow has a wide range of adjustment for width of furrow so that three horses can be used with a two-horse plow but the work done will be less than satisfactory. If you are going to do a lot of plowing you should make an effort to acquire a plow that is matched to the number of horses you will be using. And matching means beam set as well as share width. Commonly 12" to 14" shares work for two horses. 14" to 16" work for three abreast. Some models of wood beam walking plows from early this century were adjustable, at the frog end of the beam, landing, and therefore allowing conversion, to a true three horse plow from a two. (See Fig. 137).

**Plowshares** The condition of the plowshare is important in plow adjustment. Correct suction, relative sharpness, and correct width for the conditions are all critical. These subjects are covered in Chapter Two.

**Eveners.** Plow bottom widths are measured from the share wing, flat across underneath the plow, to the landside. (See Fig. 132)

The center of draft of a 14 inch plow is approximately 2 inches in from the landside and directly under the middle of the plow beam. The distance from the center of draft to the edge of the furrow wall will be 14 inches minus 2 inches. Also, since the furrow horse walks in the middle of the furrow, the doubletree clevis will be 7 inches from the furrow wall. (See Fig. 125). Thus the distance from the center of draft or the middle of the evener to the doubletree clevis should be 19 inches.

---

*You might find the beginning of your plow work to be in soil of perfect moisture but increasingly warm days dry the soil rapidly and alter the draft. Or you may find your work interrupted by rain and what was scouring nicely now comes to stick and puddle. It is your job to recognize the changes in soil condition and make appropriate adjustments.*

Please read this list and note which of these terms or phrases are unfamiliar.
    *doubletree, singletree, neckyoke, clevis, shackle, trace, tug, lines, collar, point of draft, hitch.*
If any of these terms are confusing and you are serious about plowing with horses soon, this author insists that you preferably get help from a knowledgeable teamster or at the least get copies of **The Work Horse Handbook** and **Training Workhorses/Training Teamsters.**

The doubletree then should have a total distance of 38 inches between doubletree holes for a 14 inch plow. (See Fig. 123)

For a 12 inch plow, the center of draft is approximately 10 1/2 inches from the edge of the furrow wall plus 6 inches to the middle of the furrow, or 16 1/2 inches. This makes a total distance of 33 inches between doubletree holes for eveners to be used on 12 inch plows.

**Effect of Wide Eveners.** A wide evener causes the horses to walk further apart than they should, and the horizontal hitch on the plow will have to be moved to the left to prevent the plow from cutting too wide a furrow. Moving the clevis to the left tips the plow toward the right. In other words, if a long evener is used, the handles must be constantly held up against the pull of the horses. The use of too long an evener, such as 42 inches, results in increased draft of the team and constant work for the plowman.

**Three Horses Abreast.** The correct dimension of 3 abreast eveners for walking plows utilizes the same formula as the team. Just make sure that the point at which the evener fastens or hitches to the plow places the furrow horse in the same position as the doubletree does for the team. In other words, for 14 inch share, it should be 19 inches from hitch hole to center of furrow single tree. For 12 inch share 16 1/2 inches from hitch hole to center of furrow single tree, etc.

**Good plow, Bad plow?** Unless you are fortunate enough to have purchased a brand new Pioneer walking plow with all the right attachments, there is no guarantee that the good looking old walking plow you inherited, bought at auction, or drug out of the brush is suitable for use with horses. Hard as it may be to take, there are thousands of older plows that are best used as yard ornaments and mailbox holders. That said, there are hundreds of thousands of older plows that, in the hands of the knowledge-able plowman will work, or can be made to work quite well. Chapters One and Two provided a wealth of information on how to determine the good plow bottom. With all animal drawn plows, another critical factor will be the beam. If the beam of the individual bottom is sprung upward

Fig. 125. For a 14 inch plow the width of the evener should be 38 inches and for a 12 inch plow 33 inches. The center of draft on the 14 inch plow is 2 inches in from the landside, directly under the beam on a two horse plow. The horse should walk center of the fourteen inch plow cut, or 7 inches from the furrow wall. Add that 7 inches to the 12 inches, 19 inches total and double for the length of the double tree.

it may account for the plow tip pointing downwards or sideways causing the plow to run too deep, too wide or too narrow. As in Fig. 127, with a new share on the plow and the plow on an even floor, the distance between the floor and the under side of the beam should be between 14 and 16 inches. If the beam is more than 16 inches up, it should be considered sprung. It will have to be bent back into shape at a blacksmith shop or metal shop. Before taking that step, check the beam also for sideways spring. To do this, follow the instructions in the caption to Fig. 131.

As a general rule, if you should happen upon a walking plow with a bent or sprung beam do not consider for actual work. It is tricky to get this problem corrected and sometimes a corrected beam will be weakened by the process.

Another thing to suspect in any old plow under consideration is knowledge or evidence of tampering. This author has seen several implements sold at auction which featured tractor bottoms which had been jerryrigged to go on to horse plow beams and frames. Unless the prospective purchaser has some clear evidence that this plow worked and worked well with its modifications, it may be best to stay clear.

And the last note of caution goes with fresh paint restorations. Especially with cast iron, paint can be made to

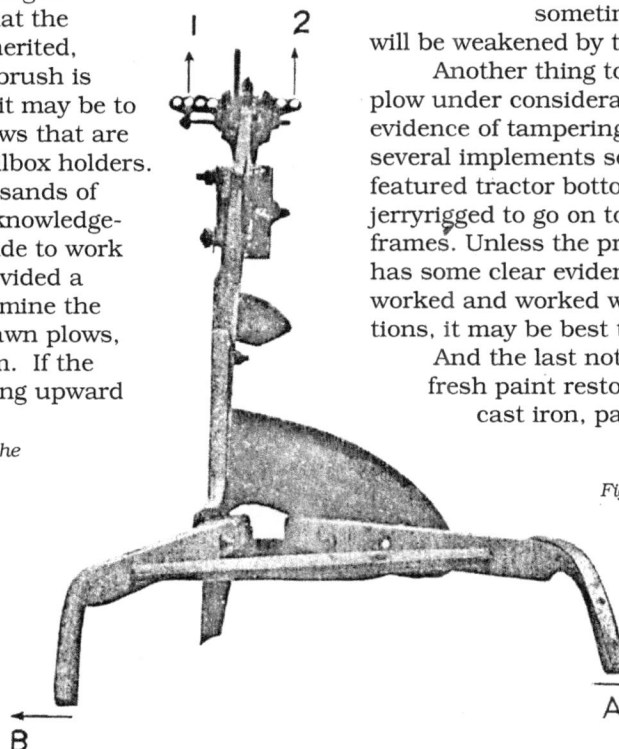

Fig. 124. The amount of wing bearing affects the adjustments of the front clevis.

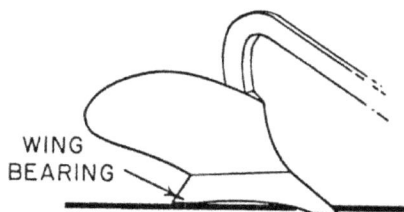

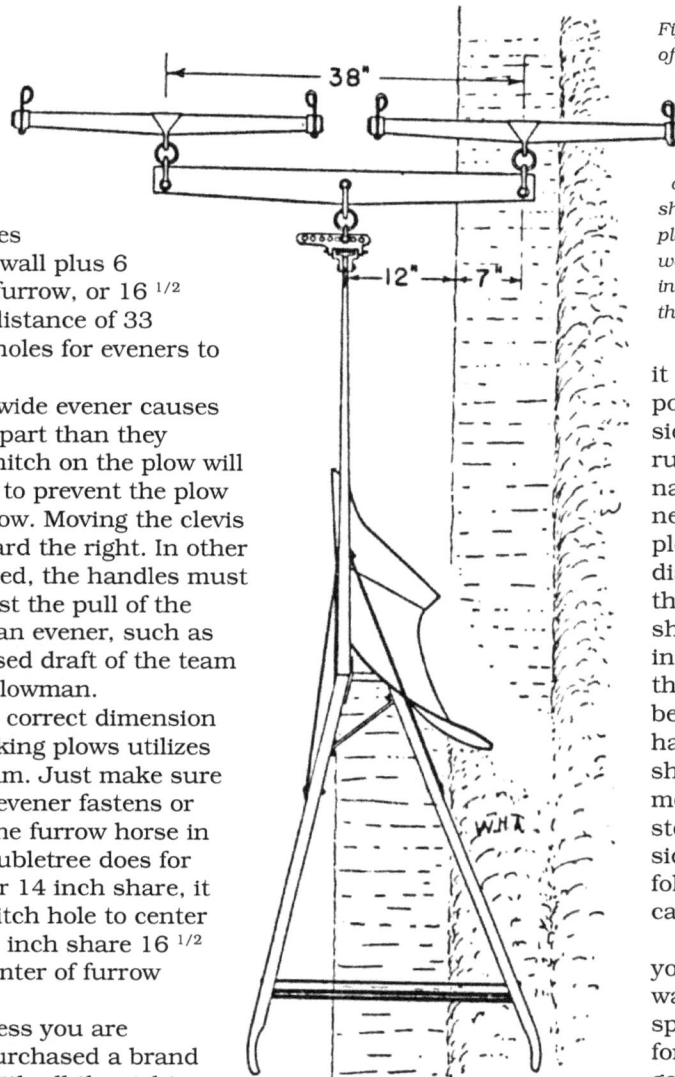

WING BEARING

Fig. 126. Plow Balance. Moving the hitch toward #1 tends to tip the handles toward direction A. Moving the hitch toward #2 tips the handles toward B. Somewhere between 1 and 2 the plow should balance.

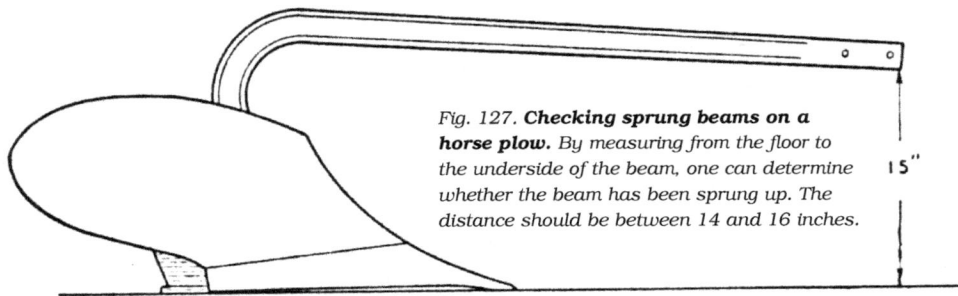

Fig. 127. **Checking sprung beams on a horse plow.** By measuring from the floor to the underside of the beam, one can determine whether the beam has been sprung up. The distance should be between 14 and 16 inches.

15"

A homemade plow shoe (Fig. 128) can be made quite easily from found materials. This piece is best constructed from a tough wood like white oak. Look for the fork of a tree limb 6 or 8 inches in diameter with just the right angle. One limb of the fork should be left 18 or 20 inches in length, and the

cover a world of ills. It is possible to braise-weld cast iron breaks, grind them smooth and paint over them. To the eye and touch there may be no evidence of the break, but beware. Cast iron is difficult or impossible to weld up strong. It is safest to assume that any cast weld will rebreak under strain. Paint can also hide excessive rust or corrosion. Poke around with the sharp end of a knife to determine if suspect areas are weak. This author prefers to purchase old iron implements that have not been "restored?" because what I see is what I get.

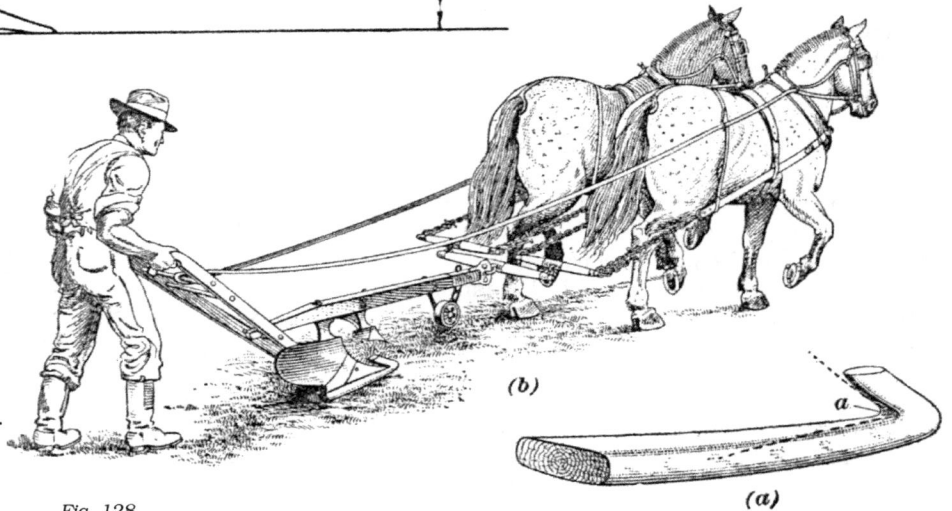

(b)

a

(a)

Fig. 128

**Moving the walking plow.** When the walking plow is out of the ground and rolled towards the moldboard surface, so that the outside edge of the moldboard rests along with the share edge on the ground, the plow can be drug short distances with some accuracy and a minimum of wear. If the plow is rolled over on to the landside, which at first might seem logical because it tends to lay somewhat flatter, it will skate dangerously side to side and rub excessively on the handle. The process of rolling the plow towards the moldboard edge and dragging the plow, once it is out of the ground, is the customary procedure, as will be seen, when coming out of a furrow and turning around to re-enter the plowing. If the plowman has a distance to go from the shed or barn or equipment yard out to the field where plowing is being done, dragging it in the fashion described above is not recommended. It causes too much unnecessary wear on an important tool and it will mark up roads and land.

Fig. 129

The ingenious farmer will come up with many clever ways to transport the plow, conveniently, to the field. Here are a couple;

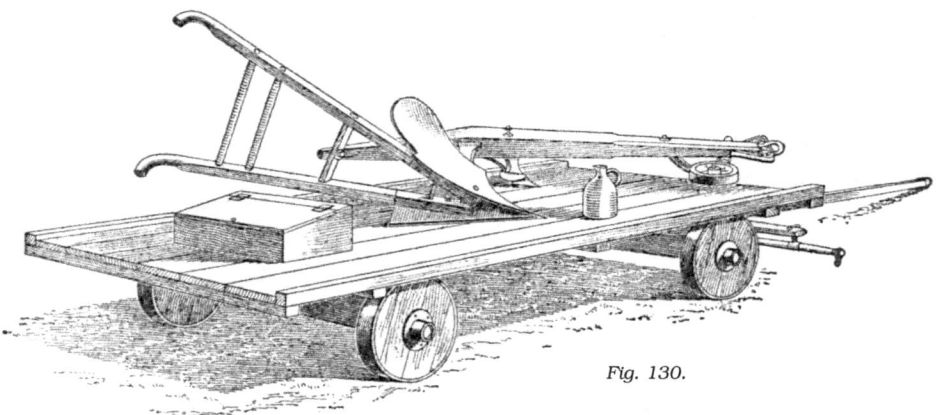

Fig. 130.

fork 3 to 4 inches. The top and the bottom of the lower one are flattened, and the point is rounded like a sled runner. If an appropriate limb cannot be found, a stout section of plank can be designed to work. A hole (see fig. 128a) is made in the elbow of the fork. This hole should be just large enough to receive the point of the share. When it is time to go to the field place the shoe directly in front of the plow, push down on the handles to raise the plow point, speak to your horses to step ahead, once the plow is started up on the shoe drop the point so that it fastens into the elbow hole. Now all is ready to proceed on the mini plow sled, or shoe, out to the field.

A stone boat or work sled is a common implement on horsepowered farms. And it is an excellent way to transport a walking plow. Make sure, if you are using a low stone boat, that parts of the plow are not dragging off the edge of the boat and on the surface of the ground. If making a sled for this purpose, a size of at least 4 foot by 5 foot is necessary. Provide a box or retainer for any tools or items that might accompany the trip. When hauling the plow on a sled, it is best to lay it on the landside.

A low set wheeled vehicle can work quite well for transporting walking plows, but keep in mind that if the plowman is alone, there will be a physical challenge to getting the heavy implement up and down from the wagon bed. Whereas with the shoe or sled, it is possible to tilt and lever the plow on and off.

**Once in the field.**

Some of the adjustments necessary in the beginning may not have to be changed again if the same plow and team are being used. For example, once the correct size of evener is selected and the hitch points are established, it is likely these may remain the same for the duration of the work. It is important, however, to fully understand these elements and be prepared to return to them if problems need correcting.

When this author first set out to learn plowing, a predictable arrogance ensued. For in the beginning I was working preset plows and ready horses in fields already correctly begun. It all seemed so easy after the first unsure rounds. Returning to home field, horses, and implements and setting out alone I quickly became stymied and frustrated for nothing worked right and I hadn't a clue as to how to proceed. Now, some thirty years later, I think back and wonder at the power of my persistance given no 'how-to' manuals or instructional videos (and they are seldom helpful). But persistance was not the whole of it for if I had not been given tough instructions by two masterful old friends, Ray Drongesen and Charlie Jensen, I'd doubtless be dreaming still about broken crooked furrows.

All that said, this author returns as a fifty-something years old horsefarmer, to a parallel moment of arrogance. I do now believe that if this book does its job and delivers to the reader an understandable presentation of the

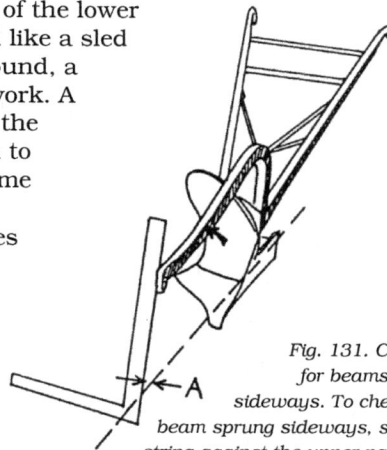

Fig. 131. Checking for beams sprung sideways. To check for a beam sprung sideways, stretch a string against the upper part of the landside. Place a square so that the edge is in line with the middle of the plow beam. Measure the distance A. For a two-horse beam this distance should be 1 1/2 inches. For three-horse beam, the string should just touch the outside edge of the square.

rudimentary basics of plow readiness and adjustment, the person who knows how to work horses will quickly learn and love the wonderful art of plowing. From a book!? But these past sentences contain a critical caveat. We said, 'the person who knows how to work horses'. If you find yourself at this point in the book and you do not know how to work horses I implore you to stop and go back to a beginning before this book's first page. At the very least please get copies of the two books which precede this one, *The Work Horse Handbook* and *Training Workhorses/Training Teamsters.*

If you proceed inexperienced and without key bits of information, to attempt to hook dependent animals to a plow, and drag it with accuracy and SAFETY through that waiting plot of land, you take enormous risk with the well-being of everything in view, animate and inanimate. See that small child watching? See that lovely fence?  See that clothes line? See that new outhouse? Look down on yourself and at those fine animals and that lovely plow. You put them all at risk. You owe it to all of them and every thing else to have, at the very least, someone knowledgeable and physically capable close at hand, and preferrably you with head and hands fresh from learning.

The reader might reasonably ask why these words of warning appear at this point in the book and not at the beginning. They did appear at the beginning, they are repeated here because this author's many years of writing and teaching have repeatedly shown that many anxious beginners fast forward to the 'action parts' and fail to digest the 'warning labels'.

It is a challenge to set these instructional words into the best and most helpful order. Almost a 'what comes first, the chicken or the egg?' question. As we move to a logical physical procedure for the adjustment of the walking plow, that procedure which needs to be done while actually plowing, we are mindful that some of the readers of this text need, first, to know how to plow before they may understand these next steps. Some readers might reasonably choose to read the sidebar "First Time Plowing" before moving to the next paragraph.

Fig. 132. Share width is measured from the wing to the landside of furrow wall.

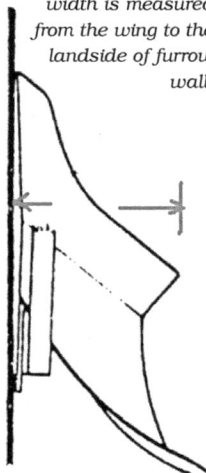

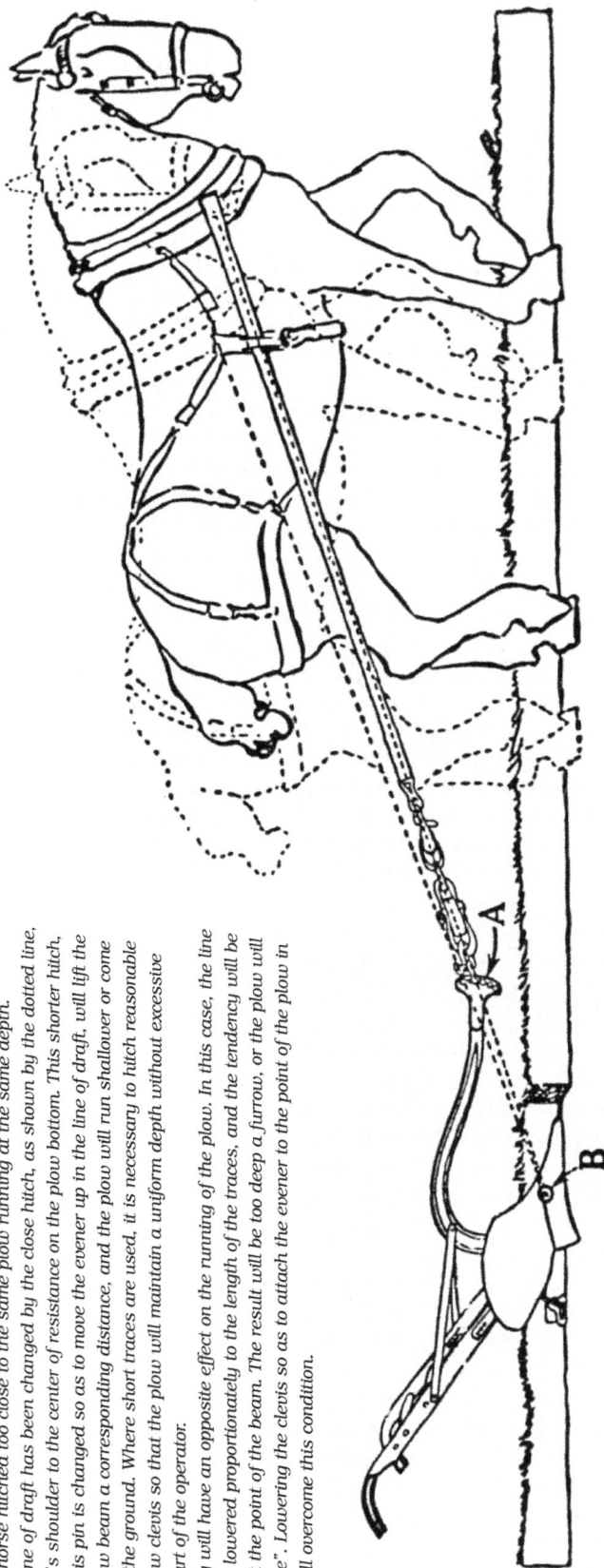

Fig. 133 illustrates:

(A) "Heavy-line horse" correctly hitched to a walking plow running at a normal depth.

(B) Dotted-line horse hitched too close to the same plow running at the same depth.

Note that the line of draft has been changed by the close hitch, as shown by the dotted line, from the horse's shoulder to the center of resistance on the plow bottom. This shorter hitch, unless the clevis pin is changed so as to move the evener up in the line of draft, will lift the point of the plow beam a corresponding distance, and the plow will run shallower or come entirely out of the ground. Where short traces are used, it is necessary to hitch reasonable high on the plow clevis so that the plow will maintain a uniform depth without excessive effort on the part of the operator.

Traces too long will have an opposite effect on the running of the plow. In this case, the line of draft will be lowered proportionately to the length of the traces, and the tendency will be to pull down at the point of the beam. The result will be too deep a furrow, or the plow will run on its "nose". Lowering the clevis so as to attach the evener to the point of the plow in line of draft will overcome this condition.

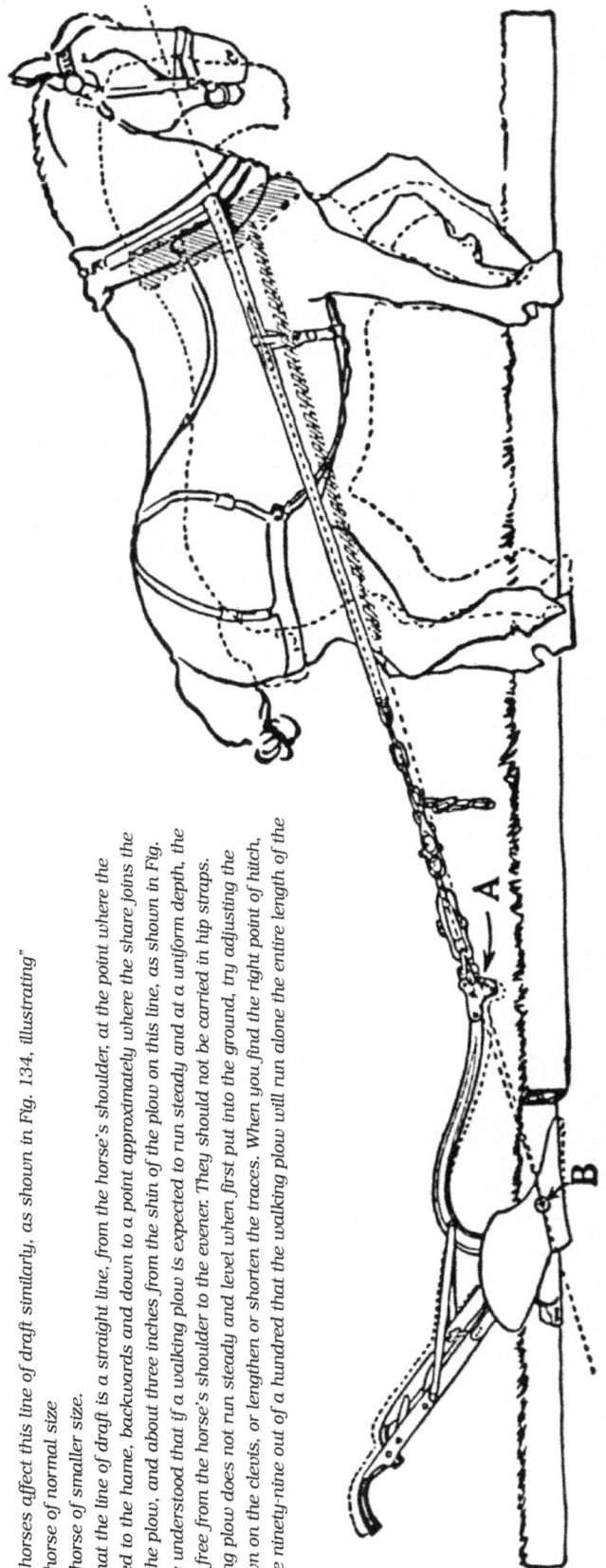

Different-sized horses affect this line of draft similarly, as shown in Fig. 134, illustrating"

(A) Heavy-line horse of normal size.

(B) Dotted-line horse of smaller size.

Bear in mind that the line of draft is a straight line, from the horse's shoulder, at the point where the trace is attached to the hame, backwards and down to a point approximately where the share joins the moldboard on the plow, and about three inches from the shin of the plow on this line, as shown in Fig. 75. It must be understood that if a walking plow is expected to run steady and at a uniform depth, the traces must be free from the horse's shoulder to the evener. They should not be carried in hip straps. If a new walking plow does not run steady and level when first put into the ground, try adjusting the hitch up or down on the clevis, or lengthen or shorten the traces. When you find the right point of hitch, the chances are ninety-nine out of a hundred that the walking plow will run alone the entire length of the land.

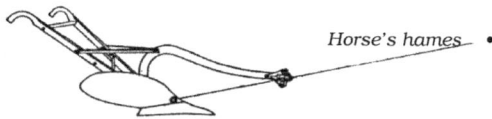

Fig. 135. The line of draft of a walking plow.

the same effect on a plow as does changing the height of the vertical clevis. Short traces raise the end of the plow beam and have the effect of a low vertical clevis. On the other hand, long traces tend to lower the plow beam and have the effect of a high vertical-clevis hitch.

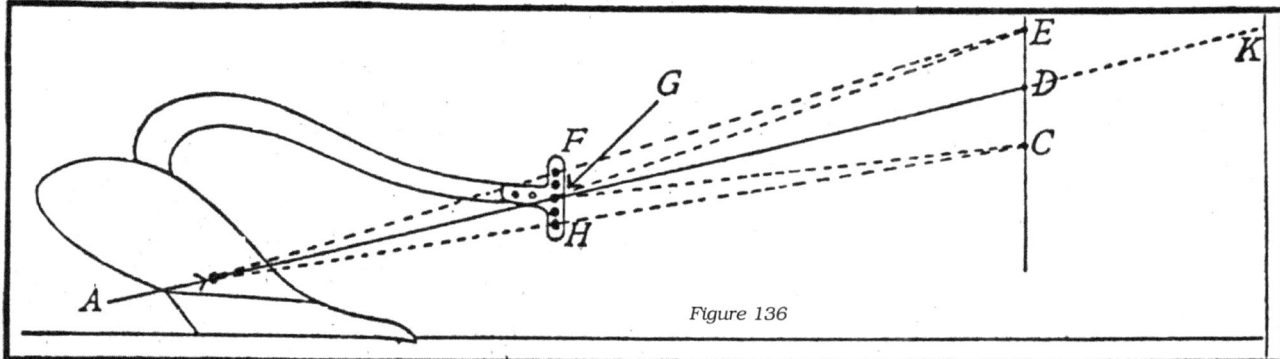

Figure 136

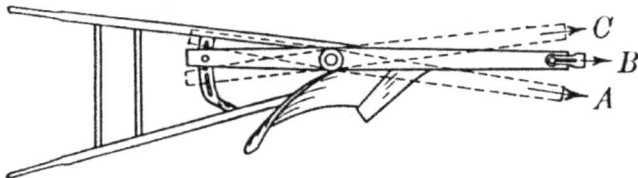

Fig. 137 Adjustment of beam. A, set for 2 horses; B set for 3 horses; C set for 4 horses. If the beam is rigid, a corresponding change must be made in the clevis at the end of the beam

Fig. 136. Illustrting the effect of height of hitch in a walking plow. The line from A to D is True Line of Draft. If point of draft, or trace joint at hame, is at C the Line of Draft will be bent up (thereby pushing down) at G and putting plow point too deep into the ground. In this case the clevis should be dropped to H. With the point of draft at E, (as in a taller horse) the clevis must be raised to F or hitch lengthened to K.

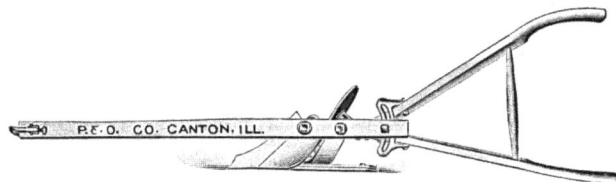

Fig. 138. A one-horse P & O vineyard plow. Note position of moldboard relative to beam and the adjustability of handles for offset.

## Getting set to plow.

To adjust the walking plow: If a depth or beam wheel is being used, raise the wheel as high as possible on the bracket. When the wheel has been raised adjust the vertical clevis to the proper height. (See Figs. 133). It will be helpful to have someone available to head your horse(s) so you may back up to view the angle of draft* . With the vertical hitch approximately right lift the plow handles, speak to the animal(s) and commence to plow. If the plow instantly runs too deep you have hooked too high (or too close or both, see Figs. 133 & 134) try first to adjust by lowering the vertical hitch point.  If the plow instantly runs too shallow you have hooked too low (or too far) try first to adjust by raising the vertical hitch point. All of this is being done with the depth wheel full up. Never try to adjust the plow depth by use of the wheel. If the clevis is too high and you try to correct this by lowering the wheel, it causes the plow to run on its nose; it is unstable; it tips over easily and it is difficult to hold. By adjusting the depth with the clevis, the plow runs evenly.

Changing the length of the traces or tugs has

If everything is as it should be, it is a good practise to begin these adjustments by hooking the tugs mid length, this allows fine tuning, plus or minus, later. This is, of course, a lost option if the harness is poorly fitted to the animals and the tugs are too short or too long.

Once the plow is running easily at the depth you want (see Chapter One for depth of plowing) lower the depth wheel until it touches the ground while plowing. This wheel is only necessary to prevent the plow from going too deep into soft or wet parts of the field. If the vertical clevis is properly adjusted the plow heel just lightly touches the ground while plowing. Part of the time it does not even turn.

As you are plowing, let go of the handles to determine if the plow is tipping right or left. If it is holding upright and plowing properly you are in luck. Tipping is due either to the horizontal hitch position or to the wing bearing of the share or to the condition of the share point. First determine if the horizontal hitch position is the problem. If the plow tips right, move the clevis to the right. If the plow tips to the left, move the clevis to the left until the handles balance in the upright position. (see Fig. 126). If this adjustment does not correct the tipping problems refer to Chapter One and Two about shares and wing bearing.

If the plow is moving ahead easily upright let

*The tugs or traces will have to have some tension on them to give anything approximating a true line of draft. But complete accuracy will be impossible without having the animals actually pulling the plow in the soil. Collars (and yokes) change position, sometimes dramatically, when under a full load. And animals carry themselves differently when pushing against the harness. Some will pull more upright and erect, thereby increasing the angle, some will pull low to the ground reducing the angle. Hopefully, if two or more animals are being used their manner of working, as well as size, will be similar although a good evener will compensate for the differences.

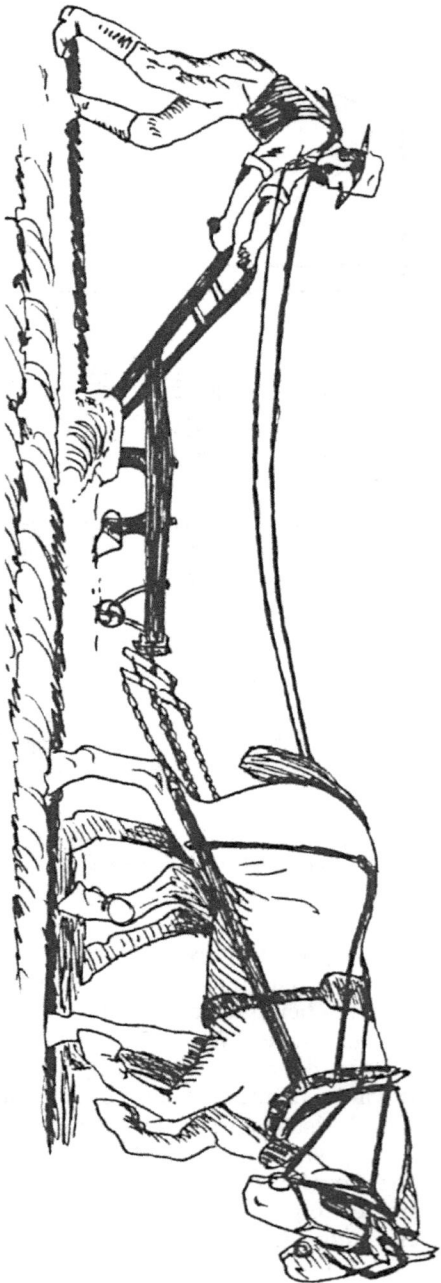

Fig. 139. If you are bent over, pushing, pulling, sweating, groaning you will be through long before that field is plowed. Something's wrong. The plow should run itself if properly adjusted and hitched.

If you are plowing for the first time without help, it will seem an impossible procedure to master. Experienced horses, a helping teamster and a clean furrow would be ideal circumstances for the beginner. If this is not possible, try at least to have some of the first furrows started for you. If that is not possible, have someone either drive or lead the team until you have a good clean furrow.

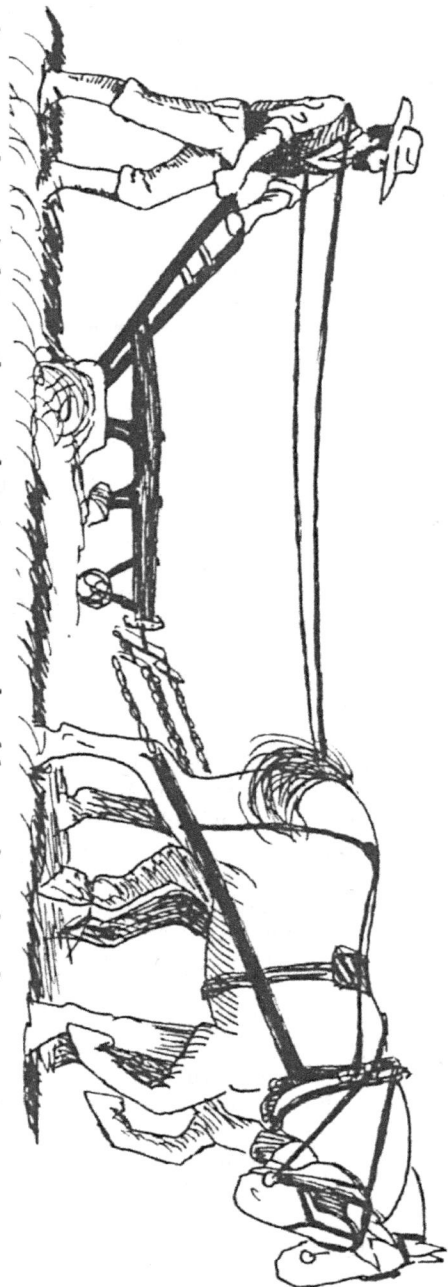

Fig. 140. The lines are best tied and put over one shoulder and under the opposite arm. If an accident of some sort should occur, the teamster, by tucking his head, can easily get free of the loop and still have lines in hand. To start plowing, lift up on the handles and have the team step ahead. The plow will suck down into the ground and find proper depth without your help. As you move ahead, slight pressure on either handle will cause the plow to move to the opposite side. Increased pressure to either side or both will cause the plow to come out of the ground. Pulling up on the handles will cause the plow to go deeper.

The rule with the walking plow is relax. Walk upright, light touch on the handles, follow the plow, speak softly to the animals and enjoy the day farming. Now this is what plowing is supposed to be.

## Lines and the Walking Plow

*The driving lines for a walking plow will not work if they are too short. The size of animals, and the nature and design of the plow, will both affect the length of lines necessary. For in depth information about this important subject, look to* <u>The Work Horse Handbook</u> *and* <u>Training Workhorses.</u>

*Team driving lines, or three abreast lines or single lines will all conclude at the plowman with two straps. These two straps will need to be fastened together for plowing. These may be set up with a buckle on one end and holes perforated on the other for convenient buckling. This system works fine but this author prefers to keep the ends of the lines free of anything that might hang up or hurt the hands. Illustrated is an age-old knot which works exceedingly well for tying two leather lines together. This knot is handy because, besides holding firm, it is easy to untie and, by a simple push of the fold, easy to loosen for quick adjustment. This knot, when placed at the center of the back between shoulder blades, gives a comfortable position to the lines.*

*Throughout this book the reader will see myriad variations of how plowmen deal with the driving lines. To include all of the visual information necessary it was impossible to edit certain images of how lines are dealt with. For the* <u>beginner</u> *it is, in this author's opinion, critically important to offer this one catagorical absolute.* <u>NEVER</u> *put a loop of tied or buckled driving lines under both armpits or around your waist. Here is the reason: should an accident occur that would have, heaven help you, your animals take off at a run you would not be able to get free of those lines and you would be drug and hurt badly. It is this author's contention, and many exceptional plowmen disagree, that the lines should always pass over one shoulder and under the opposing armpit. Here is the reason: should an accident occur and the animals take off by lowering your head away from the shoulder line the lines will roll off of you and free without hanging up and dragging you.*

*It needs to be noted that the precaution described has little or no bearing on the quality of plowing you might accomplish. It is strictly a safety concern and one which most particularly benefits the inexperienced plowman.*

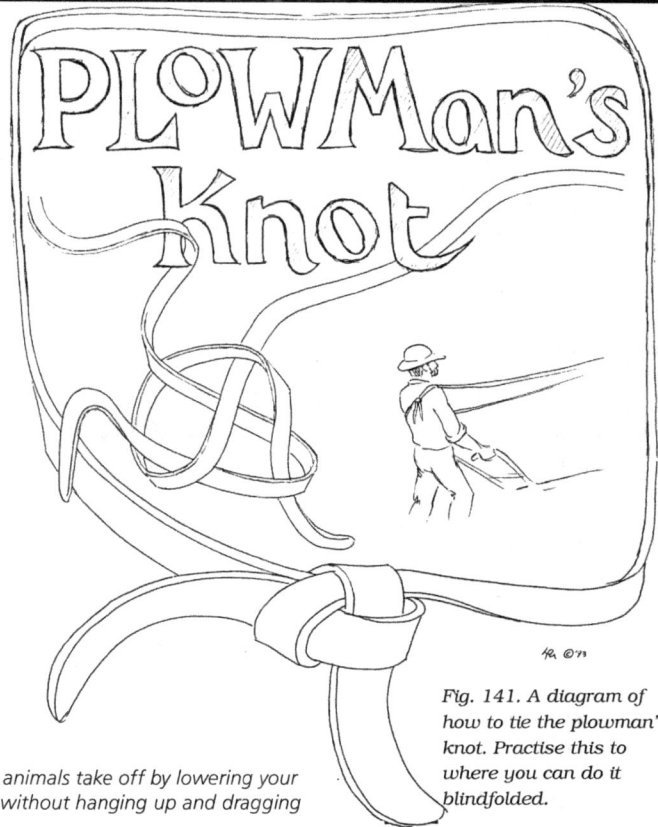

Fig. 141. *A diagram of how to tie the plowman's knot. Practise this to where you can do it blindfolded.*

Walking Plows

go of the handles and notice whether the plow is cutting too wide or too narrow. The width of the furrow should be corrected, not by the horizontal adjustment, but by the spacing of the team and the width of the evener. Make sure that you have matched the plow share width to the correct evener width. (See Fig. 125).

At this point you will be ready to adjust the jointer or coulter for a clean even furrow. See chapter two, page number 32.

## First Time Plowing

If you've never plowed before, trying to do it on your own, with a plow you know nothing about, and with horses or mules you might be unfamiliar with, is asking for frustration at the least. The best way to learn to plow is with competent help. The way to learn is to start with little introductions. If you know someone who plows with horses, or know of a plowing demonstration or competition, or a work horse workshop which includes plowing, you owe it to yourself to sign up, invite yourself, check in, plead for a taste. The best way to learn to plow, and this is especially true of the walking plow, is by having someone else worry about the animals while you try your hands on the handles and in the furrow.

If you cannot do that and insist on trying to do it on your own, here are some tips. Assuming you have paid attention to the warnings and have learned about working animals in harness, you should practise ground driving (driving horse[s]

which are not hooked to anything - not pulling anything) while the lines, tied or fastened together (see Lines sidebar), are over one shoulder and under the other. Practise maintaining a perfect light tension on the lines without touching them with your hands, only by slightly leaning back and/or slowing yourself down. You'll find that it is tricky at first and you'll be reaching, with both hands, for the lines. Try not to. The horse(s) should start out smoothly on a quiet voice command, slow when you apply pressure to the lines, and stop when you say "whoa".

When you think you are ready to actually hook to a walking plow, it would be helpful if you knew the plow was ready and right. And it would also be helpful if you had a person available to drive or lead the horse(s) for you (see Fig. 143) so that you could deal

Fig. 142. *The first time you plow, use common sense and do not put yourself in a tight space with lots of obstacles. Also, try to have someone available to help lead or drive the horses so you can concentrate on learning about the plow.*

just with the plow.

Know what you are going to plow and how. (See Chapter 8) As in Fig 142 set a marker or a flag to where you will be heading. With the plow properly hitched, lift up on the handles slightly, as the animal(s) step(s) ahead the plow will draw down into the ground. Stop lifting on the handles. The properly set plow will find its own level or depth. To steer the plow, you will do the opposite of what seems natural. Push down on the right handle to go left. Easy, a very little pressure goes a long way. Push down on the left handle to go right.

Fig. 143. Whether you are new to plowing or the horses are or both, a little help getting started will pay big dividends.

The beginner's first sensation is that the plow, regardless of what speed its going, is going too fast and is out of control. The beginner struggles to slow the plow by pulling back, and to hold the plow in the ground by pulling up on the handles. It is natural to want to compensate, to correct. But your efforts are less than worthless. You aren't strong enough to slow the plow against the power of the draft animals. And you aren't strong enough to hold the plow in the ground if it doesn't want to stay in. Remember that if the plow is properly set you should be able to let go of the handles and have it run true without your help. Working the walking plow is not about strength, its about adjustments and finesse. Learning to relax while plowing is a difficult but necessary challenge. If you are

going down your first furrow and you feel out of control simply say whoa, stop and take a breather. Start up again. Easy. Do it in short runs until you feel calmer. When you get to where your furrow should end, slow your horses and prepare to say whoa but don't say it yet. Push down on the handles with a slight lean towards the moldboard. As you feel the plow come out of the ground say whoa. The first few times, the release of resistance that comes when the plow leaves the ground may have the animal(s) stumble slightly or even naturally speed up. This is why it is important that you prepare them and yourself by slowing and stopping as soon as the plow is out of the ground. With the plow out of the ground, lay it over on the moldboard side and drive around to prepare for the return pass.

In an ideal situation the beginner would have a clean straight furrow, made by someone else, to enter into for his or her first rounds. The crown, or opening furrows, are always a challenge. Sometimes we don't get ideal situations. If you're struggling double because of the crown (see Chapter 8) don't be disheartened if it doesn't go well to begin with. It will get better.

## Teaching Horses to Plow

*Pulling a plow is amongst the best and most honest work you could ask a horse, mule or ox to do. And, contrary to what the uninitiated might think, if the plow is matched to the soil, sharp and properly set, hitched right and matched for size to the number of animals, the job is not an overly difficult one.*

*There is only one aspect of the plowing that a good work animal might need some help getting used to and that is walking in the furrow. It is surprising how quickly most animals learn this. But until they do, stepping in and out of a furrow will create a messy nightmare of the plowing. This is why a simple repeated excercise can help.*

*Again, an ideal situation would be the availability of an actual furrow to practise in. But barring that any narrow ditch will work fine. And the excercise is simple. Just ground drive the animal back and forth in the ditch, without being hooked to the plow. Keep doing this, patiently returning him or her to the furrow or ditch whenever they step out. With some animals it will take one pass, with others it will take several. Eventually it will become second nature.*

*If you are going to attempt plowing with animals which are not well trained you MUST have plenty of teamster experience. But know that the repetition of the work and*

*procedures will be quickly accepted by the new animals.*

*A particular circumstance which can create difficulties has to do with the animal's experience with actual work, or pulling. If the horse or mule has always pulled a wagon or light implement without steady resistance it may take some careful persistance to get it to pull the plow. The utmost concern should be taken not to make the animal balky or sullen, refusing to pull. It is best for an experienced horse or mule man to deal with such a circumstance. If no help is available, try working your animals first on a sled for long steady walks before hooking to a plow.*

*A specific problem of the greatest difficulty arises with animals which have been used in pulling competitions and not much else. These animals will often want to 'jerk' start the plow and pull it very fast and hard for a short distance. This is because they associate a resistance with the contest routine. The only way to correct this is with lots of slow steady repetition using a work sled or stoneboat until the animals have learned a new routine and are ready to take on the plow in a calm and safe manner.*

*This book is not about how to work horses or how horses or mules work, it is about plows and plowing. You are encouraged to refer to* The Work Horse Handbook *and* Training Workhorses

# Plowing with the Single Horse

Fig. 144. Verical adjustment of hitch on one horse walking plow. Note horse stands on land (unplowed ground).

Fig. 146. The three photos on this page are of Alan Slavick plowing with Noah, a belgian gelding. Note safe position of the lines over one shoulder and under opposing armpit.

Fig. 145. Demonstrating in-line hitch on the single horse. Note the beam is set further over the moldboard than with 2 and 3 horse walking plows.

All other aspects being equal the primary difference in plowing, comfortably, with a single horse is that the animal walks on unplowed ground immediately adjacent to the previous furrow, rather than in the furrow. This will cause the point of draft at the shoulder to be somewhat higher and will dictate hitching longer and/or higher than with the animal walking down 5 to 8 inches lower in the furrow. The single horse plow customarily has an 8 to 10 inch share width and adjustable handles.

Fig. 147

Fig. 148

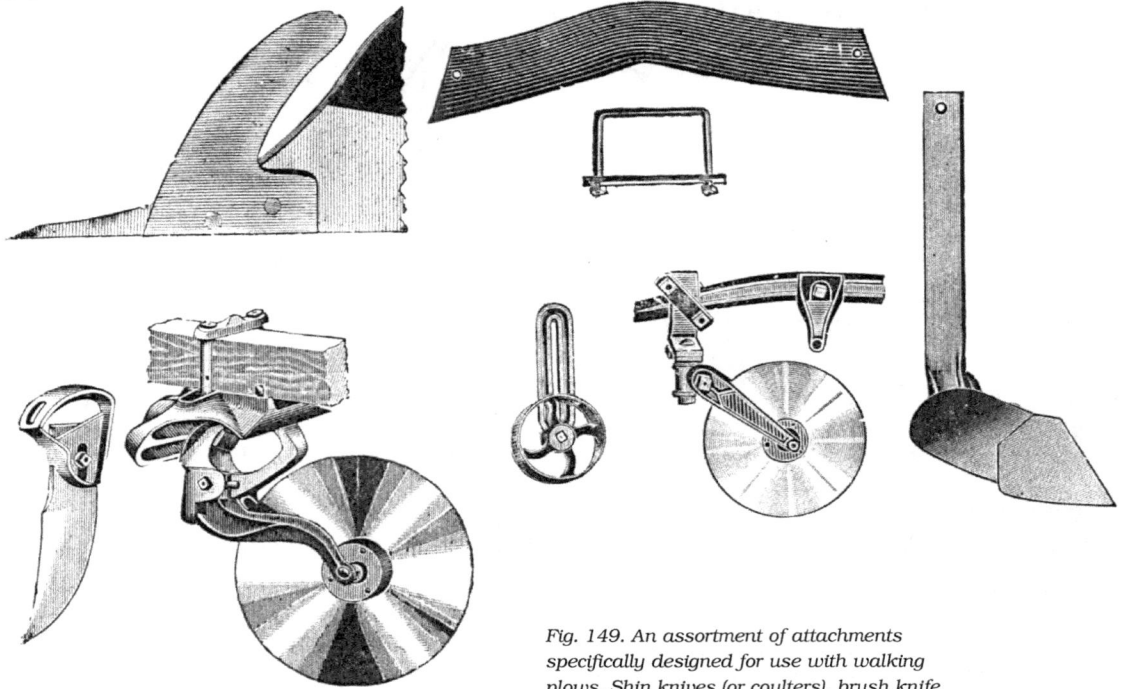

Fig. 149. An assortment of attachments specifically designed for use with walking plows. Shin knives (or coulters), brush knife, rolling coulters, jointers, and bracket systems.

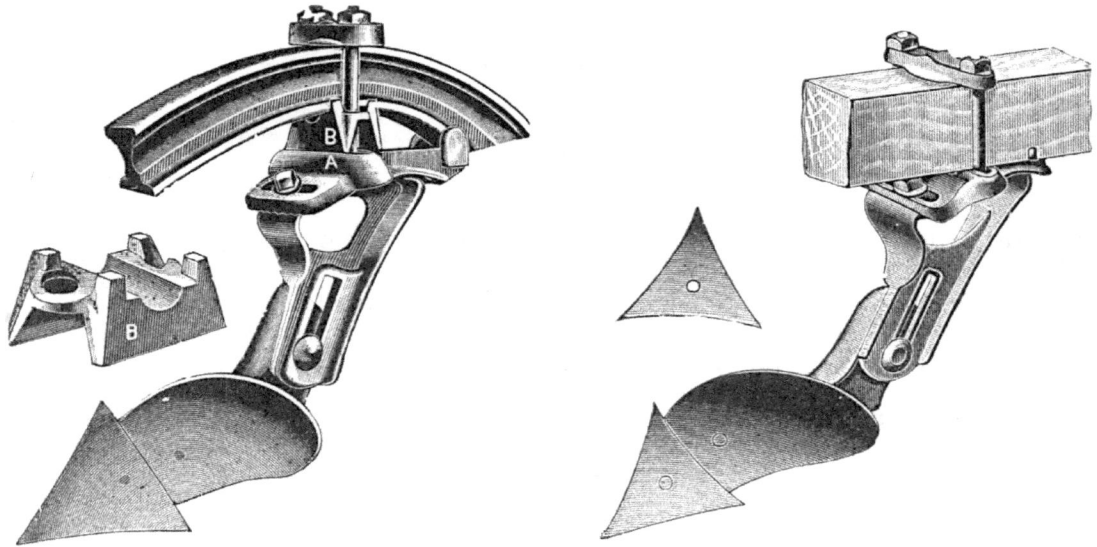

Fig. 150. This walking plow has been equipped with a drag or weed chain set to fold or bend over tall vegetation as the furrow is turned (see Fig. 108).

Fig. 151. (right) A walking subsoil plow following the regular plow and cutting a deep slice in the lower subsoil reaches of the field. The most time efficient system for subsoiling with walking plows would require two plowing outfits following one another. One doing the regular depth work while the subsoiler follows immediately after and before the next furrow covers.

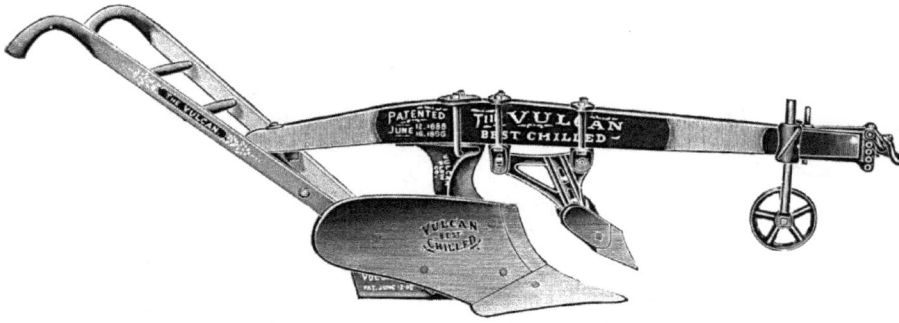

*Fig. 152. Classic Vulcan heavy duty general purpose walking plow with detachable shin, adjustable jointer and depth wheel.*

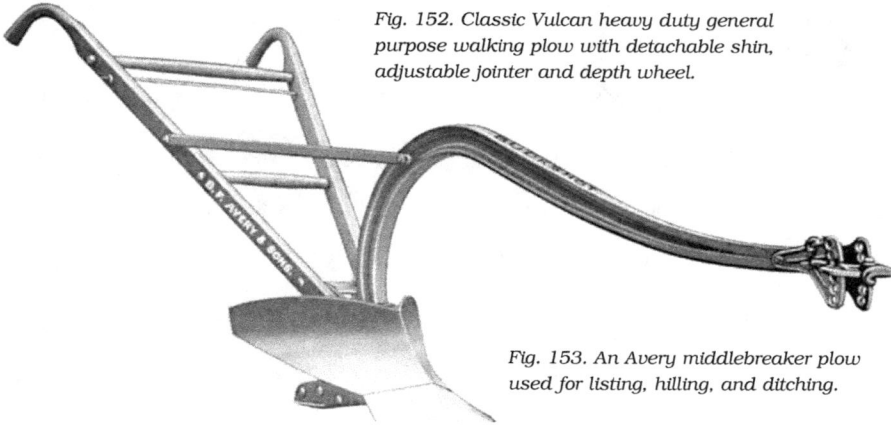

*Fig. 153. An Avery middlebreaker plow used for listing, hilling, and ditching.*

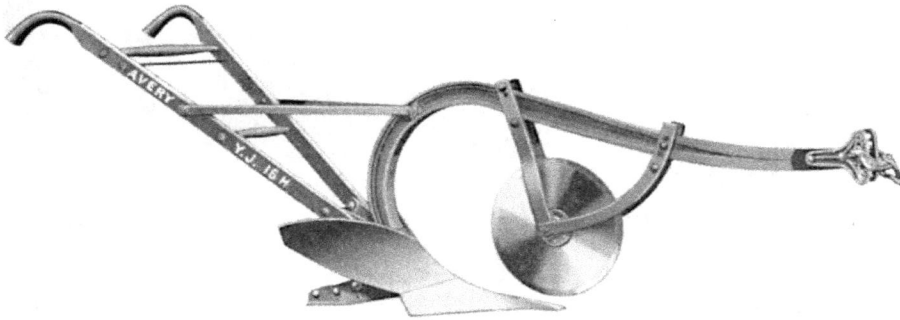

*Fig. 154. An Avery breaking plow with reinforced rolling coulter.*

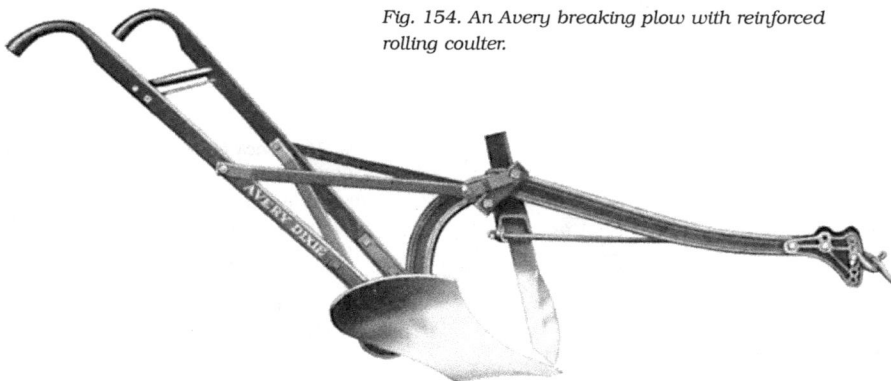

*Figs. 155 & 156. Two views of an Avery New Ground plow equipped with a reinforced root knife (coulter) and a wheel in the landside to reduce draft. This plow is designed for working in forest-type soils with lots of roots.*

## What to take with you to the field.

There have been occasions when this author has had to travel a half mile or more to the plowing field and be best prepared to spend an entire day. In this particular case, whether a work sled or wagon is employed, space is provided to take tools, spare clevis and bolts, drinking water, lunch, feed, buckets, coat, halters and lead ropes, notepad, pencil.

The object is to stay a full day and get the plowing done. This is my itemized checklist. The reader will and should customize his own:

1.) crescent wrench, channel-lock pliers and small hammer.
2.) spare clevis with spare heavy steel ring.
3.) 2 or 3 spare bolts for repairing evener or hitch.
4.) 4 gallons of water for animals and people. (suggestion, use plastic gallon jugs if available).
5.) plowman's lunch.
6.) small jag of hay, small quantity of grain. Water bucket for animals.
7.) plowman's coat & hat.
8.) halters and ropes to fasten animals during lunch.
9.) something to write on & with. optional: reading material for lunch break & a pillow for a nap.

Walking Plows

982

Fig. 157. Most plow manufacturers sold assemblies, such as these, which converted the walking plow into a frameless riding plow. This particular design employs an 'axle' to which the beam is fastened.

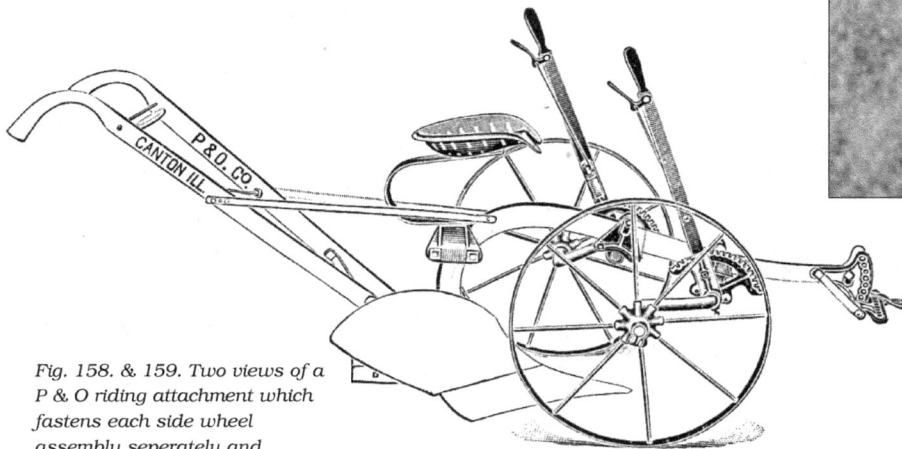

Fig. 158. & 159. Two views of a P & O riding attachment which fastens each side wheel assembly seperately and eccentrically.

Fig. 160. As this turn of the century photo illustrates, clean true wide plowing can be accomplished with a minimum of physical effort if all aspects of adjustment, hitching and sharpness are understood and respected.

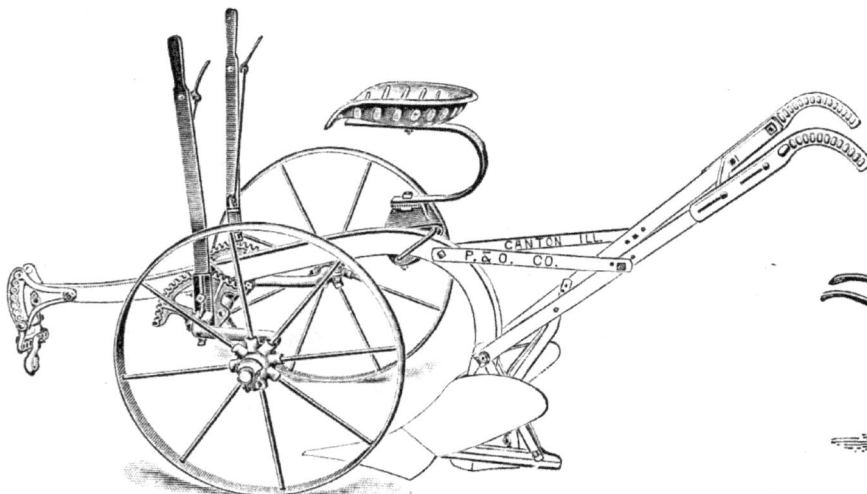

Fig. 161. Back in the mid 1800's illustrations appeared of wild imagined plows, such as this four bottom walking affair. With it's straight wooden beams and centered large depth wheel this author imagines a most difficult time plowing even one complete furrow.

## Three Abreast on a Walking Plow

Ample information has been presented in this chapter on how to determine and adjust a walking plow for three abreast work. In particularly tough soil conditions, or when able to use a large bottom 16 inch share, or when wanting to employ three small animals for the work, three abreast is a good option.

For the right size evener, remember that a three abreast walking plow should have it's beam landed (see fig. 123). If you are using a 16 inch bottom on a 3 abreast plow, add 8 inches (midfurrow) to the 16 inch cut for a total of 24 inches which is the distance from center hole of the evener to center hole of the furrow horse single tree. There will be some adjustment available on the horizontal plate of the plow clevis but remember that this adjustment also affects the vertical stability, in motion, of the plow.

Fig. 162. At a 1997 rainy Oregon Draft Horse Assoc. plowing meet, novices get to try their hands at operating a walking plow while the owner/teamsters handle the lines. At planned for events it is not always possible to have the ground in ideal conditon

Fig. 163 Three abreast makes tough work look easy.

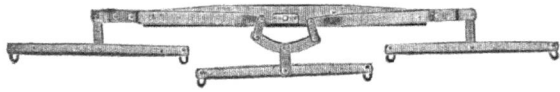

Fig. 164. A John deere 3 abreast plow evener.

**In conclusion**: There is a great deal of information in this chapter which applies to all plows. This author hopes that, if the walking plow is your sole interest, you have found many valuable pieces of information here. But more important, this author hopes that the reader goes out to plow, discovers a world of delight and satisfaction and becomes an enthusiastic member of the horsefarming community. The world needs more good farmers! And horsefarmers are the stuff of good farmers!

### Getting the walking plow to work right.
#### The short sheet.

*Note: Hopefully this will mean nothing to you unless you've read the text up to this point.*

1. right plow for the soil
2. broke, properly harnessed, animal(s)
3. sharp share
4. proper suction
5. right size evener
6. hitch in straight line from point of draft (Fig.133) to evener hitch point
7. hitch at proper height without pushing down on, or pulling up on, beam
8. adjust horizontally to have plow stand up straight while moving ahead
9. there you go!

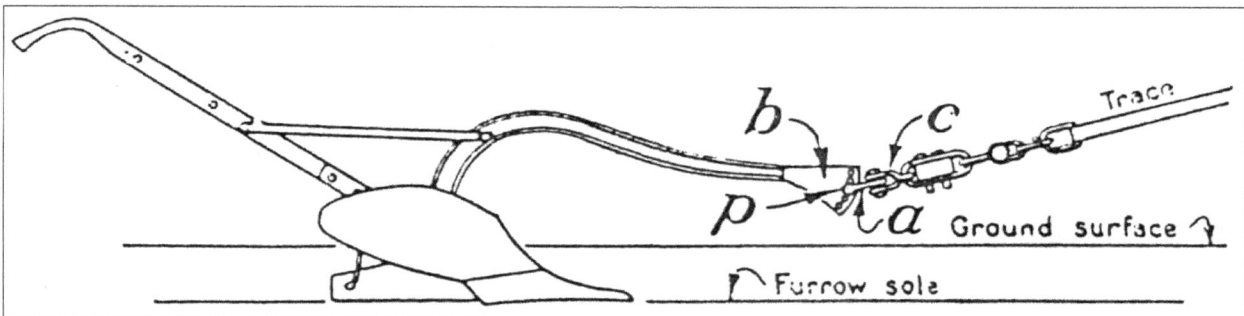

Fig, 165. At the risk of obscene redundancy we offer yet another view of the all important vertical hitch of the walking plow. If the pin at P is too high the effect will be a pushing down of the plow beam. If P is too low it will serve to lift up on the beam. Get the hitch point at the right vertical position and your plow will almost run itself! (A. equals the cross clevis, B equals the beam clevis and C equals the twisted hitch clevis fastened to the evener.)

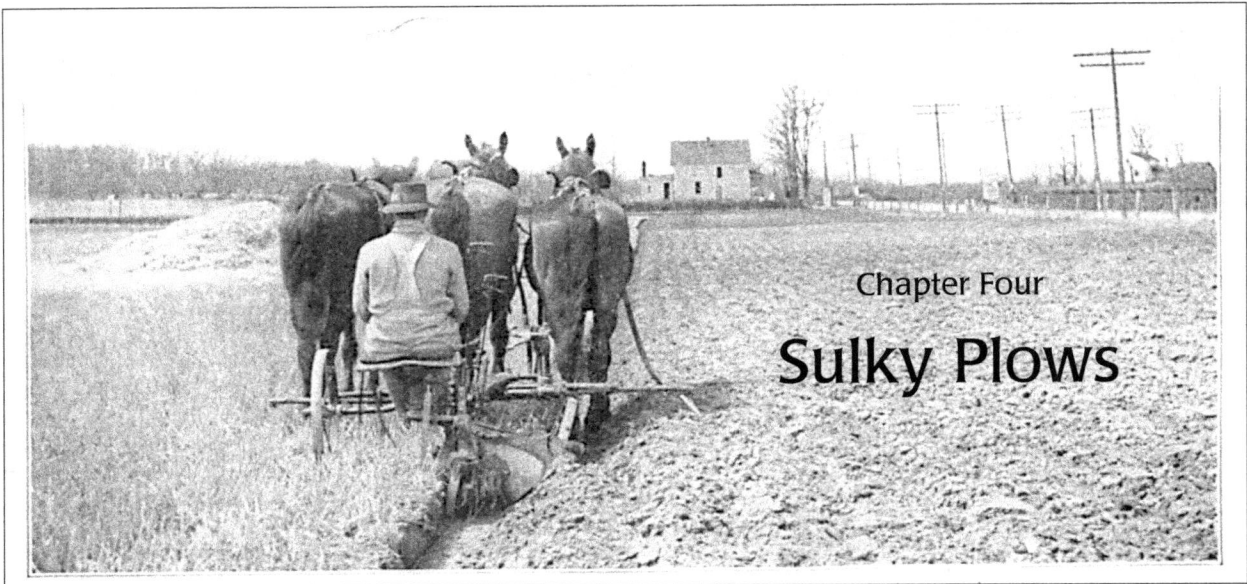

Chapter Four

# Sulky Plows

Fig. 166. The big selling factor for the first sulkies was 'plowman comfort' but a careful review of all aspects of the plow suggests additional benefits to this style of two and three horse plow.

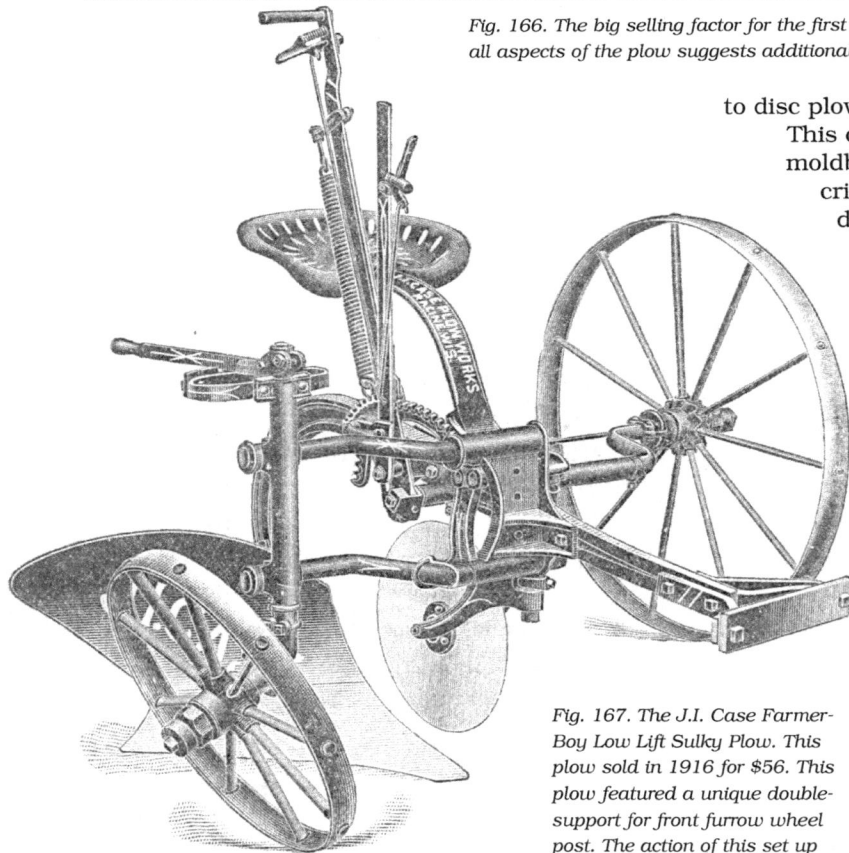

Fig. 167. The J.I. Case Farmer-Boy Low Lift Sulky Plow. This plow sold in 1916 for $56. This plow featured a unique double-support for front furrow wheel post. The action of this set up was timed to raise and lower the front wheel at the same time as the land wheel, keeping the plow level at all times. Tucked in behind the bottom is a rolling landside.

Horse-drawn plows which the operator rides on are made with one or more bottoms. When it is made of one bottom it is called a sulky plow. When made of more than one bottom plowing in the same direction it is called a gang plow. When it is made of two bottoms facing opposite directions it is called a two-way riding plow. (All these aspects generally apply to disc plows as well.)

This chapter will cover basic sulky plows of the moldboard type, though the reader will find critical information here which applies to the different plow designs in the next three chapters. We have seperated out all disc plow information into Chapter Seven.

The sulky plow may be classified as *frameless* (or 'low lift') and *frame* (or 'high lift' also known as 'foot lift'). The first successful sulky riding plow was patented by S. F. Davenport in 1864.

Literally millions of sulky plows representing hundreds of designs were manufactured by dozens of companies across North America. Though they are generally felt to be an improvement over walking plows for the sake of operator convenience, some, this author included, do love the walking plow most. Serious horsefarmers will find appropriate need and use, in most cases, for both walking and riding plows. All riding plows were designed to receive each style of bottom (see Chapter Two).

The sulky plow is the result of an effort to reduce friction as well as to provide a way for the plowman to ride. The size of the plow is usually either 14 or 16 inches, and it is essentially a 3-horse plow, although two are commonly used. The plow bottom, together with the frame if it has one, is mounted on 2 or 3 wheels, a feature which changes into rolling friction, the sliding friction of the walking plow. Hence its possible to add the weight of the frame and the driver and still not materially change the draft.

**The Low-lift (Frameless) Sulky Plow.** - The frameless sulky plow consists of little more than a walking plow minus the handles, with a truck under

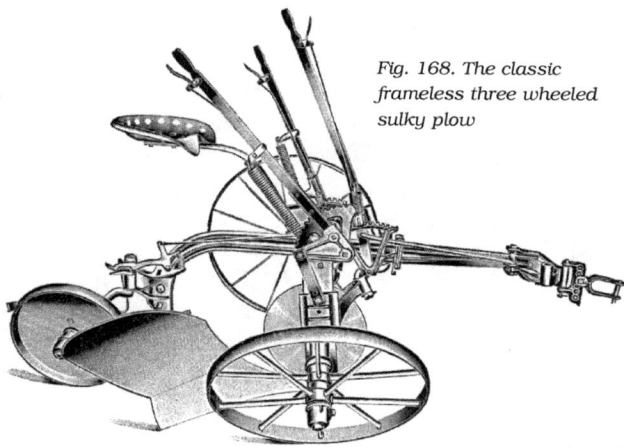

Fig. 168. The classic frameless three wheeled sulky plow

Fig. 171.  Three Fjord draft ponies hitched to a new Pioneer two wheeled frameless sulky plow in Ohio.

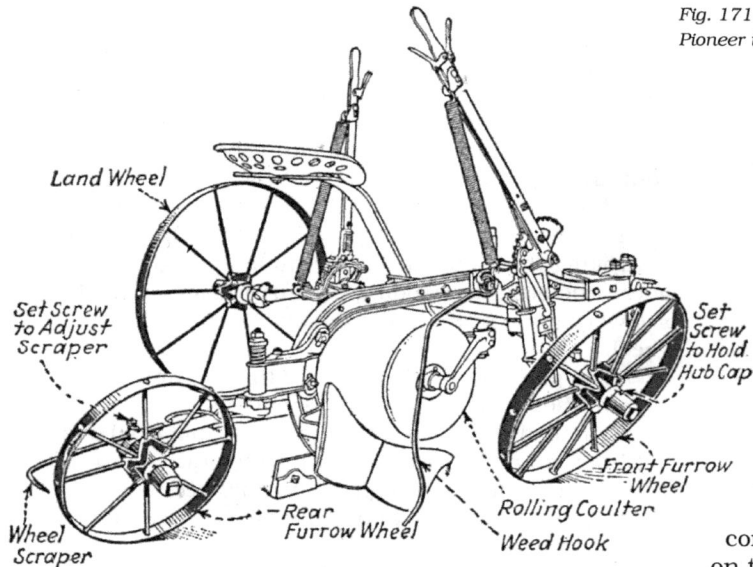

Fig. 169. Frameless Sulky.

Land Wheel

Set Screw to Adjust Scraper

Wheel Scraper

Rear Furrow Wheel

Weed Hook

Rolling Coulter

Front Furrow Wheel

Set Screw to Hold Hub Cap

Fig. 172. Jim Sackett and his Suffolks in eastern Oregon on Gene Westberg's JD footlift sulky.

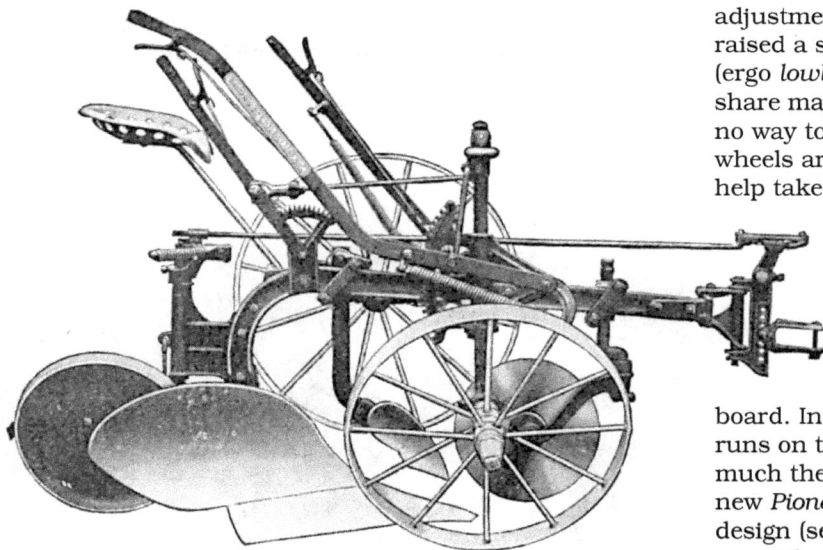

Fig. 170. Rock Island No. 3 Sulky. This is a fancy frameless plow that was especially designed to make square corners without taking the plow out of the ground (note that the rear wheel is connected by rod to the hitch clevis thereby activating a turn as soon as the team steps around. The rear wheel also featured a brake lock to keep the plow from rolling up on the horses when in transport. The brake would unlock automatically if and when the horses were turning.

the end of the beam, a seat attached, and a wheel substituted for the landside. There are two levers to control the bottom - one to raise the end of the beam on the front furrow wheel, the other to level the plow by controlling the land wheel. These levers have a range of adjustment only sufficient to allow the bottom to be raised a short distance above the surface of the ground (ergo *lowlift*). By means of these levers the point of the share may be lifted free from the ground, but there is no way to lift the rear end. The front and rear furrow wheels are inclined from the land (or tilted) in order to help take the pressure of the furrow slice on the moldboard. The front furrow wheel is always castered to swivel; the rear one is usually castered, but in some types is rigidly fastened, and so becomes in reality a rolling landside. These rear wheels are inclined to counteract the side pressure of the furrow slice as it moves backward over the mold-board. In some two wheel models a built-up drag heel runs on the far end of the extended landside acting in much the same manner as a rear furrow wheel. The new *Pioneer* sulky plows are of 2 wheel frameless design (see page 77).

The frameless plow is usually operated without a tongue, although in some plows the tongue may or may not be used. In case a tongue is not used, the front wheel may be controlled by a hand lever for convenience in turning, or by a connection from the hitch at the end of the beam. If the rear wheel is castered (and allowed to swivel) it is connected to the front wheel, so that both are controlled by the same lever or by the

Fig. 173. A highly unusual Gilpin (John Deere) sulky plow from circa 1860. This is basically a variation on the frameless plow with the beam running over the top of an eccentric axle arrangement. No historical evidence of how successful this plow proved to be. It is the author's contention that this represents one of thousands of inventor's 'works in progress'.

Fig. 175. Charlie Jensen of Halsey, Oregon drives his Percheron team, out to a plow competition land. He's pulling a highly unusual walking plow rigged with seat and wheels.

or foot-lift sulky allows the bottom to be lifted higher than in the low lift. The high-lift sulky plow framework is supported by and carried upon three wheels, two running in the furrow and known respectively as the front and rear furrow wheels, and the other running upon the land. The frame may be raised or lowered upon the wheel-supports by means of two levers, one controlling the land side of the frame through the land wheel, the other controlling the furrow side by raising or lowering the frame upon the shank of the furrow wheel.

The plow bottom is attached to the frame by bails, the plow being designated as a single-bailed or a double-bailed plow, according to the number of bails. A foot lever is provided which, by means of the bails, lifts the bottom sufficiently high for turning purposes, so that it is not necessary to use the hand levers after the plow is properly leveled.

It is interesting that so many present day owners and operators of this style of plow do not realize that it was designed to work in rocky and rooty ground. Whereas the frameless riding plow might buck its

same hitch connection. Because this plow is short coupled - the rear wheel, especially in the rigid form, running close to the bottom - and because of the short and rigid connection between the bottom and the wheels, it is possible to turn either a right-hand or a left-hand corner with equal ease. Moreover it is not necessary to lift the plow from the ground in turning in either direction. (see Fig. 169)

**High-lift (Framed) Sulky.** The high-lift

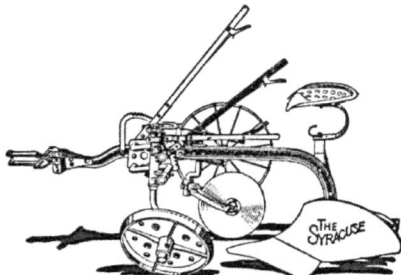

Fig. 174. Two wheel Low-lift, or Frameless style of John Deere sulky plow.

Fig. 176. The same plow as shown in Figure 170.

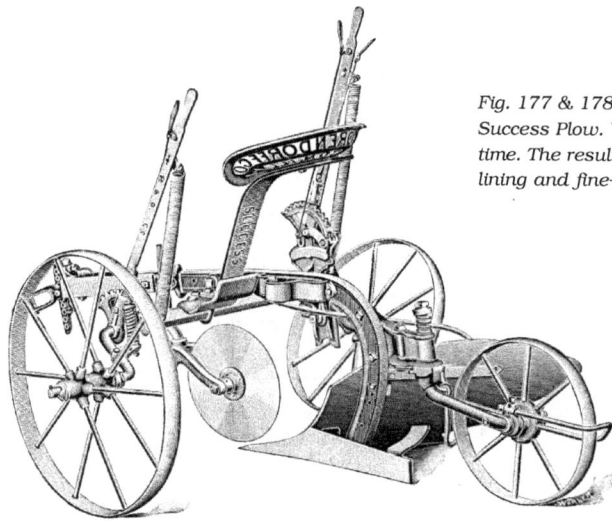

Fig. 177 & 178. (right & left) P & O
Success Plow. Very popular in its
time. The result of years of stream-
lining and fine-tuning

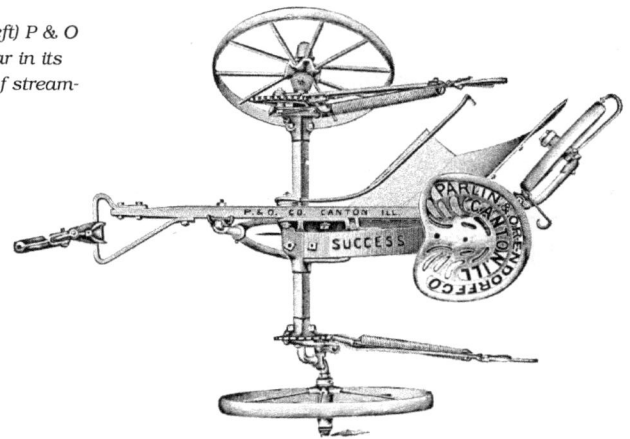

operator off when hitting a subterranean obstruction, many frame plows were designed to allow bail adjustments which might spare the plowman this unsched-

Fig. 179. P & O No.1 Diamond Sulky Plow. A pinnacle of frame plow design. Simplicity, strength, light draft, high lift, easy operation, auto rear wheel control, adjustable seat. This plow included unique turning features evidenced in the next two figures.

uled dumping. This style of plow can be made to float. This means that it will automatically come out of the ground if the bottom strikes an obstruction. When set to float, the bottom is not locked in the ground. When locked, should the bottom strike an obstruction, it will not come out of the ground as in the case where it is floating.

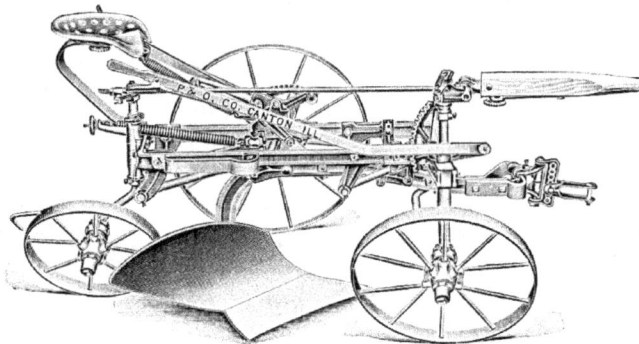

The furrow wheels are on casters and inclined to the vertical, so that they may run in the furrow without undue friction and may better take the pressure of the furrow slice on the moldboard. The front furrow wheel, which is connected to the rear furrow wheel by a flexible rod, is guided by the tongue, which is connected directly to the front wheel. The tongue or pole extends forward between the furrow

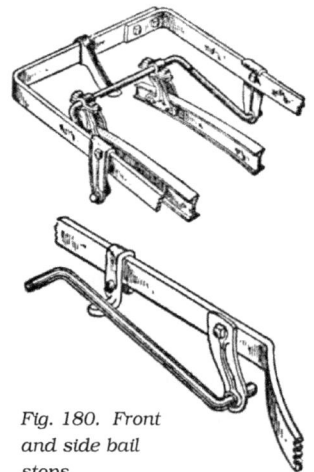

Fig. 180. Front and side bail stops.

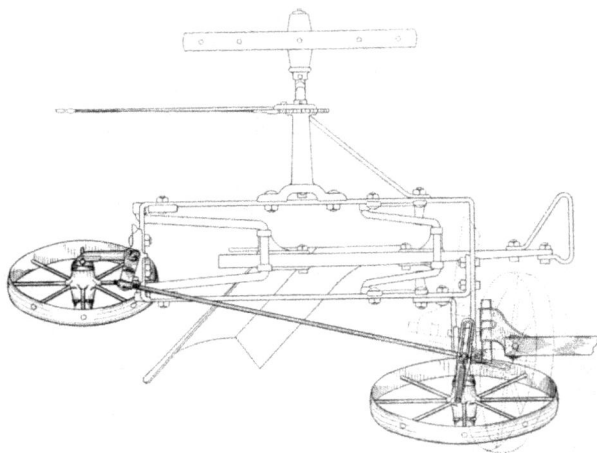

Fig. 181. P & O No. 1 Diamond Sulky Plow frame showing the position of the highly sophisticated turn-controlling rod and position of wheels while plowing and preparing for a turn. The dotted lines illustrate how far the front furrow wheel can be turned without acting on the rear wheel. The rear wheel will not castor or swivel until the frame is actually being pulled around at the end of a furrow, it then immediately responds and as quickly resumes its normal position when the turn has been made.

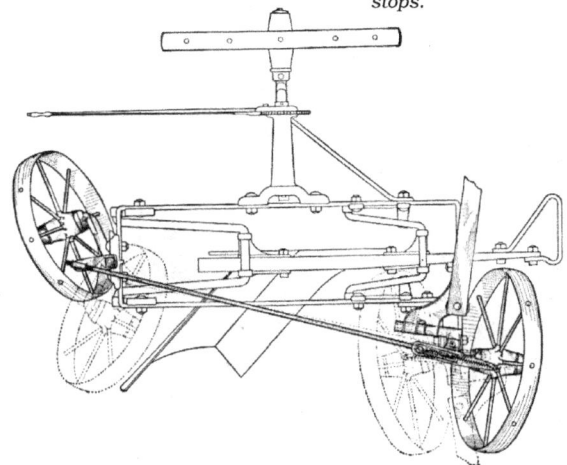

Fig. 182. Showing action (bold lines) of controlling rod and position of wheels while a left turn (or turn around the land) is in progress. The dotted lines show the position of the wheels when turning to the right or back furrowing.

horse and the one next on the land.

**The Gang Plow.** Not a true 'sulky' plow by definition, the gang plow has two or more bottoms. There is no material difference between the horse-drawn gang plow and the various sulky plows except that a gang plow has more bottoms. The gang plow may be either frame or frameless. The gang construction and arrangement of the wheels and methods of control are practically the same. Of course, it takes more horses to pull the extra bottoms and, when this is the case, greater potential trouble will develop from side draft, which will be discussed in Chapter Six.

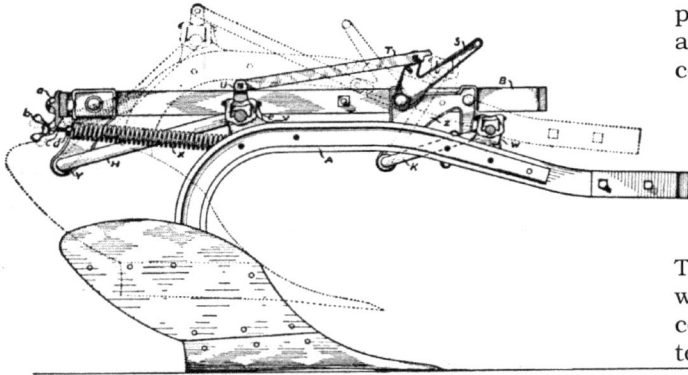

Fig. 183. The P & O Diamond Sulky, plow featured on the previous page, incorporated a perfected foot-lift system. In this drawing both positions are shown. The heavy lines, bottom lowered, dotted lines raised.

Fig. 184. A landside view of the P & O Diamond foot-lift system. The hand lever is always ready for use when necessary, but the foot levers are used almost exclusively. By releasing the hand lever from the quadrant (lock) it remains idle. This feature, of allowing the hand lever to remain idle, prevents it from flying up and possibly injuring the driver by striking his arm when the plow strikes an obstruction, an occurrence which can happen on plows where a rigid lever is used.

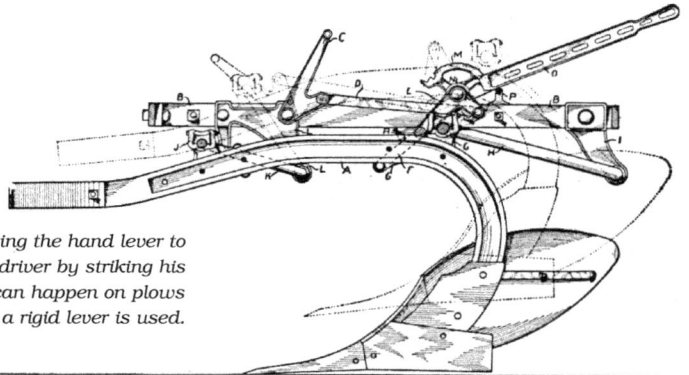

Fig. 185. The P & O Diamond Sulky Plow shown from behind with the landside removed for structural view.

## The set of the sulky plow.

The set of the sulky plow differs in many important ways from that of the walking plow. The point of the share is the only part of the bottom that touches the floor when the plow is properly leveled. The wing of the share just swings free of the floor, while the heel of what little landside there is does not touch at all. The suction, then, is not measured by the amount the point dips, but rather by the amount the heel of the landside is raised from the floor. Since the length of landsides varies, the suction cannot be stated in a formula involving terms of distance of the heel from the floor unless the distance back to the point of the share is given. In general, at a distance of 18 inches from the front, the landside should be ½ inch from the floor. Except in a few instances there is a way provided in the plow's design for regulating the suction from zero to about 1 inch. This device, which is usually found in connection with the shank of the rear furrow wheel (see Fig. 198), provides a means for raising or lowering the rear end of the frame. *

**The furrow wheels** are given lead either toward or away from the land. The front wheel is frequently given lead toward the land in an effort to throw the line of lead as far to the land as possible. This is particularly true of gang plows. The rear furrow wheel is given lead (or advance) from the land. There is considerable pressure on this wheel, a pressure which tends to crowd it into the furrow wall, and the wheel

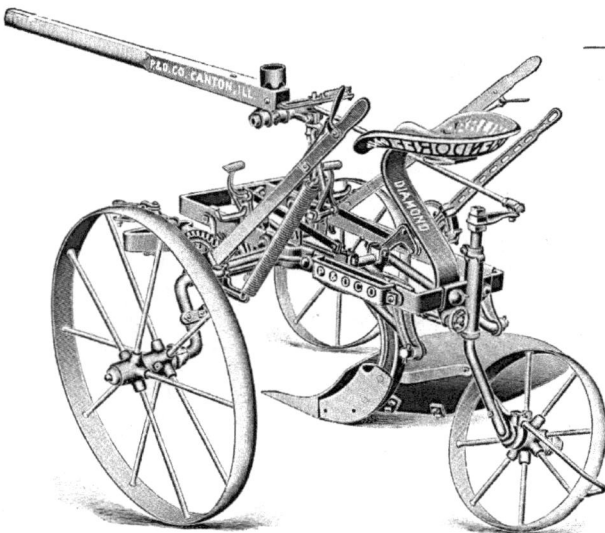

properly set leads away in order that the plow not struggle to climb over the furrow wall. Different devices are provided for giving these wheels lead (or advance).

**The rear wheel** should be set in the corner of the furrow and carry the land side-pressure. Adjust at *Fig. 197 "F"*, moving the bracket towards the land until the rear end of landside sets about ½" away from the furrow wall when the plow is at work. If you can place your fingers between the landside and furrow wall, the setting is all right.

Rear wheel should carry the weight at the rear. The landside on the rear bottom should set about 5/8" above the furrow bottom.

Adjust the collar at Fig. 197 "E" up or down until

*(In some frameless sulkies with a rigid rear wheel it is impossible to alter the suction in this way. For this reason these plows must be treated more like a walking plow with wheels attached. The guidelines for bottom set in this case will be seen to duplicate the walking plow.)*

the landside sets about the thickness of your finger (5/8") above the bottom of the furrow.

**The front furrow wheel** must be properly adjusted to the width of the furrow; that is, this wheel should always be kept running in the bottom of the furrow and against the furrow bank. If the hitch is changed to cut a wider or a narrower furrow, a corresponding change must be made in the wheel.

**Float.** It is possible, too, to set the high-lift plow to float. Ordinarily when the bottom is lowered by the foot lever, it is locked in position and cannot be raised unless the frame is lifted with it. In nearly all cases a set-screw is found in connection with this foot lever, which may be so adjusted as to prevent its locking. In this way if the bottom strikes a stone or root, it alone will be thrown out of the ground, swinging upward on its bails. This device is useful in the plowing of stoney or stumpy ground.

## Attachments.

*Much of the information in the next few paragraphs is a rehash of particulars from Chapter Two but nonetheless important to touch on once again in the context of sulky plows.*

**Rolling coulters and jointers**. The function of the rolling coulter in loose ground is to insure a clean furrow and cut trash so it does not accumulate in the throat of the plow. The adjustments necessary to get this job done are comparatively few and simple. To cut sod and roots and thereby lighten the plow in breaking, have these things in mind:

First: To insure a clean furrow the coulter should be set to cut at least one-half inch wider than the cut of the plow, or toward the unplowed land from the shin of the plow. This is clearly illustrated in Fig 66. The coulter in this position will make a clean cut, and the plow following will turn the furrow away from this cut, leaving the furrow wall standing clean.

Second: When plowing loose land, or doing what is commonly termed, "stubble plowing", where there is very much trash, the coulter should not be set deeper than about one and one-half inches. There seems to be more or less of a general idea that the more trash there is to cut, the deeper the coulter should be set, or to quote a remark often heard, "Set her good and deep so that it will cut through all of it." This is wrong. The rolling coulter set deep for trashy conditions causes the coulter to push the trash rather than cut through it. Fig. 103 shows the coulter properly adjusted, running shallow, crowding the trash into a wedge or scissor shape, where it is easily cut and passed on through with the furrow slice.

Third: In breaking, the coulter should be set deep because here its function is to cut the sod and roots, and it can be run well into the hub of the coulter when plowing six inches or more deep.

Fourth: The fore and aft location of the rolling coulter should be with the hub of the coulter about three inches back from the point of the plow. This is the best location for average plowing conditions. In extremely loose and trashy soil it is sometimes advisable to move further forward so the coulter has the chance to cut against the firm soil before the plow

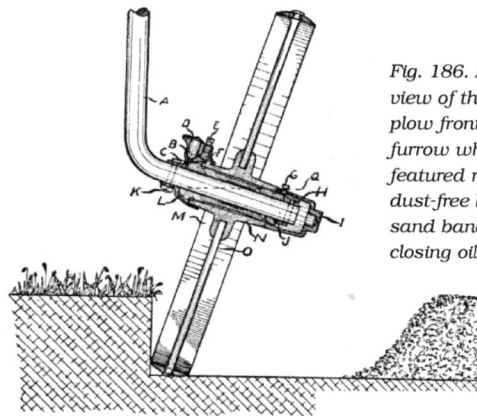
Fig. 186. A cross section view of the P & O sulky plow front and rear furrow wheels which featured removable dust-free boxes and sand bands and self-closing oil cups.

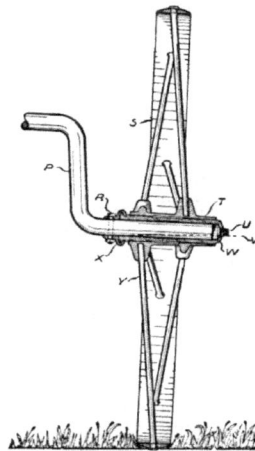
Fig. 187. P & O land wheel cross se The manufacturer touted that this wheel could be designated as having a 1,000 mile oil-carrying capacity.

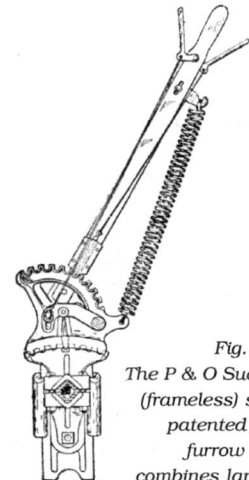
Fig. 188. The P & O Success (frameless) sulky patented front furrow lever combines landing and raising (or lowering) all in one lever.

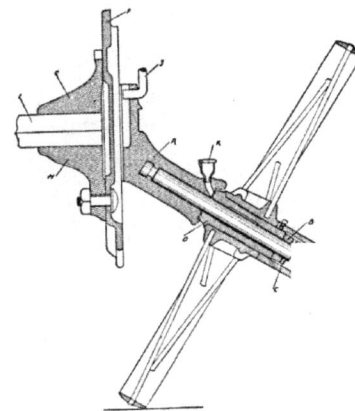
Fig. 189. (left) Sectional view of front furrow wheel for the P & O Success sulky

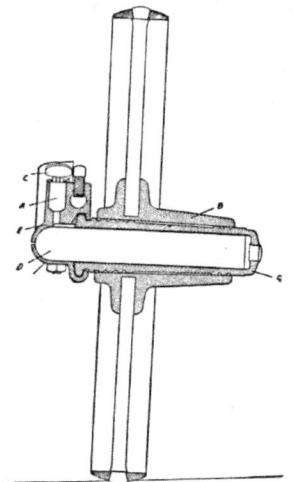
Fig. 190. Sectional view of P & O Success (frameless sulky) rear wheel . This rear wheel has a brake hook, which can be used when at work on steep hillsides or on inclines, to prevent the plow from running down the team. The Success is a plow which can be easily handled on hillsides or uneven places, as the levers are all counterbalanced and within easy reach (see Fig 177 & 178).

starts to lift and loosen it.

Just remember two things: In loose, trashy land, the coulter must cut the trash on top of the soil. This requires a shallow adjustment. In sod or breaking, the coulter must cut the roots beneath the surface of the ground, and this requires a deep adjustment.

Another point bearing on the necessity for shallow adjustment in trash is the fact that the shallower the coulter is set, the more clearance there is between or below the coulter stem and yoke and the less danger of clogging.

*Now about jointers*: The function of the jointer is to cut and turn a small furrow at the extreme landside edge of the big furrow. In loose, trashy land, this puts the trash well away from the furrow edge, covers it up, and insures a perfectly clean job of plowing. In sod the same thing is accomplished, so there are no ragged edges of the furrow protruding and no grass left near enough to the surface to start growing and interfere with crop raising and cultivation. The jointer is also a great aid in pulverizing the soil. Many farmers claim that a field plowed with the jointer attachment is in a better state of pulverization than the same field plowed without it and harrowed once.

The adjustment of the jointer is very simple. It should be adjusted so the point of the jointer sets approximately over or slightly ahead of the point of the plow, and toward the unplowed land from the shin of the plow from one-half inch to three and a quarter inches, so that it will cut and lift the small furrow and leave a clean standing furrow wall.

Fig. 104 illustrates the proper location of the jointer in relation to the point of the plow.

**Clean plowing** consists of covering all trash deeply and completely, leaving the surface of the plowed field absolutely clean and free of trash.

*Rolling Coulter*: The rolling coulter is of much importance in clean plowing. The coulter should be sharp so it will cut down

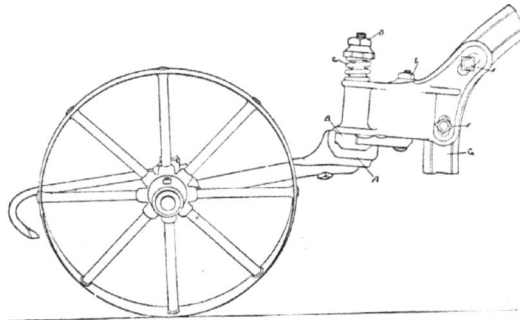

Fig. 191. The P & O Success sulky has this rear wheel with an automatic clutch which can be adjusted and compels the plow to run absolutely straight. The land friction can be relieved by manipulating one bolt.

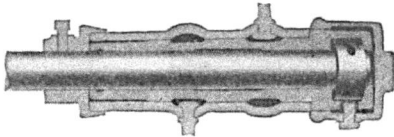

Fig. 192. Wheel boxing, showing dirt-proof and oil-tight construction. Collar on outer end of boxing holds wheel on axle and takes end thrust. Screw cap provides means of applying lubrication.

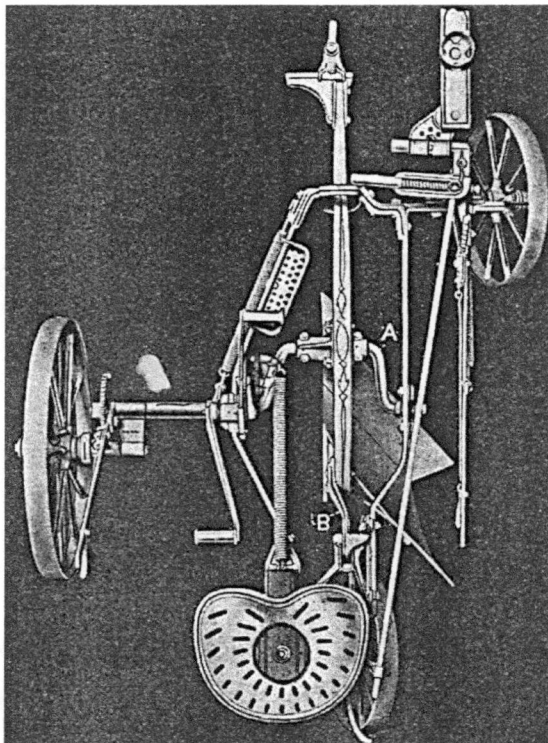

Fig. 193. Top view of a single-bailed high-lift sulky plow. The single bail is at 'A'. At 'B' is shown a leveling device; suction is also altered at this point.

through the surface trash.

*Independent Jointer*: The independent jointer is ideal equipment, as the jointer can be set at a definite and invariable position, and, being rigidly attached to the beam, it goes through all conditions. Again, it has no effect on the coulter, allowing the coulter to do its work of cutting without interference.

*Moldboard Extension*: The object of a moldboard wing extension is to control the furrow slice after it passes the wing of the moldboard. It helps in securing clean plowing, as it insures close lapping of the furrow slices and thus eliminates openings between the furrow slices.

*Trash Wire*: A No. 9 wire about 10 feet long, which can be clipped to the coulter shank, makes a good trash wire for holding down the loose trash. It never clogs, regardless of conditions.

*Weed Rod*: The addition of a weed rod may sometimes be helpful. It should be clipped to the beam above the shin of the moldboard.

## Tuning up wheel plows - *before and after*

Before plowing it is critical to assess the readiness of the plow. If the reader has purchased a new plow he or she may reasonably assume the plow is ready for work. But this author recognizes that most will be dealing with a relic plow in various states of repair. What follows is a good checklist of things to do to the plow before attempting to use it. And this list serves an additional purpose

There is no better time to get the wheel plow (or indeed any farm implement) in shape for another season's work than immediately after you have concluded this season's work.

In either case, whether in preparation for that first plowing or at the close of a season, the following tips will be useful.

**First** - Examine the wheel boxes. *(See Fig. 197 A)* If they have been properly lubricated, they should last as long as any other part of the plow. However, in summer and fall-plowing particularly, the ground is frequently dry and dusty, and if overlooked for only a short time, the wheel boxes

Fig. 194. Amy Beyer drives 3 draft mules on a sulky plow to win the ladies sulky class at the 1999 Heritage Farm Plow Days in Hudson, Iowa.

Fig. 195. Jim Sackett of Oregon using his foot-lift, frame plow with two excellent Suffolk Draft Horses.

will begin to wear very rapidly. If the boxes are badly worn, they should be renewed; if not, they should be slipped off the axle and both axle and box washed clean with kerosene, and a fresh supply of grease applied. The wheels are fastened to the axle on practically all wheel plows in three distinct ways; a clamp or hinged collar on the inner end of the box, a collar in the middle box (the box in this case being in two pieces), or a collar and linchpin on the end of the axle. In any case, if the collar has become badly worn so as to allow excessive end play of the wheel box on the axle, it should be replaced with a new one.

**Second** - Examine your shares. (See Fig. 197 B and refer to Chapter Two). No other part of a plow so quickly affects its good running and good working qualities as the share. Shares must be not only reasonably sharp at all times; they also must be properly shaped and set when heated for

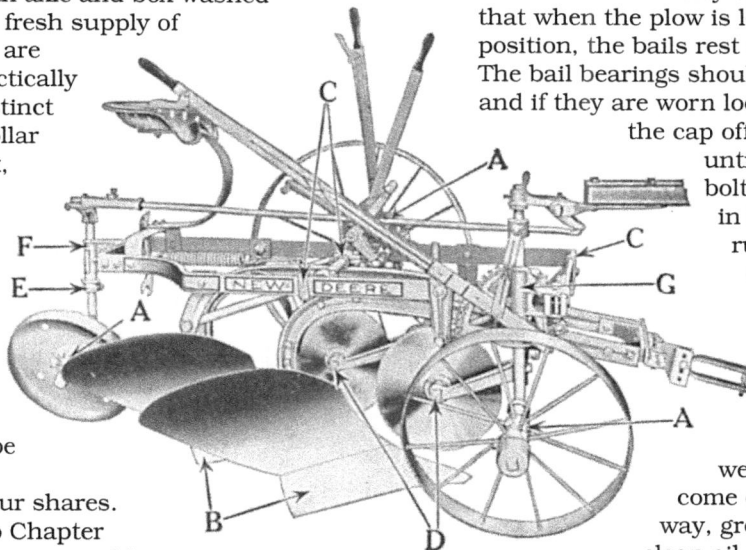

Fig. 197. Though this is a gang plow it is a good representation of a foot-lift frame with parts identified. See text.

sharpening, and they should not be allowed to rust.

**Third** - On high-lift, foot-lift plows, suspended by one or two bails, examine the bail stops. (See Fig. 197C) These are located on the right frame bar, and on the front frame bar. They should be so adjusted that when the plow is locked down in plowing position, the bails rest securely on the stops. The bail bearings should also be examined, and if they are worn loose and sloppy, take the cap off and file or grind it until it fits snug with all bolts tight. This will help in keeping the plow running steady and quiet.

**Fourth** - Examine the rolling coulters and hub bearings. (See Fig. 197 D) Coulters should be sharp and well polished. If they come out of the field that way, grease them with some clean oil or grease. A dull rolling coulter acts like a gauge wheel, prevents the plow from penetrating properly, and increases the draft. Coulter hub bearings, if badly worn, should be renewed. The rolling coulter should not be permitted to get loose enough on the hub bearings so that it cannot be kept running true and steady.

**Fifth** - Check on the set of the rear axle collar. This collar on both high lift and low lift plows should support the rear end of the frame and transmit the weight of the entire plow and rider to the rear wheel. If it has slipped down on the axle, this weight will be carried on the bottom of the plow landside. This will increase the draft and throw the plow out of level.

**Sixth** - Rear axle frame bearing carries the vertical part of the rear axle, and, in addition to carrying the weight of the plow and rider, it transmits the side pressure created by the moldboards to the rear furrow wheel. If this bearing becomes badly worn, the landside of the rear plow will have to carry this pressure, in place of the rear wheel. This will increase the draft and wear out the landside prematurely. On the plow illustrated, this bearing is provided with a take-up

Fig. 196. Liza Howe with 2 American Creams and 1 Suffolk on a sulky plow. She won 'best youth teamster' at the Heritage Farm 1999 Plow Days

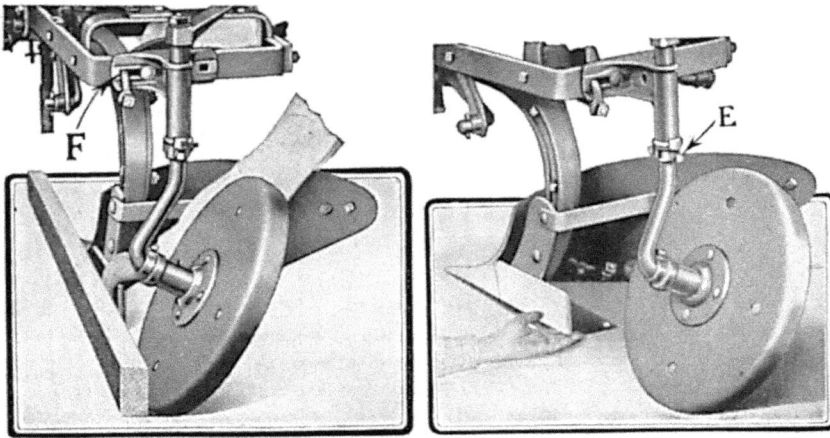

Fig. 198. *On the left*: *showing test for clearance between landside heel and furrow wall to determine if furrow wheel is carrying landside pressure. If there is not room for the fingers between straight edge and landside, adjust at "F". Adjust at connection rod to set rear wheel to run straight.* **On the right:** *There should be enough clearance beneath heel of landside to permit fingers to pass as illustarted, between floor of furrow and heel of landside, when plow is at work. Adjust at "E" for proper clearance.*

at "F" (Fig. 197) by means of a slot in the specially-designed casting with heavy bolts at both the upper and lower end. This makes it easy to keep a snug fit on the bearing.

**Seventh** - Front furrow axle frame bearing should fit snugly, in order to keep the wheel running at the proper angle, and the front furrow at the proper width. This, however, is not quite so important as the rear axle frame bearing, as in most sulky or gang plows the manufacturers provide an adjustment at the right front corner of the frame, or where the furrow axle bearing is attached to the frame, so that this may be moved in or out, to widen or narrow the furrow.

And last, but not least, do not overlook loose nuts or badly-worn bolts. Go over the plow from stem to

Fig. 199. (below) John Deere was proud of its detachable shares. The literature includes these notes: 1. No trouble to remove share - only one nut to take off. 2. Eighty per cent time saved. 3. No danger of damaging share. 4. Share is drawn up closer. 5. Share is stronger - not weakened by bolt holes. 6. Resharpened or sprung share can be drawn into place - no drift punch necessary. 7. No unequal strain on share. 8. No danger of injuring hands in taking off share.

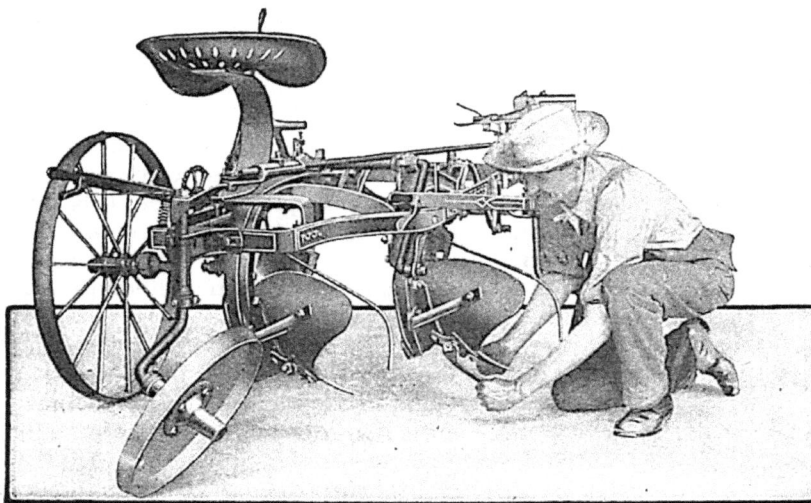

stern, tighten up every loose bolt and replace those that are badly worn. Fit all new bolts with lock washers - it is well worth while. You can save horseflesh by carefully checking up the foregoing adjustments on your horse-drawn plow. Make the wheels carry the load - that is what they are intended for.

## Correct Hitch on Riding Plows

We must assume, in discussing the correct hitch on a three-wheel plow, that all adjustments of the plow are correct and the share properly sharpened. If the plow be a new one, erected according to factory instructions and none of the adjustments are changed, you can proceed with the assurance that with a correct hitch, the plow will work as it was intended. If it is an older plow a careful reading of Chapters One and Two is important.

The wheels are attached to a three-wheel plow for the purpose of carrying the entire load, regulating the depth and width of furrow or furrows to be turned. Keep this point firmly in your mind.

The load consists of the weight of the plow, the weight of the operator and the weight of the soil being lifted and turned. All of this weight should be carried on the wheels and none of it carried by the team, on account of improper hitch.

Fig. 203 illustrates the true line of draft when a three-wheel plow is pulled by horse power. This line of draft you will notice is from a point indicating the center of resistance on plow bottom to the point where the tug or trace attaches to the hame at the horse's shoulder. The correct place to hitch, then, would be at a point where this line passes through the vertical part of your plow clevis. The dotted lines show the effect at this point on your vertical clevis of too short a hitch when horses are worked abreast. The natural tendency is to hitch too close to a three-wheel plow.

Understand, then, that if you hitch too close or too low on a three-wheel plow, your team will carry a

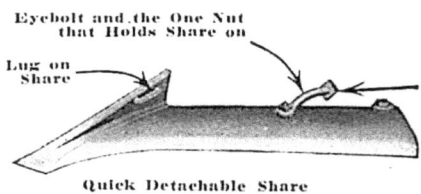

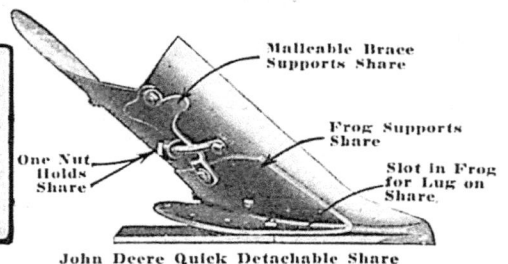

Eyebolt and the One Nut that Holds Share on

Lug on Share

Quick Detachable Share

Malleable Brace Supports Share

Frog Supports Share

One Nut Holds Share

Slot in Frog for Lug on Share

John Deere Quick Detachable Share

# Pioneer Plows

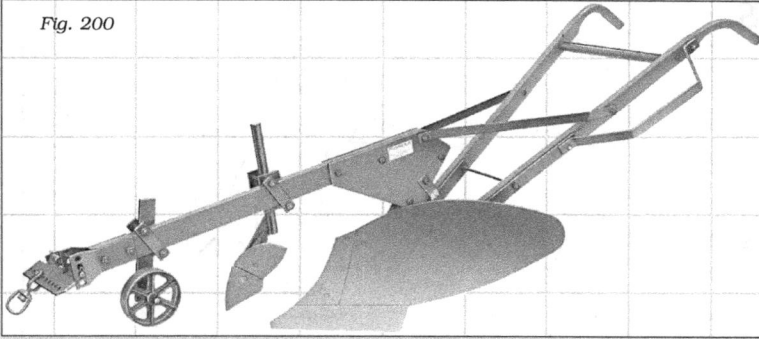

Fig. 200

*The good news is that exceptional new horsedrawn plows are being manufactured today. One of the companies hard at work is the Amish family business PIONEER Equipment of Dalton, Ohio. They build walking plows, sulky plows and a new two bottom gang. In this sidebar we feature pictures and information on two models. The PIONEER Walking Plow and Sulkies are both available in 12, 14 and 16" widths right or left handed. The beams are a two piece reinforced high carbon steel. The hitch features horizontal adjustment for two or three horses. They adjust vertically for any size animals. They feature Oliver raydex-type chilled steel bottoms. The sulky features tapered roller bearings. Shares, heels, shins, and landsides are all replaceable and readily available.*

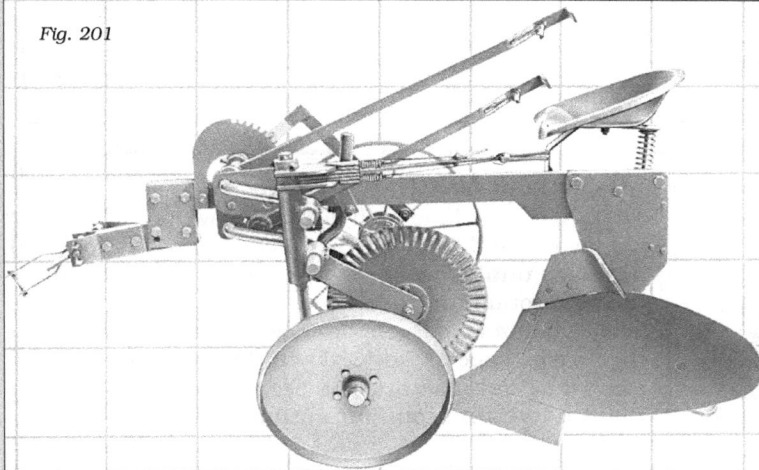

Fig. 201

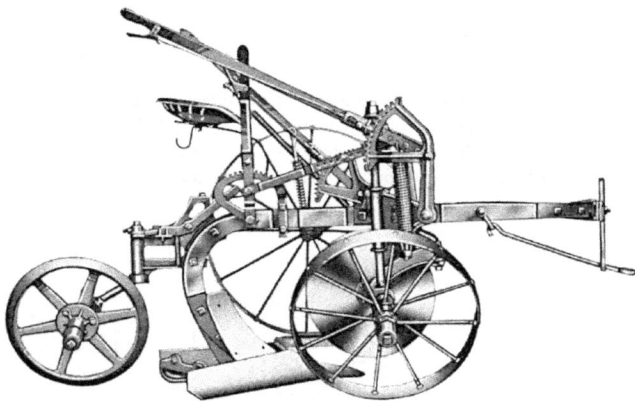

*Fig. 202. The Oliver No. 81 Corn Borer Sulky plow equipped as a corn stalk lifter. Note the moldboard and jointer have been removed. When equipped this way the No. 81 is used for lifting the first 12 rows of stalks around the outer edge of fields heavily infested with the European Corn Borer. Several different makes of plows were built to receive special application options.*

good portion of the load on their traces rather than allow it to be rolled along on the wheels of the plow. The weight they carry will be taken mostly from the front furrow wheel.

By hitching too close or too low, you have not only added this weight to your team but you have taken away the means of controlling the plow, because without proper weight on the front furrow wheel, it will not be possible for you to control accurately the width or depth of furrow.

A good rule to go by on this point is to hitch long enough and high enough so that you have as much weight on the front furrow wheel as you have on the rear furrow wheel. A long hitch is better to accomplish this, rather than a high hitch, as the long hitch affords more room for your horses to walk, making it all the easier on them.

You can test this very nicely after you have opened up your land and your plow is running at the depth you want it to run. Slip off the seat and grab the front furrow wheel and see how much effort it takes for you to slide it. Then do the same thing to the rear furrow wheel. Adjust your hitch until you have as much resistance on the front furrow wheel as you have on the rear furrow wheel.

You will be surprised at the difference it will make to your team when you relieve them of carrying part of the load of the plow and lengthen their traces enough so they simply roll the load along.

If you will follow these simple instructions, you will have no trouble in getting a proper hitch so far as the vertical adjustment of your clevis is concerned. With this part of your hitch right, move your evener clevis sidewise on the cross clevis of the plow, whether it be a sulky or a gang plow, until your team walks comparatively straight. Do not insist on hitching directly in front of the point of the beam, on a single-bottom plow, or between the point of the two beams on a two-bottom plow. If you do and use fairly good-sized horses, they will have to walk sidewise and the traces will chafe their legs. With the first hitch adjustment proper, you will have sufficient weight on the front furrow wheel, so that you can get over far enough on the cross clevis to allow your horses to walk away straight and free.

***Take care of your horses***; hitch them so they will be the most comfortable at work, and you will be surprised at the additional service they will give you, and it will also improve the operation of your plow.

# Hitching Riding Plows

**Fig. 203. Hitching Riding Plows.** *The satisfactory performance of a riding plow depends to a great extent upon correct hitching. If both horizontal and vertical hitch adjustments are correct, the plow will run smoother, pull lighter and do a better job of plowing than when carelessly hitched.*

*The drawing above illustrates the correct up and down adjustment on the vertical clevis. The correct hitch at "A" is the place where "A" is in a true line between "B" and point of hitch at the hame. When plowing deep or using tall horses, hitch at "A" should be higher than when plowing shallow or using small horses. When hitching horses strung out, the hitch at "A" must be lower than when using four horses abreast.*

*The results of improper adjustment of the vertical hitch are easily noticed. If the hitch is too high at "A", there is a down-pull on the front end of the plow and the rear end will tend to come up. If the hitch is too low at "A", the draft will tend to lift the front end of the plow . By changing the position of the clevis up or down one or two holes at "A", a trial will generally show which hole places the clevis in a true line of draft.*

*Hitches are adjustable horizontally for the purpose of accommodating the position of the horses and the various sizes and types of eveners. Consequently, the cross hitch is very long and has a large number of hitch positions.*

*The operator should aim to get the horizontal hitch as near as possible in direct line between the center of draft and the center of power when the plow is running straight and horses are pulling straight ahead. (See page number 37).*

## Hitching Horses or Mules to a frameless, poleless, sulky.

*As the preceding information has indicated there are a myriad of variables with the frameless style plows including whether or not a tongue is used and whether or not there is any braking system for the rear furrow wheel. It is important to keep in mind that driving a frameless plow out of the ground is like driving a walking plow with wheels. With any unevenness to the ground (or a sloppy furrow wheel) the implement will skate back and forth as it moves forward. This would be reason for caution if green or nervous horses are employed. Also, use of this style plow on steep terrain is risky unless either a tongue is employed or some sort of brake is involved with the third, or rear furrow, wheel. Two wheeled sulkies such as the Pioneer employ drag on the landside heel as a brake. With this style of plow it would be smart to think about leaving it in the field until the plowing is done rather than using it as a conveyance back and forth from the barnlot. (Please also read 'Hitching to Frame Plow' pg 79.)*

**CAUTION: This author recommends DO NOT use a poleless framelss sulky with horses or mules suspected to be runaways! (Almost all frameless sulkies can be custom equipped with a pole or tongue which will prevent the implement from running up into the legs of the animals.)**

## Opening A Land

To open a land with a riding plow, both the land wheel and front furrow wheel levers should be raised until plow opens up to depth desired. On the next round, the landside lever is adjusted to permit bottoms to cut depth desired to plow, and the furrow wheel is set level with the bottoms. Once the correct setting has been made, the plow will run level and continue cutting at uniform depth.

Many plowmen prefer to turn the first furrows in each land with a walking plow, thereby simplifying the adjustment of the riding plows.

When plowing around a field, the plow is not lifted at the corners. The driver stops a short distance from the end, turns square and does not permit his team to start ahead until completely turned.

*Fig. 204. A diagram showing the net effect on hitch point of greater distance from the plow. The correct vertical hitch in BGC for a short hitch and AHC for a long hitch. With horses hitched it is important to remember that too low is okay, too high is a problem.*

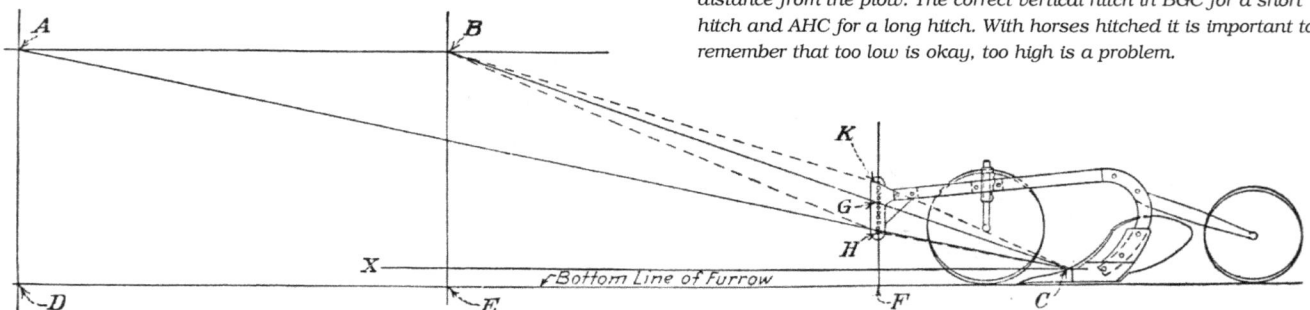

## Floating Over Rocks!

The foot lift on a riding plow acts both as a lift and a lock. The plow is locked down when in plowing position by pushing forward on the upper lift pedal until the lock goes over center. In plowing stony land, the set-screw on the lock can be screwed down to set the strap anti-lock. When so set, the lifting spring should be loosened sufficiently to prevent bottoms lifting excepting when they strike an obstruction. The bottoms will then maintain their depth, but will automatically come up when the shares strike an obstruction.

## Lifting spring: a hint.

*The lifting spring aids in raising the bottoms and should be adjusted with enough tension to make the plow lift easily. If left too loose, the plow will lift heavily.*

Fig. 206. A rare 1880's J Thompson & Sons 'Ole Olsen' frameless featuring tongue.

## Hitching Horses or Mules to a Frame-style, sulky plow.

*As the preceding information has indicated, there are variables within the frame type plows but less so than with frameless. All of the frame, hi-lift, plows feature a pole or tongue which functions to articulate the steering. This translates to greater stability when driving with plow out of ground.*

*Driving a frame plow out of the ground is like driving a vehicle from which hangs a plow. In other words there is less reason for concern than with the frameless.*

*One word of caution: until all aspects of the sulky plow are completely understood it is safest to secure the neckyoke to the end of the tongue (if not already done so).*

*When hitching, always remember to fasten neckyoke FIRST! Only after the neckyoke is fastened, is it safe to hook tugs. If the reader is uncertain about the horses it is IMPERATIVE that help be available when hooking. Do not attempt to hook horses to a plow unless you know how to work horses. The riding plow, in knowledgeable hands, is a comfortable, relatively safe implement. The riding plow in inexperienced hands can be an EXTREMELY HAZARDOUS tool. Know what you are doing, excercise common sense and think safety. If the reader is looking for kicks and thrills, this author suggests setting aside the riding plow and considering a life in real estate sales, plastic surgery, or politics. Leave the riding plow for people who want to farm.*

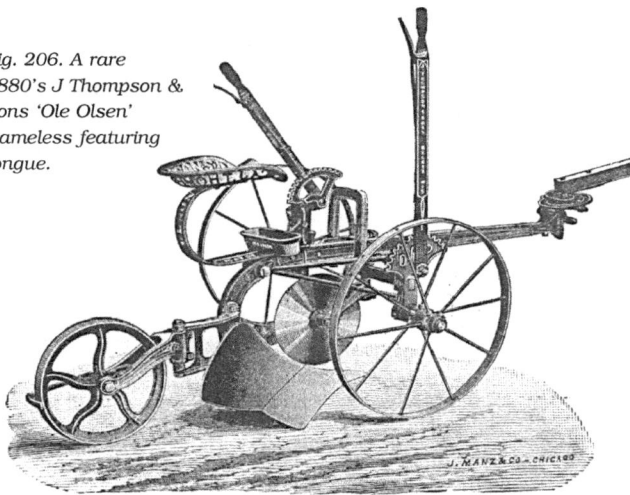

Fig. 207. A P & O cut from 1889 showing their early 'Clipper Tricycle' frame-style sulky.

## How Many Horses & What Evener

*Sulky plows can employ anything from 12" to 16" bottoms. Depending on soil conditions, and the physical condition of the working animals, the plowman might need two draft horses or mules - or possibly three abreast.*

- *For a team on a twelve inch sulky, try a 34" doubletree.*
- *For a team on a 14" sulky, try a 38" doubletree.*
- *And, though it is unlikely a team would be employed, for a 16" bottom try a 42" doubletree.*

*As a rule of thumb (always subject to overrule); 12" use two full-size horses, 14" and 16" use three full-sized horses. This author believes strongly in using more horses than the job requires for several good reasons. For a 14" or 16" sulky in open field plowing it is hard to beat a strung out four-up. Because of the extra horse, none of them need be completely exhausted at day's end. And a green horse hitched landside wheel (nearest the teamster) is handy to get to and learns quickly about team work. And, as will be covered in Chapter Six, for the experienced teamster driving four strung out on a plow, is actually every bit as easy as driving a team of two and more fun.*

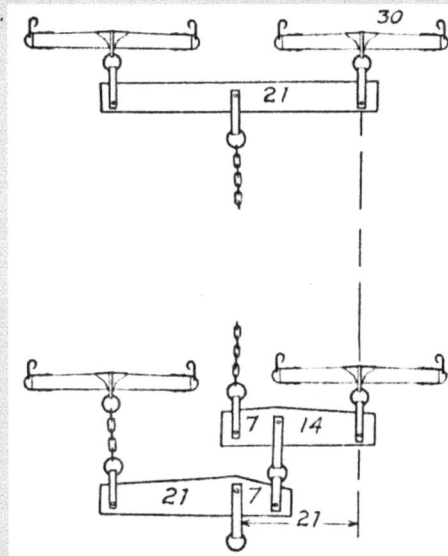

Fig. 205. A diagram of a serviceable homemade four-up evener. Notice that the distance from center of furrow horse to hitch point is 21 inches. With a 16" bottom 21" to 22" inches will be perfect, with a 14" bottom 18" to 19" would be perfect. With a slight adjustment of the horizontal position the above evener will work fine for either size plow. (See Fig. 125).

Sulky Plows

Fig. 208. In 1889 the Gale Manufacturing company advertised its 'Big Injun Three-Wheel' Sulky Plow with this cut demonstrating how a square corner could be made by a one-armed plowman without ever having to lift the plow.

For in depth information on working horses and mules in harness please refer to preceding texts in this series, The Work Horse Handbook and Training Workhorses/Training Teamsters.

***Comfort to horses*** when plowing depends a great deal upon all aspects of harnessing. The fit of the collars should be perfect and this author recommends the use of sweat pads. Hames should fit the collars properly. All leather (or nylon) should be strong and smooth. Buckles should face away from the animal's hide and be strong. The lines should be correctly adjusted so the horses cannot spread out too much, or they will not be well under control of the driver.

Long traces give the horses more room and tend to make the plow run steadier. Short traces do not lighten the draft and may add much discomfort to the horses.

If hip straps are used, they should be adjusted so that the loops hang free. If loops pull up on the traces, they will change the line of draft, making the plow run unsteadily and cause weight to be carried on the horses' backs.

Comfortable animals will be ready to return to work. Uncomfortable work animals won't.

The plow operator should constantly keep in mind the welfare of his horses as well as his plow.

# Oliver 'Horse Lift' Sulky Model 26A

This author's first experience with the riding plow was with Ray Drongesen's customized Oliver 23B two way plow. Ray had removed one bottom when he discovered that the unique two-wheel frame and lift system of this plow made it a superior implement for plowing out the last furrows and cleaning a dead furrow. The foot-lift tongue-articulated steering sulkies all bounced in and out of the dead furrow passes while Ray's beaut held straight

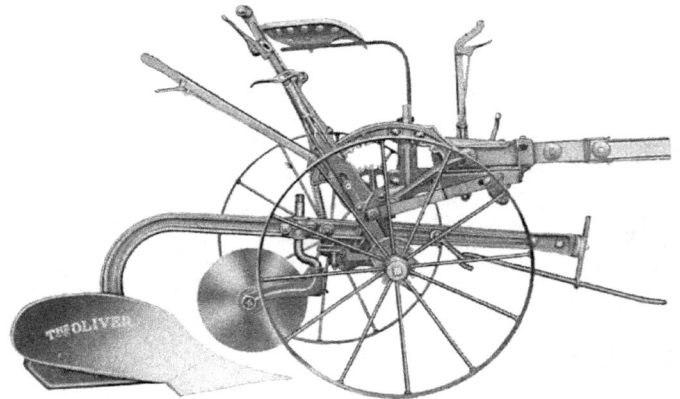

Fig. 209. Oliver 26A equipped with rolling coulter.

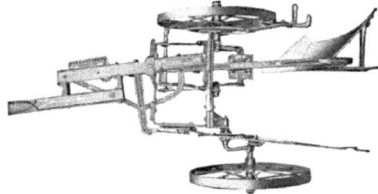

Fig. 210. The above illustration shows how the tongue can be shifted by the convenient guiding lever which enables the operator to straighten crooked furrows or to aid in making short turns at the end of the furrow.

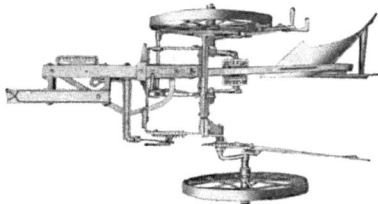

Fig. 211. The illustration shows the pole brackets placed on the left side of the stub pole thus fitting the sulky for use with three horses. It is equipped for two horses when the brackets are on the right side of the pole.

every time. Ray and this author won dead furrow classes at plowing matches with this outstanding plow. I was surprised while doing my research for this book to discover that Oliver knew the same thing and literally built almost the same plow. It was called the 'Horse Lift' 26A. The company literature contains this quote:

*"The Oliver 26-A Sulky is a horse lift plow. Both foot and hand trip are provided for putting the lift into operation. ...The land lever is equipped with a cushion spring which permits the land wheel to pass over uneven ground or stones without affecting plow depth, or riding comfort of the operator.... Easily adjustable for two or three horses."*

If this plow works anything like Ray's custom unit it is one to seek out and capture.

# James Oliver No. 11

The Oliver Plow Company considered its finest design to be the James Oliver No. 11 two wheel frameless sulky with the rolling landside. Considered by many at the time to be a masterpiece of engineering it is illustrated here with some of its features.

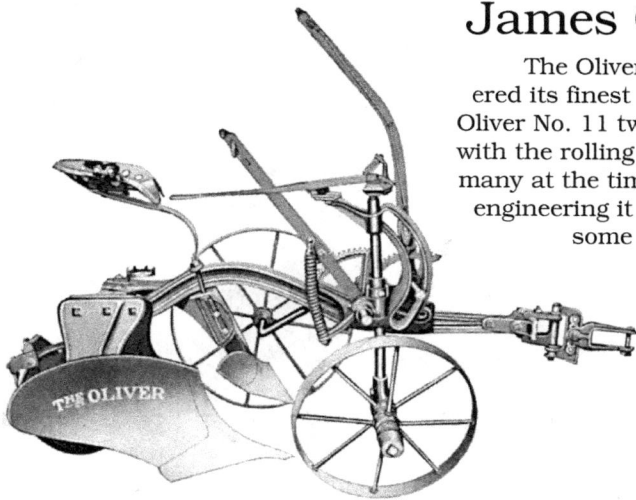

Fig. 212. The James Oliver No. 11 Improved

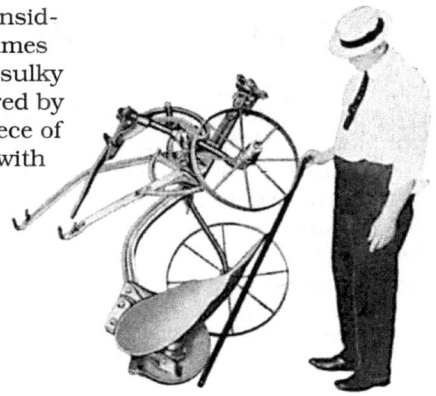

Fig. 213. Light and well balanced. The entire weight is carried on wheels; friction is reduced to a minimum. While working, only the tip of the share and the rear furrow wheel tire come in contact with the furrow bottom.

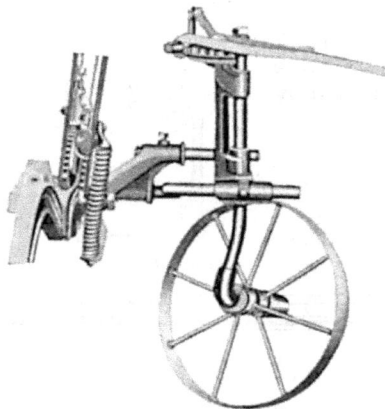

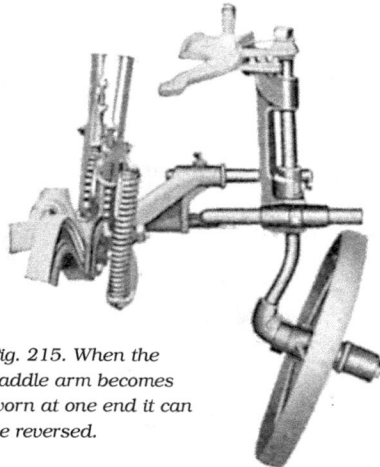

Fig. 215. When the saddle arm becomes worn at one end it can be reversed.

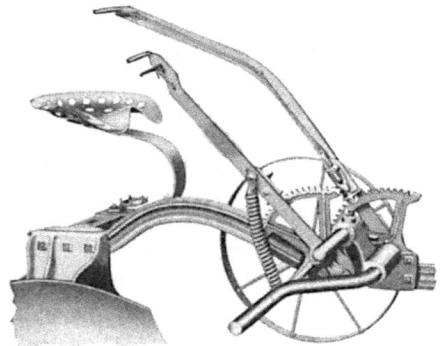

Fig. 214. In turning corners or straightening crooked furrows, the guiding lever operating the front furrow wheel is easily controlled. Square corners can be made while plowing at full depth. Interlocking teeth in the lever and lever bracket hold the wheel axle securely in whatever position it is set.

Fig. 216. The low, comfortable seat is easily reached. The levers are well placed and the quadrants feature a wide range of fine teeth for exacting adjustments.

Fig. 218. (below) The construction of the land wheel axle makes it possible to slide the land wheel over close enough to the bottom so as to drop it into the furrow for finishing the land without necessitating the use of a walking plow.

Fig. 217. (left) Sturdy rolling landside is angled to work in the furrow corner. It will not climb or tear down the furrow wall. Three scrapers keep it free of dirt.

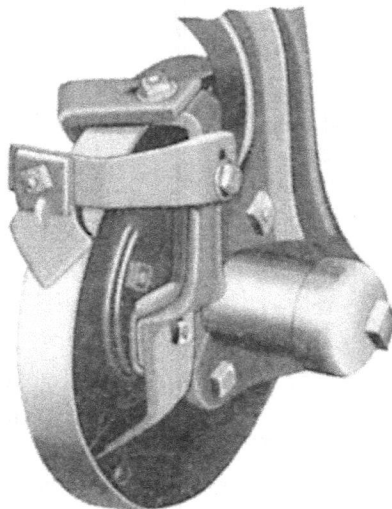

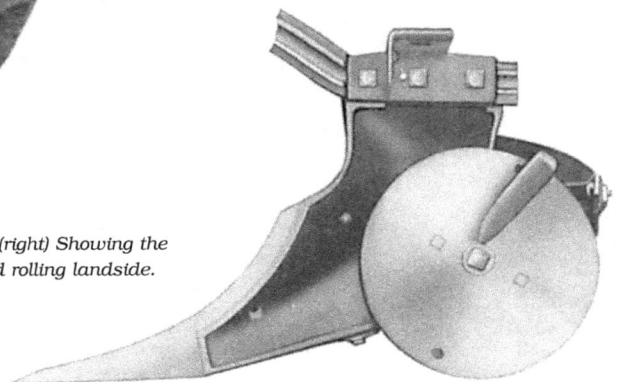

Fig. 219. (right) Showing the share and rolling landside.

# Moline's 'Good Enough' Low-Lift Sulky Plow

*Moline Flying Dutchman Farm Implements* built a long line of horsedrawn plows. In this sidebar we showcase their economy model (they called it 'Good Enough' - doubt that modern advertising agencies would go along with such honesty these days). It was a three wheeled frameless with accomodation for a tongue and an ingenious 'steering clevis' affair. The plow also featured an adjustable beam position to facilitate use of different size bottoms.

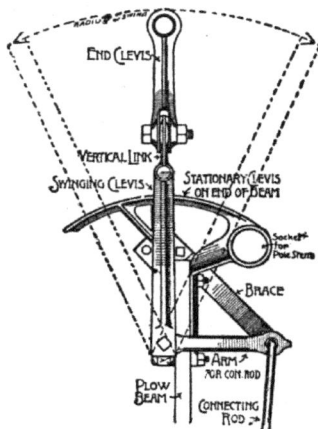

Fig. 220. GE3 Sulky by Moline.

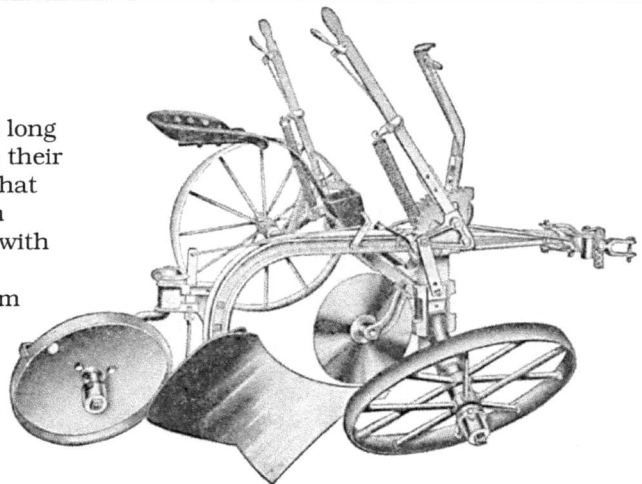

Fig. 221. The hitch clevis, as well as the front furrow wheel, is connected to the guiding lever. With this one lever both wheel and hitch work in conjunction with each other, permitting the plow to be moved quickly to the right or left, to straighten crooked furrows or overcome drifting on hillsides.

**Guiding Device**

A foot controlled lock was provided which allowed the plow to swing to the left without tripping or to make a sharp right turn by releasing lock.

Fig. 222.

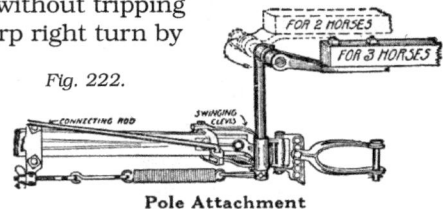

**Pole Attachment**

**Harrow attachments** used to be available for both horsedrawn sulky and gang plows. They were of either the spike-tooth, rotary or the disc type and of a width designed to harrow just the furrow(s) turned. They were fastened to the plow by a more or less rigid connection and could sometimes be made to raise from the ground by convenient levers. Many farmers felt they were useful, particularly in the fall, when the upturned soil can be rather lumpy and dry quickly. They insured excellent immediate harrowing. Such attachments might be made to order in farm shops. This author believes that there would be a ready market for such an attachment if it could be designed to be somewhat universal in its application with different model styles.

Fig. 223. The 1916 Kramer rotary harrow attachment for sulky and gang plows

# Getting the sulky plow to work right.
## The short sheet.

*Note: Hopefully this will mean nothing to you unless you've read the text up to this point.*

1. *right plow for the soil*
2. *broke, properly harnessed animal(s)*
3. *bottom in order (Chapter 2)*
4. *check all nuts/bolts - grease*
5. *right size evener*
6. *hitch in straight line from point of draft at shoulder (Fig. 203) through evener hitch point to point of draft on bottom*
7. *correct vertical hitch position (go long/never short)*
8. *adjust horizontally for animal comfort*
9. *after plowing is started, fine tune wheel positions for objectives*
10. *you are on your way!*

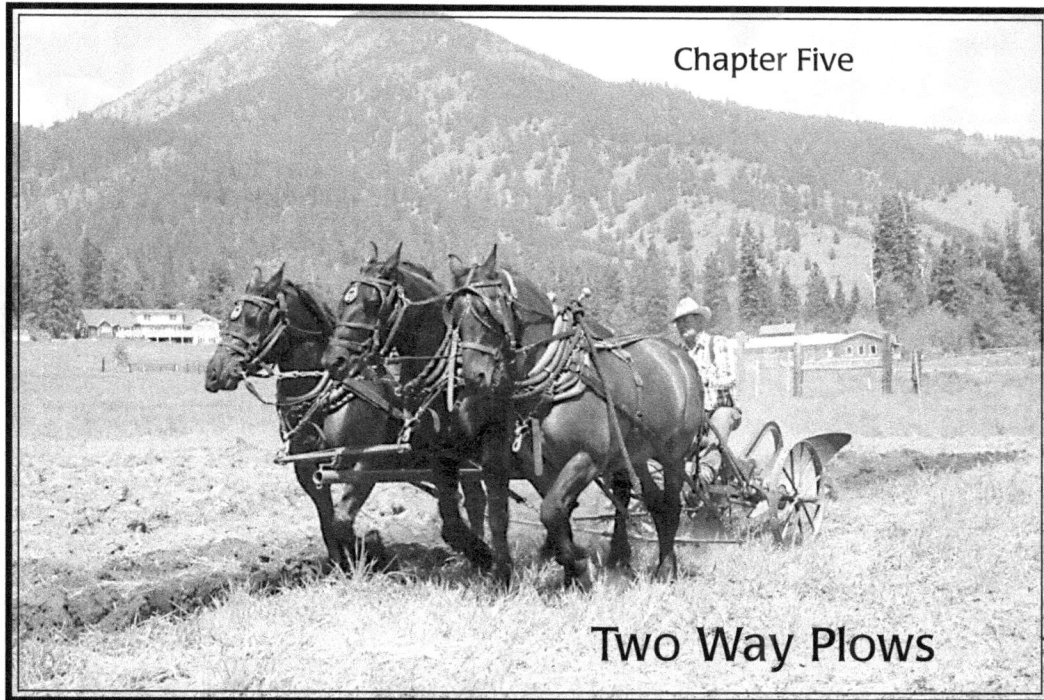

## Chapter Five

## Two Way Plows

*Fig. 225 Gene Westberg with Peggy, Meagan and Charlie on his J.I. Case Model 4F 16"
two way plow near Baker City in Eastern Oregon*

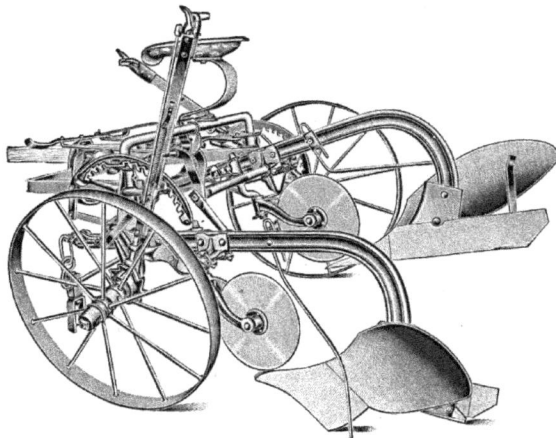

*Fig. 224. A Moline Dutchman No. 2 Two way plow. This plow
featured a pedal-articulated steering which allowed the
plowman to fine tune the furrows quite like a straddle-row
riding cultivator does. The Dutchman's wheels are toed in
slightly to hold the furrow wall.*

Conventional plows throw or roll the soil in one
direction as they travel. As will be seen in
Chapter Ten, arrangements need to be made in
field layout to account for the ridge, or crown, from the
first two passes and the ditch, or dead furrow, which
results when plow lands meet or at the edge of plowed
fields. Conventional plows are also difficult to use for
contour plowing on hillsides. Rolling soil up hill on one
pass and then down hill on the next, results in a mess.
As early as the 1840's successful designs of walking
hillside, rollover, plows were being used. Late in the
nineteenth century riding two way plows were well on
their way to perfection.

The two-way sulky plow has two bottoms, a left-
handed and a right-handed bottom. Only one of these
bottoms is used at a time. One bottom is used going
across the field in one direction and when, at the end
of the furrow, the team is turned around, this bottom is
raised and the other bottom is lowered into the soil.
Therefore, by the use of the right- and left-handed
bottoms, separately, all the furrows are thrown in the
same direction. This makes it a particularly good plow
to use on hillsides, terraced fields and on irrigated
lands. It also works good in small fields of irregular
shape. No dead furrows are left when a plow of this
type is used. This would be advantageous to the
irrigated farmland because the dead furrows left by the
other types of plows will hold more water than is
necessary and will likely cause a drowning out of
plants. On terraced fields the soil washes down between terraces and partially fills the channel. Throwing
the furrows up hill with a two-way plow somewhat
offsets the down hill movement of the soil, helps in
keeping the channel clear, and prevents the formation
of bench terraces. Also, when a true level contour
furrow is planned for and maintained, soil moisture is
retained and erosion held in check.

Pressure on a pedal causes the 'power lift' to raise
the working bottom when the end of the field is
reached. The 'power lift' consists of an ingenious
eccentric on the axle which when engaged causes the
forward motion to literally raise the designated bottom.
Some of the manufacturers referred to this as 'horse
lift' but it would be more accurate to refer to it as a
ground-drive activated lift. Once the designated bottom
has been lifted and the turn around completed, the set
of a lever puts the idle bottom to work for the return
trip down the field. On this style of plow when one
bottom is raised and the other one lowered, the hitch

Fig. 227. Because of the illustrated bail lock adjustment, straight beams can be used because no stress is put on them to pull them out of shape. They are adjustable on the blocks for either 12, 14 or 16 inch bases.

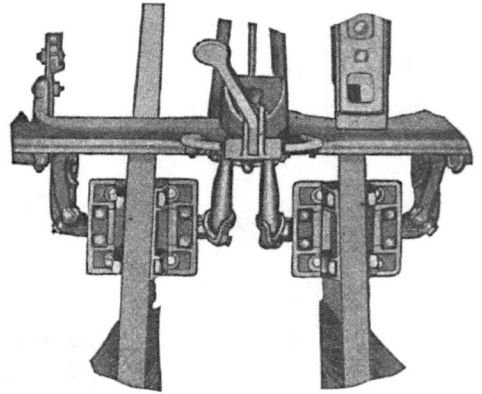

Fig. 226 The Oliver 23 Two way sulky

# Oliver
## 23A and 23B
## Two Way Sulky Plows

Fig. 229. Showing the optional converting roller evener clevis.

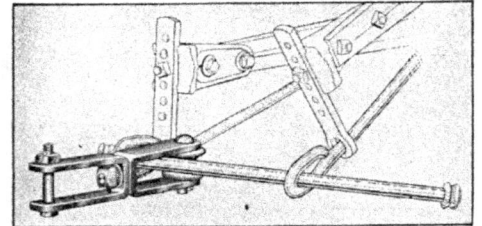

Fig. 230. (left) Rear view of the seat tilting device for maintaining a level seat when plowing on hillsides. Regardless of the slant of the hill, the operator can by shifting a conveniently located lever, keep the seat in a comfortable location.

OLIVER CHILLED PLOW WORKS

Fig. 228. Showing detail of the convenient location of both the hand and foot trip on the 'horse lift', the turn of the engaged wheel pulls the beam up. This mechanism can be used when walking behind or riding the plow.

Fig. 231. Here the pole bracket is placed on the right side of the stub pole adapting the plow for use with two horses. When the bracket is placed on the left side of the stub pole three horses are used.

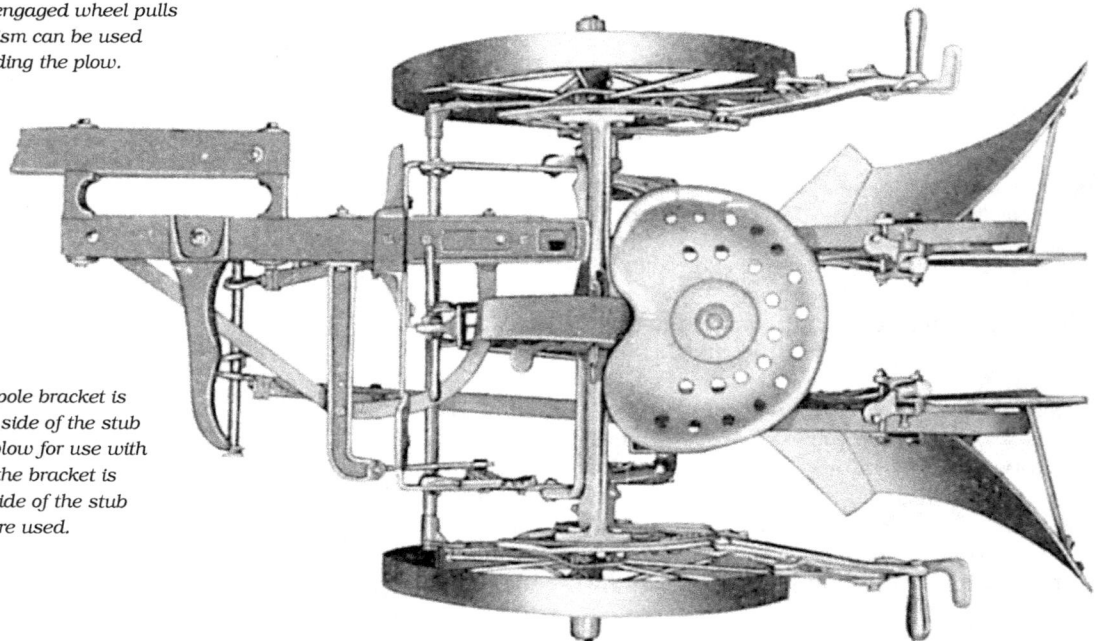

Fig. 232. Two turn of the century three abreast outfits pull two two-way sulky plows across a long field.

automatically shifts to the proper position. The seat can also be tilted for use on hillsides, which assures comfort for the driver.

All the major plow makers built their own version of the two-way plow. The basic design is amazingly similar throughout. Out in the western U.S. the most popular 2 way model has easily been, and remains, the Oliver 23B. This is not to say that it is superior to the other models. But certainly, from the standpoint of those who choose to farm with the older implements, the 23B is easiest to find by dint of the millions which were sold. For utility purposes the P & O (McCormick/ International), the Syracuse (John Deere), the J. I. Case, the Le Roy, and the Moline are all excellent models.

### Adjustments

The adjustments of the two-way sulky plow are the same as those for the walking plow. In other words the two way plow, with two in-line (ground-drive)

Fig. 233. An Oliver model 23 plow waiting for bid at a horsedrawn equipment auction.

wheels and the nature of the beam suspension, is quite different from most other sulky plows and more akin to the walking plow with regards to sensitivity to hitch. The fact that the plows are supported by a frame makes the adjustment easier in some respects and more difficult in others. Incorrect adjustment of a walking plow is evident to the plowman because it makes the job of holding the plow much harder, while the poorly adjusted sulky plow which is mounted in a frame has no effect on the operator. If the sulky plow is not correctly adjusted, draft is heavy and the plowing unfortunate.

**Vertical clevis.** The depth of plowing is determined by the height of the vertical clevis. ***The vertical clevis is kept as low as possible.*** If this clevis is too high, the end of the beam will be too low, the plow will "ride on its nose," and the back end of the landside will not touch the bottom of the furrow. Such a hitch will cause a broken furrow, rapid wearing of the plow point, and a hard-pulling plow. If the clevis is too low, the plow will not go as deep as is desired.

**Width of furrow**. The width of the furrow for a two-way sulky plow is determined by:
- the position of the plow in the frame;
- the position of the tongue-shifting lever;
- and the position of the horizontal clevis.

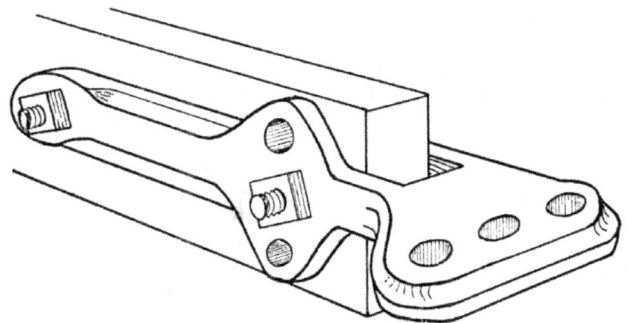

Fig. 234. The beam clevis for a two-way sulky plow.

*The plow should be set in the frame 15 inches from the inside edge of the wheel to the landside face for a 12-inch plow and 17 inches for a 14-inch plow.*

Sulky-plow wheels should run about 3 inches out from the furrow wall. The plow beam should be bolted to the frame so that the distance from the landside of the plow bottom to the wheel is the width of the plow bottom plus 3 inches. That is, for a 12-inch plow, the beam should be attached 15 inches from the wheel; for a 14-inch plow, it should be 17 inches. This measurement should be the same for both the right and the left plows, and this position determines the width of furrow. The series of holes in the frame are not for adjustment, but are for mounting various widths of plows in the same frame.

*The tongue-shifting lever should be adjusted so that the wheels run parallel to the furrow wall.*

On level ground, the wheels should run straight and parallel to the landside of the furrow. On side hills, further adjustment of the tongue will maintain a correct width of furrow.

*The horizontal clevis should be so adjusted that the plow pulls straight.*

The horizontal clevis should be adjusted after the plow has been set in the frame and the tongue-shifting lever has been adjusted. To do this, one should walk behind the plow and observe the plow beam to determine whether or not it is pulling in a straight line with the furrow. If the plow pulls sidewise or cornerwise, so that the front end of the beam tends to pull toward the right, the clevis should be moved to the right. If the end of the plow beam pulls to the left, it should be moved to the left.

**Horizontal clevis for two or three horses**. The position of the hitch for the horizontal clevis should be set according to the number of horses to be used. If the hitch is not correct, the plow will pull sidewise or cornerwise and it will not run straight.

*For two horses, the horizontal hitch should be from a point approximately 1 ½ inches toward the plowed land from the landside. For three horses, this hitch should be out toward the unplowed land about 1 inch from the landside.*

There are two different hitch types for two way plows. On one of the two types of hitches a rod is attached to the beam about 18 inches from the end. The height and the width adjustments are controlled at the front end of the beam by shifting the position of the draw end of the rod to the right or left.

For the rod type of hitch when two horses are used, the clevis should be so adjusted that the pull will come from a position approximately 1 ½ inches in from

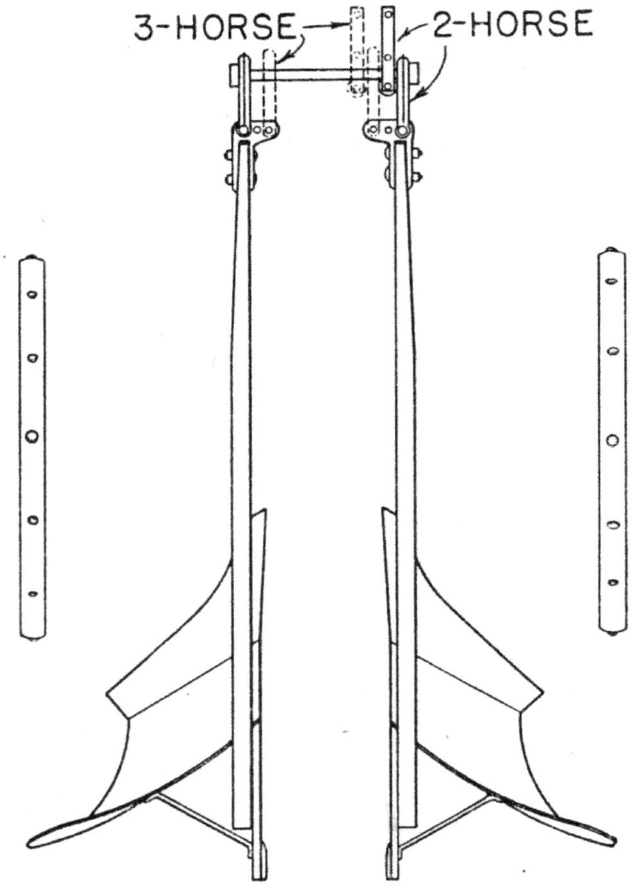

Fig. 235. Setting clevises for straight beams. Note that these beams are parallel to the landside of the plow (two-horse beams). The clevis projections are turned in toward each other for this type of beam. For two horses, the outside holes are used; for three horses, the inside holes are used.

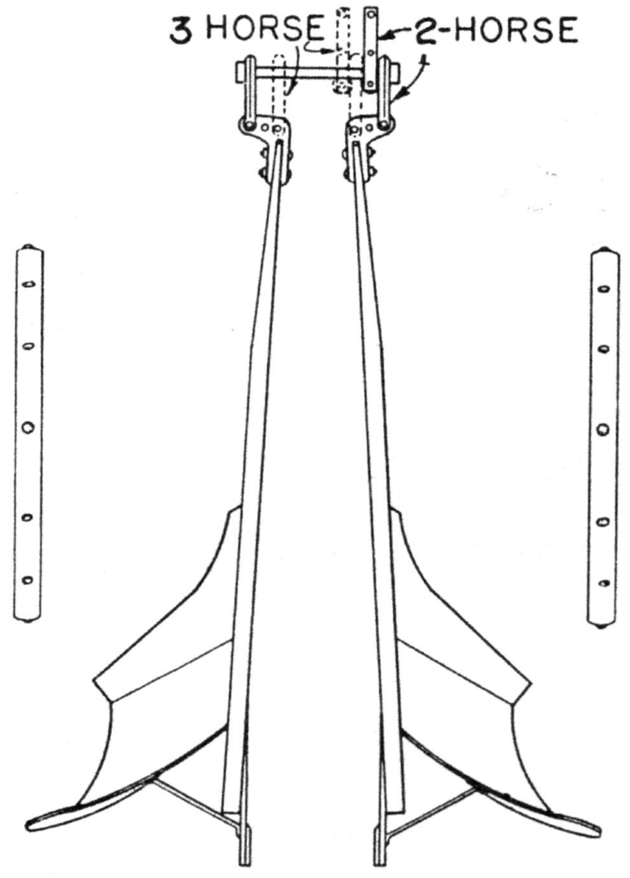

Fig. 236. Setting clevises for landed beams. Note that these beams are not parallel to the landside of the plow (three horse beam). The clevis projections are turned out for this type of beam. For two horses, the outside holes are used; for three horses, the inside holes are used.

a line parallel to the landside of the plow. When three horses are used, the front clevis should be so adjusted that the pull is about 1 inch out from the landside.

The second type of hitch consists of a clevis with vertical adjustments and three holes for horizontal adjustments (Figure 234). This clevis is offset and it can be turned over so that the offset projects either in or out according to the landing of the beam (Figure 236). The three holes are for two or three horses.

*With clevis hitches, the offset should be turned inward for plows with straight beams; for landed beams, the offset should be turned outward.*

The correct type of adjustment for the clevis type of hitch is shown in Figures 235 and 236. The offsets of the front clevises should be set according to the landing of the beam. If the beam of the plow is parallel to the landside and not landed (this is commonly called a two-horse plow), the offsets of the clevises should project inward, or toward each other (Fig. 235 ). If the beam is landed (this is commonly called a three-horse plow), the offsets of the clevises should project outward, or away from each other (Fig. 236). The actual number of horses to be used on the plow does not affect the position of the clevis offsets.

*For two horses, the hitch should be in the hole toward the plowed land. For three horses, the hole toward the unplowed land should be used.*

The position of the draw clevis and the hole to be used depends on the number of horses. For two horses, the hitch should be in the outside hole, or the one toward the plowed land. For three horses, the inside hole, or the one toward the unplowed land, should be used.

**Uneven furrows.** A very common trouble with two-way sulky plows is an uneven appearance of the furrows. Of the several possible reasons for this, the underlying cause is that one plow is running deeper or is cutting a wider furrow than the other, because of the following:

1. One of the beams or frogs may be sprung.
2. The plows are not set alike in the frame.
3. The jointers are not set alike.
4. The tongue-shifting lever is not shifting the frame the same amount in each direction.

# McCormick Deering No. 1 Two-Way Success Plow

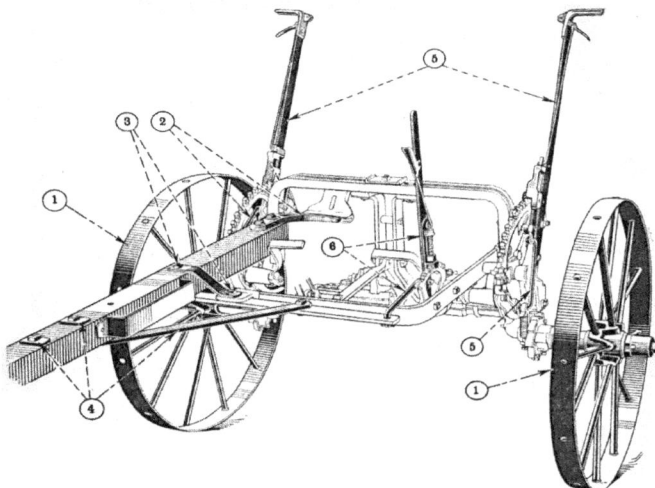

As late as 1943 International Harvester Company built the *McCormick Deering No. 1 Two Way* fine tuning its design and operation. The literature which this company put out to send along with purchasers of their implement, is amongst the best. We have chosen to present some of the cuts from that material to offer insights into the structural peculiarities of most of the two way sulkies.

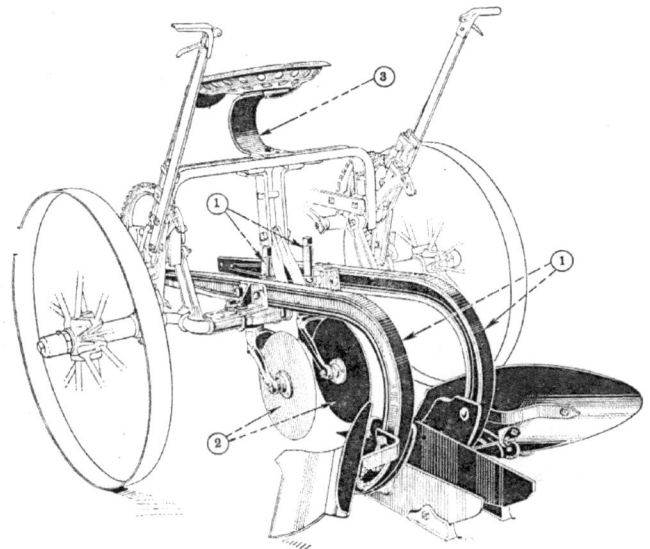

*Fig. 237. This cut shows the beam, coulters, bottoms and seat placement of the McD No. 1.*

*Fig. 238. (left) With all else stripped away this illustration shows wheel, lever and tongue placement. #1, screw hubcaps force grease into the wheel hubs. #2 is the sliding end bracket for the tongue. #3 an important tongue support bracket. #4 reinforced diagonal tongue brace. #5 depth adjusting levers. #6 tongue shifting lever.*

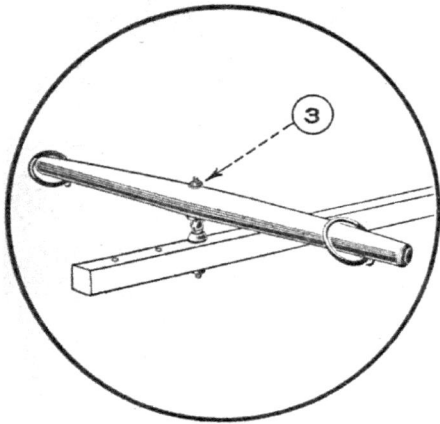

Fig. 239. Showing neckyoke and evener setup. Note that the neckyoke is bolted direct to the end of the tongue at 3. (Below) note how #1 illustrates the three points where draft pulls the McD No.1 beams. The three abreast evener is attached to the sliding roller-loaded clevis. When one beam is lifted and the other set the evener naturally slides to the appropriate point.

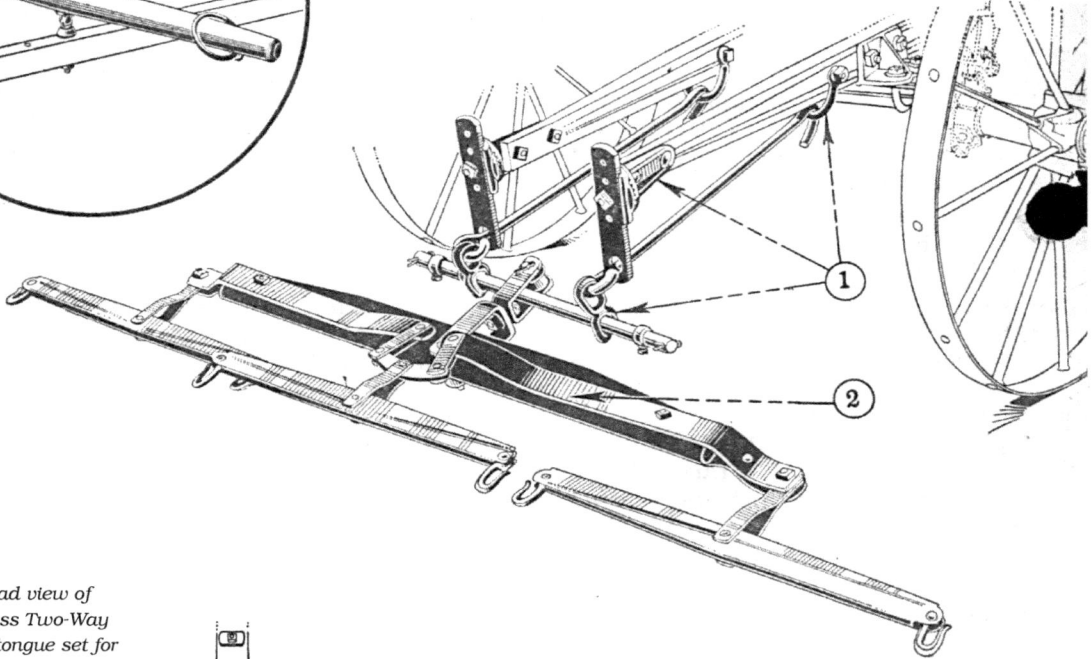

Fig. 240. An overhead view of the McD No.1 Success Two-Way Sulky showing the tongue set for three horses.

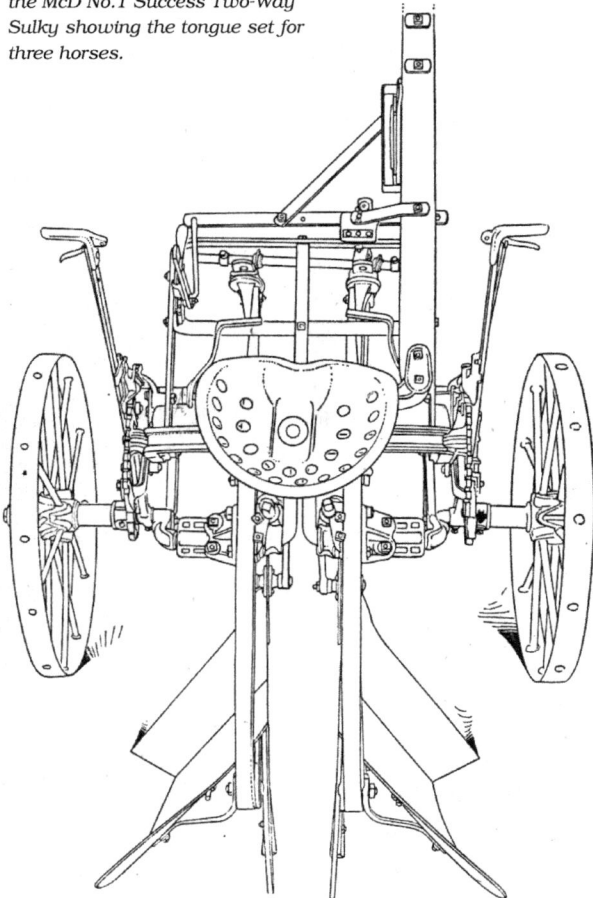

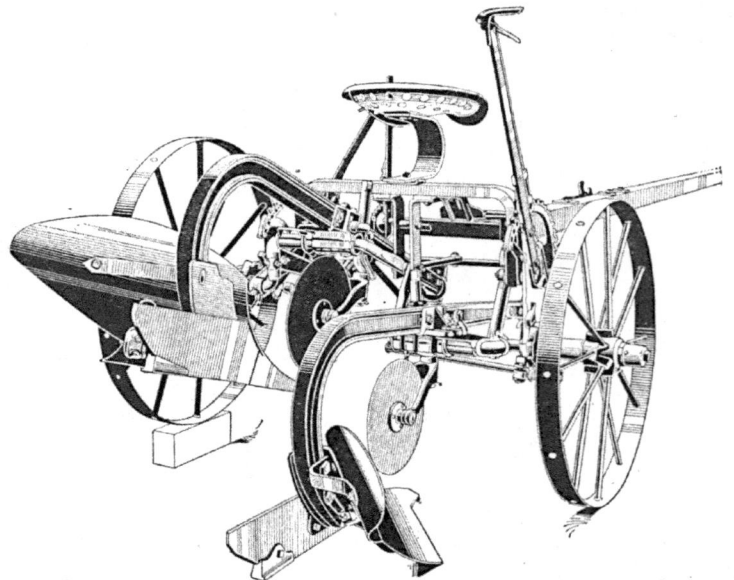

Fig. 241. Here the plow is all set in the plowing position with the land wheel up on a block to approximate level offset.

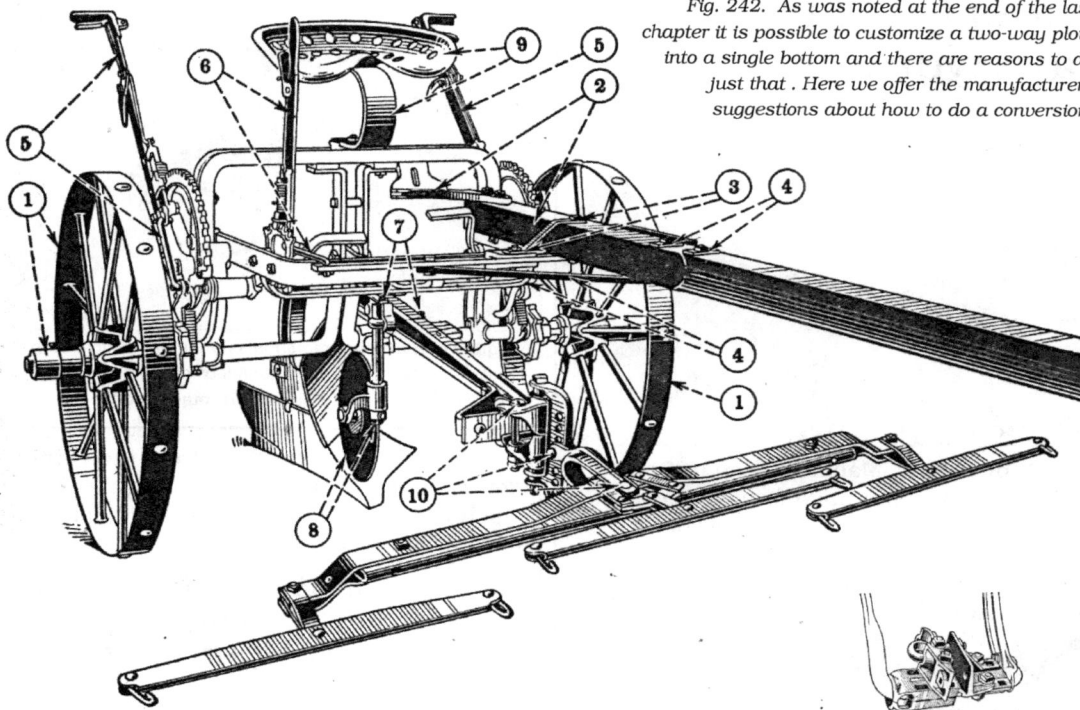

Fig. 242.  As was noted at the end of the last chapter it is possible to customize a two-way plow into a single bottom and there are reasons to do just that . Here we offer the manufacturers suggestions about how to do a conversion.

## SETTING UP THE No. 1 TWO-WAY SUCCESS AS A LEFT-HAND ONE-WAY SULKY PLOW

NOTE: Illustration shows the tongue set for use with three horses. For use with two horses, reverse the side arm and tongue brace.

1.  Put on wheels and replace linch pin collar and linch pin. Fill wheel cap with grease and screw on to wheel, forcing grease into the wheel bearing. Repeat until bearing is well filled, then secure cap with set screw.

2. Remove side arm from side arm bracket and bolt same to rear end of tongue with tongue clip on underneath side of tongue.

3. Force frame into an upright position, slide tongue side arm into bracket, and bolt tongue arm and top tongue brace to tongue. Remove cotter from pin through pivot bracket and put on top tongue brace. Replace pin and cotter.

4. Bolt tongue arm brace to tongue with clips on top of tongue. Bolt other end of tongue brace to tongue arm. Put on tool box.

5. Bolt on depth adjusting levers. Attach raising lever link to right-hand or furrow lever and secure with cotters.

6. Attach tongue shifting lever complete. Connect link to lever and tongue arm and secure with cotters.

7. Remove lower saddle plates and clamp the upper and lower saddle plates, complete with beam brackets to the axle by means of "U" bolts and carriage bolts, setting the beam brackets on the saddle plates, as shown in the small illustration (above). Remove wooden block and bolt beam on top of bail between the brackets; at the same time put on coulter shank, then tighten bolts.

8. Attach coulter; see that the lugs on collar are set so the coulter will pivot over point of share; then secure with cotter.

9. Attach seat spring and seat.

10. Slide clevis completely over draw bar and tighten bolts through clevis clamps. Attach evener by means of clevis pin. Replace cotter through end of pin.

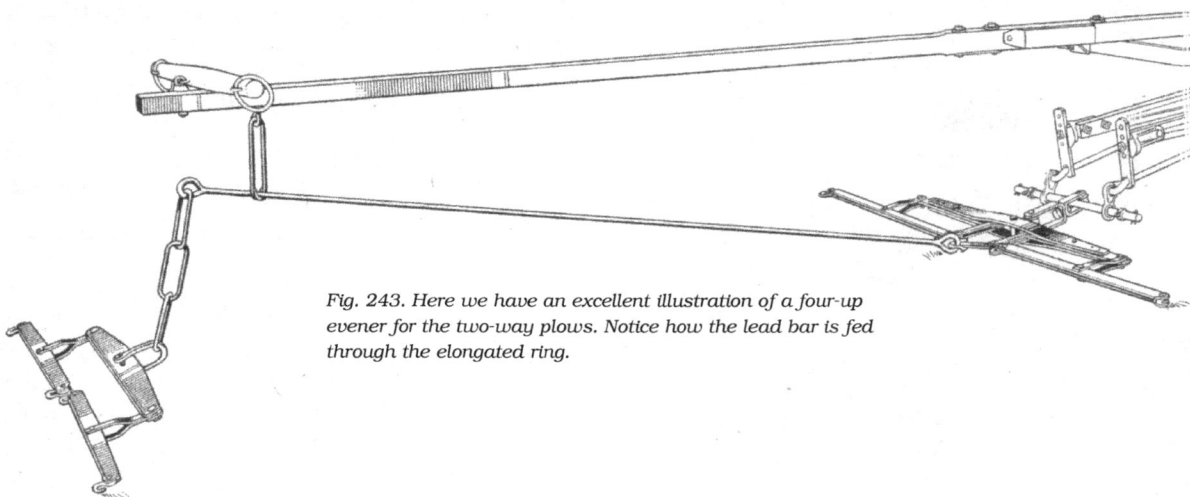

Fig. 243. Here we have an excellent illustration of a four-up evener for the two-way plows. Notice how the lead bar is fed through the elongated ring.

# White Horse Heavy Duty 2 Way Hydraulic Plow

Fig. 244.

The folks at White Horse Machine Shop in Gap, Pennsylvania have designed and build for sale the only modern horsedrawn 2 way plow we know of at press time. It features a hydraulic reset. Each bottom and the tongue/hitch combination are moved individually by ground-drive accumulator-plumbed hydraulic cylinders.

Fig. 245 & 246 (above & below) At the Ohio Horse Progress Days the author caught these two images of the excellent White Horse 2 Way plow at work. No, it doesn't take two men to operate, but at Horse Progress Days it was pretty near impossible to have an implement seat go empty. Note the beautiful furrow this plow delivers.

Fig. 247. This old photo shows how it is possible with the two way sulky to plow along the side-hill or contour thereby avoiding the erosive ditching that downhill furrows create.

Below is an extraction from the 1896 Le Roy Plow company literature regarding their two-way sulky which is pictured, along with David Jones, on the inside front color cover.

## THESE ARE REASONS WHY FARMERS LIKE THE LIGHT DRAFT TWO-WAY Le ROY SULKY

*The Light Draft Le Roy Two-Way Sulky ... will save driving around the ends, leave no dead furrows or back furrows, permit the operator to drop dead furrows where it is necessary to drain the land, and also be so easy to handle that a small boy like David Jones or an old man can operate it without any trouble...*

*The Le Roy is very light draft, steady, durable, reliable, pays for itself quickly, saves walking 7 miles to the acre, and does good work in all the various conditions of sod, stubble, clay, hard, stony, or gravelly soils. It will work equally well where there are stumps, fast stones or large loose stones, where the ground is uneven and where the fields are small or of irregular shape....*

**READ WHAT DAVID'S FATHER, MR. OWEN J. JONES, SAYS**

*Mr. P. N. Shoemaker:--Welmore, PA., May 9, 1914.*

*Dear Sir:--The Le Roy Sulky Plow is giving me good satisfaction. I consider it the lightest in draft of any that I have seen. My boy 13 years old does good work with it.*

*It is my choice of them all.*
*Yours truly,*
*Owen J. Jones, Prop. Cherry Ridge Farm.*

**Horse in the Furrow**. Most horsefarmers know what is meant when referring to this or that mare or gelding being a good 'furrow' horse. With the Two-way plow the 'honored' position switches from side to side. Quite by accident this switching becomes an excellent training excercise. Changing between furrow and land teaches restless animals patience and acceptance. The author goes so far as to suggest that, if three animals are being used, positions be rotated so all animals have time in the furrow.

Fig. 248. At the Ohio Horse Progress Days a White Horse Two Way plow was demonstrated hooked to six matched Haflinger Draft ponies utilizing White Horse's rope and pulley hitch (see Chapter Ten for details). The plow didn't require so many ponies but, as was remarked on before, these Haflingers won't be overworked at the end of the day.

## Getting the two-way riding plow to work right. The short sheet.

Note: Hopefully this will mean nothing to you unless you've read the text up to this point.

1. right plow for the job
2. broke, properly harnessed animal(s)
3. bottoms in order (Chapter 2)
4. check all nuts/bolts - grease
5. right size evener
6. beam clevises set for number of horses.
7. correct vertical hitch position. (make sure plow is not running on its nose.)
8. smile, this is as good as it gets!

John Erskine of Monroe, Washington with two of his Shires hitched to an International Two-way Plow. John has helped a great many people get started with horses. Photo by Heather Erskine.

Fig. 249. Jiggs Kinney of Columbus Junction, Iowa with ten head of home-raised Belgians on a 3 bottom frame plow.

The 'gang' in gang plows refers to a cluster of two or more staggered bottoms built to plow simultaneously and in the same direction. The purpose of ganging bottoms is to permit the single plowman to cover more acres in the same time frame. The two bottom gang doubled the number of acres one man could plow in one day and possibly with just one or more additional draft animals.

Small farms with fields of just one to four acres may not find justification for gang plows, as these

Chapter Six

# Gang Plows

Fig. 250. Jiggs Kinney, a legendary draft horseman, has spent a lifetime working with the outstanding Belgians horses he has bred, raised and trained. This ten-up features two stallions.

*Fig. 251. Four abreast of turn-of-the-century farm chunks walk away with a frame-style two bottom gang in easy-to-plow old ground.*

*Fig. 252. A magnificent set of 6 Percheron horses effortlessly pull a new Pioneer two bottom frame-style plow at '98 Ohio Horse Progress Days.*

the sulky and the gang plow are the same make, should an exchange of coulters or shares be necessary to keep farming. Obscure models may be hard to get parts for.

Horsedrawn gang plows can be divided into four categories of design: frameless riding, frame, frameless walking, and pull-type or 'tag" plows. In Chapter Four we covered the design aspects of the frame and frameless riding plows. These aspects are identical in the similar gang categories.

All aspects of the plow bottoms are well covered in Chapters One & Two.

The big differences, unique to gang plows, all relate to the draft power requirements and the mechanics of big hitches.

Some interesting, but non-critical, design variables in gang plow construction are covered in the next pages.

Though we go into some detail in this chapter on big hitch options, from four-up to eight-up, additional variables appear in Chapter Ten.

**How Many Horses.** If it takes a three abreast to pull a 14" sulky plow, under the same soil conditions a strung-out five horse hitch will provide adequate power for a 14" two bottom gang. Adding a sixth horse improves the chances that all horses will be less tired at end of day.

Eight horses will handle a three bottom gang nicely.

Nine to twelve head will handle a four bottom gang.

As the following information will indicate, the exact number of horses or mules used will be determined by

* soil conditions,
* plow,
* size of field to be plowed,
* number of trained animals available,
* teamster experience and comfort level.

Keep in mind that these bigger hitches will require considerable headland room (see Chapter Eight). For example a six-up (two teams of three abreast) and frame or frameless gang plow will string out up to 32 feet long. (A tag plow or tractor plow hitched behind a forecart could add six to ten additional feet of length. With a nine horse hitch (three teams of three) and plow, 40 to 45 feet could be consumed. When factoring hitch length into what width headland to plan for it's a good idea to plan for one and a half times overall length. With good horses and an experienced plowman/ teamster it is possible to make headlands the same

hitches tend to dictate certain spacial concerns to plowing layout (see below 'how many horses' and see Chapter Eight). Having four to ten animals strung out ahead of the plow requires that the plow be taken out of the ground a good distance before the edge of the field, leaving enough 'headland' to allow for a turn and positioning for the return pass. Five to ten acre fields may well justify the use of a two bottom gang plow along with a walking plow or sulky plow (perhaps even a two-way plow). The serious horsefarmer with 40 to 160 acres of cropland will enjoy having a good plow of each size/style. Careful shopping and trading might result in a lineup of affordable plows but an important suggestion is in order; if the reader is planning on using older implements some intelligent specialization is called for. Find out which makes were most prevalent in your area. Select one of those makes of plow (i.e. P & O, John Deere, Moline etc.) and stick with it. It will come in mighty handy if

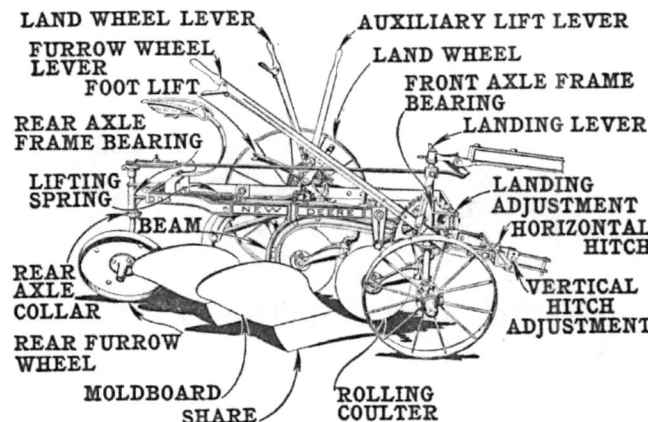

*Fig. 253. The parts of a New Deere (JD) frame-style foot-lift gang plow are here identified.*

width as the length of the plow outfit as the horses can

be asked to sidestep or side pass as the plow leaves the ground.

**Understanding relative draft efficiencies of certain larger hitches**. Above and beyond the issues of side-draft, which are addressed in following paragraphs, it is important to understand how larger hitches of horses and mules can be more or less efficient in their work output. When driving a team of two horses the knowledgeable teamster has many ready indications of whether the two animals are pulling equal. When driving four, five, six, eight or more horses, this becomes more difficult. And the difficulty can be more or less, depending on the mechanical nature of the equalizers employed. In almost every case, if the teamster is not paying close attention it is possible, with most equalizers, to have the lead animals pulling the entire load. Understand that when we speak of equalizers we are referring to the devise or system employed to allow a balanced distribution of load between spans or teams. There are three basic types of equalizers. Whether a *pulley-style, bar-style,* or *evener-style* equalizer is employed (See Fig. 274) the animals need to be 'driven' (read controlled) otherwise the teamster may expect that some animals will lay back while others will attempt to take the whole load.

One of the challenges for the teamster is to have all the animals comfortably pulling their fair share. And though 'driving'' the horses is an important piece of the puzzle, there is work ahead of the actual plowing which can lessen the challenge.

**Lead horses versus wheel horses.** When driving horses strung-out the teamster quickly values the placement of certain equine temperaments and intellect in certain spots within the hitch. When the sole plowman, seated 30 plus feet back from the heads of the leaders, speaks to his hitch, it is imperative that the animals in the lead step ahead promptly. If the lead horses are slow to respond and the *wheel horses* (those closest to the teamster) and/or the *swing horses* (the mid span) step ahead on command it is possible that eveners and tugs will get tangled up. The result can be bothersome at best and tragic at worst. Look at the

Jiggs Kinney hitch in Figs. 249 and 250 and imagine the bother if the lead three failed to step ahead as the swing and wheel spans started out. Is it any wonder then that plowmen highly value 'lead horses' and especially 'lead furrow horses'?

Understanding that the quicker more willing animals may be placed in the lead it should be clear that the plowman will need, through voice commands and line pressure, to steady those leaders and thereby require the other animals to pull their own shares. It was mentioned in a previous chapter that this author likes to place trainee animals in the wheel land position. One of the reasons is that should this green horse or mule choose to hesitate or lay back, he or she is within easy reach for prompting by touch.

Along with considerations about slow versus quick, smart versus dull, the teamster should also pay attention to personality and the day's temperament when setting up the hitch placement order. For example; obviously a stallion should not be worked directly behind or next to a mare in heat (some might argue that the mare in heat should not be worked at all though this author confesses to enjoying the challenge). If the stallion tries to mount the mare, harnesses will tangle and break. If the mare tries to kick the stallion, the same result can be expected. Beyond those sorts of obvious concerns all horsefarmers have had experiences with difficult animals, shy animals, and stupid animals which when poorly organized in a big hitch have caused no end of squabbles and nervousness. Think of a big hitch as you might a big dinner party. The hostess knows not to seat Sheila next to Brenda and to keep Roger away from Jill or the party will be a bust. Same way with horses and mules. Trust what you know about who likes or dislikes who. And definitely trust what you have observed about any mare's mood of the morning. A little planning will usually result in a 'seating arrangement' that results in harmony - and harmony in the hitch will make the plow run smoother.

*Fig. 254. John Erskine at Duvall, Washington, plowing with six of his and Heather Erskine's homegrown Shires.  Photo by  Heather Erskine.*

# ROCK ISLAND NO. 8 GANG

Fig. 255.

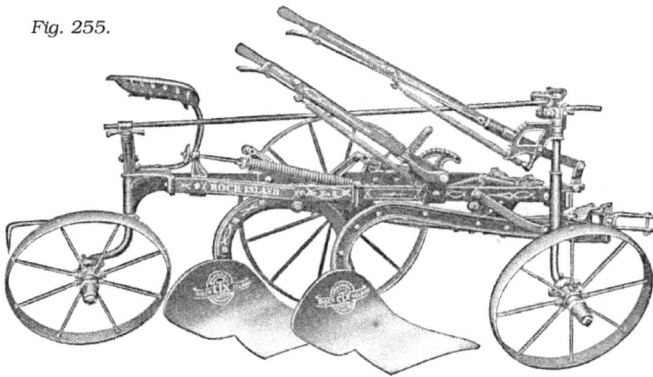

In researching this book we discovered 36 different companies which designed and manufactured gang plows at the beginning of the twentieth century. There is not space to do a detailed showcase of each in this book though Chapter Eleven offers illustrations of hundreds. Though most all gang plows fall into the design categories we have described, many had unique little variables. Here we present some of the company literature on the Rock Island No. 8 Gang.

HEAVY FRAME. Notice the heavy, one-piece frame opening at the front right hand corner. It is the stoutest frame ever put on a plow. We don't open it in the rear, but in the front where the least pressure is. The two ends come close together and are not separated by an axle bracket, but are firmly bolted together. It is the strongest construction possible.

The wheels of the No. 8 Gang are higher all around. The big land wheel is 36 inches, and the front and rear furrow wheels are 24 inches. It is the largest rear wheel used... Each wheel is equipped with a dust-proof hub. Unscrew the large nut on the end of the hub, fill it with hard oil and it forces the grease the entire length of the hub.

LIGHT DRAFT... a long wheel base yet a short hitch. The long wheel base steadies the plow and holds it to its work. It also makes it possible to seat the driver well back over the rear wheel so that he can see the work that is being done while the ground is being turned over.

Special consideration has been given the hitch or evener. In order to give the horses plenty of spread behind, we use a 28-inch singletree on the two inside horses and 32 inches are on the two outside.

FOOT LIFT. When you are seated on the Rock Island No. 8 Gang, your left foot is resting right on the foot lift (See Fig. 256). Then, by putting your right foot on the

upper treadle and pushing forward easily, you will see that the point of the plow drops downward until the two points on the front bail catch on top of frame. This naturally causes the down suck of the plow to load itself on the two back wheels and not on the horses' shoulders, as on a single bail plow.

The Rock Island is designed perfectly to throw a large portion of the load on the front wheels. Naturally, a greater part of the weight of the dirt and bottom falls on the rear of the plow. We, however, distribute the load on all the wheels, putting a great deal on the front but leaving enough on the rear wheel to keep the plow from jumping. It will stay in the ground perfectly and hold right to the work without the weight of the driver. That is why we have such an unusually light draft.

EASY TO OPERATE. This gang is also equipped with the famous Rock Island foot lift, which has a double, compound lever. With this lever you don't have to raise your knee up under your chin to handle it. The foot lift is underneath the frame--the pivot is below. We start the lift lower down and finish higher up. Any small boy, with his own weight, can lift the plow out of the ground very nicely and he doesn't get away from his seat, so there is no danger of being thrown off or carried over the front on the wheels.

We have both levers on the right side, which gets them out of the way, making it easier to get on and off and avoids catching your clothes. With this construction the operator can hold the lines in one hand and operate either lever without changing lines from one hand to the other. This gives him perfect control over team and plow.

UNIFORM CUT. Look at Fig. 258 Take a straight edge, hold it central with the upright or vertical portion of front furrow wheel axle, dropping to the floor. Notice that the direct pressure of weight of plow rests on the outside edge of front furrow wheel. This forces wheel out at top and in at bottom. Therefore width of cut remains the same at all times. This construction means a smooth sliding axle, as the load is carried directly under center of sleeve. Set a straight edge in the corner of the frame and you will see that you have a straight down pressure. It will be from 2 ½ to 7 inches inside of the furrow wheel. Think how the wheel gets out from under the load.

LEVERS ON RIGHT SIDE. Take the other side of the plow. By looking at Fig.259, you can see how we run the main axle from the land wheel up, then over and across the frame, putting the land lever on the right side. This braces the frame and supports the down pull of the plow on the axle. On other plows where the axle is sawed off short, and is simply bolted to the left side of the frame, you will readily see that the down draft of the plow will cause these bolts to give, or the frame to spring or twist. It is bound to do it, and when it does, the large land wheel will spread out at the bottom. It gets out of plumb.

Take hold of any other plow, pull

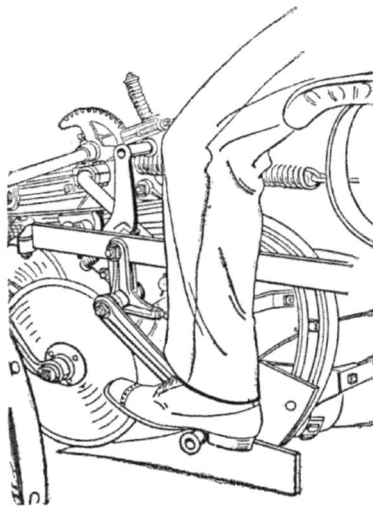

Fig. 256.

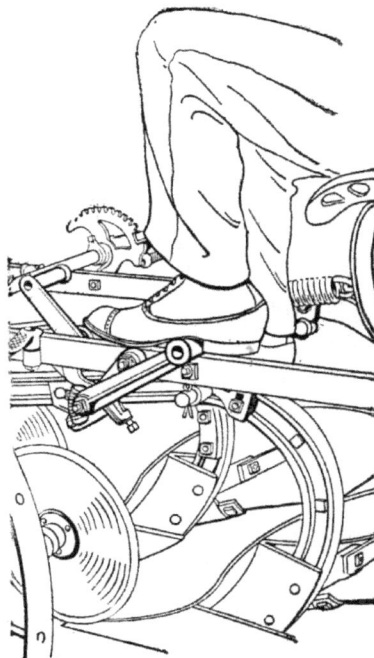

Fig. 257.

Gang Plows

out this wheel at the bottom and see how easily it gets out of line. But you can't do this with the Rock Island construction.

In order to put the strain on the frame directly in the center, we use short bails and short hangers. Then the side pressure of the dirt against the moldboard is thrown square in the center of the frame. With long bails and hangers a long leverage is formed, which accounts for so many twisted frames on plows other than the Rock Island.

STEERING ROD. Still another important feature that makes the Rock Island superior is the steering rod. The steering rod is put on the plow to adjust the rear wheel so as to carry the pressure of the moldboard to the right on a right hand plow, relieving the landside of all pressure except what is necessary to keep it scouring well. At the same time the rear wheels must be controlled by the horses.

See by Fig. 258 how we place this rod within one and one-half inches of the straight line of the axle. The pressure of the dirt against the moldboard, unless checked, will turn the rear wheel around. It would be just the same as if you put your foot on the rear wheel and gave it a shove.

NO DRAGGING AT CORNERS. If you take hold of the pole of a Rock Island and give the rear wheel a good, hard push on the front edge, you will see that no matter how hard you push, it hardly affects the pole plate. A great advantage where you have a team that isn't steady. It lessens the danger of uneven furrows, for the uneven movements of the team don't affect the rear wheel.

Notice how we bolt the rear wheel

casting on to the frame with two heavy bolts through slotted holes in the casting. Without dropping the rear of the frame or the heel of the plow, you can now move the wheel to the left so as to prevent the landside from touching the dirt. You don't change the suck of the plow in any way. If we were to use a common rosette, like many, on the rear of frame, you would loosen the bolt, tip the axle to a different position and tighten. The result would be, the top of the wheel drops further to the right, drops the point of the axle closer to the ground, also rear of frame. This would drop the heel, and that takes the suck out of the plow.

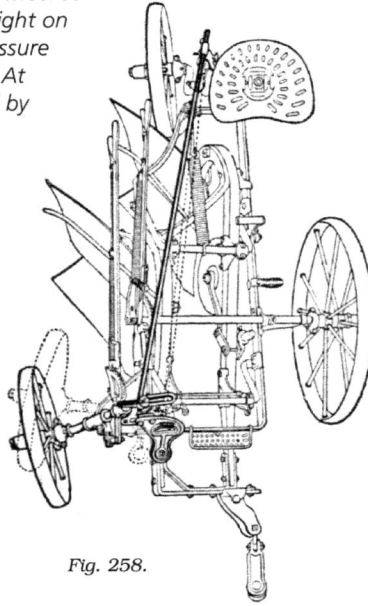

Fig. 258.

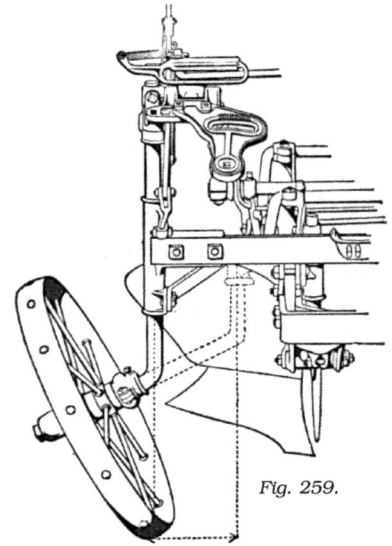

Fig. 259.

Fig. 260. The reigning queen of Pacific Northwest Lady teamsters, Donna Anderson, drives six of her family-raised Shires hitched to a two bottom plow. The setting was the plowing demonstration at the Bohnet Ranch in Washington.

# Oliver  No. 1 Improved High Lift Gang Plow

The Oliver company specifically built this plow, with landed beams, for use with four abreast with minimized side draft.

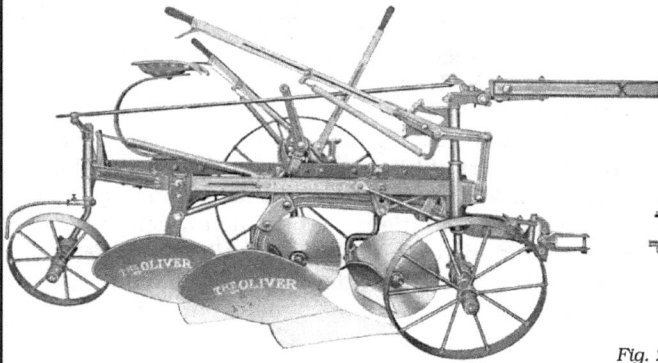

Fig. 261

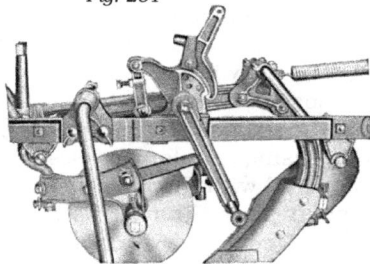

Fig. 262. Detailed picture of the foot lift mechanism. An adjustment was provided to fit the foot levers for different size operators. When lowered to full depth the bottoms are automatically locked. An auxiliary hand lever was provided to assist, in extremely hard ground, getting the bottoms up.

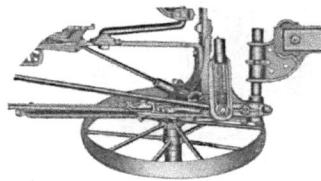

Fig. 263.  The rod which connects the furrow wheels is pivoted directly over the front wheel axle. A spring absorbs all movement to the front wheel and the course of the rear wheel remains unchanged.

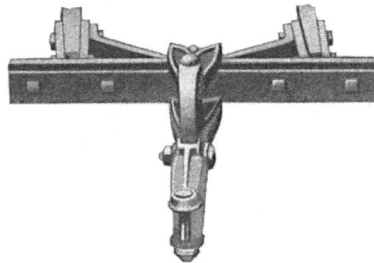

Fig. 264. The hitch can be shifted right or left to secure the correct line of draft with small or large horses.

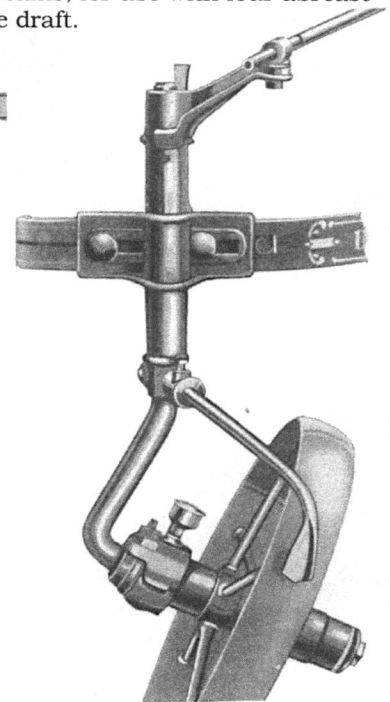

Fig. 265. An adjustable rod connects the rear furrow wheel to the tongue plate on the axle arm. The rod could be adjusted so the wheel would run at the correct angle to the furrow wall when either three or four horses are used.

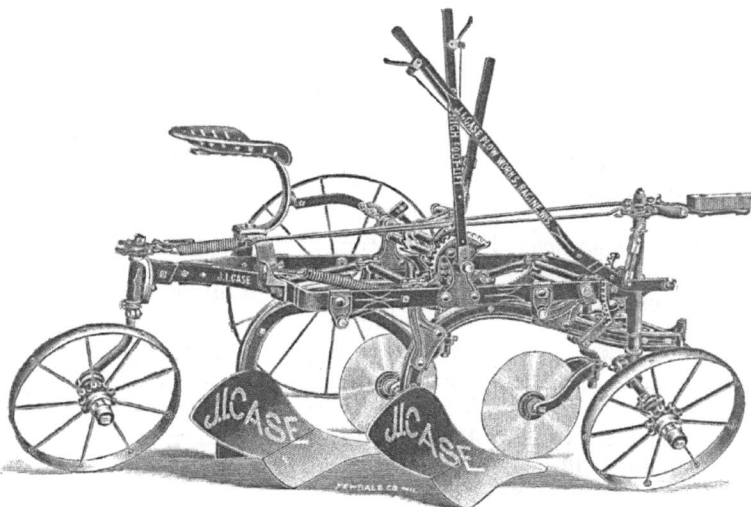

Fig. 267. (Below) The Emerson Brantingham Plow Company  used this photo to demonstrate what they claimed to be the enormous lifting power of their foot-lift gang plows. The man on the seat was said to be 110 pounds. While the big man, standing just back of the hitch on the beams, is supposed to weigh 506 pounds. The smaller man is lifting plow and big man with the foot pedal.

Fig. 266. The J. I. Case High Foot Lift Gang Plow. The penetration and suction of the plow bottoms are uniquely adjustable in this plow by brackets securing the rear bail to the frame. They  have a slot through which the bolt passes. By moving the brackets ahead in the slot the heel of the landside is raised and the share point is lowered. By moving the brackets towards the rear the reverse effect is obtained.

Gang Plows

Fig. 268. This cavalier attitude with a four abreast on a walking gang plow is hard to accept. How long could that left arm stay in that position?

Fig. 269. Dale Hendrickson drives John Erskine's Shire six-up on the JD Wheatland plow. 'Bout the only time one sees John sitting on that seat is when someone else is driving. Photo by Heather Erskine.

**Hitches.** The greatest and most common objection against the use of gang plows is that of side draft. And that objection stems from the fact that many horsefarmers are uncomfortable with the idea of stringing out their horses preferring to work abreast. While in some soil conditions and with some two bottom plows it is possible to plow with four abreast it is never as efficient as tandem or strung-out hitches because of the tiring effect of side draft on the horses.

Fig. 270 shows clearly just why there is side draft when four horses are hitched abreast to a two-bottom gang plow. Here the line of draft is shown to fall 16 inches outside the line of load. The line of load is found by locating the theoretical points of resistance on the two bottoms as shown and described as "Line of draft" on page 37 (Fig. 75) and then dividing this line at the middle.

Anything that brings the horses closer together will alleviate side draft but will not eliminate it. Fig. 271 shows how the beams of the gang plows of certain companies (like Oliver and its *Model No. 1 Gang* illustrated on page 96) are set, in an effort to bring the line of draft and the line of load together. It should be

> **Please Note:** In the research and compilation process for this book we discovered many slightly different illustrations of big hitches with important variations in how an idea is presented and in the mechanical information. We've elected to present several of them even though they seem to repeat similar information.

noted, however, that the change in the beams does not alter the line of load. The only way to eliminate side draft is to hitch tandem. In Fig. 272 a tandem hitch is shown in which longer doubletrees are used than shown in Fig. 205. That side draft is virtually eliminated is shown by the fact that the line of draft falls inside the line of load. Fig. 274 shows two methods of equalizing this sort of hitching. The pulley and cable is satisfactory for 2-horse teams, but the lever must be used for other combinations. While the tandem hitch makes the handling of the teams only slightly more awkward and only in the beginning, especially in turning, it's merits in other ways make one wonder why it is not more commonly used on gang plows.

Returing to the argument in favor of comfort for the working animals, the ideal hitch for a two-bottom gang plow is the five-horse strung-out hitch shown. Most of us do not take into consideration the real load we are putting on the horses when using only four horses on a gang plow, as compared with three horses on the ordinary 16" sulky plow. Suppose both plows are running six inches deep. The three horses on a 16" sulky plow are each cutting and turning 5-1/3" of the furrow slice, whereas, the four horses on a 14" two-bottom gang plow, cutting and turning a total of 28", are each cutting and turning 7" of soil, or an increase over the sulky plow team of approximately 30 per cent per horse. Figuring the other way, two horses on a 16" sulky plow would each pull only 1/7, or about 14 per cent more than is required of each of the four horses on the gang plow.

If the reader has never driven horses hitched tandem, or strung-out, on a gang plow, he or she may object to the team not being quite so convenient to handle. This objection will melt away once the system is tried. And the horses will do the work so much easier becoming quickly accustomed to the strung out manner of being hitched. Keep in mind, again, that in practically every team of four or five horses, some of them will work better in one position on the team than another. After a little practice in shifting the horses around with the goal of contented work, better results will follow. A horse won't render efficient service when nervous and fretted, any more than human beings will.

The third set of cuts (Fig. 278) illustrate the common four-horse-abreast hitch on a two-bottom gang plow. The heavy outlines on this illustration represent four horses hitched to work to the best possible advantage. The dotted outlines represent the same team working to great disadvantage, both to the horses and to the plow.

Note the difference where the dotted line crosses the plow clevis and evener, showing clearly the added side draft caused by working the horses with their heads too far apart.

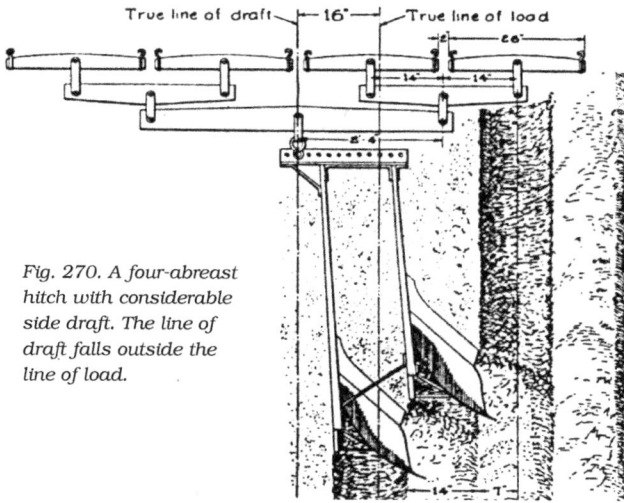

Fig. 270. A four-abreast hitch with considerable side draft. The line of draft falls outside the line of load.

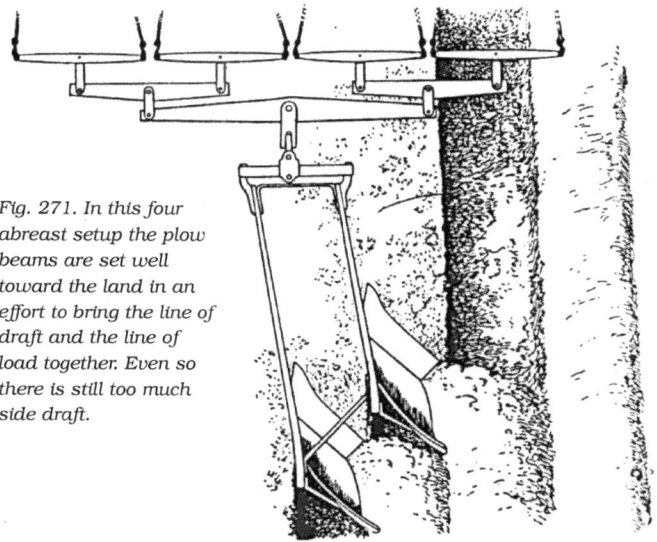

Fig. 271. In this four abreast setup the plow beams are set well toward the land in an effort to bring the line of draft and the line of load together. Even so there is still too much side draft.

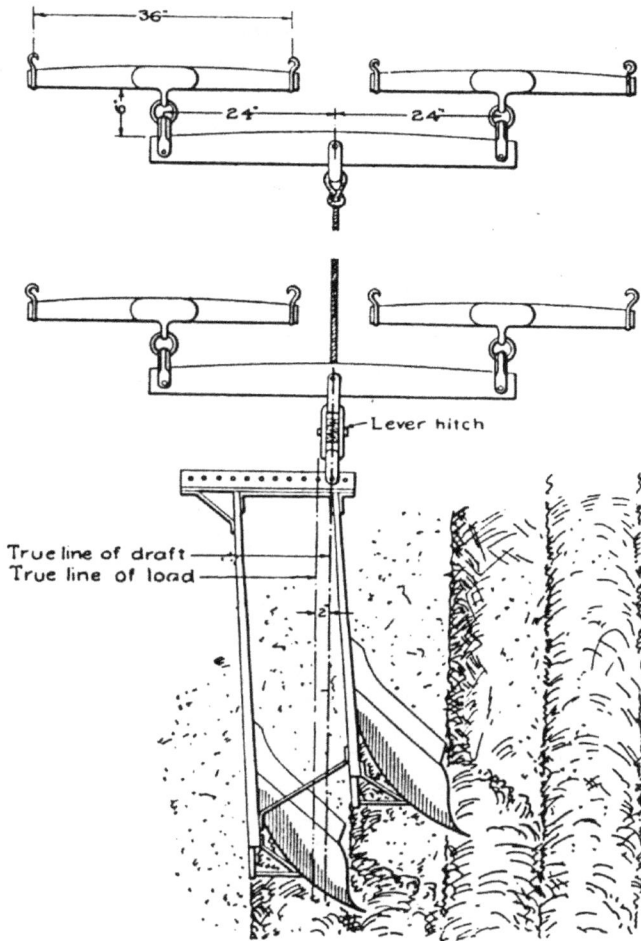

Fig. 273. A Five horse tandem hitch, demonstrating good proximity between line of draft and line of load.

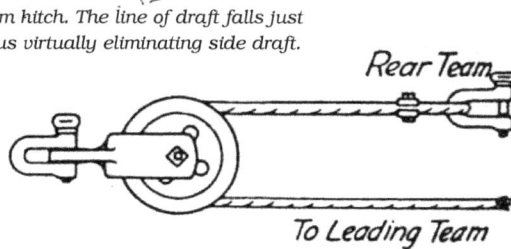

Fig. 272. A four-up tandem hitch. The line of draft falls just inside the line of load, thus virtually eliminating side draft.

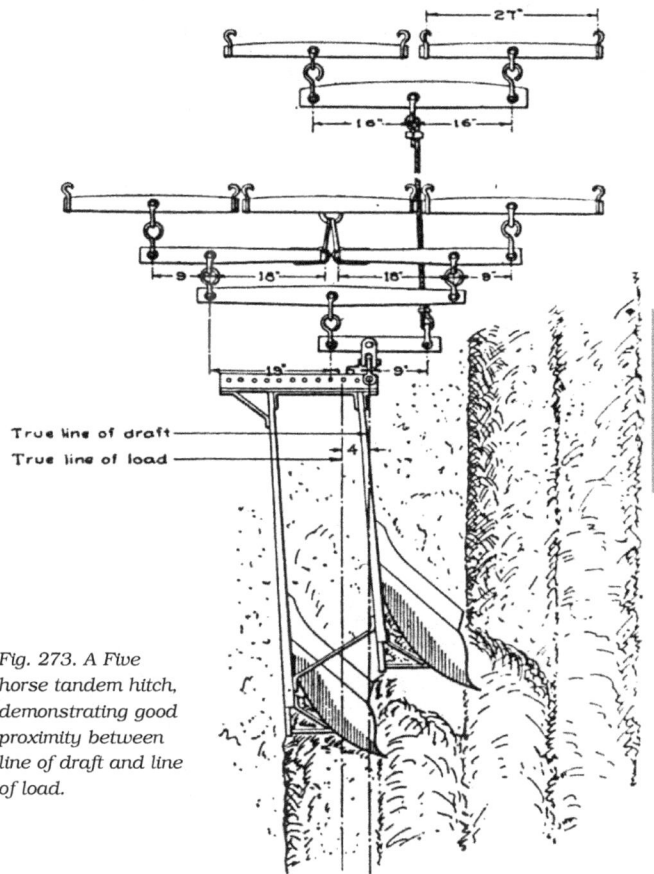

Fig. 274. Tandem-hitch equalizer devices illustrating two methods, the pulley and the bar, for facilitating an even distribution of load between wheel and lead teams. Note bar is drilled for two working with three.

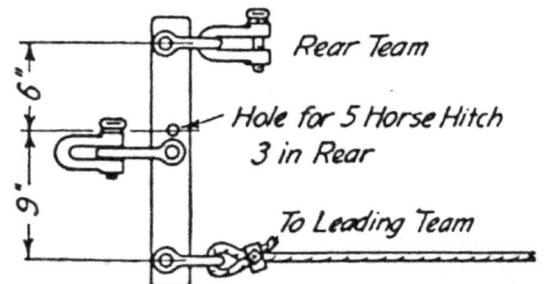

Gang Plows

The most favorable four-horse abreast hitch that can possibly be secured creates considerable side draft, because the center of the team is well to the side of the center of the plow. Note that the horse next to the furrow is almost exactly straight ahead of the center of resistance in the plow, and every inch that the three land horses swing away from the furrow horse, not only increases the side draft, but increases the direct draft as well.

Carrying this to the extreme, imagine that the third horse from the furrow be permitted to swing around far enough to pull straight from the end of the evener, at a right angle to the line of travel of the plow and the furrow horse. This would not help to move the plow forward.

Work the horses tandem, or strung out, and they will have all of the room they need, and will be working more nearly in front of the line of draft of the plow; consequently, they will pull it easier.

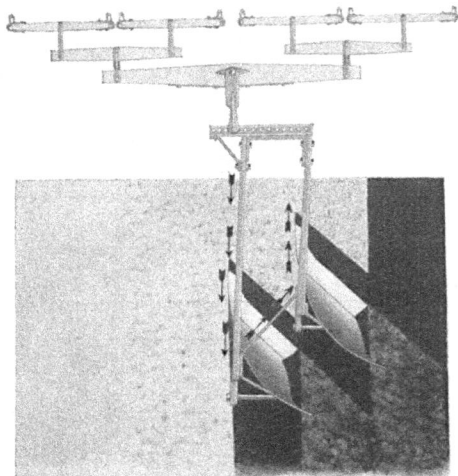

*John Deere's recommended four-up plow hitch*

Fig. 275. Another view of the straight ahead four abreast hitch, the same one as in Figure 270, and in this case showing how the side draft transfers the pull back up the beam proving highly inefficient.

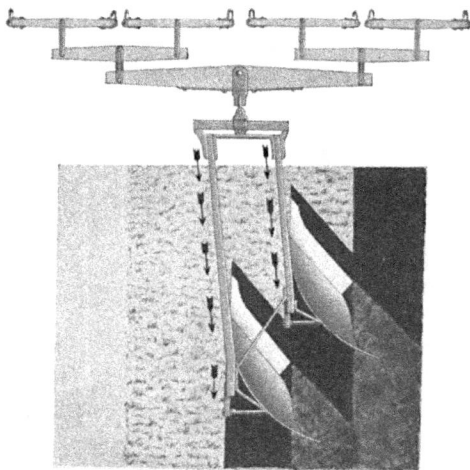

*John Deere's recommended five horse tandem plow hitch*

*The John Deere four abreast plow evener*

Fig. 276. In this view it can be seen how landed beams and a centered evener will pull equally on the beams. However it should be obvious that a serious side draft remains and that the correct adjustment of the front furrow wheel is critical.

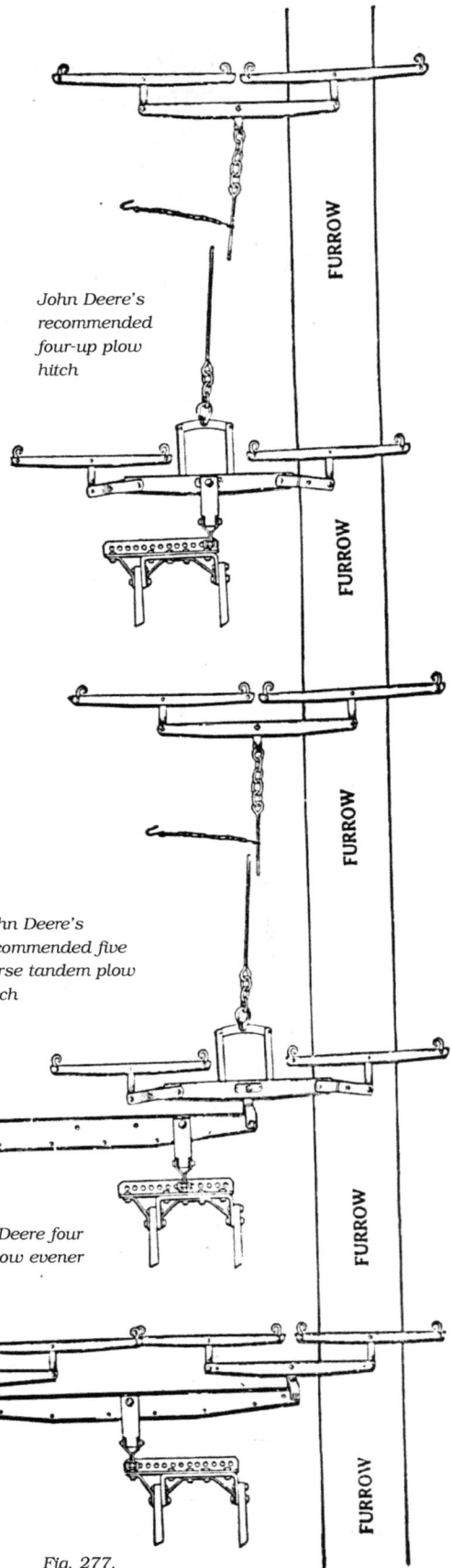

Fig. 277.

**To the Right—Four-horse abreast hitch on two-bottom gang plow. Note crowded condition of horses.**

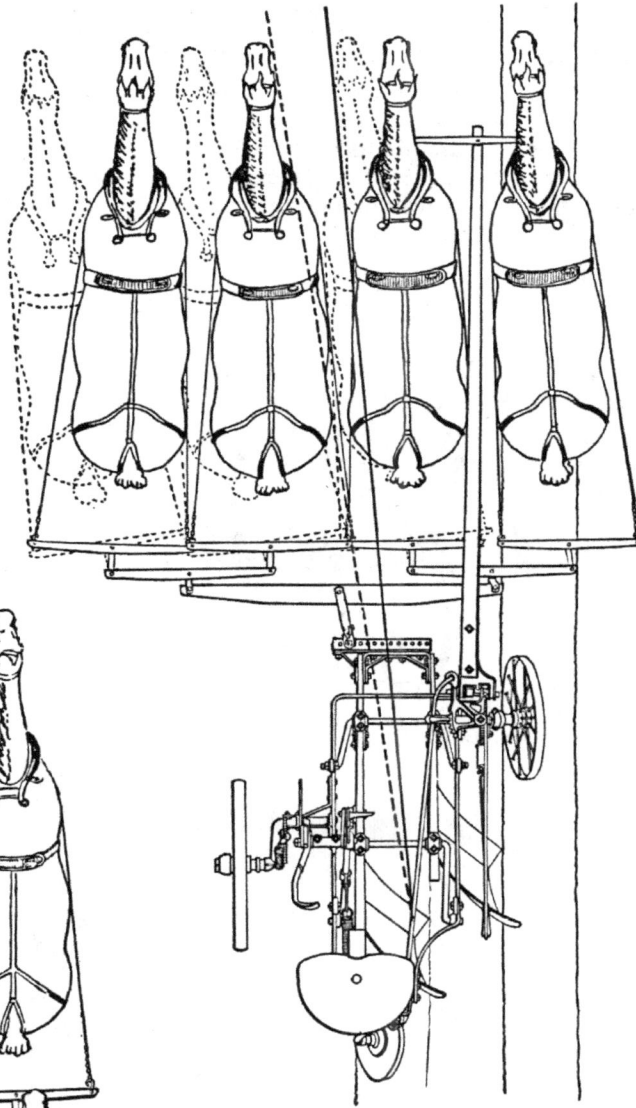

**Five-horse strung-out hitch.**

*Fig. 278. These two good diagrams were originally published in John Deere literature further demonstrating the clear advantages of a tandem, in this case a five horse, hitch over the four abreast. Many plowmen, including this author, prefer the five horse tandem which puts the three abreast in the lead (see Fig 294) and spreads the wheelers out for better visibility and air circulation.*

**To the Left—A five-horse strung-out hitch on a two-bottom gang plow. Each horse has plenty of room.**

# P & O No. 2 Diamond Gang Plow

The illustrations on this and the next page are taken from a setup and operation pamphlet sent with the new owners of P&O gang plows around 1920. We present them here to allow an inside construction view of one popular make of gang plow.

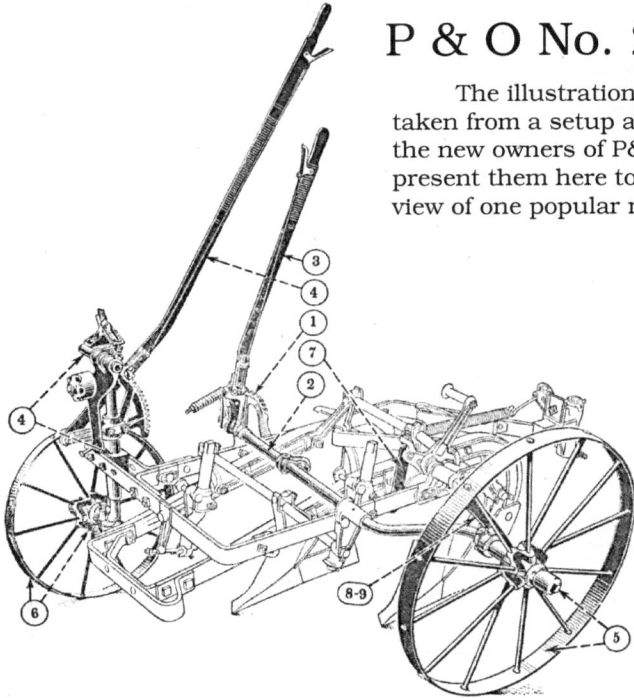

Fig. 280. Here the seat and bracket, the steering rod and the coulters are shown.

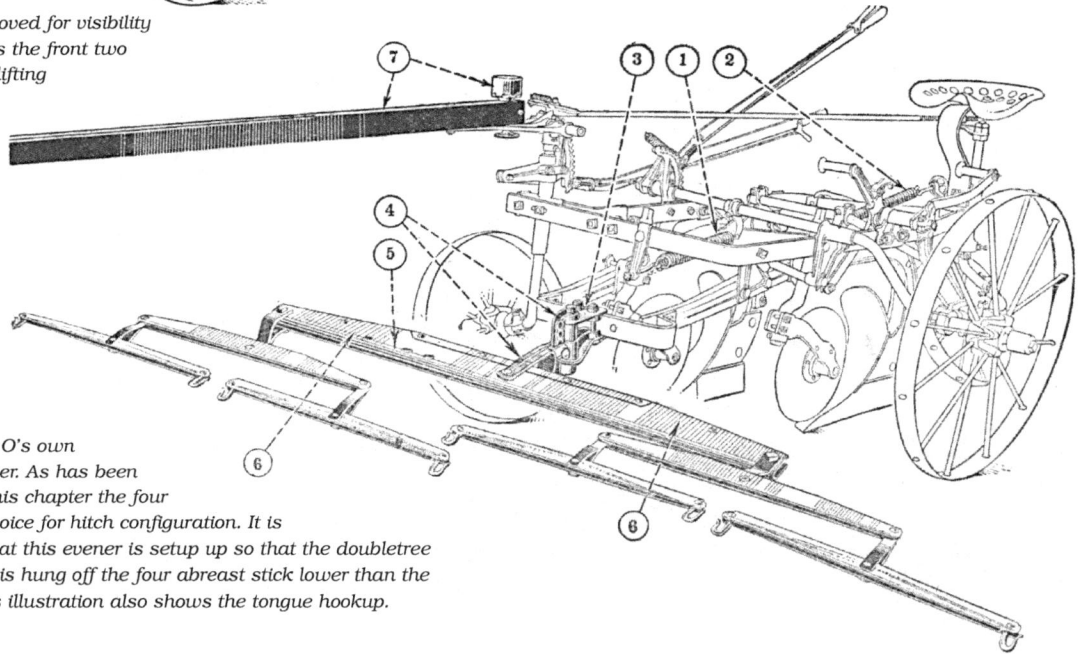

Fig. 279. With the seat removed for visibility the darkened portion shows the front two wheels and their axle and lifting mechanisms.

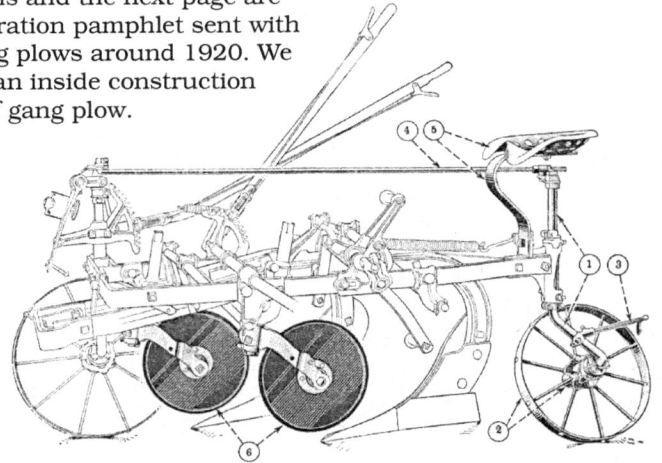

Fig. 281. Illustrating P & O's own factory four abreast evener. As has been argued exhaustively in this chapter the four abreast is not the first choice for hitch configuration. It is interesting to point out that this evener is setup up so that the doubletree serving the furrow horse is hung off the four abreast stick lower than the landside doubletree. This illustration also shows the tongue hookup.

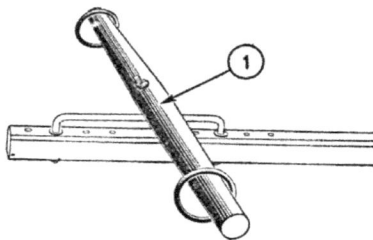

Fig. 283 shows how the weed hooks are attached. These help to turn tall vegetation and trash in under the furrow.

Fig. 282. Note that neckyoke is set, by use of a center eyebolt, to slide on the reinforced rod on the front of the tongue. As noted elsewhere, backing these plows up is difficult or impossible beyond a couple of inches.

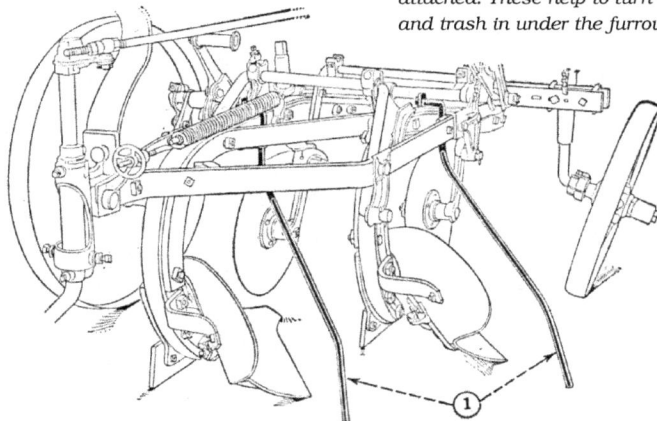

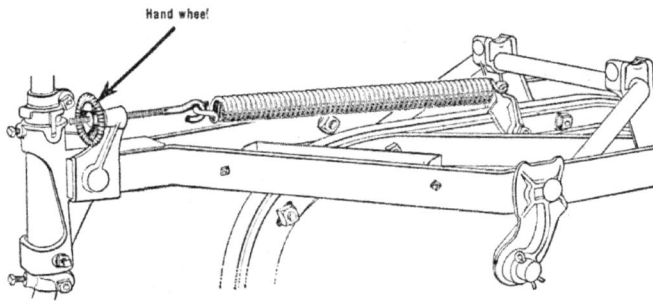

Fig. 284. The P & O No. 2 Diamond Gang featured a handy wheel for adjusting the Raising Spring tension.

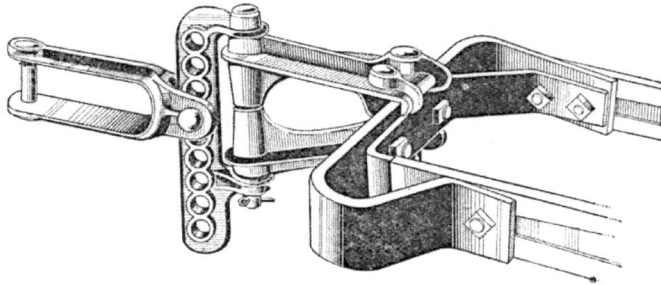

Fig. 285. An excellent close up of the P & O Gang hitch adjustable up, down and sideways. Note that the vertical hitch can be removed and flipped over if additional height is required. Remember it is always better to hitch too long (as in 'too high').

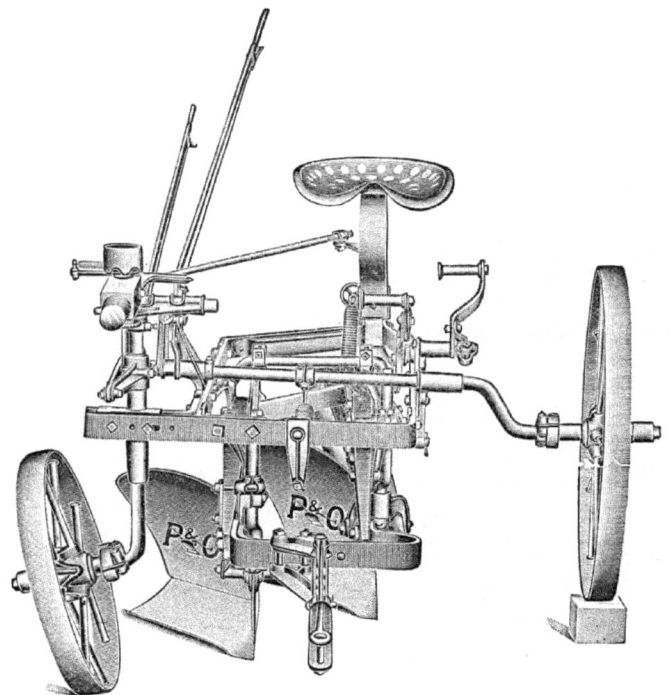

Fig. 286. The P & O Diamond Gang Plow in a front view with the land wheel on a block to show furrow wheel and bottoms down and frame level as is desired in plowing. Notice line of coulters in relation to shares.

**Backing gang plows.** It will become clear when hitched that having the animals back up a three wheeled frameless, tongueless, riding plow is impossible. What is less clear is that the frame-style, with articulated steering off the tongue, is only slightly more amenable to backing. Due to jack-knifing and rear furrow wheel complications do not expect to back more than a couple of inches. Many is the time that this author has had to get off and yank the plow back six inches, back the horses one step and yank the plow six inches, etc., etc. On one early occasion, with a team well trained to back easily and quickly, this author, younger then and dumber, was unceremoniously dumped as the plow folded up.

# Opening a land with a gang plow

Fig. 287 & 288. As has been previously noted, many plowmen prefer to open a land (see Chapter Eight) with a walking plow and after the crown is complete and a clean furrow awaits, then commencing with the sulky or gang plow. As these old photos indicate it is altogether possible for a knowledgeable teamster to adjust the riding gang to do a highly respectable job of the crown furrows.

Fig. 289. John Erskine in front and Donna Anderson behind, each with six Shires on two bottom gangs in eastern Washington state in 1998.

## Correct Hitch on Riding Gang Plows

Repeating from last chapter we must assume, in discussing the correct hitch on a three-wheel gang plow, that all adjustments of the plow are correct and the shares properly sharpened.

The wheels are attached to a three-wheel plow for the purpose of carrying the entire load, regulating the depth and width of furrow or furrows to be turned. It's important to keep this point firmly in mind.

The load consists of the weight of the plow, the weight of the operator and the weight of the soil being lifted and turned. All of this weight should be carried on the wheels and none of it carried by the team. Improper hitching can result in the horse's lifting the front end of the plow and causes many problems.

Please refer to the drawing in Fig 203 on page 78. It illustrates the line of draft which comes from that point indicating the center of resistance on plow bottom to the point where the tug or trace attaches to

the hame at the horse's shoulder. The correct place to hitch, then, would be at a point where this line passes through the vertical part of your plow clevis. The dotted lines show the effect at this point on your vertical clevis of too short a hitch when horses are worked abreast. *The natural tendency is to hitch too close to a three-wheel gang plow*. This often results in poor plow performance and will always result in the animals having to work too hard.

If the hitch is too close or too low on a three-wheel gang plow (just as with sulkies), the team will carry a good portion of the load on their traces rather than allow it to be rolled along on the wheels of the plow. The weight they carry will be taken largely from the front furrow wheel.

By hitching too close or too low, weight is added to the team and the means of controlling the plow is robbed. Without sufficient weight on the front furrow wheel, it will not be possible to control, accurately, the

## Oliver No. 22-A Gang Plow

This Oliver gang was the height of simplicity with just one lever for all operation. Advertised for general field work as well as for orchard and for vineyard plowing, the rear furrow wheel was a free castor so that the entire plow turned like a two wheeled cart. With the handle length this plow worked for either walking or riding. Oliver offered an attachment lever for the front furrow wheel for when the plow was used in vineyards and orchards.

Fig. 290.

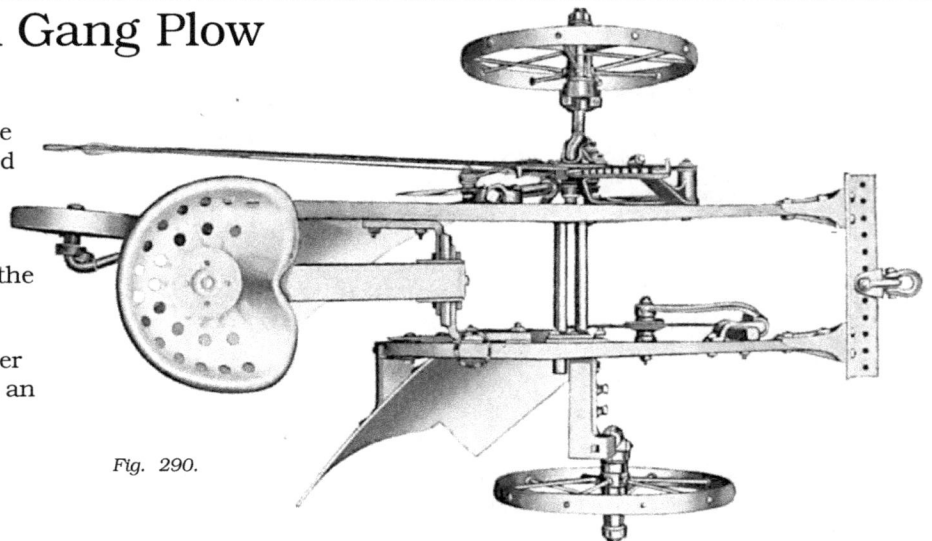

Fig. 291. An excellent worm's eye view from 1918 of six head of Percheron cross chunks pulling two bottoms down a looooong furrow.

Fig. 292. Three big outfits plowing roaring twenties Eastern Washington wheatlands.

width or depth of furrow. When the horses are lifting the front end of the gang plow it will tend to float and bounce making accurate plowing difficult or impossible.

A good rule to go by is to hitch long enough and high enough so that you have as much weight on the front furrow wheel as you have on the rear furrow wheel. A long hitch is better to accomplish this, rather than a high hitch, as the long hitch affords more room for your horses to walk, making it all the easier on them.

As was said before, this is easily tested. After having opened up the land, and with the plow running at the depth wanted, get off the seat and grab the front furrow wheel and see how much effort it takes to slide it. Then do the same thing to the rear furrow wheel. Adjust the hitch until there is as much resistance on the front furrow wheel as there is on the rear furrow wheel.

**The goal is to have the animals simply roll the plow load along.**

Fig. 293. The old Draft Horse and Mule Association of America published pre WW II, excellent information advocating big farm hitches. This diagram of six, bucked back and tied-in (see page 110), hooked to a two bottom plow also features a "tag harrow".

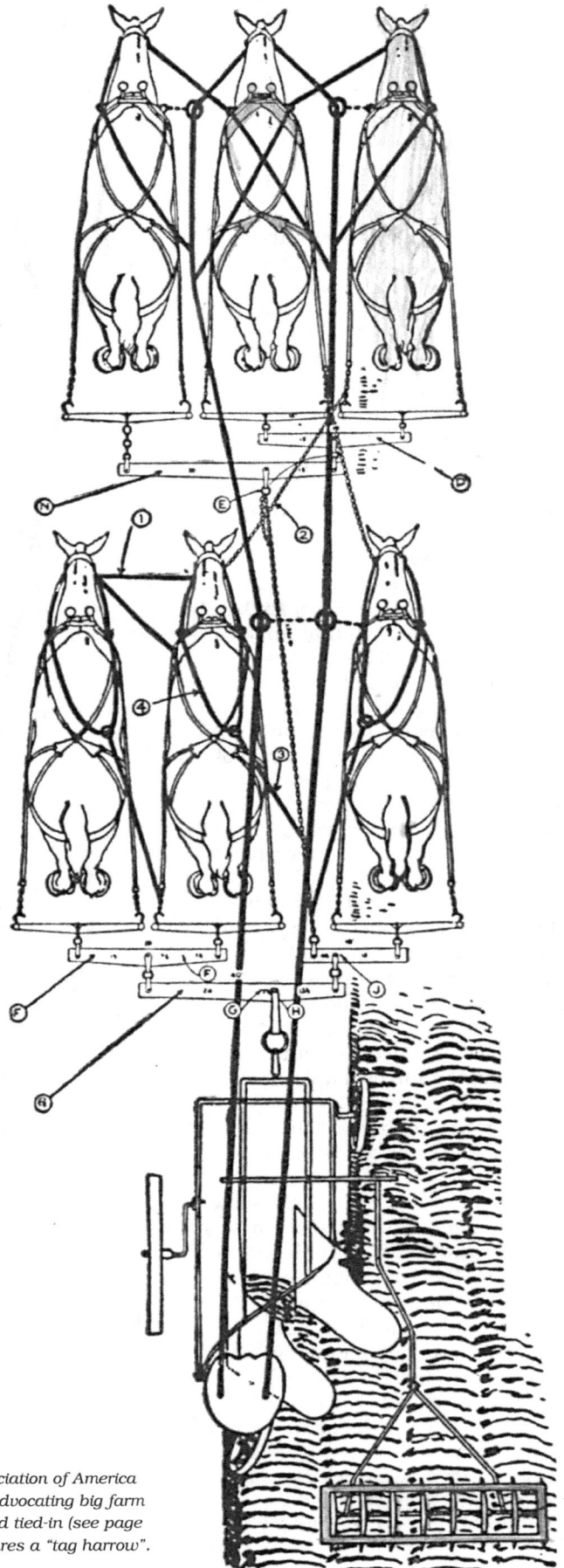

Because it is better to be hitched too long, getting a proper hitch so far as the vertical adjustment of your clevis is concerned is much simpler than with walking plows and some sulkies. Move the hitch clevis sidewise on the cross clevis of the plow, whether it be a sulky or a gang plow, until your team walks comparatively straight. Its a mistake to insist on hitching directly in front of the point of the beam, on a single-bottom plow, or between the point of the two beams on a two-bottom plow. If you do and use fairly good-sized horses, they will have to walk sidewise and the traces will chafe their legs. With the first hitch adjustment proper, you will have sufficient weight on the front furrow wheel, so that you can get over far enough on the cross clevis to allow your horses to walk away straight and free.

**Getting the Furrows Right**. A common complaint is that furrows are not uniform. This applies particularly to gang plows with more than one bottom. This complaint may be due to any one of the following faults: First, frame of plow not level; second, improper rolling coulter adjustment; third, improper front furrow wheel adjustment; and fourth, incorrect hitch.

Speaking now of frame-style plows, the frame should be parallel with the surface of the field. If the field being plowed has a slope, the gang-plow frame should slope accordingly; otherwise one plow bottom will run deeper than the other and the furrows will not be uniform.

It is very important that both rolling coulters be adjusted an equal distance ---- outside or to "land"---- from the shin of the plow. For instance, if you are using a gang plow with two 14-inch bottoms, and the front rolling coulter is set flush with the shin of the plow, this robs the front plow bottom of from one-half to five-eighths of an inch of it's furrow slice. With the rear rolling coulter set from three-fourths of an inch to

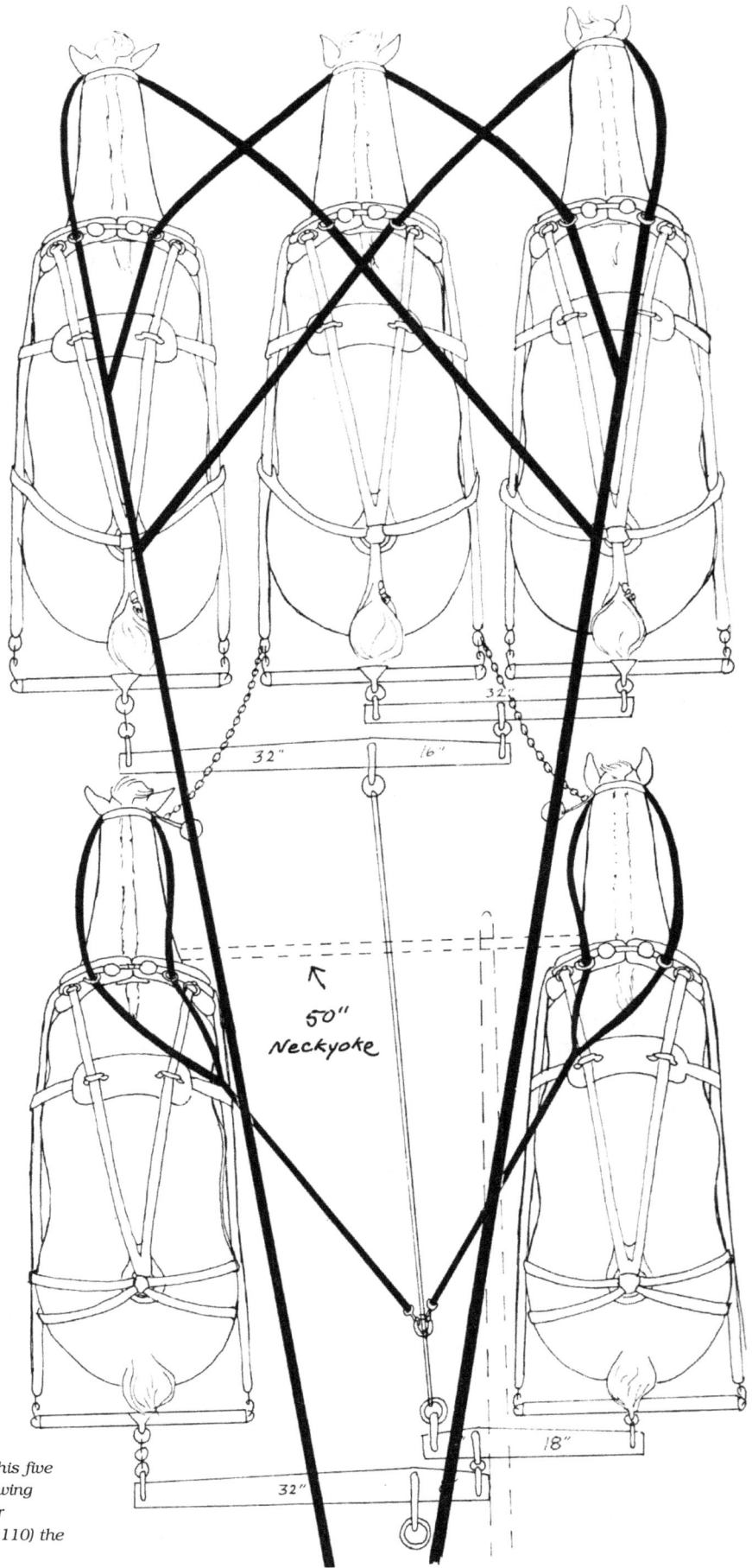

Fig. 294. Taken from _The Work Horse Handbook_ this five horse tandem is an excellent gang plow hitch allowing tremendous visibility for the teamster and good air circulation for the horses. Once again (as on page 110) the buck-back system is employed.

*Fig. 295. Jack Eden of Corvallis, Montana plowing in 1998 with six head of his all purpose draft mules. Photo by Helen Eden.*

one inch to "land," it adds that much to the rear plow furrow slice. This causes the front plow to turn a 13-1/2-inch furrow slice and the rear plow 15-inch, and as a result the furrows are not uniform.

It is also important that the front furrow wheel be adjusted so that the front plow can freely cut a full-width furrow. If the front furrow wheel is adjusted so that when it is run in the corner of the furrow the front plow cannot cut freely a full-width furrow, whereas the rear plow, owing to the spacing of the beams, must always cut a full-width furrow ---- here again the furrows will not be uniform. With rolling coulters properly adjusted, measure the furrow slice from the rolling coulter to the edge of furrow wall. This, on a 14-inch bottom gang plow, should measure 14 inches. Adjust the front wheel "in" or "out" until this measure-

*Fig. 296. The author back in 1976 plowing with two Percherons and two Belgians and an Oliver frame 14" sulky still in use today*

ment is obtained.

And here again comes importance of correct hitch. With the frame level, rolling coulters properly adjusted, and furrow wheel properly adjusted, it is still possible to hitch so low or so far out of line on the clevis of the plow that the front furrow wheel will not run snugly in the corner of the furrow. Lengthen the traces and make adjustments on the cross clevis of the plow until the front furrow wheel runs snugly in the corner of the furrow.

Whether standing or sitting on a well-adjusted, well-hitched, horsedrawn gang plow the plowman will always enjoy an immense satisfaction as he or she watches, listens to, smells and tastes the soil roll and crumble into a waiting seedbed. It is a living window on creativity.

**Uneven Furrows**. When the furrows turned by a gang plow do not lie alike, it is usually due to the furrow slices not being the same width. This may be corrected by leveling the bottoms (which is done in a frame plow by leveling the frame - with a frameless gang it is done by leveling beams) or adjusting the plow so both bottoms cut the same width furrows. The width of cut of the rear bottom is regulated by adjusting the rolling coulters. The same applies to the front bottom unless the plow is old and worn, in which case it is necessary to set the front furrow wheel in, thereby narrowing the cut of the front bottom. This is done by moving the landing adjustment casting in on the frame.

# Pioneer Gang Plows

After years of field testing, Pioneer Equipment of Dalton, Ohio recently began selling its all new frame-style two bottom gang plow. This author had occasion to witness, at the Horse Progress Days, field tests of this plow and reports without hesitation that it is a superior piece of modern horsedrawn engineering. It features, as does the Pioneer Sulky Plow, Oliver Raydex bottoms. A bar-style foot lever and well-balanced assist arm make easy work of lifting the plow out of the ground. The platform, with expanded metal floor for traction and excellent safety headboard design, features a high seat for excellent visibility. As more modern horsefarmers get the chance to see this beauty in action, it's distribution is sure to expand.

Fig. 297.(above) and Fig 298, (below) are two views of the Pioneer two bottom gang plow on display at Horse Progress Days. The white sheet was information. Spread throughout this book are several photos of this plow at work.

Fig. 299. (below) Wayne Wengerd at Pioneer Equipment had his shop modify an Oliver 4 bottom tag or trail plow to be used with horses at the Horse Progress Days in Ohio ('98). The author had occasion to plow with this implement and twelve head of Percherons. Certainly, if demand warrants, Pioneer has proven it is ready to offer a three and/or four bottom horse plow.

# White Horse Gang Plows

The folks at Gap, PA's White Horse Machine Shop build excellent hydraulically operated gang plows which, to date, have been primarily seen in the eastern half of the U.S. One model is direct hitch of a three point plow and the other is the combination of a trail plow with a forecart.

Fig. 300. The White Horse 3 point hitch Forecart with Ground drive hydraulic system and 2 bottom detachable 3 point Hydraulic Reset Plow.

Fig. 301 (above) A view at Horse Progress Days of the White Horse three point Gang doing good work

Fig. 302. (left) and Fig. 303. (below)  Two views of a combination of White Horse Hydraulic forecart with a White Horse two bottom trail plow working at Ohio HP Days.

Fig. 304. An earlier version of the White Horse three point hydraulic gang plow at work at the 1996 Indiana Horse Progress Days.

Fig. 305. Six magnificent Belgians walk out with ease pulling a White Horse two bottom hydraulic plow at the Ohio HP Days in '98.

Fig. 306. Six beautiful Percherons make easy work of plowing with the Pioneer frame-style two bottom gang plow in Ohio. This photo does an excellent job of illustrating how the optimum hitch for a gang plow has the hitch far enough ahead to "pull it along" without lift.

# Buck-back Tie-In system
## of hitching

Though many teamsters choose to drive a strung out hitch with a pair of lines to each span, it is possible to drive tandem hitches with just two lines. On this page we've borrowed an illustration from *The Work Horse Handbook* illustrating how an eight up can be

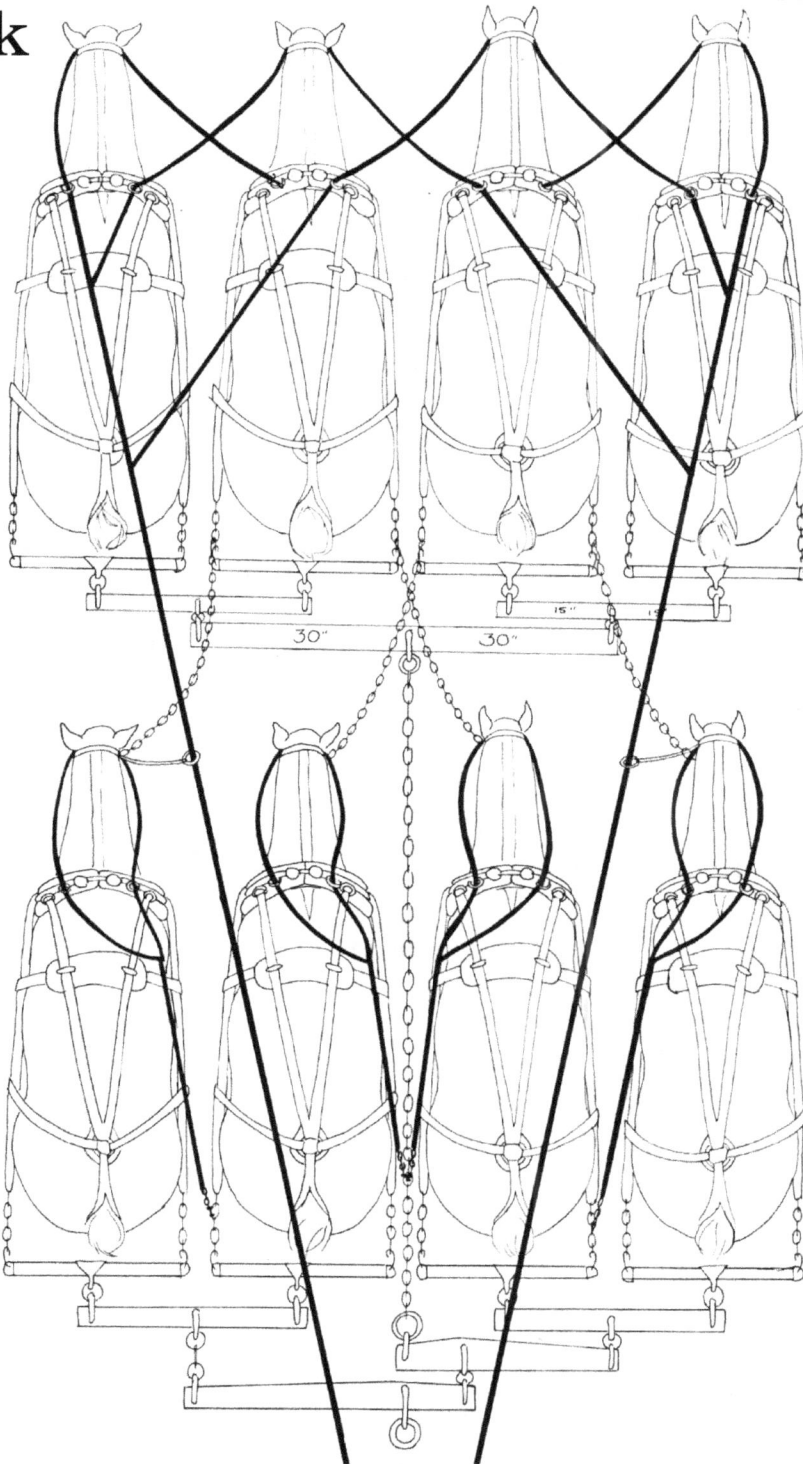

*Fig. 309. (below) A buck-back strap with a snap at each end.*

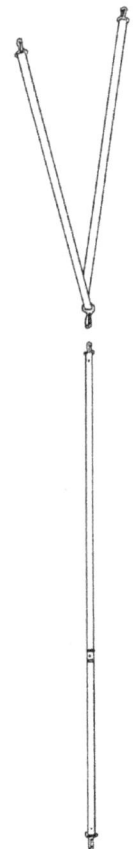

*Fig . 308. Tie-in chain with a snap on each end.*

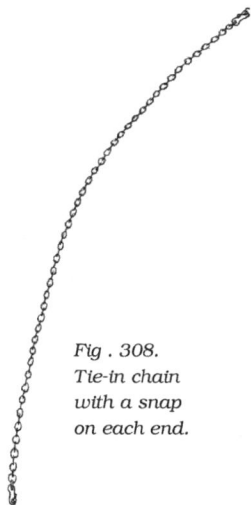

driven with just two long lines employing the *'buck-back tie-in'* system. For a more indepth explanation of this setup the reader is encouraged to go to *The Work Horse Handbook*.

For the purposes of this book, here is an abbreviated explanation. The wheel (and swing team if used) are fastened into the hitch by *tie-in* chains snapping halters to the trace chain ahead and *buck-back* straps which fasten from bit rings over the same horse's withers and back to the neighboring horse's trace chain or the lead bar or chain. In this way these buck-backed horses must go when the leaders go and they must stop when the leaders stop. By following the lines of action (*note series of arrows along lead chain in the middle of the hitch*) it

*Fig. 307. A good diagram, taken from The Work Horse Handbook, of an eight-up showing the lines setup and the buck-back tie-in system*

will be seen when the lead team stops, if the wheelers step ahead they will be pulling the entire lead team through the buck back straps.

Though this system may seem awkward, this author can attest to it working comfortably for teamster and animals; so comfortably that once this system is tried for a couple of hours, the teamster will likely not return to the old way.

# Rope and Pulley Hitches

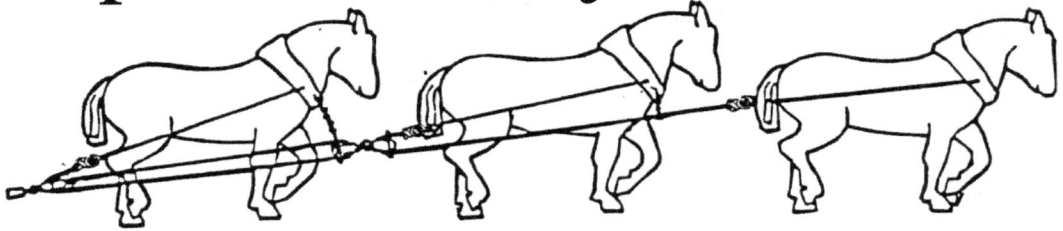

Fig. 312. Demonstrating how the rope and pulley hitches work.

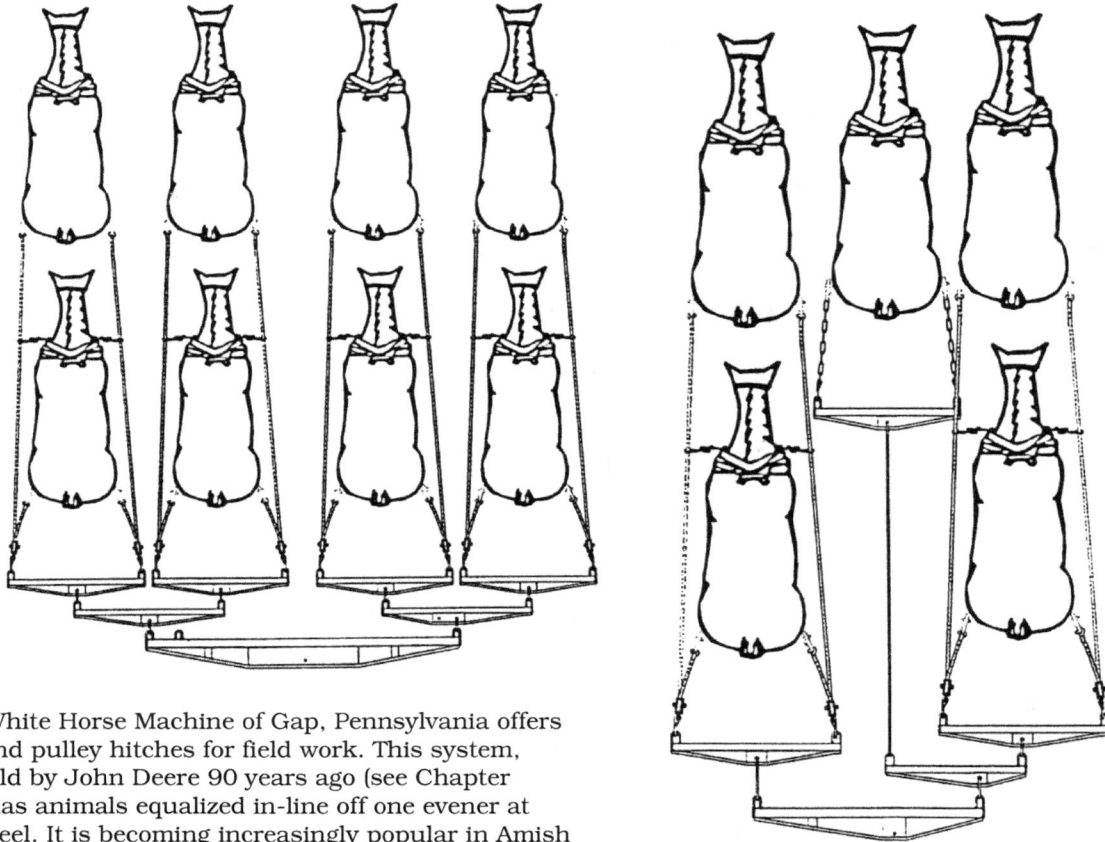

White Horse Machine of Gap, Pennsylvania offers rope and pulley hitches for field work. This system, first sold by John Deere 90 years ago (see Chapter Ten), has animals equalized in-line off one evener at the wheel. It is becoming increasingly popular in Amish circles where big hitches are employed for field work. The White Horse gear features heavy duty nylon rope and Nylatron Pulleys. These drawings, done by Doug Groff and Sam Moore, were borrowed from White Horse's sales literature.

Fig. 310. (Below) A closeup of the rope and pulley setup at the wheel evener.

Fig. 311. The author entrusted with the lines of the Yoder 12 Percheron plow hitch utilizing the rope and pulley system.

Fig. 313. John Erskine driving his six Shires on a John Deere two bottom gang plow with Marianne Frank along for the ride. Heather Erskine took this picture at the Thomas Ranch in Waitsburg, Washington.

Fig. 314. Bill Anderson plowing with eight of the family Shires and a three bottom plow. Bill's mother Dorothy Anderson is riding. Following hitch is Ross Frank driving the John and Heather Erskine six-up of Shires. Photo by Heather Erskine.

Gang
Plows

Fig. 315. In front is Donna Anderson driving the Anderson's eight Shires on the Oliver three bottom plow with mom, Dorothy, on the seat.

Fig. 316. Bill Anderson and a good side view of the eight Anderson Shires and Oliver three bottom.

All photos on this page by Heather Erskine.

Fig. 317. At the Waitsburg, Washington Thomas Ranch, John Erskine again with his six Shires, Cathy Lee-Haight riding. To quote John, "Its always nice to have company on the plow, but it is a blessing if something goes wrong and you need a third hand."

Chapter Seven

# Disc Plows, Trail Plows, & Misc Plows

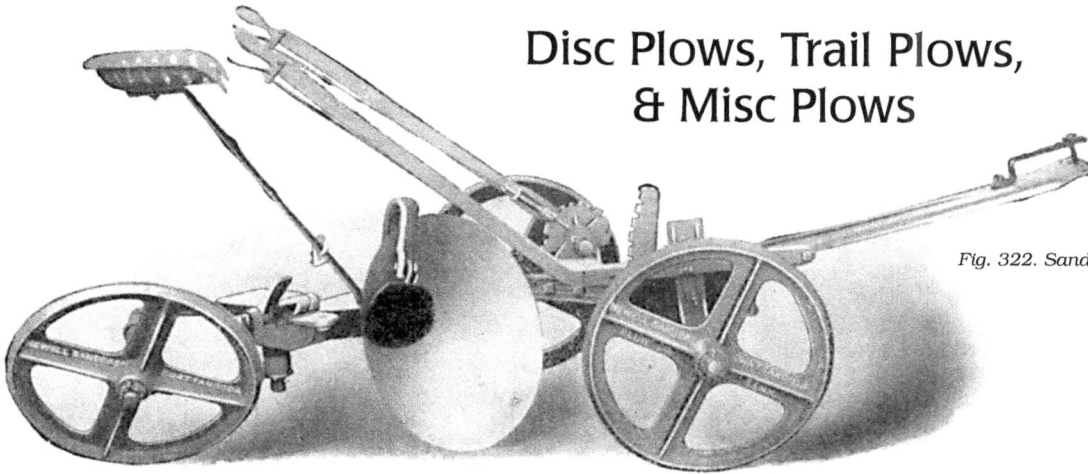

*Fig. 322. Sanders sulky disc plow.*

The disk* plow was invented in an effort to reduce friction by making a rolling bottom instead of a bottom that would slide along the furrow. It has not been proven that, after the extra weight is incorporated into the plow, it has any less draft than that of the moldboard type. The results of the disk plow usage, however, show that it is adapted to conditions where the moldboard will not work or works with difficulty. Some of these conditions are as follows: first, the disk plow can be used in hard ground that is too hard for the moldboard; second, it will scour using a scraper, in most soils; third, it does not form a hardpan; fourth, the angle of the disk can be changed for hard or

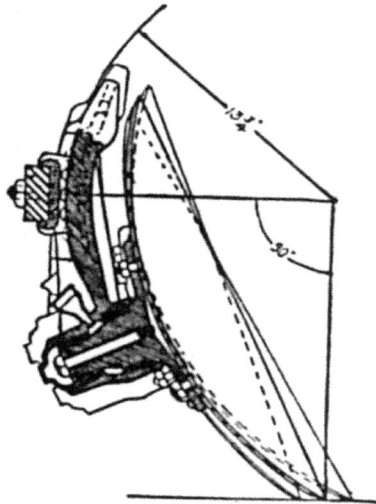

*Fig. 323. Vertical angle of disk can be easily changed.*

loose land. fifth, the disc plow will roll over big rocks.

This type of plow is used in the South and North, and very extensively in the Southwest and the semi-humid regions of the Middle West. It is of special value in Texas because of the large areas of soil having a close texture which will not scour on the average moldboard plow. Texas is called by the plow manufacturers, a disk plow state. There are large areas, however, where the moldboard plow does work satisfactorily.

**The Sulky Disk.** The sulky disk plow as in the case of the sulky moldboard plow, is a disk plow with only one bottom. The disk plow bottom is a perfectly round, concave, disk of steel, sharpened on the edge to aid in the

*\*Please note: Across the history of horsedrawn plows the accepted spelling of Disc (Disk) plows went back and forth between makers. This book will reflect that inconsistency.*

*Fig. 324. Sanders two bottom riding disc plow.*

penetration of the soil. There are several holes for bolting this disk to the malleable casting upon which it fits. These disks are set at an angle both to the plow sole and to the furrow wall. This allows the disk to have a sort of scooping action. The use of heat treated disk plow bottoms assures longer life, a smooth cutting edge and easier penetration. A malleable iron bracket to which the disk proper is bolted has an extension forming an axle projection, which fits into a hub in such a manner as to give a close fitting bearing. These parts are usually chilled to increase their lasting and wearing qualities. Some plows are equipped with ball and roller bearings. The majority of plows, however, use plain cone bearings. The bearing allows the disk free action as far as turning is concerned.

When the plow is pulled forward the disks turn due to the action of the furrow slice upon it. The top of the disk is revolving to the operator's left. The furrow

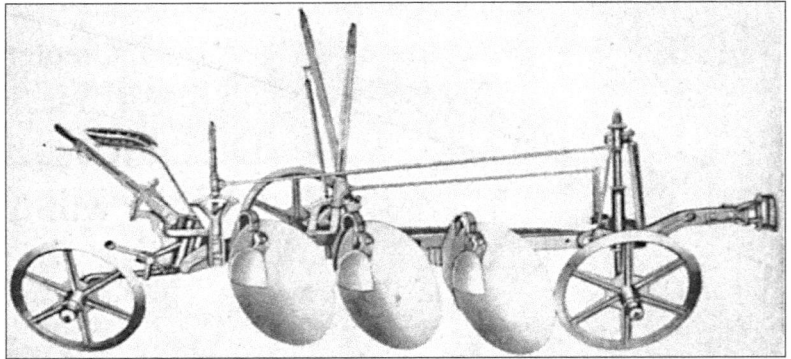

Fig. 325. A three bottom disc plow.

Fig. 326. Overhead view of a reversible disk plow with seat removed for visibility. Stopping the plow and swinging the team causes the disk bottom to pivot to the opposite direction.

slice, then, is cut by the left edge of the disk, brought under and up to the right, and then thrown out to one side. The furrow slice is pulverized to some extent when carried over the concave surface of the disk.

All disk plows should be equipped with a scraper which can be adjusted to work from the center to the edge of the disk. With the aid of the scraper it is possible to get greater pulverization of the furrow slice. It is also possible to invert the furrow slice much better.

The disk plow can be made to penetrate more easily by setting the disk more in a vertical position. The flatter it sets, the less tendency there will be for it to penetrate. To further enable the disk plow to take the soil properly, weight is added to the frame and wheels to force the plow into the ground. There is one great difference in moldboard and disk plows: the moldboard plow is pulled into the ground by the suction of the plow, while the disk is forced into the ground by adding weight and by the suction of the disk due to the angle at which it is set. The frame of this plow is made of very heavy steel with many large castings to give plenty of weight.

The wheels, instead of being made of light rolled steel as in the moldboard, are cast and are smaller and heavier. The rim of the wheels, instead of being flat, is

usually flanged or V-shaped. This construction aids in preventing the wheel from slipping sideways. Provision is made for additional weight by means of weights which can be bolted between or on the sides of the spokes of the wheel, usually on the rear wheel. This may be necessary if unusually hard ground is encountered.

Another difference of construction, in comparison with moldboard plows, is that instead of the plow beams curving over the top of the plow and attaching to the back of the plow bottom, they come from the side. This, of course, does not allow enough clearance and often gives trouble where a large amount of trash is on the land.

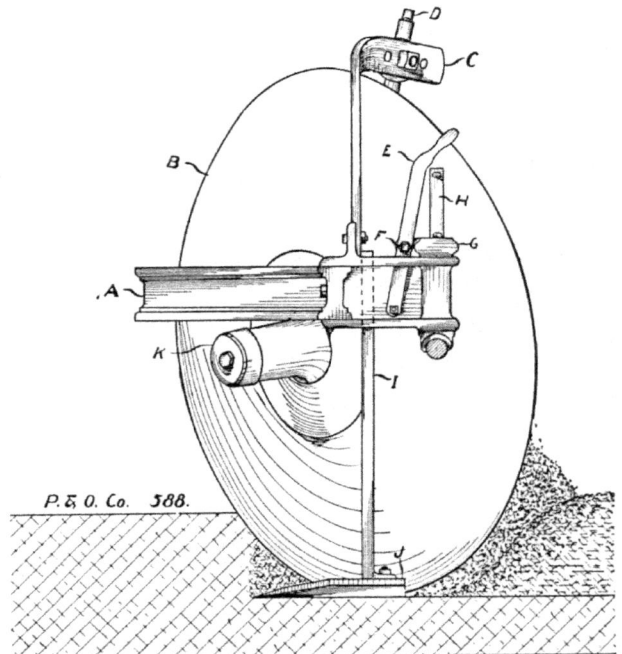

Fig. 327. P & O's highly unusual Standing Cutter is here demonstrated cutting a square furrow. The standing share cutter is placed back of the disk, directly in line with the furrow bank or wall, and is adjustable on the bracket for the three sizes of disks with which P & O plows are furnished. The upright has a sharp edge and runs up close to the bank, while the horizontal share cutter is bolted to the upright and runs on the bottom of the furrow, cutting several inches under the bank towards the landside. This cutter squares the bottom of the furrow, holds the plow in line, and cuts out the ridge. The standing cutter always cuts under the ground towards the bank, and when the plow makes the next revolution of the field, the disk turns over a perfectly flat furrow, turning up and pulverizing all the soil the entire depth.

The furrow wheels of the disk plow, like those of the moldboard, are inclined. This is to aid the plow to overcome the side pressure created by the furrow slice upon the plow which is increased by the rolling of the bottom itself, causing the rear end of the plow to swing around to the left. There are levers for each of the wheels for adjusting and leveling the plow. There is a special lever for landing the front wheel, that is, it can be given more or less lead to or away from the furrow wall.

**The Gang Disk Plow**. The gang disk plow differs from that of the sulky disk plow in that there are two or more bottoms. Many of the sulky plows are so constructed that they can be changed into a gang plow by adding another bottom, making either a two disk or three disk plow. For this reason, this type of plow is sometimes called a multiple gang plow. The construction of the disk, frame, wheels, and arrangement of the levers is practically the same as for the sulky plow.

**Angle of Disks**. To successfully meet the conditions outlined, the disks' blades must be so arranged on the frame that they will function properly. From experience it has been found that the disks should be placed at an angle both vertically and horizontally. This angle depends on the proper distribution of the entire weight of the plow, which is necessary to hold the disks in the ground. Weight is required because the disks do not have the suction that the moldboard has. By referring to Fig. 331, it is seen that the vertical angle can be varied from an abrupt angle to one that is quite flat. The more vertical the disk is set, the greater the tendency to penetrate.

The horizontal angle of the disk influences the width of the furrow slice and the tendency to roll. Disk blades set more perpendicular to the direction of travel cut

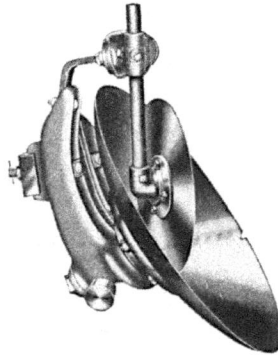

Fig. 328. An Oliver rotary disk scraper adjustable for every condition and ideally suited for sticky soils. It can be set in or out, up or down and close to or away from the disk. It helps to turn the soil and crumble it.

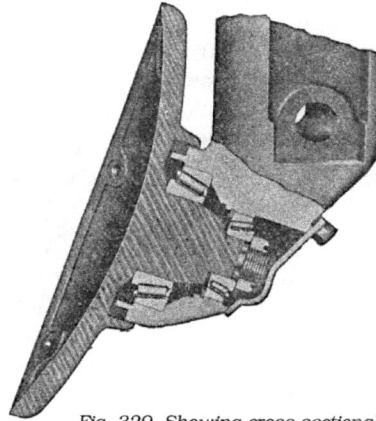

Fig. 329. Showing cross-sectional view of roller bearing disk bearing with disk removed. Tapered roller bearings installed at spindle end and shoulder carry the pressure of plowing with a minimum of draft. Expanding felt collar and overlapping flange on the disk spindle make this bearing dirt-proof.

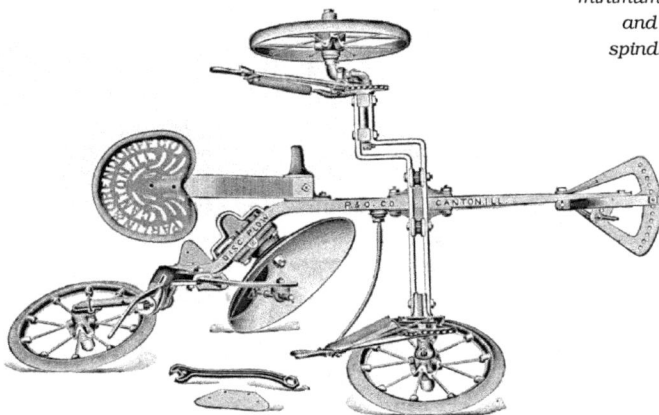

Fig. 330. This overhead view shows many of the features that made the P & O Canton Disk Plow unique

wider furrows and do not turn so freely as when more parallel to the furrow.

The disk angle, or the angle the face of the disk makes with the line of forward travel (Fig. 331), can be regulated on most plows. This is usually done by changing the adjustment of the land-wheel bracket in relation to the plow frame. As the disk angle is increased, the width of cut of the gang is decreased; the smaller the angle, the greater is the width of the cut. The narrow cut at the greater disk angle is used in hard ground. Some plows have a wedge in the disk bearing mounting for adjusting the angle of each disk individually, leaving the total width of cut substantially the same.

The tilt of the disk (Fig. 331) may be adjusted in various ways for different plows - by arrangement of holes for the bolts connecting the bearing standard to the frame, by use of wedges of eccentric washers in connection with the bearing support, or by a pivoted bearing support. The tilt should be increased for sticky, waxy soil and decreased for loose, sandy soil and in hard, dry soil. Decreasing the tilt puts the disk in a more nearly vertical position.

For extremely hard ground, where penetration is difficult, set the plow at its narrow width with the disks in their most nearly vertical position. If this does not give the depth desired, add weights to the plow.

**Disk Scrapers.** Scrapers with brackets are furnished as regular equipment with most standard disk plows. The three most common types (Fig. 332) are the moldboard, or universal, the hoe and the rotating. The brackets are designed to give a considerable range of adjustment for the scraper.

The moldboard scraper is used in soils that offer no scouring difficulties. Under these favorable conditions and with proper adjustment it assists greatly in covering trash and vegetation. Rotating and hoe scrapers are best in sticky soils where the moldboard type will not scour. Regardless of the type used, the scraping edge should be close to the disk face but with sufficient clearance to avoid the friction in case the disk does not run true.

**The Center of Resistance**. The center of resistance is closer to the furrow wall than on moldboard

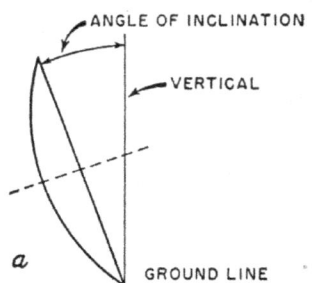

Fig. 331.
Demonstrating disk angle and tilt.

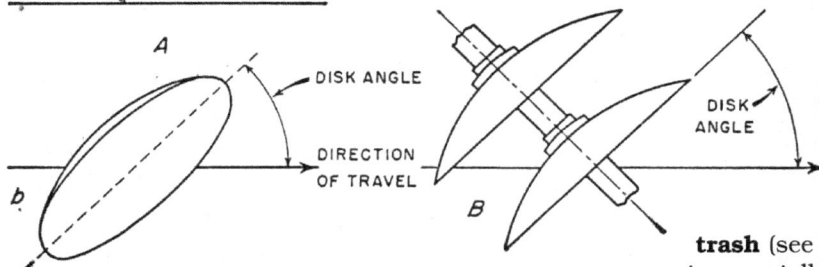

plows. Its location is to the left and below the center of the disk blade. The point varies with the vertical and horizontal angles, the depth, and the amount of concavity of the blade.

**Disk Blades**. The size of disk blades ranges from 20 to 28 inches. The average thickness for disk plow blades is 3/16 inch. The amount of concavity varies with both the different diameters and the same diameter.

### Disk Plow Troubles

There are troubles and adjustments of the disk plow which are different from those of the moldboard plow. Yet there are many troubles that are common to both classes.

**Failure to penetrate**. This problem may be caused by lack of weight and improper angle to the disk. Weight can be important to forcing the disk into the soil. Provision is usually made on most disk plows

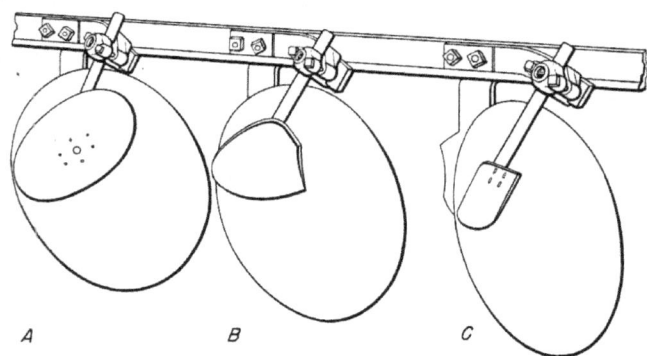

Fig. 332. Types of disk scrapers: A, rotating; B, moldboard; C, hoe. In heavy sticky soils it is necessary to use scrapers on the disks to avoid increasing the draft and having to add weight to keep the disk in the ground. Hoe-type and revolving-disk scrapers are more effective in very sticky soils. The moldboard-type scraper aids in doing a more thorough job of inverting surface trash and cover crops in soils that do not adhere to the scraper. The moldboard scraper tends to turn the furrow slice downward as it leaves the disk.

for the placing of additional weight, usually at the rear wheel. Changing the angle of the disk to set nearer to perpendicular will increase the tendency to penetrate.

Under most conditions, the front furrow wheels should run straight forward, parallel to the line of the furrow. If the plowing is exceptionally hard, it may be given a slight lead to the furrow wall. Usually, the wheels are given a lead toward the plowed ground to counteract the side pressure of the furrow slices. Since disk plows do not have landsides, the wheel must hold the plow in position.

If the bottoms **tend to trail** it may be due to the hitch being too far to the right. The hitch set this way will have a tendency for each of the bottoms to cut a narrow furrow width due to their trailing behind one another.

Disk plows do not always **cover trash** (see page 32 as well as do moldboards. This is especially true when they are operated without scrapers. If the disk is set rather flat from the vertical, it will not cover trash as well as when set more nearly straight up and down. When set straight up and down, the furrow slice is in use, the furrow slice will be taken from the disk and turned. The scraper having a curved surface will turn the furrow slice better than the straight type; however, the straight type will shed

Fig. 333. The author struggling to make a P & O two bottom Disk Plow line out properly in sandy desert soil

soil better and give less trouble when sticky soils are being plowed. As a general rule, the scraper should be set low and at an angle of about 35 degrees with the disk. It is also tilted to throw the soil toward the furrow.

**Care of Disk Plows.** Standard and vertical disk plows are sturdy, but they should have good care. To maintain proper alignment, keep all nuts tight and promptly straighten or repair bent or broken braces, hitch parts, or levers.

Wheel bearings and the thrust bearing on a vertical disk plow must be kept well greased and properly adjusted when wear occurs. The radial

Fig. 334. A four abreast evener of the type sold with two bottom Disk Plows

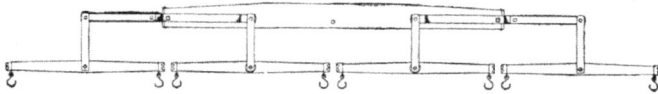

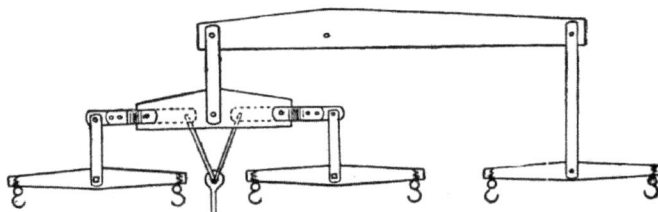

Fig. 335. A five horse tandem hitch sold specifically for use with Disk Gang plows.

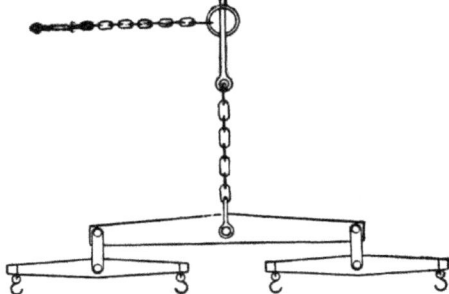

bearings on the disk gang may or may not need lubrication. If they are replaceable and of maple wood the manufactuers usually recommended that they not be lubricated. A metal bearing should be greased or oiled.

The bearing of each individual disk of the standard disk plow must be kept in adjustment and well-greased whether it is of the plain or the anti-friction type. These bearings should be disassembled at least once a season and thoroughly cleaned and the dust seals cleaned or, if necessary, replaced. Bearings should be lubricated at least twice a day when the plow is in use. All screw adjustments, axle sleeves, and other moving parts need to be oiled to permit ease of adjustment.

The disk blades must be kept sharp, ground on the same side as the original level. Rolling the blades to sharpen has not proved satisfactory for the heat-treated disks. When not in use, the disk blades should be coated with oil or grease.

**Uses of the disk plow**. In certain territories where plowing is extremely difficult, disk plows are used. The conditions adaptable to disk plow use are enumerated as follows:

1. Sticky, waxy, gumbo, non-scouring soil, and soils having a hardpan or plow sole.
2. Dry hard ground that cannot be penetrated with a moldboard plow.
3. Rough, stony, and rooty ground, where the disk will ride over the rocks.
4. Peaty and leaf-mold soils where the moldboard plow will not turn the slice.
5. Clay and sandy loams.
6. Deep plowing.

Fig. 336. Overhead view of a heavy duty three bottom Moline Disk Plow.

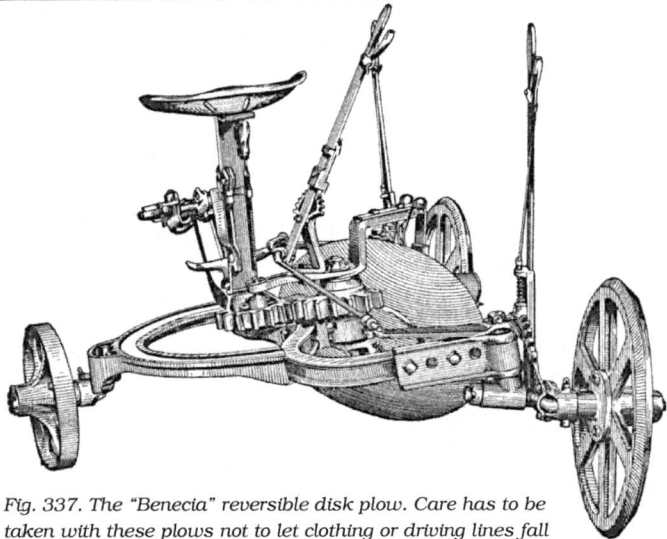

Fig. 337. The "Benecia" reversible disk plow. Care has to be taken with these plows not to let clothing or driving lines fall into the gear mechanism. As the horses side pass to change direction that gear could tear a shirt or wind up a line dangerously

# Reversible Disc Plow

**Reversible disk plows.** This disk plow consisting of one bottom is so constructed that the disk can be reversed and the soil thrown in the same direction at all times. The change from a right to a left handed plow or vice versa is accomplished by a beam that is pivoted at the center on one end of which is hitched the team, while at the other end there is a large semi-circle gear meshing with the gear on the bracket of the disk. If it is desired to turn the plow in the opposite direction, a latch is kicked loose releasing the beam and the team turns without turning the plow frame. As the beam revolves, the disk is also turned by the action of the gear. The seat is attached to the beam and moves with it. The furrow wheels are adjusted automatically, changing the rear furrow wheel into the front furrow wheel and giving them the proper lead. This can be further adjusted by a special lever for that purpose.

Fig. 338

Fig. 339.

These four photos were sent to us by Michael, Terry and Simone Denton of Clarksville, Virginia of their reversible Chattanooga Reversible Disk Plow. The Dentons started with a rusted pile and completely restored the plow to this condition. These photos show excellent details.

Fig. 340

Fig. 341

Fig. 342. At the Indiana Horse Progress Days a big hitch of Belgians pulled a Pioneer forecart hitched to a tractor pull-type plow.

Fig. 343. Two models of White Horse Forecarts. The seat swivels, and the hitch and tongue are fully adjustable. White Horse Machine is located in Gap, PA.

# Trail Plows & Forecarts

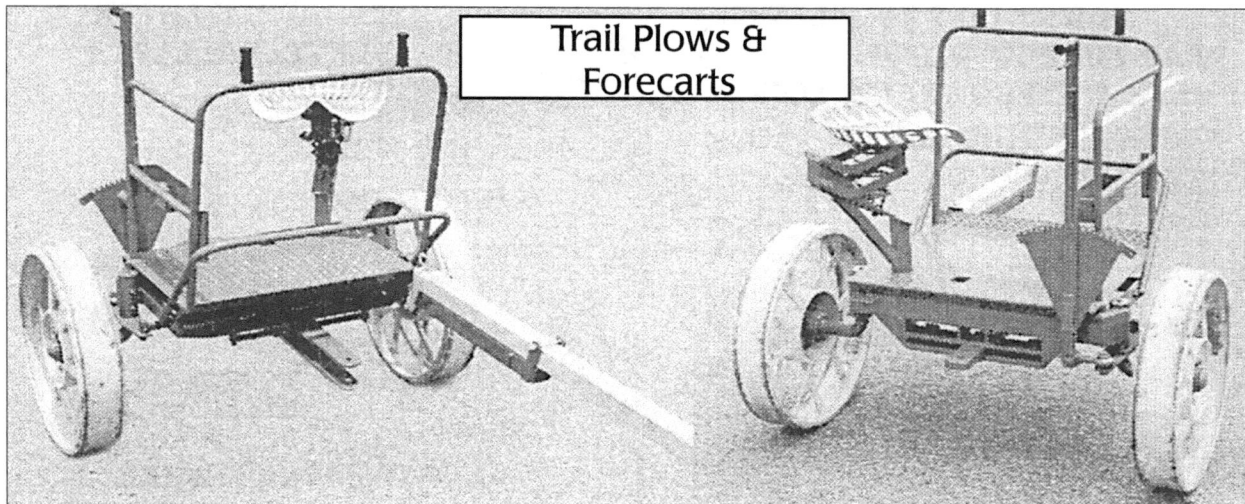

With a basic forecart any tractor pull-type plow of four bottoms or less can be reasonably used with horses or mules. In some cases the plow can be hitched to direct, without a forecart, and a platform or seat rigged for the teamster.

For those uninitiated, forecarts are wheeled implements which act as hitching conveyances between the work animals and assorted appropriate implements. Far and away the more numerous are those with two or three wheels and a simple adjustable draw bar set up. The horizontal adjustability of the draw bar can be important when working to set the outfit up with the forecart wheel running in the previous furrow.

This book is meant to deal primarily with those plows which were designed specifically for animal power. The important variables in dealing with modifications to make tractor plows work for animals are deserving of a book on their own.

Fig. 344. At the 1996 Indiana Horse Progress Days twelve Belgians pulled this behemoth four bottom tractor plow.

Disc, Trail &
Misc. Plows

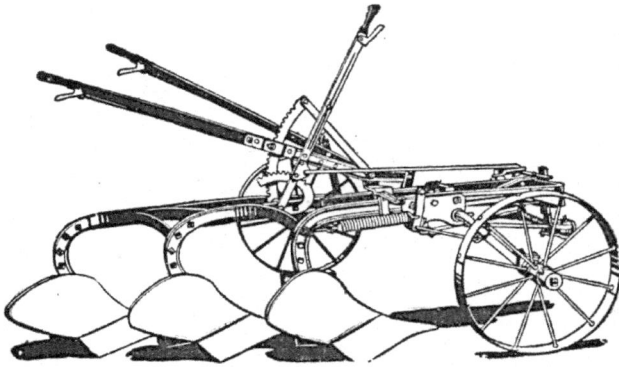

Fig. 345. This was classed as a three bottom walking plow by John Deere, however many western farmers used this beauty and fastened a platform and seat on.

THE PIONEER FORECART

Fig. 346. The popular Pioneer forecart. The tongue will fit right or left and the seat swivels.

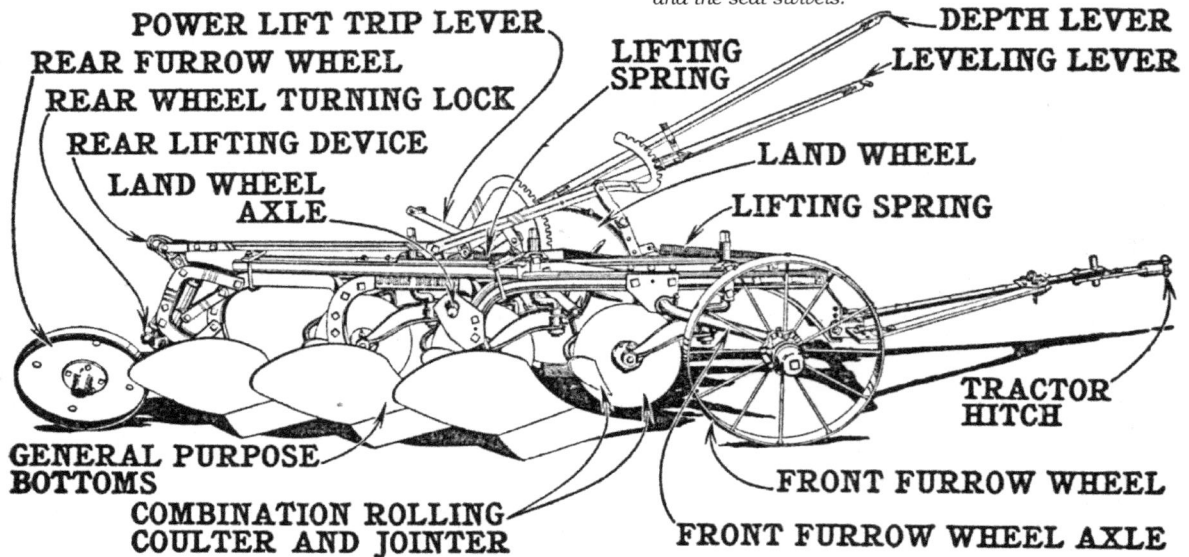

POWER LIFT TRIP LEVER
REAR FURROW WHEEL
REAR WHEEL TURNING LOCK
REAR LIFTING DEVICE
LAND WHEEL AXLE
LIFTING SPRING
DEPTH LEVER
LEVELING LEVER
LAND WHEEL
LIFTING SPRING
TRACTOR HITCH
GENERAL PURPOSE BOTTOMS
COMBINATION ROLLING COULTER AND JOINTER
FRONT FURROW WHEEL
FRONT FURROW WHEEL AXLE

Fig. 347. A John Deere three bottom pull-type tractor gang showing parts.

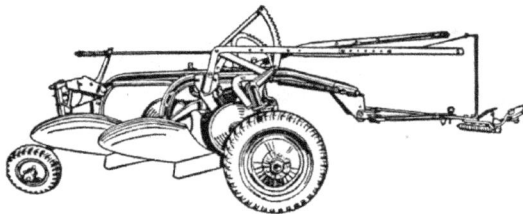

Fig. 348. Allis Chalmers two bottom tractor plow. This little plow would work well with a forecart as the levers and trip cord are far enough forward to reach the teamster.

Fig. 349. (Right) A Pioneer motorized forecart demonstrating the incredible sophistication of modern horsedrawn equipment. Forecarts are available from a variety of cottage industries and bigger companies. And these carts can have live hydraulics, full live PTO, three point hitch capability, adjustable load dispersion, and steerable wheels and axles.

Fig. 350. This highly unusual Secretary combination disk and subsoil plow with tongue articulated steering was sold by John Deere company.

Fig. 351. A slightly more modern subsoil Disk Secretary by John Deere. Notice the tongue is gone, the hitch simplified (over simplified?) and the subsoil shank is beefed up.

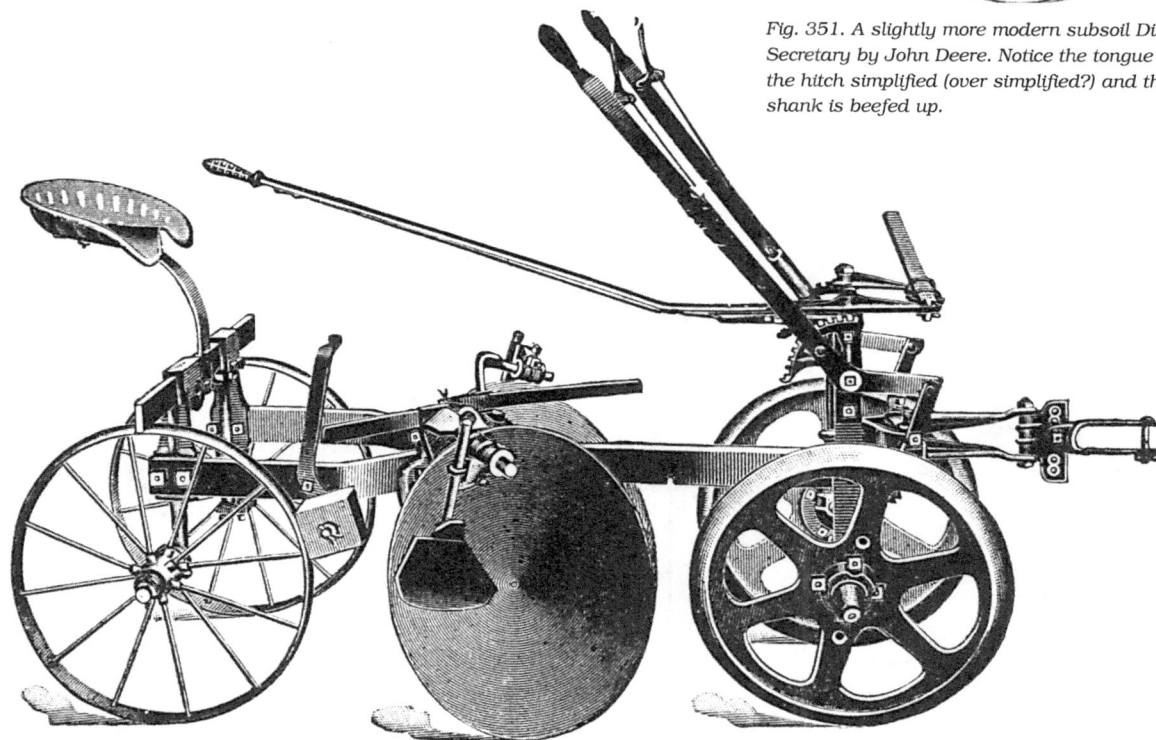

Fig. 352. John Deere Disk Bedder. Wish we could come up with more information and pictures of this beauty. It looks like a row croppers dream implement.

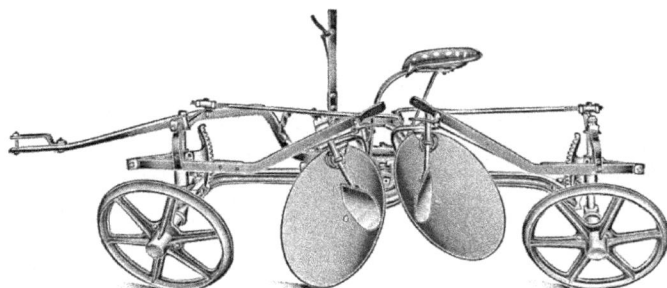

Fig. 353. Moline No. 2 Reversible Disk Plow. This design is different from most reversible Disk plows as it incorporates two bottoms which raise and lower alternately. The beam and seat pivot. The lever raises the one bottom while lowering the other.

Disc, Trail & Misc. Plows

## Front Furrow Wheel Lift

## High and Level Lift

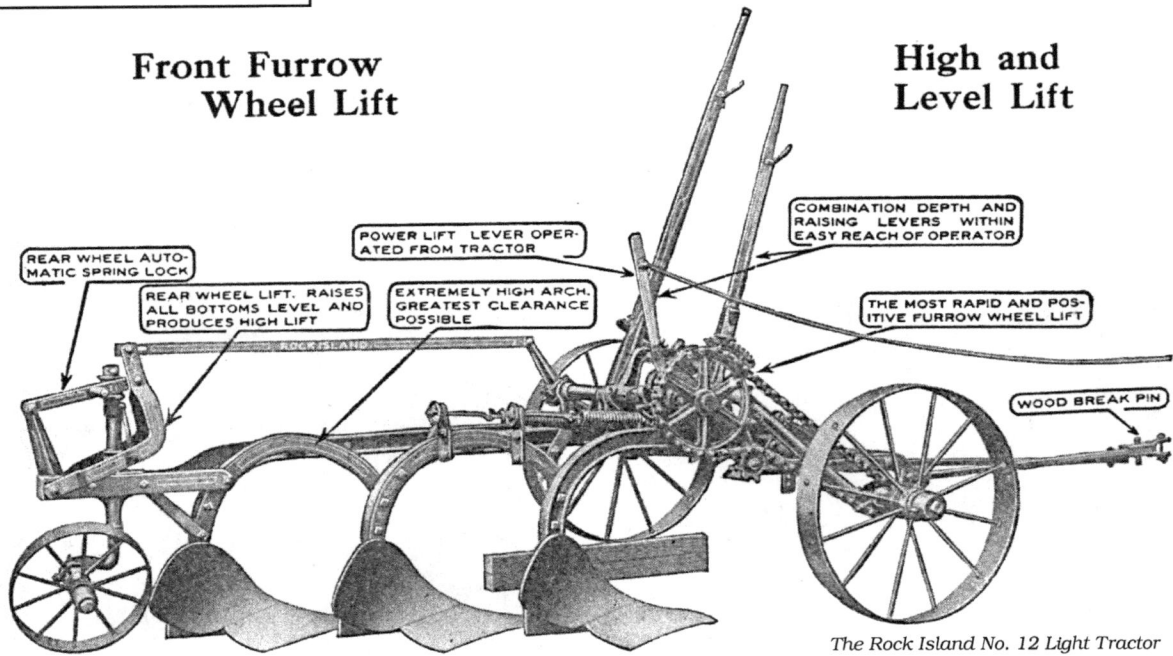

REAR WHEEL AUTO-MATIC SPRING LOCK

REAR WHEEL LIFT. RAISES ALL BOTTOMS LEVEL AND PRODUCES HIGH LIFT

POWER LIFT LEVER OPERATED FROM TRACTOR

EXTREMELY HIGH ARCH. GREATEST CLEARANCE POSSIBLE

COMBINATION DEPTH AND RAISING LEVERS WITHIN EASY REACH OF OPERATOR

THE MOST RAPID AND POSITIVE FURROW WHEEL LIFT

WOOD BREAK PIN

*The Rock Island No. 12 Light Tractor Plow featuring a chain lift.*

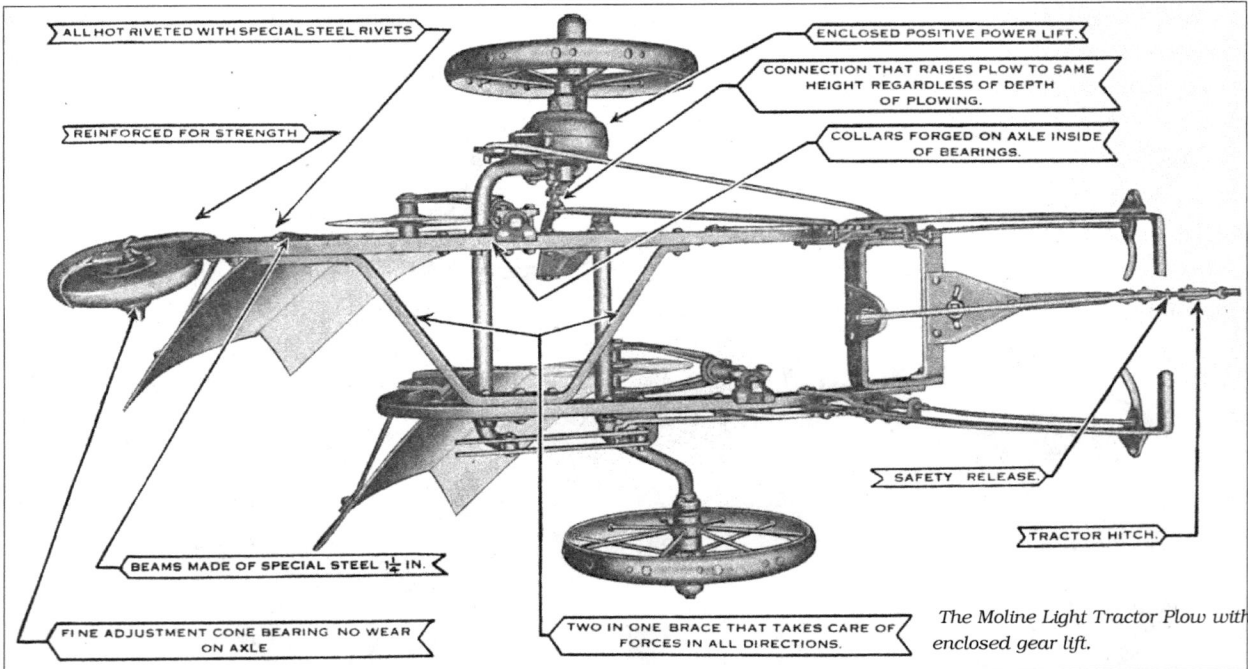

ALL HOT RIVETED WITH SPECIAL STEEL RIVETS

REINFORCED FOR STRENGTH

ENCLOSED POSITIVE POWER LIFT.

CONNECTION THAT RAISES PLOW TO SAME HEIGHT REGARDLESS OF DEPTH OF PLOWING.

COLLARS FORGED ON AXLE INSIDE OF BEARINGS.

SAFETY RELEASE.

TRACTOR HITCH.

BEAMS MADE OF SPECIAL STEEL 1¼ IN.

FINE ADJUSTMENT CONE BEARING NO WEAR ON AXLE

TWO IN ONE BRACE THAT TAKES CARE OF FORCES IN ALL DIRECTIONS.

*The Moline Light Tractor Plow with enclosed gear lift.*

*The McCormick Deering Telephone Wire Laying Plow No. 17 - T.*

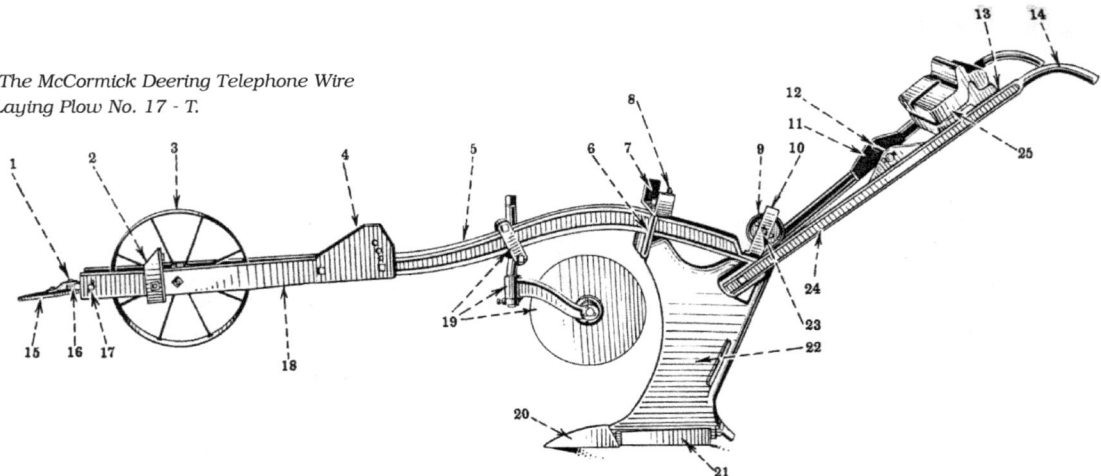

## Chapter Eight

# Laying Out Fields
# For Plowing

Common crown
L = land, A = opening furrow sole, B = Crown
(or back) furrow, K = unplowed ground.

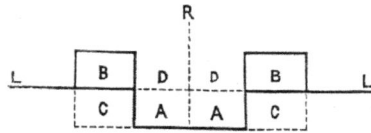

Split crown
B = furrows turned back right and left from A. C =
Second furrow (B & C) turned back together to fill R

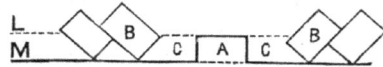

Last furrow
A = 1st furrow, BB = last furrows turned right and left. CC =
furrow soles, L = surface of land, M = bottom of furrow.

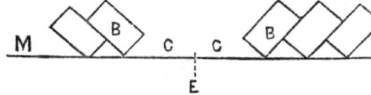

Open finish
BB = last furrows turned, CC = dead furrow, E = middle of
dead furrow, M = bottom of furrow.

Fig. 355. Showing the parts and systems of a plowed field

Understanding the equipment for plowing with horses, it is time to move on to the geometry of setting up a well-plowed field.

**Parts of a plowed field.** As shown in Fig. 355, the plowman needs to understand a handful of basic terms; *crown* (the ridge formed with the first facing furrows - in some circles this is referred to as the back furrow), *furrow* (that ribbon of turned earth), *furrow sole* (some call it the furrow as in 'walking in the furrow' but the more accurate term is furrow sole), *in-and-out* (literally that edge line where the plow goes in and comes out of the soil), *dead furrow* (the low trough that results when a field is plowed to an open finish), *back furrow* (the turn, up against the first furrow, the result of which is the crown), *headland* (the portion of the field often like a frame of unplowed ground left for turns and plowed out when the main portion is complete), *land* (refers to that integral unit of plowing which might be a given field in its entirety but which more

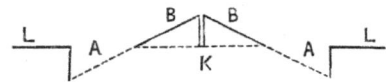

Fig. 354 Demonstrating the simplest approaches to layout. 'A' is with the use of a two-way walking or sulky plow where the plowman starts at one edge and works all the way across alternating bottoms and leaving no dead furrow. And 'B' demonstrates throwing furrows out to the edge until a certain width remains and then leaving off the short side until the dead furrow is completed

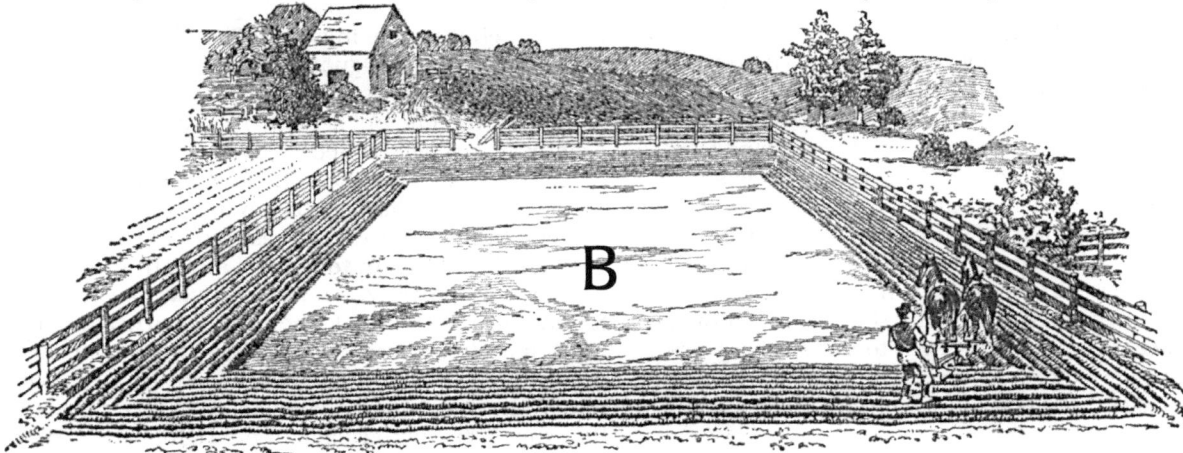

often refers to the portion of a plowing which stretches from dead furrow to dead furrow).

Lands are finished in various ways; open furrow (or dead furrow), closed furrow (furrow back), flat finish (as with two way plows eliminating dead furrows).

Much of this nomenclature and geometric abstraction will become clear with the following illustrations and explanations.

**Length of Hitches.** As was mentioned in Chapter Six page 92, when preparing to lay out a field for plowing it is important to know the length of the hitches which will be used to plow. The reason is that enough room must be allowed for turns. The reader may choose to actually step off the length or do a more accurate job of measuring but here are a few rough guidelines:

• team and walking plow or sulky plow (including two-way plows), 16 to 18 feet.

• tandem hitch of two spans with sulky or gang plow, up to 32 feet. (add 6 to 10 feet for tag plow and forecart.)

• tandem hitch of 3 spans and gang plow, 40 to 45 feet.

**Headland widths.** When factoring hitch length into what headland width is needed, it's a good idea to plan for one and a half times overall length.* And to make the headland width a combined factor of the share width so that the plowing out of the headlands comes even. Here are a few examples taking into account particular plow setups:

• 4 acre field, team and 12" walking plow = headland width 24 feet.

• 10 acre field, team and 14" two-way plow or sulky = headland width 23'4".

• 40 acre field, tandem and 14" gang = headland width 46'8".

The text which follows came, for the most part, from USDA bulletins, and has been edited to suit this book. What follows first is a set of simple approaches after which comes the heavy duty variables, some of which are not recommended but still of interest.

**Simple Approaches**

There are four general plans, or methods of plowing fields. These are: (1) to plow from one side of a field to the other; (2) to plow around the field; (3) to plow a field in lands; and (4) to start the plowing in the center of the field.

To plow from one side of the field to the other is the quickest method of all, but it has some disadvantages that make it used only occasionally. The reversible hillside walking plow and the two-way riding plow are used in this method of plowing. (See Chapter Five)

Fig. 354A illustrates the method of plowing from one side of a field to the other. The plow team is started at one corner and a furrow turned along one side of the field. When the end of the field has been reached, the team is turned as shown in the figure, the plow reversed, and the next furrow thrown against the previous one. This process of reversing the plow each time and throwing the new furrow next to the one that has

just been made is continued until the entire field is plowed. There is little loss of time in this method of plowing and no tramping of the plowed portion at the ends. Either a rectangular, triangular, or in fact, any shaped field may be plowed in this way, but when fields shaped other than a rectangle are plowed by this method, the last furrows will generally be short.

The next quickest way is to plow around a field, as illustrated in Fig. 354B, using the landside plow. The start is made at one corner and furrows are turned outward on one side of the field. As each corner of the field is reached, the team is turned and a furrow made along the adjoining side to the place of beginning, where a second furrow is turned inside the first. This process is continued until the center of the field has been reached. This plan of plowing is a good one in many respects; but it is often objected to on account of the tramping of the plowed ground at the corners when the team is turned. The land at the corners is, in this way, often made as hard as before it was plowed.

If a field has been plowed in the manner just described for several years, it is likely to be a little dish-shaped and to have a hollow in the center, on account of the furrows having been thrown from the center toward the outside of the field so many times. When such a condition occurs, water is likely to collect in the hollow.

A very general method is to plow a field in lands. In farming terms, a land is a space on a field a certain number of steps in width. Farmers, as a rule, measure by steps, or paces, rather than by using a tape. It is a good plan for each farmer to know the length of his pace, as he can then readily approximate the distance he paces. (A more accurate system would involve using a measuring device, i.e. wheel gauge or a triangle as in Fig. 362.)

To lay out a field to be plowed in lands, a farmer proceeds as follows: the width of the field is first paced off, no attention being paid to measuring the length. Suppose, for example, that a field is found to be 106 steps wide. It is divided into 5 lands of 20 steps each, with a width of 3 steps left on each side and also at each end. This border or headland, as it may be called, is to be plowed last. Fig. 356 is a diagram of the field and shows the different lands. The lines *ab, bc, cd, da* represent the outside edge. The lines *ef, fg, gh, he* are each 3 steps from the outside lines. The space between these sets of lines forms a 3 step border to the field. The lines *ef, ij, kl, mn, op* and *hg* divide the field into lands, each 20 steps in width. The different lands are, for convenience, designated 1, 2, 3, 4, and 5.

To begin the work after the field is laid out, stakes are set on the line *op*. If the field is long or hilly, three or four stakes will be necessary; if short, two, or even one at the end *p*, will be sufficient. The first furrow is turned along the line *op*, beginning at the end *o*. By keeping in line with the stakes that have been set, it is easy for the plowman to make a straight furrow. At each end of a furrow, if a plow is used that throws the furrow slice to the right, the team is turned to the right; if a plow is used that throws the furrow slice to the left, the team is turned to the left. In the explanation of the diagram it is considered that a plow is used that turns the furrow slice to the right. When the end *p*

---

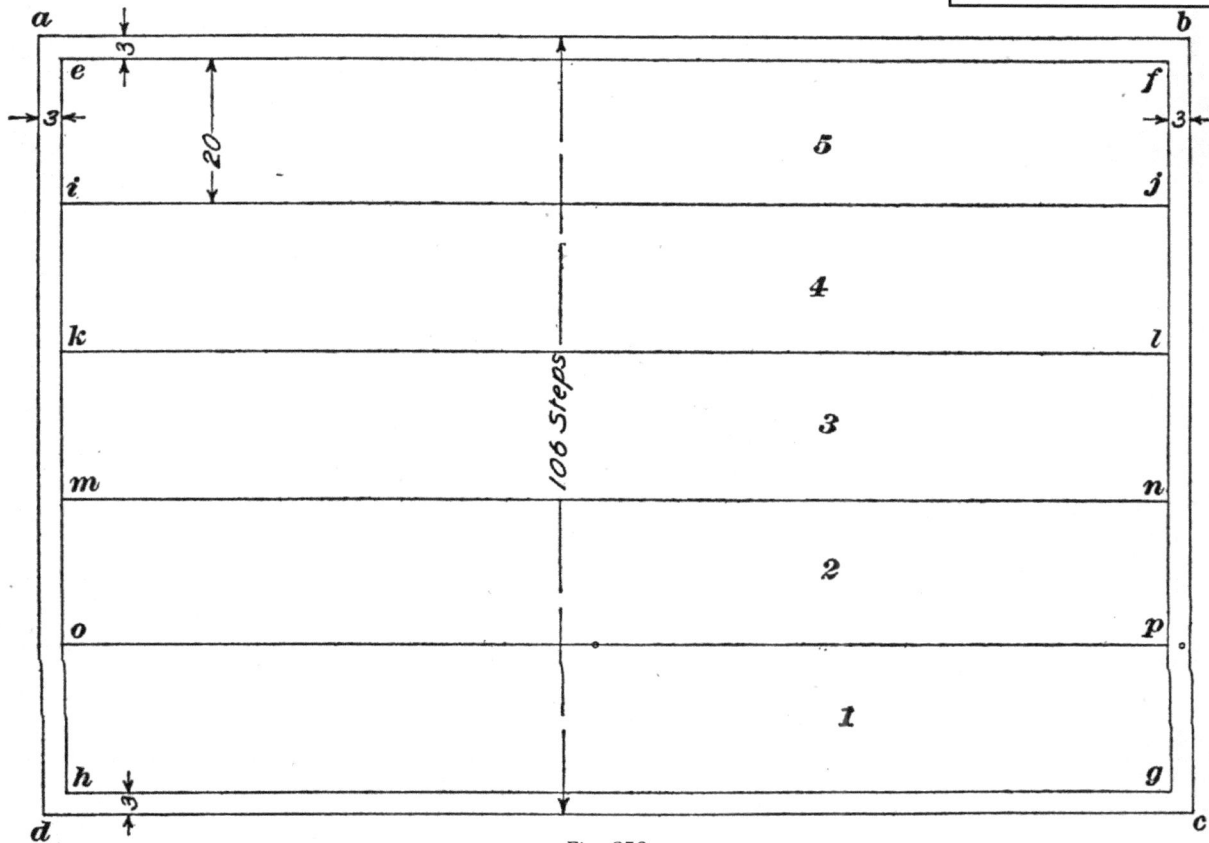

Fig. 356

has been reached, the plow is thrown out of the ground, the team turned to the right, and a second furrow slice thrown against the first one. No stakes are needed this time, as by making one horse of the team walk on the edge of the furrow slice, the new furrow will be straight. These first and second furrows should be shallow, no deeper than it is necessary to plow in order that the furrow slice can be made to turn over. When the end *o* has been reached, the plow is thrown out of the ground, the team is again turned to the right, and the third furrow is thrown against the first. This going back and forth, up one side of the plowed portion and down the other side, is continued until lands 1 and 2 have been plowed. The plow is then taken to the point *I* and lands 4 and 5 are plowed in the same manner.

The 3 step border and the middle land bounded by the lines *kl* and *mn* is all the ground left to be plowed. The middle land 3 is plowed by turning a furrow to the right by driving up *mn* and down *lk* and the plowing continued until the center is reached. In using this method there is sure to be one dead, or open furrow along the middle of the field. To avoid making it too wide, the last few furrows should be narrower and deeper than those on the rest of the field. Furrows are made narrower and deeper by regulating the plow at the clevis. A narrow dead furrow is more easily closed by harrowing than a wide one.

After the five lands are plowed and the plow readjusted to cut the regular furrow width, the plowman takes the team to one of the corners, and begins to plow around the outside of the lands, continuing

until the 3 step border has been plowed. The furrows of the 3 step border may be thrown either toward the fence or toward the lands, depending on which direction will keep the field level. After the 3 step border has been plowed, the field is finished.

When the same field is plowed next time the start should be made at the open furrow, and lands plowed on either side, as this will tend to level the field. It is a good plan if lands have been plowed toward the center one time, to plow them toward the outside the next time, to keep the field level.

Many farmers prefer this method of plowing to any other. Whether or not it can be used depends, however, to a great extent, on the shape of the field, an irregularly shaped field not being easily divided into lands. One advantage of this method is that the horses never step on the plowed ground; an objection is that at least one open, or dead furrow is necessarily left in the field.

It is often an advantage to start the plowing in the center of a field, throwing all the furrows toward this point. When fields have been plowed by throwing furrows away from the center, as just described, it is usually a good plan, if the shape of the field permits, to plow by starting at the center. Perhaps the reason that this plan is not used more is that it is necessary to carefully stake out the field before beginning the work and that all calculations for the staking must be very exact, otherwise too much unplowed land is left on the ends and sides of the field.

The staking out and calculating is not especially difficult; however, if care is exercised. Suppose, for

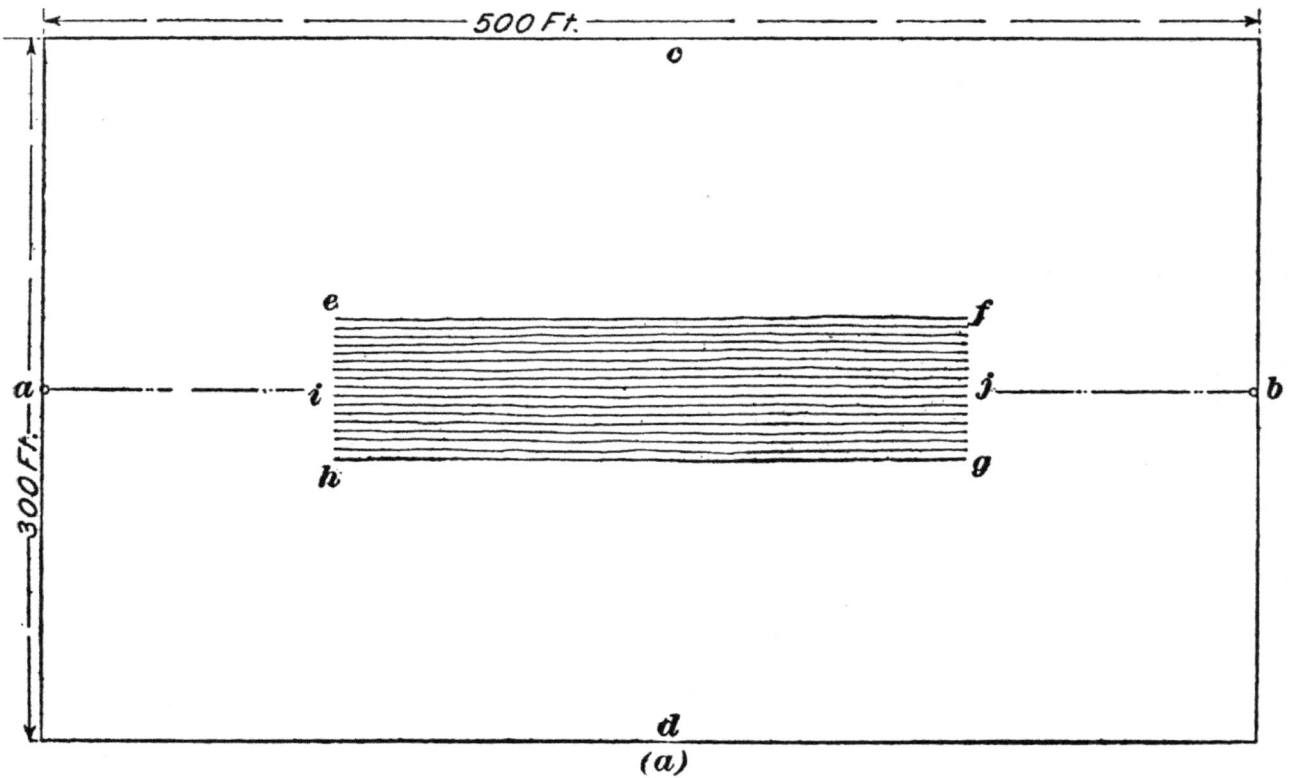

*Fig. 357*

(a)

(b)

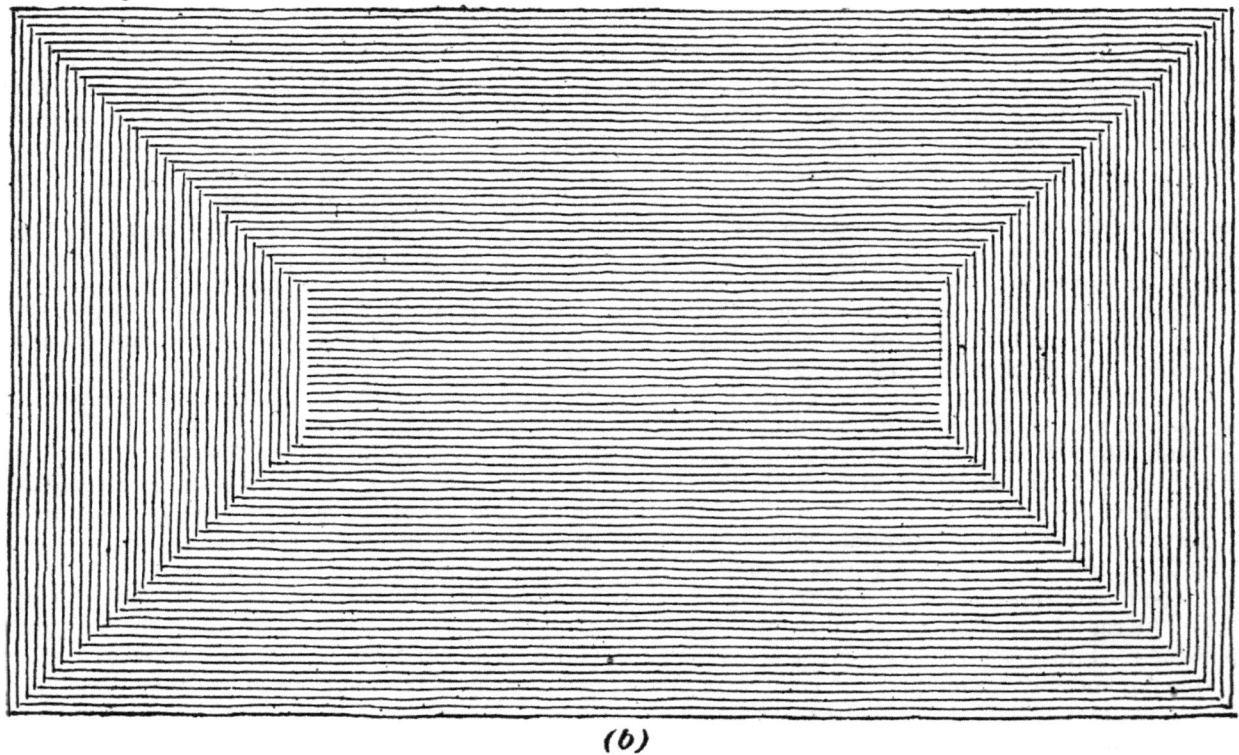

example, that a rectangular field as represented by the diagram of Fig. 357(a) is to be plowed by this method. The first thing to do is to find the length and breadth. On the farm, this measuring, as already explained, is generally done by stepping off the ground. In this example, the length, measured from a to b, is found to be 167 paces, and the width from c to d, 100 paces, or the field is approximately 500 feet by 300 feet. It is a good plan also, to pace both ends and place a stake at the middle of each end - that is, at the points *a* and *b*. The farmer is then sure that the line from *a* to *b* is through the center *f* of the field. It is generally custom-

Figs. 358. 359, 360. These great photos by Helen Eden, of her husband Jack and his six mules in Corvallis, MT, demonstrate dramatically the need for adequate headland room to turn the bigger hitches at furrow's end.

This also demonstrates that a good teamster needs nothing more than the overall length of plow and animals as he or she can get the hitch to 'sidepass or side step basically having the plow pivot in place.

This maneuver is not as easy as it looks because the teamster must gauge with absolute accuracy how the swinging position of the hitch and the steering of the plow will come out. The only way to get a second chance for correct starting position on that next furrow is to pull out and drive a circle and try again.

ary, with this method, to plow a land in the central portion of the field, as indicated by the rectangle ef gh of the diagram. This land is of such a size that both its ends and sides are equal distances from the he ends and sides of the field. The remaining portion of the field, outside of the land, can then be plowed by turning furrows consecutively around the plowed land. The width of the land, of course, will depend on the size of the field. In the example under consideration, it is 20 steps, which is 10 steps each side of the center line ab. Since one-half of the width of the field is 50 steps, there remains 40 steps to be plowed on each side of the land. The distance from a to the end of the land is 40 steps, and the distance from b to the end of the land is the same. These distances the farmer measures by stepping them off from a to I and from b to j. The plowing is then continued until the land is plowed. Care should be taken when plowing the land to start and to end the furrows along the land on a straight line, otherwise the first furrow plowed along the end of the land will be irregular.

After the land at the center of the field has been plowed, the plow should be taken to a point just outside the edge of the center land, and furrows turned all around the land, plowing along the ends as well as the sides and turning the furrow slices toward the center of the field. The plowing should be continued until the last furrow near the outer edge of the field has been turned. The field then is plowed as shown in Fig 357(b) and all the furrows are turned toward the center.

The four methods described are applicable under different conditions. In deciding which one to use on a given field, a farmer should be guided by four considerations: (1) the way the field has been previously plowed; (2) the nature of the soil; (3) the natural inclination of the land; and (4) the kind of a crop to be grown. The kind of plow to use is also taken into consideration. The two-way or reversible plow is the only kind, as previously explained, that can be used to plow a field from one side or end to the other. Both landside and two-way plows are; however, applicable to the other three methods.

Head land

Dead furrow

Back furrow

Dead furrow

Head land

Head land

Head land

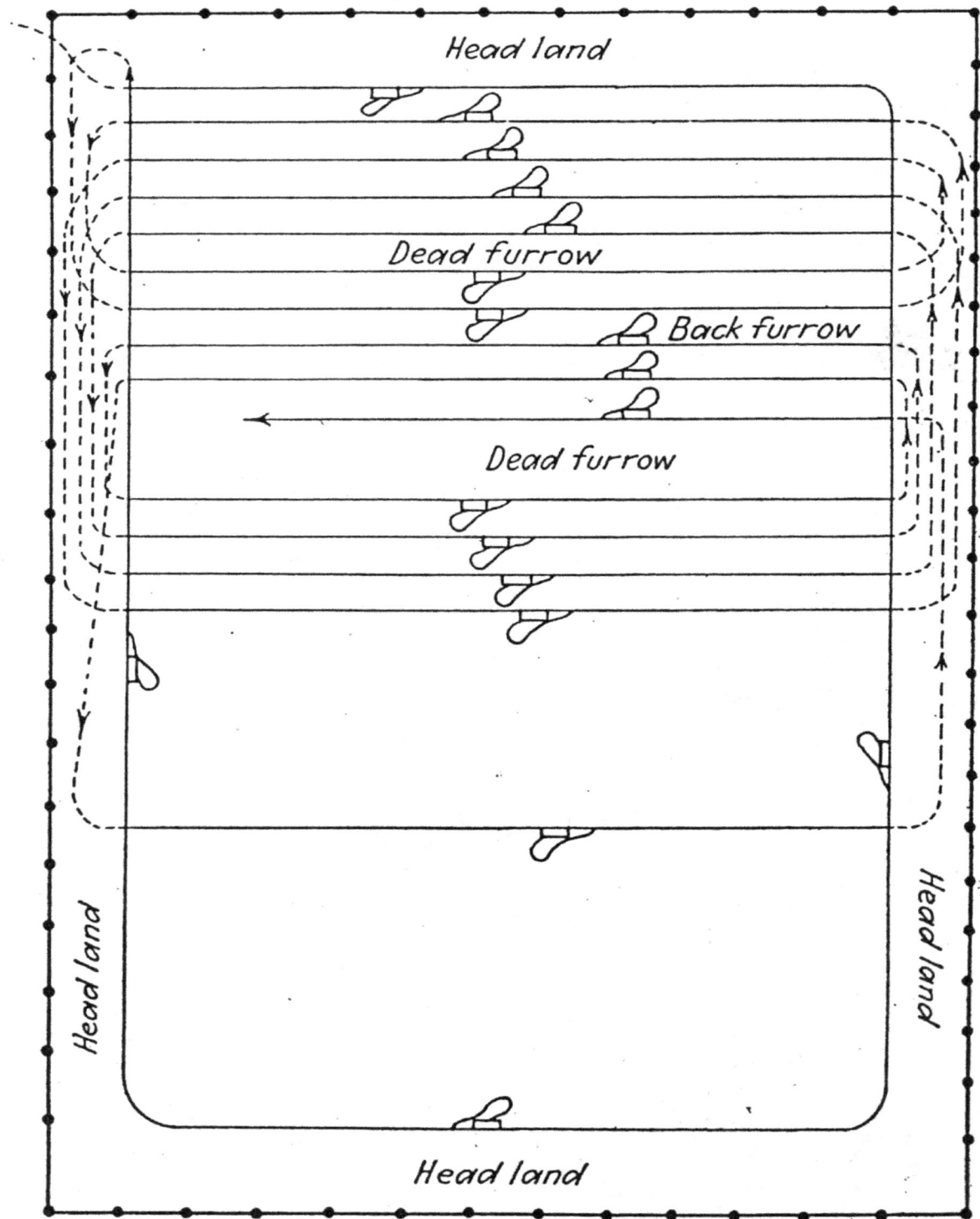

*Fig. 361. A basic diagram of the most common method for laying out a field for plowing.*

**Gang Plows and plowing layout.** If precision plowing is the goal this type of plow is perhaps best adapted only to the method of plowing around the field, the furrows being always thrown toward the edge of the field. The reason is that a gang plow arrangement has the bottoms in a slanting line. When the forward motion stops, the one plow that has plowed to the end is the first one of the gang. The more plows there are in the gang, the greater the strip of unplowed land at the end. These unplowed ends, or the *ins-and-outs*, form a zig zag which is best plowed by using a walking plow or sulky plow.

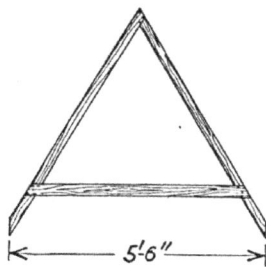

Fig. 362. An A frame for measuring fields in place of stepping off.

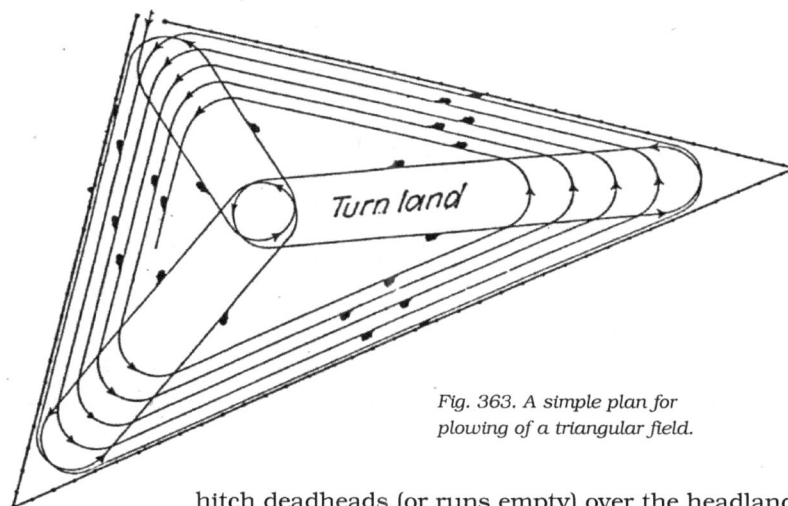

Fig. 363. A simple plan for plowing of a triangular field.

The following information, as previously noted, is an edited version of a USDA bulletin and expands on the geometric possibilities of field layout. For the purist, it might make for interesting thought, but it is not necessary or vital for the serious horsefarmer who will be well served by the information immediately preceding. LRM

**Heavy Duty Variables**

Farmers, whether they plow with tractors or horses, want to do a good job of plowing over the entire field. And they need to reduce to a minimum the time spent in turning and in traveling with plow bottoms out of the ground. Many circumstances must be considered in deciding just what method is best for a particular field with a particular outfit. No one method can be considered best for every size and shape of field.

Advanced methods of laying out fields for plowing might fall into two classes: 1) The method in which fields are laid out so that when the field is plowed the plow bottoms are taken out of the ground at the ends of the field and thereby creating *headlands*; 2) those methods in which fields are laid out so they can be plowed without lifting the bottoms out of the ground in crossing the ends and possibily doing away with headlands.

The advantages of the method of the first sort are that short turns are eliminated, except in some cases at the beginning and ending of the lands, and that it is generally possible to do plowing of a little higher quality at the corners or turns. Making short turns can be awkward and the plowman often has difficulty in getting the bigger hitches in the correct position for starting the furrows after such turns have been made.

The advantages of methods of the second sort are that little or no time is lost in traveling with the bottoms out of the ground and that ordinarily the number of dead furrows and back furrows will be considerably less. The longer the time spent in turning or running with the bottoms lifted, the smaller the acreage that can be plowed in a day. Though it may pay to make some additional effort to avoid short turns when using a large hitch, the loss in time due to making long idle runs across the ends of the field is just as serious with large hitches as it is with smaller, more easily handled hitches.

Wide lands increase the number of times the

hitch deadheads (or runs empty) over the headland, thus increasing the tendency to pack the soil. These factors should be taken into consideration in deciding on the most desirable size and number of lands. From the standpoint of time lost in running empty (or with bottom[s] up), the size of the hitch should be considered only with reference to the relative difficulty in making short turns.

In deciding on the method to use, the ease of handling the hitch and plow is not always the most important consideration. In areas of heavy rainfall it may be best to make narrow lands with frequent dead furrows and back furrows as an aid to drainage; in dry areas the reverse may be true. In other cases the contour or shape of the field may be such as to determine almost entirely the method that must be followed.

If a field is rectangular, relatively level, and not farmed on the contour, the choice of the method of laying it out for plowing will usually depend on how short a turn can be made with the hitch and plow and how objectionable the additional back furrows and dead furrows are.

**Method in Which Bottoms are lifted at the ends**. If it is decided to lift the bottoms out of the ground in going across the ends of the field, it must then be decided into how many lands the field should be divided, how wide to leave the headlands, and where to set guide stakes or markers.

**Width and number of lands**. The wider the lands are made the fewer will be the dead furrows and back furrows (crown), but the greater will be the time consumed in idle running across the ends. The effect that lands of different width has on idle travel across the ends can be illustrated by considering three alternatives for dividing a field 40 rods* (660 feet) wide. In the following examples idle running time has been calculated for divisions of 5, 11, and 3 lands. Calculations are for a 3-bottom, 14-inch plow.

When the field is laid off into 5 lands of 132 feet each, it will take about 38 trips lengthwise of the field to plow out each land as well as 19 turns on each headland. The average length of travel across each headland is half the width of the land, or 66 feet. The idle travel in turning for each land is therefore 2,508 feet, or almost half a mile. The total idle travel in plowing the entire field will be almost 2-1/2 miles.

If the field is laid out in 11 lands, each 60 feet wide, the unproductive travel at the ends would be reduced to approximately one mile. This reduction would be largely offset by the greater number of figure-8 turns necessary in starting the extra lands and also by the probability of the plow running at less than its full width of cut for a considerable distance in finishing the extra number of lands.

If the field were laid out in only 3 lands the travel across the ends would be increased to about 4 miles, but there would be only 2 dead furrows to finish out

with the possibility of the plow not cutting its full width.

Decreasing the width of lands increases the number of dead furrows and back furrows or crowns. The time necessary to make the difficult turns at the ends of each dead furrow or back furrow reduces the advantages of narrow lands to a certain degree.

The length of the field is also important in deciding the width of the land. The turning time in proportion to the total plowing time is greater on short fields than on long fields. For this reason wider lands are usually selected for long fields than for short fields.

The dimensions of the field will determine whether the saving in time in making narrow lands is sufficient to offset the disadvantages of the extra deadfurrows and back furrows and any difficulties of

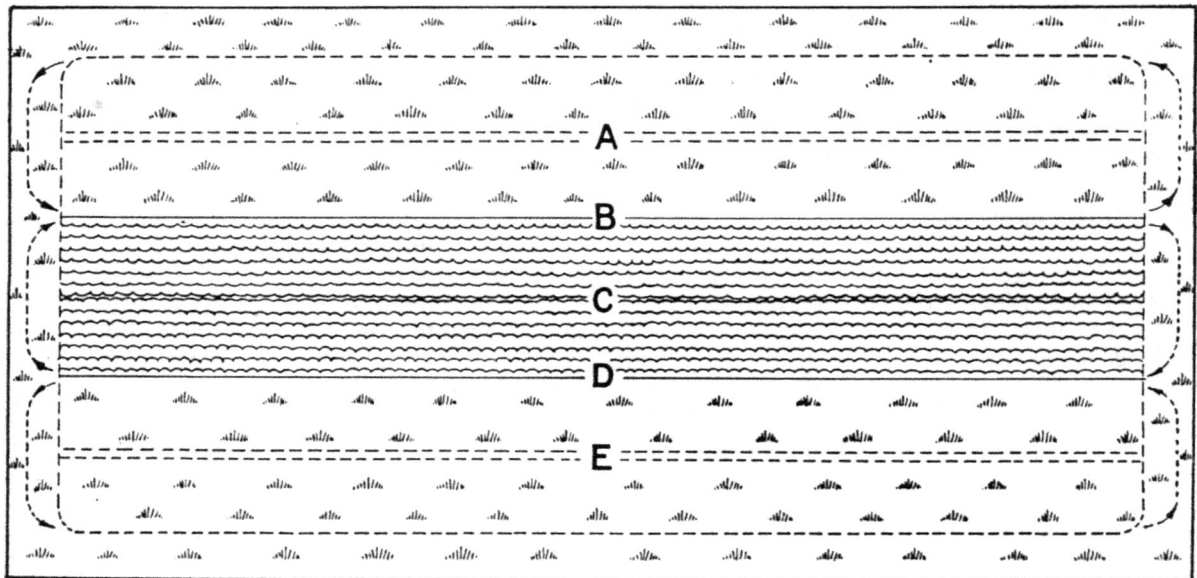

Fig. 364 Method 1, plan A, at the end of the first step. The locations of the two dead furrows are shown by the double dotted lines, A and E. The direction of travel across the ends is indicated by dotted lines and arrows.

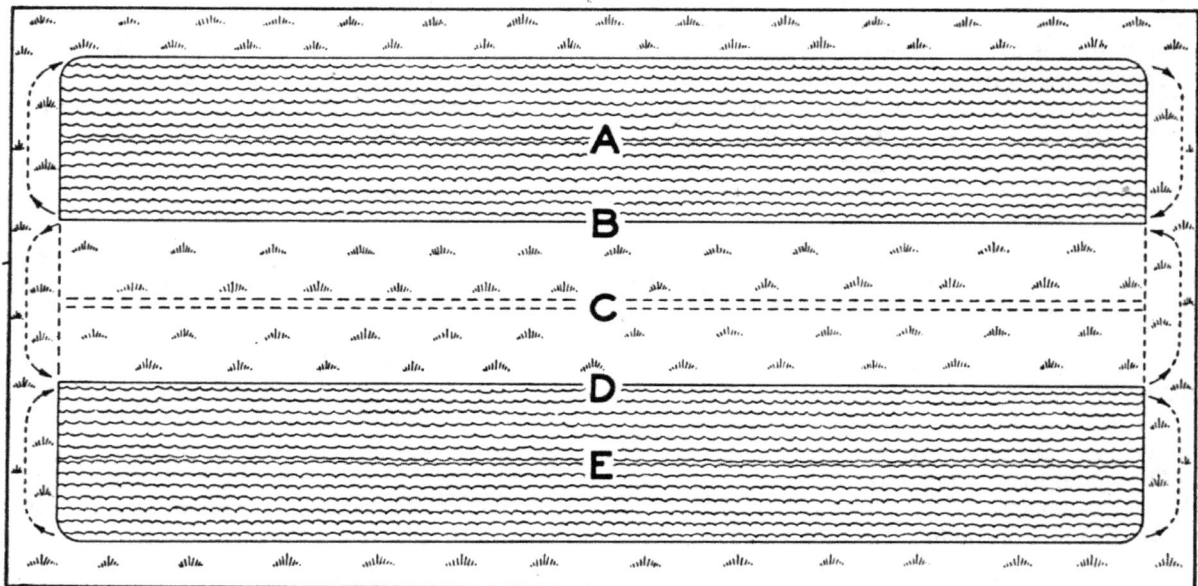

Fig. 365. Method 1, plan B, at the end of the second step. Note that the two back furrows at A and E are where the two dead furrows were in plan A, and the dead furrow, at C is where the back furrow was at the previous plowing.

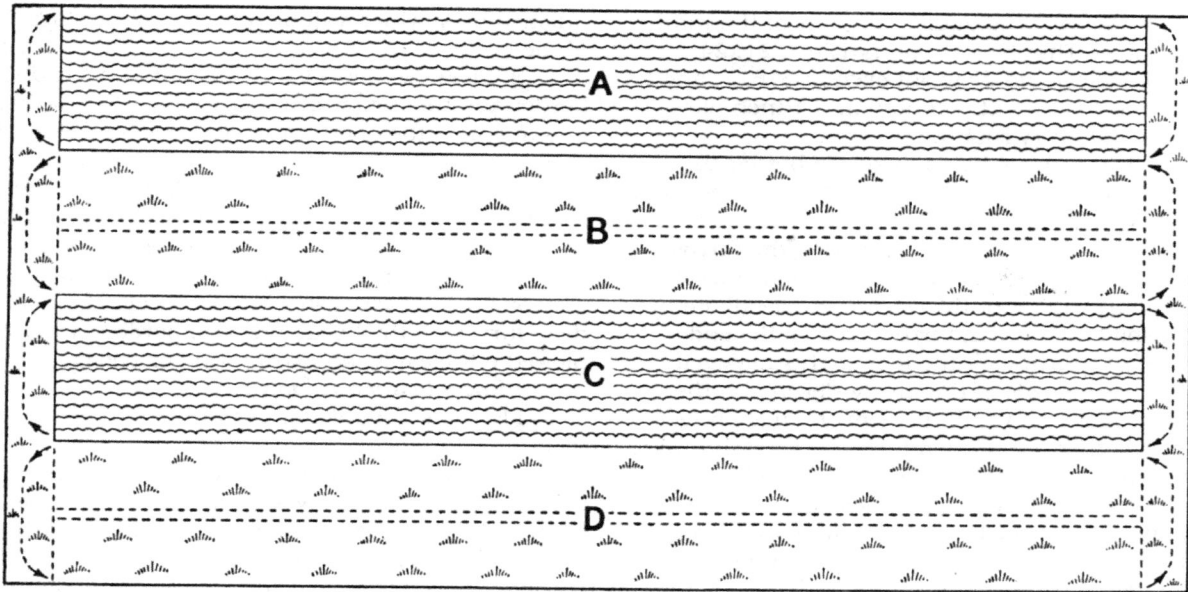

Fig. 366. Method 1, plan C, at the end of the second step. The order of plowing might also be C, D, A and B. Each headland is plowed seperately after the body of the field is plowed.

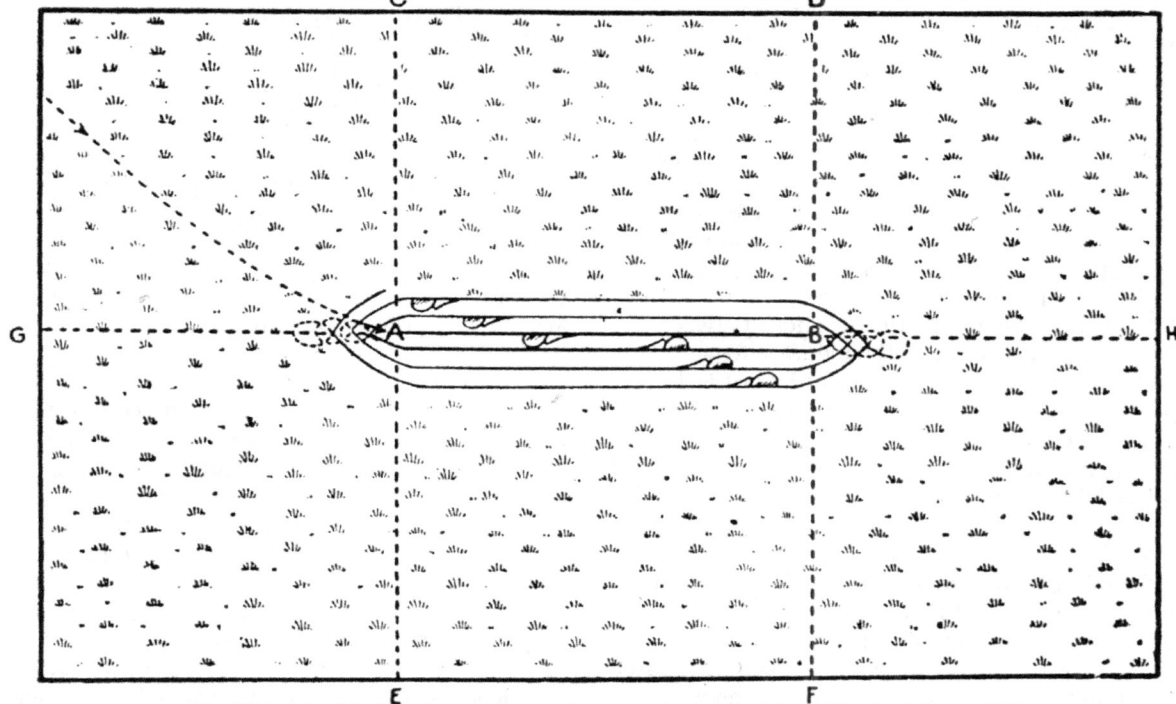

Fig. 367. Method 2, first stage. To begin plowing in the center of the field, a backfurrow (AB) is laid out in the center of the field. The continuation of this plan is shown in Fig. 368.

making short turns. The most popular width under average conditions seems to be about 100 feet for a 2 or 3 bottom plow. If the field has no irregularities, however, its entire width should be measured and divided into lands of approximately equal width.

**Headlands.** The width of the headland will depend largely on the total length of the hitch and plow and the turning radius of the hitch. Some farmers having outfits which are short in length and handle easily do not leave more than 10 to 15 feet. However, any extra ground in the headland can be plowed just as quickly

as if it were plowed with the body of the field, and plenty of room should always be left to allow easy turning and to get the outfit headed in properly at the beginning of the furrows. The wider the headland, the less the tendency to go over the same ground repeatedly in turning at the ends. Soil packing therefore tends to be less intense in wide headlands.

Headlands 15 to 35 feet wide may be suitable when one of the smaller outfits is to be used. With most outfits a headland 1-1/2 times the total length of the hitch and plow will give plenty of room for turning. It is a good idea, particularly with the larger outfits, to make the border or headlands a multiple of the width cut by the plow. That is, headlands 17-1/2, 21, 24-1/2, and 28 feet wide would plow out even in 5, 6, 7, and 8 rounds, respectively, with a 3 bottom, 14-inch plow taking a full cut every round.

If the field is fenced on all sides, a border the same width as the headlands may be left on each side, and it will be possible to finish the field neatly by plowing around the entire field, throwing the furrows either in or out as is required to keep the field level.

Many farmers mark the edge of the headland by plowing a shallow "scratch furrow" across the end of the field before starting on the lands. This makes it easier to keep the ends of the furrows even and the headlands uniform in width. It seems to make little difference whether the scratch furrow is thrown toward or away from the edge of the field, but throwing the furrow away from the edge of the field seems to be the more common practice. This furrow across the end of the field sometimes helps the bottoms to enter the ground more quickly at the beginning of each round.

**Setting stakes and markers**. To finish up a field without having to plow irregular or wedge shaped strips it is essential that the lands be started straight and parallel, and that the headlands be kept uniform in width. If a field is once laid out accurately and marked permanently, it will not be necessary to measure off lands at each plowing. In fields that are fenced, the locations of dead furrows and back furrows may be readily marked by setting stakes along the fence. After this has been done, old dead furrows can readily be found when plowing the new back furrows even if the field is covered with tall weeds.

Most farmers "step off" the distances between lands. This method is sufficiently accurate for a layout in many cases, but when a permanent layout is to be made or a large number of narrow lands are to be laid out it is advisable to use a tape or some other accurate method of measurement. (See Fig. 362)

### Method 1

An outline is given below to show how Method 1 applies to a 20-acre field 40 rods wide and 80 rods long. For plowing by this method the field would have headlands 2 rods wide and a border on each side 2 rods wide, leaving the body of the field 36 rods wide and 76 rods long. Two plans should be used alternately on any field plowed by this method. The plans will be called Plan A and Plan B.

### Plan A

1. Lay out a back furrow through the center of the field, as indicated at C (Fig. 364), and plow a strip 12 rods wide about the back furrow, lifting the bottoms 2 rods from each end of the field. When plowing a back furrow, it is desirable to turn the first furrow back over its original position on the return trip so that all of the soil is worked. The crown will be less pronounced if the first furrow is made with the plow operating at less than the depth to which the rest of the field is plowed.

2. Plow along one side of the land already plowed, at B (Fig. 364). Turn to the left and on the return trip across the field plow with the first furrow 2 rods from the edge of the field to allow for the border. Continue

plowing, turning to the left until the dead furrow is finished at A, which will be 6 rods from the inside edge of the border.

3. Plow the last third of the field between the border and the furrow at D, turning left at the ends and finishing the body of the field with the dead furrow at E.

4. Plow the borders and headlands, traveling around the field to the left and turning the soil toward the outside of the field. This will leave an open furrow 2 rods from each edge of the field. This open furrow is only one furrow wide, so it is not so objectionable as a dead furrow. If the headlands and borders are not plowed out cleanly, an extra round or two can be made.

### Plan B (Alternate with Plan A)

Plan B is suitable for use at the next plowing after Plan A.

1. There will be old dead furrows at A and E (Fig. 365). Plow a back furrow at A and, turning to the right about this back furrow, continue plowing until the plowed strip is 2 rods from the side of the field. The border is to be the same width as in plan A. The land should then be 12 rods wide.

2. Lay out a back furrow in the old dead furrow at E (Fig. 365) and plow, turning to the right, until the plowed strip is 2 rods from the side of the field.

3. One-third of the body of the field, a strip 12 rods wide, remains to be plowed between the two lands plowed in the first and second steps. Plow this out, turning to the left and finishing with a deadfurrow at C, through the center of the field.

4. Plow the borders and headlands by traveling around the field to the right and throwing the soil toward the center of the field. Thus the lands around back furrows A and E are finished to the field boundary, leaving only an open furrow around the edge of the field.

### Plan C

If one end of the field is unfenced and the outfit can be pulled out into a road or adjacent field for turning, it may be preferable to plow up to the fence on each side of the field and plow to the end of the field on the unfenced end. It may also be desirable to have a headland on each end of a field, but to plow to the fence on each side (Fig. 366). This plan of plowing without leaving borders is called Plan C. Back furrows and dead furrows in the lands are alternated with each plowing, as explained for Plans A and B. Back furrows and dead furrows are also alternated in the headlands with each plowing.

### Variations of Method 1

The number and width of lands may be changed as required to adapt Method 1 to fields of different widths. It should be noted that in Method 1 half the land between any two dead furrows in the body of the field is first plowed by turning to the right about the back furrow. The other half is then plowed by turning to the left until the dead furrow is finished.

Plowing fields by Method 1 reduces to the minimum the time spent in unproductive travel across the ends of the fields. With this method, where a complete border is to be plowed, there should always be a plan similar to Plan A, with an odd number of back furrows, and an alternate plan similar to Plan B, with an even

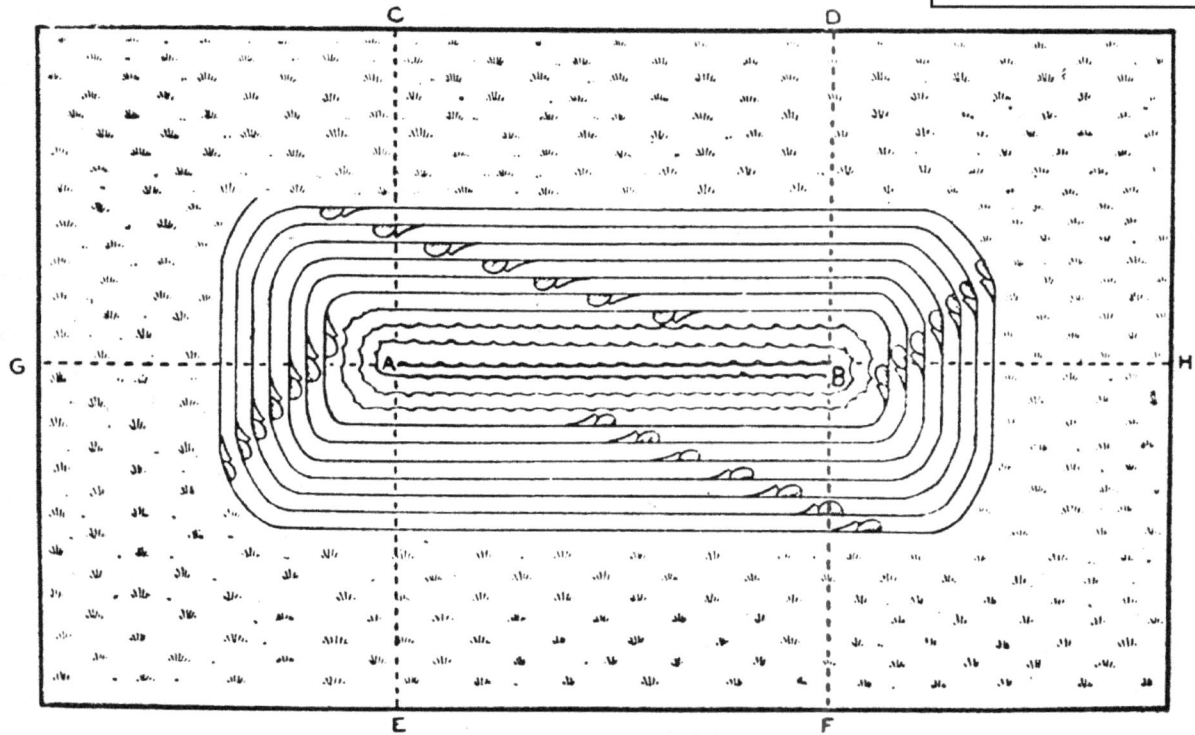

Fig. 368. Method 2, second stage. Plowing is continued around
the central back furrow crown until the field is finished.

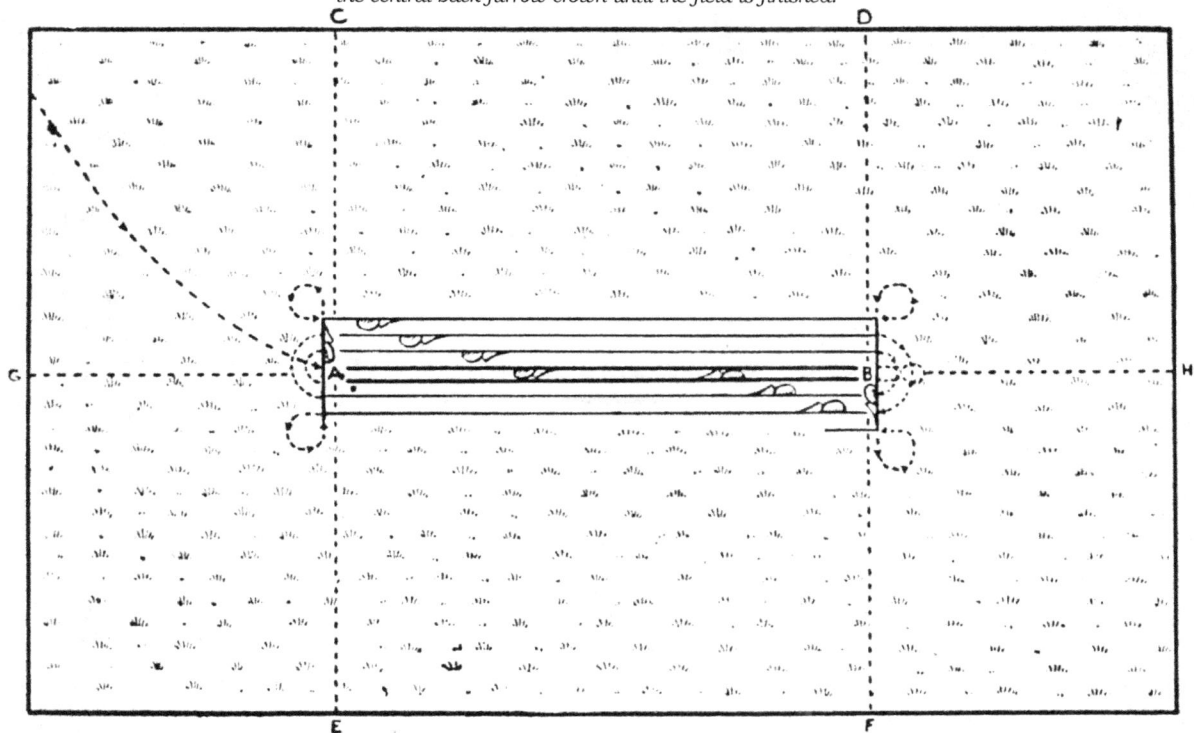

Fig. 369. Method 3, first stage. The field is plowed in one land by starting a back furrow in
the center of the field by the method illustrated in Fig. 367. The corners are kept square by
making short turns to the left and swinging around so as to plow across the ends.

number of back furrows.

**Methods in which Bottoms are left in the ground at the ends.** The objections to the method already described are that it necessitates considerable travel with the bottoms idle and that there are many dead furrows and back furrows if an attempt is made to reduce the mileage of this idle travel. The use of Method 1 usually results in a somewhat better job of plowing than use of a method involving an attempt to keep the bottoms in the ground all the time the hitch is traveling. However, many farmers think that the possible reduction in quality of the work is not sufficient to offset the time

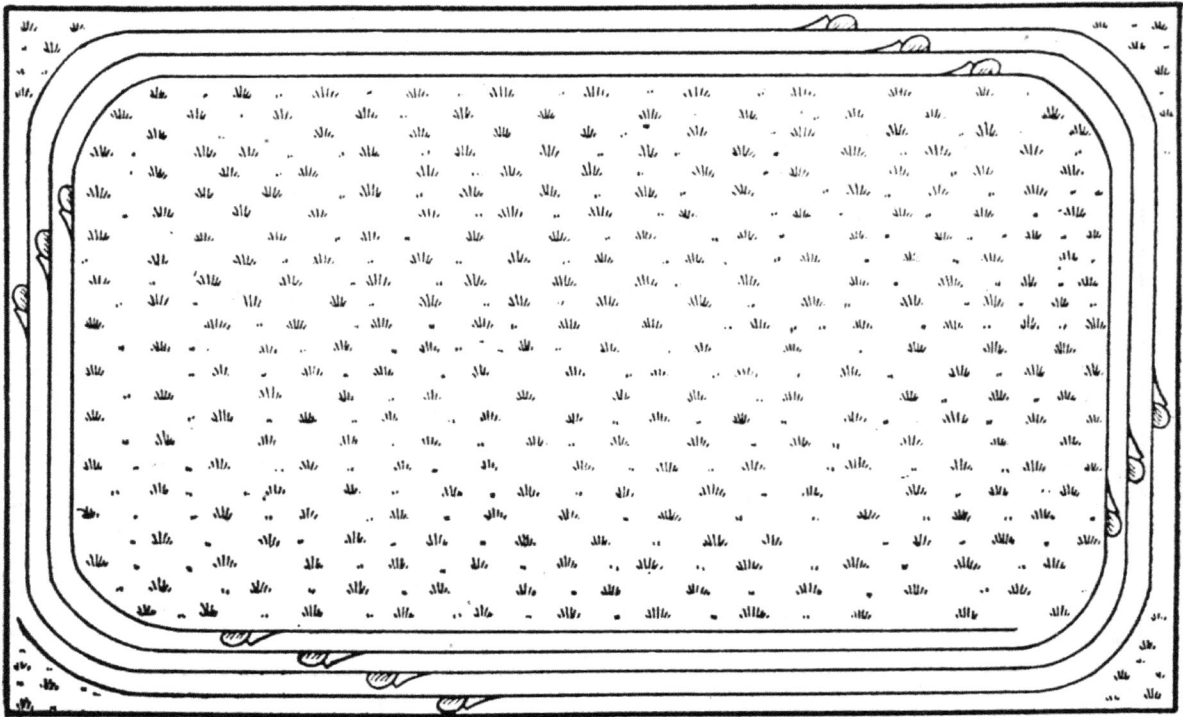

Fig. 370. Method 4, first stage. Plowing starts at the outside of the field with this method. The furrows are turned toward the fence, and the corners are rounded off enough to permit turning without lifting the bottoms.

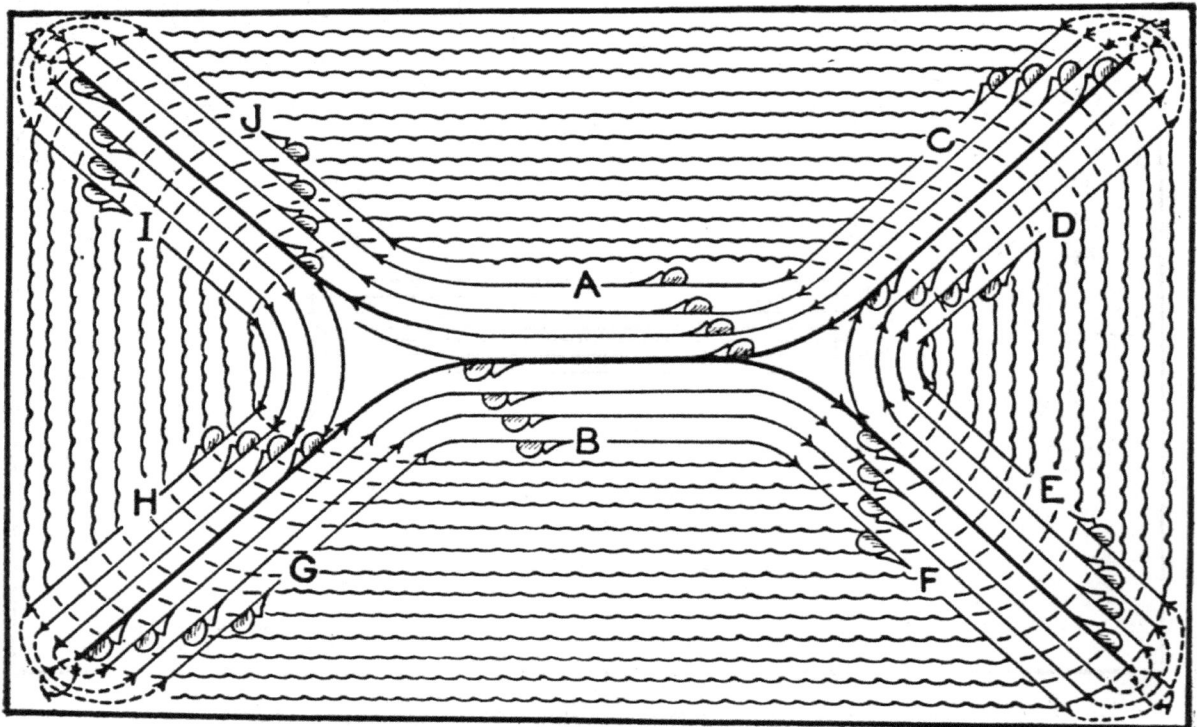

Fig. 371. Method 4, final stage. The diagonal strips that are left imperfectly plowed at the turning points may be finished by following this plan. A field plowed by this method has a dead furrow along each diagonal.

saved by eliminating idle travel.

## Method 2

*(Note: not particularly recommended. LRM)* By Method 2 a rectangular field is plowed around a single back furrow in the middle of the field (Fig. 367). The bottoms are lifted only in making the comparatively few short turns on the first few trips across the ends of the plowed land. The corners are rounded off by pulling the outfit over to the right as the end is approached. The turn at the ends on these first few rounds is made by making a complete circle to the left.

After the plowed land becomes wide enough for

the outfit to turn around the ends, the bottoms are not lifted from the ground until the field is finished. For some large outfits the land may have to be 75 feet or more wide before this can be done. A small outfit with a short turning radius may be able to plow the ends on a strip half as wide.

To determine the position of the back furrow at the center of the field, make the distance from A to C equal the distance from A to E. The distance from A to G should be enough shorter than the distance from A to C so that when the land is rounded off at the ends and plowing around the ends is begun the furrows on the ends and sides will be an equal distance from the edges of the field. The distance from B to H should be the same as the distance from A to G.

Some care will be necessary in navigating the turns after the land becomes wide enough to permit leaving the bottoms in the ground continuously if the turns are to be kept abrupt. The shorter the turns are kept the smaller will be the triangular pieces left in the corners of the field at the finish.

If the field is square, or nearly so, it can be plowed in two or more lands, each one laid out according to this method. Unplowed pieces, each approximately twice as large as the unplowed pieces at the corners, will be left at the ends of the field between the lands.

### Method 3

To plow by Method 3 a rectangular field is laid out, just as in Method 2, and the entire field is plowed in one land about a single back furrow. The back furrow is laid out along the line from A to B.

The first few rounds are made by plowing furrows the full length of the back furrow, and lifting the bottoms at the ends. When the plowed strip is wide enough to plow across the ends, the bottoms are lifted at the ends and a turn is made to the left until the outfit is headed across the end of the field (Fig 369). The corners are kept square by turning in this way until the furrows get so near the fence that not enough room is left to make such a turn. Then the corners must be rounded and the outfit turned to the right.

This method of plowing keeps the corners square, except for the last few rounds. The unplowed pieces in the corners are about the same size as those left in a field plowed by Method 1 (Plan A or B).

Except for these last few rounds, the net result of plowing a field by this method is the same as is ordinarily attained by using Method 2. The greatest objection is probably the time and travel necessary to make the turn to the left at each corner. This travel at each corner will amount to just about a complete circle. Thus the loss of time in many cases would be too great for this method to be advisable.

### Method 4

In Method 4 the plowing starts at the outside of the field (Fig. 370). The furrows are turned toward the fence, and left turns are made at the corners without lifting the bottoms.

A rectangular field like that shown is plowed in a single land with one dead furrow. The corners will have to be rounded to a certain extent on the first trip around the field and kept this way throughout the plowing. This permits the outfit to make the turns

without encroaching too far on the plowed ground or getting the furrows irregular and crooked near the corners. The plow will be pulled away from the last open furrow to a certain extent in making the turns, and the diagonal strips running from the ends of the dead furrow to the corners of the field will usually have to be replowed (Fig 371).

It is not necessary to measure any distances when this method of plowing is followed, and omitting the measuring will make this method quicker than any of the methods heretofore described. The bottoms are left in the ground from the time the field is entered until the dead furrow at the center is reached. A field with slightly irregular or crooked boundaries can be plowed very satisfactorily by following this method.

The body of the field can be plowed to a dead furrow in the center, and the diagonal strips running in from the corners replowed one at a time. When this method is used, it is necessary to turn on plowed ground when the unplowed strip becomes narrow. It is also necessary to turn on plowed ground at the center of the field when the diagonals are plowed.

Usually, it will be preferred to plow out the diagonals at the same time the dead furrow is finished (Fig 371 & 372). The method illustrated in Figure 371 leaves a dead furrow along each diagonal, and the method illustrated in Fig 372 leaves a back furrow down the middle of each diagonal and an open furrow on each side.

The following plan is used if it is decided to leave a dead furrow at the center of each diagonal: When the unplowed strip in the center of the field (A to B, Fig 371) becomes the same width as the strips that are to be replowed along the diagonals (C to D, E to F, etc., Fig 371), a right turn is made from the furrow next to A. The plowing then continues along the line indicated through J, I, H, G, B, etc., in Fig. 371 until the diagonals and center are finished. The outfit travels over very little plowed ground if this system is used. If the distances are correctly judged, the whole field is finished at the same time, except the parts left for making short turns at the corners. The only places where the bottoms are lifted are on the few short turns at the corners when the diagonals are plowed.

If it is desired to have backfurrows down the middle of the diagonal strips, the method illustrated in Fig. 372 will be used. This method is similar to the one shown in Fig. 371, except that the backfurrows are made on the first trips along the diagonals. A right turn is made at the corners of the field to get in position for the return trip along the diagonal. The bottoms are taken out of the ground in going between two diagonals at the same end of the field, and the outfit will have to travel over plowed ground at these points.

### Method 5

Method 5 is similar to Method 4 except that the bottoms are lifted each time at the corners in plowing the body of the field and the diagonals are left entirely unplowed until the finish of the field. Care must be taken to get the width of all the diagonals - C to D, E to F, G to H, and I to J in Fig 373 - the same if either of the methods shown in Figures 371 and 372 is to be used in finishing the field. The width should be ample

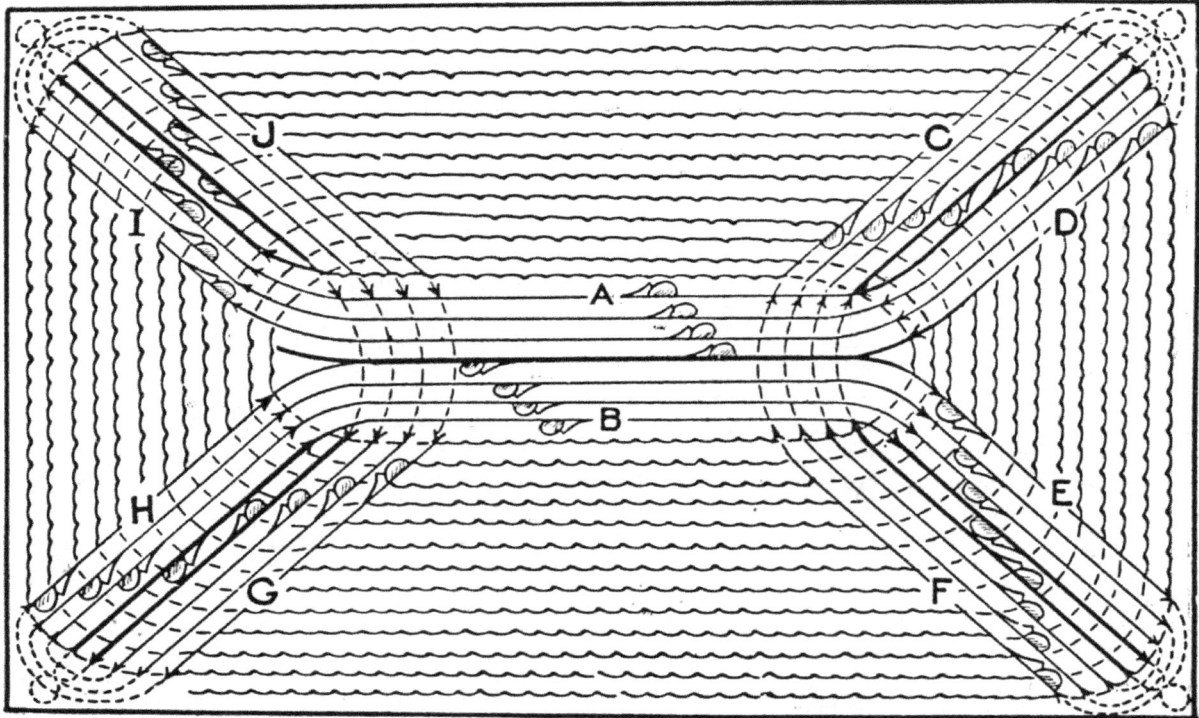

Fig. 372. Method 4 alternate final stage. When a field is plowed following Method 4 and the diagonal strips are finished following the plan illustrated in this figure, there will be a backfurrow along each diagonal.

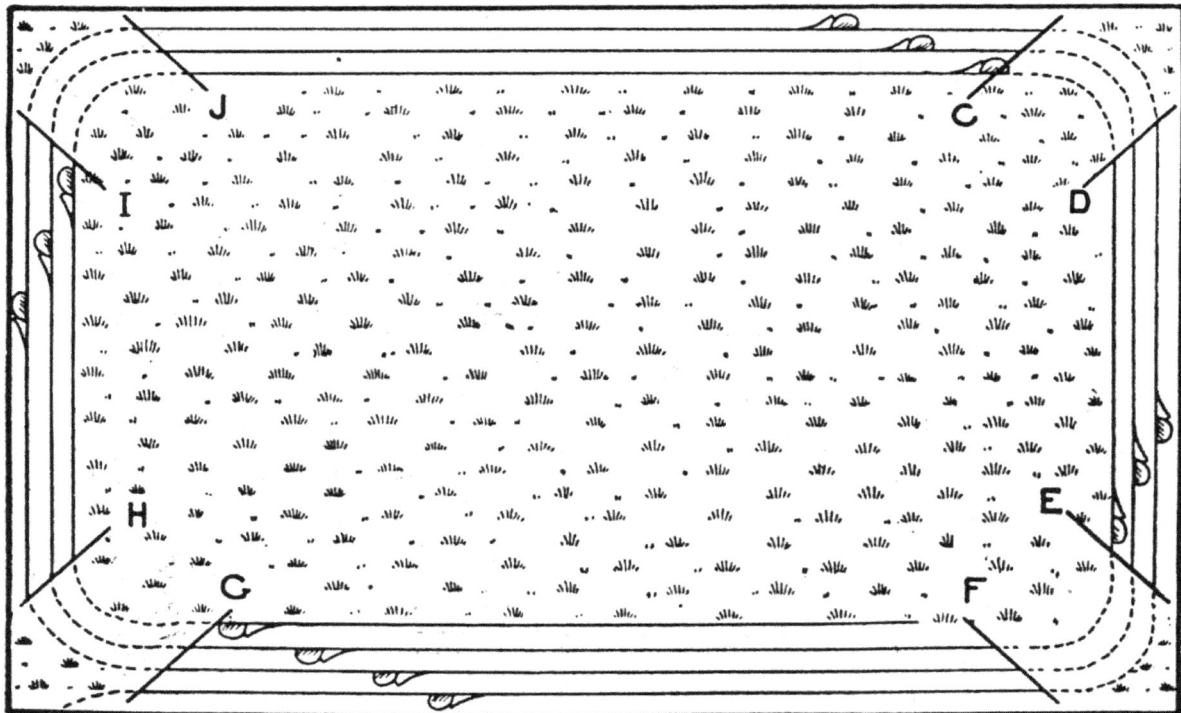

Fig. 373. Method 5. Bottoms are lifted at the corners with this method; otherwise, it is the same as Method 4. The unplowed diagonal strips may be finished by one of the plans shown in Figs. 371 & 372

for turning the outfit and getting it in line with the furrow before the point is reached where the bottoms are to be put into the ground again. It will be better to make an extra round in plowing out the diagonals than to be cramped for space at every turn in plowing the body of the field.

## Irregular Fields

Fields that are irregular because of topography or soil conservation practices may be of such a variety of shapes or present such a variety of conditions that it is impossible to give any definite directions applicable to all. If the field is comparatively level and the irregularities are confined to the boundaries on one or two sides,

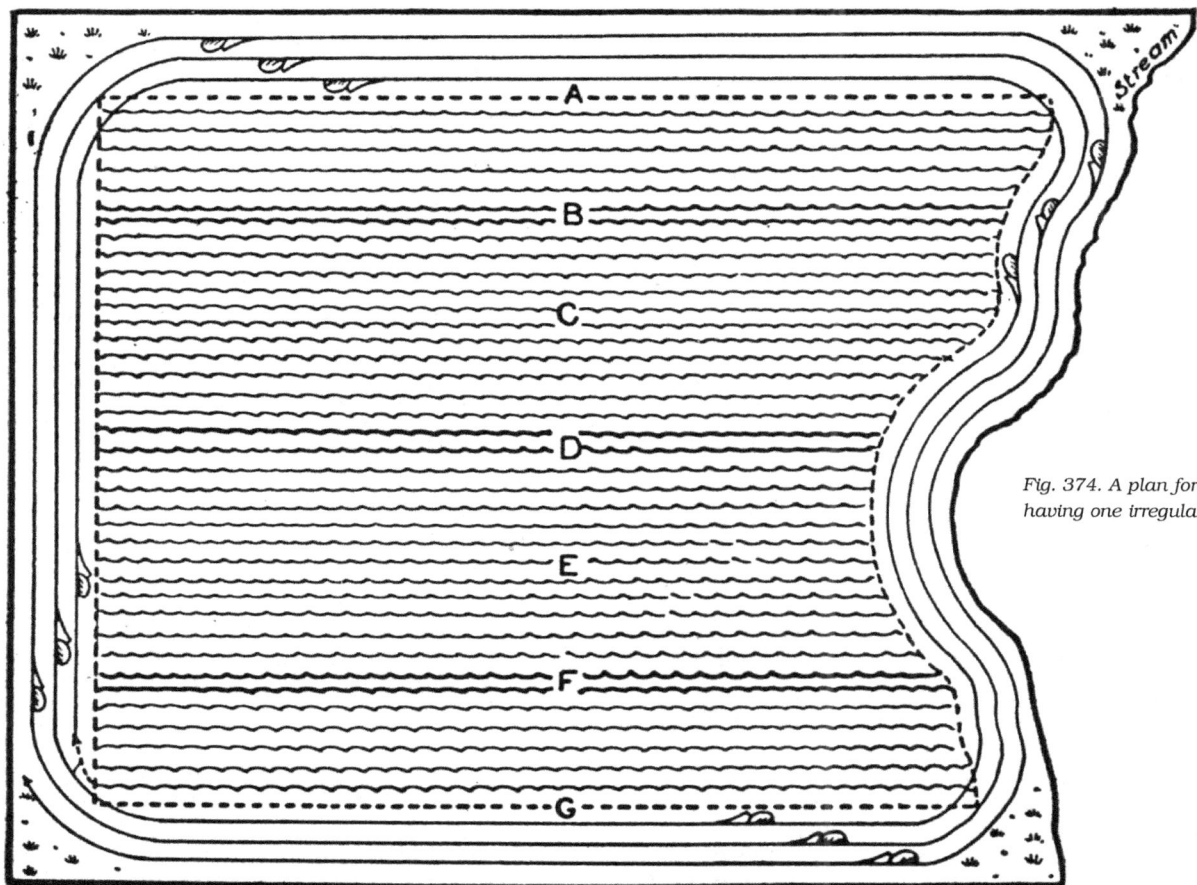

*Fig. 374. A plan for a field having one irregular side.*

usually one of the methods described for rectangular fields can be adopted.

Method 1, Plan A, B, or C, can be readily adapted to fields with two long parallel sides and one or both ends irregular. It is obvious that the lands should be plowed in the direction of the parallel sides of the field. Fields having only one long straight side can also be plowed by using Method 1 and making the obvious adaptations. In fields where the ends are far from being a right angle to the direction of plowing it is suggested that the lands be made rather narrow in order to reduce the unproductive travel across long angular ends.

A field with the irregularities confined to a stream that forms the boundary at one end usually can be plowed satisfactorily by using one of the methods in which the bottoms are lifted in traveling across a field except in laying out the headland across the end adjacent to the stream. There the line for lifting the bottoms and letting them into the ground must be made parallel to the stream if the field is to be finished without undue loss of time in plowing the headland along the stream. If the headland is plowed by turning to the left so that the first round will take in the irregularities along the stream, it will probably be less difficult to finish it satisfactorily than if it is plowed by turning to the right, as shown in the figure. Corners can be plowed out more completely with walking plows.

If the irregularity is simply due to a road, railroad, or a farm boundary which is a straight line but does

not run at right angles to the other boundary lines that join it, the problem of laying out and plowing the headland will be little if any more difficult than in a rectangular field.

A triangular field can be plowed by using a variation of Method 5, described earlier. The body of the field is plowed by starting next to the fence and going around the field by turning to the left. The bottoms are lifted at the corners. The strips A to B, C to D, and E to F (Fig. 375) left by lifting the bottoms at the turns should all be the same width. The strips should be wide enough to permit easy turning at the most acute angle of the field. That is, in the field shown in Figures 375 and 376 the distance from E to F should determine the distance from A to B and from C to D. When the body of the field is finished there will be three unplowed strips, all the same width and extending into the center of the field from each corner. These strips are plowed in the manner indicated in Figure 376.

A four-sided field in which one of the long sides is not parallel to the other can be divided into two parts, one a rectangular plot and the other a triangular plot. The two plots can be plowed separately.

If a field that would otherwise be rectangular has a square or rectangular piece taken off one corner for an orchard, a feed lot, the farmstead, or for any other reason, it will usually be better to make two separate fields in laying it out for plowing. The two fields should be divided by an imaginary line extending from the

Fig. 375. Triangular fields may be plowed by this variation of Method 5. The bottoms are lifted at the turns and the unplowed strips are finished by the plan shown in Fig. 376,

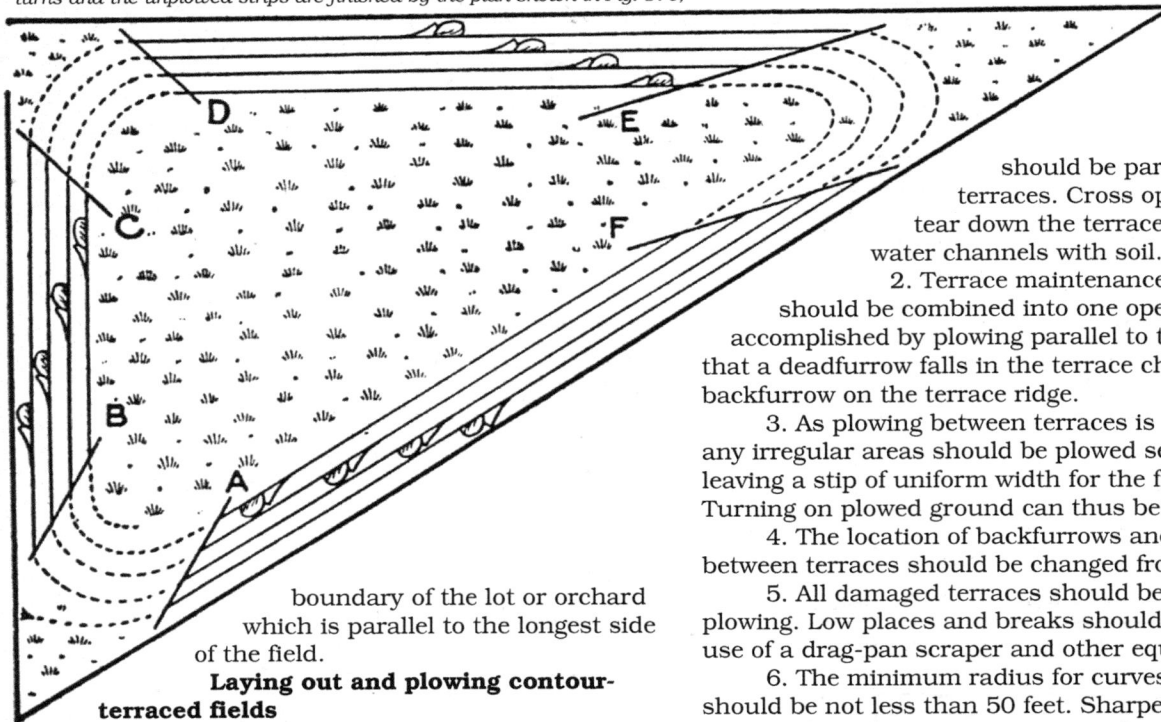

boundary of the lot or orchard which is parallel to the longest side of the field.

**Laying out and plowing contour-terraced fields**

The problems of laying out contour-farmed and terraced fields are more complicated than for level fields. Field layouts and methods of plowing must be designed to utilize the contours most effectively and to help maintain the terraces. Though the exact methods to be used will vary with different topographic, climatic, and managerial conditions, the following general principles should be followed:

1. Plowing, planting of rows, and cultivation should be parallel to the terraces. Cross operations tend to tear down the terraces and fill the water channels with soil.

2. Terrace maintenance and plowing should be combined into one operation. This is accomplished by plowing parallel to the terraces so that a deadfurrow falls in the terrace channel and a backfurrow on the terrace ridge.

3. As plowing between terraces is being finished, any irregular areas should be plowed separately, leaving a stip of uniform width for the final plowing. Turning on plowed ground can thus be minimized.

4. The location of backfurrows and deadfurrows between terraces should be changed from year to year.

5. All damaged terraces should be repaired before plowing. Low places and breaks should be built up by use of a drag-pan scraper and other equipment.

6. The minimum radius for curves in terraces should be not less than 50 feet. Sharper curves will also result in damage to row crops during cultivation. The radius should be kept greater than 100 feet if at all possible.

7. Plowing should never be done across the ends of terrace intervals. This produces natural water channels running up and down the slope.

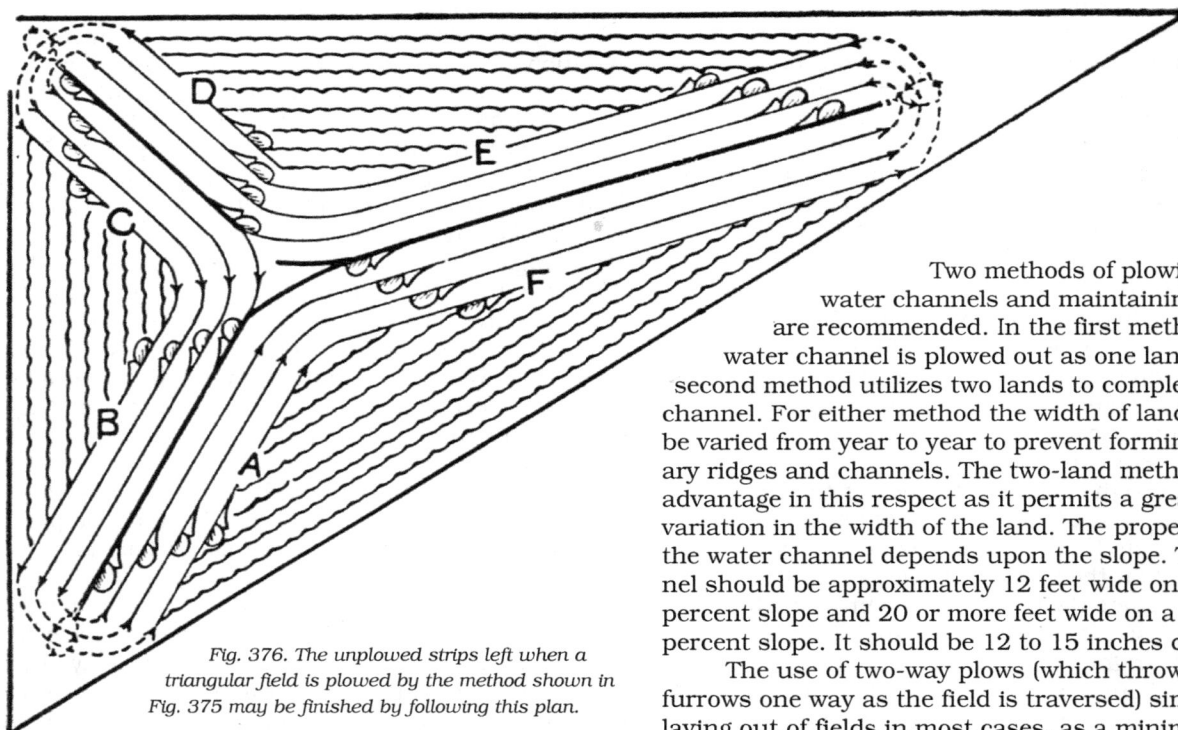

Fig. 376. The unplowed strips left when a triangular field is plowed by the method shown in Fig. 375 may be finished by following this plan.

Two methods of plowing out water channels and maintaining terraces are recommended. In the first method the water channel is plowed out as one land. The second method utilizes two lands to complete the channel. For either method the width of lands should be varied from year to year to prevent forming secondary ridges and channels. The two-land method has the advantage in this respect as it permits a greater variation in the width of the land. The proper width of the water channel depends upon the slope. The channel should be approximately 12 feet wide on a 12-percent slope and 20 or more feet wide on a 2 to 3-percent slope. It should be 12 to 15 inches deep.

The use of two-way plows (which throw the furrows one way as the field is traversed) simplifies the laying out of fields in most cases, as a minimum of

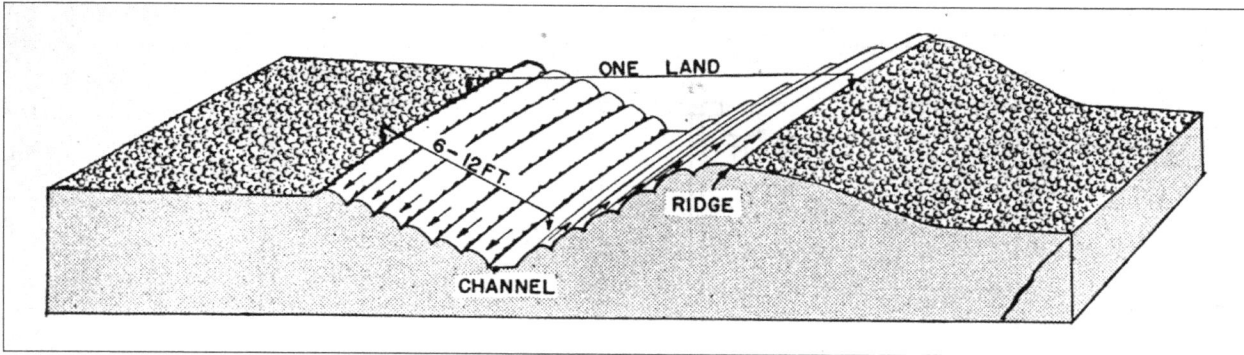

Fig. 377. In the one-land method of plowing terraces, the furrows are turned so there will be a dead furrow in the channel.

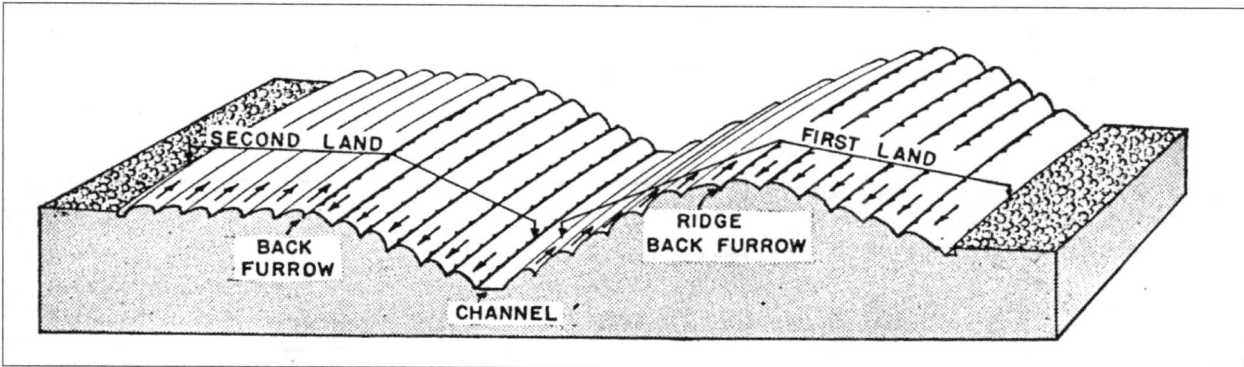

Fig. 378. In the two-land method of plowing terraces, a dead furrow is made in the channel and a back furrow is made on each ridge.

back furrows and dead furrows is required. This type of plow is well adapted to contour and hillside plowing, strip-crop farming, and particularly for fields to be irrigated.

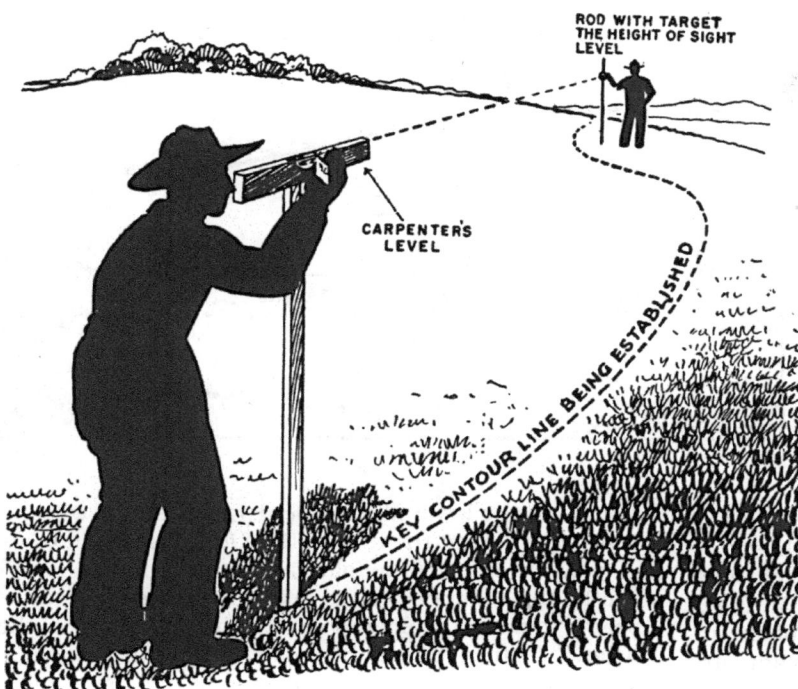

Fig. 379. Sighting a line across a field, alternating foresights and backsights for greater accuracy.

## Another View of Contour Plowing

In plowing a field on the contour, one of two common practices is to continue driving around the back furrows, which mark each contour line, until the plowed land extends half way up or down to the next guide line, at the narrowest point in the area between lines. Then the process is repeated around the succeeding guide lines. Finally, any angular or odd areas in between the two finished lands are plowed out. This places the dead furrow half way between the contour lines, and makes it follow closely the true contour at this point.

Another method is to plow around each back furrow until the plowed land extends from one back furrow a quarter of the way to the next: This is repeated around succeeding back furrows, leaving a full-width land to be finished between the two plowed areas.

## Contour Row Crops

In planting row crops on the contour, the rows are run parallel to the first contour guide line in both directions from it - up to the top of the slope; and for one-half the distance down to the second guide line at the narrowest point between the two. Then planting is started at the second contour line and continued up toward the first and down half way to the next. This places point rows near the center of the area between the two contour lines. It gives the greatest number of "through" rows and puts all of the short

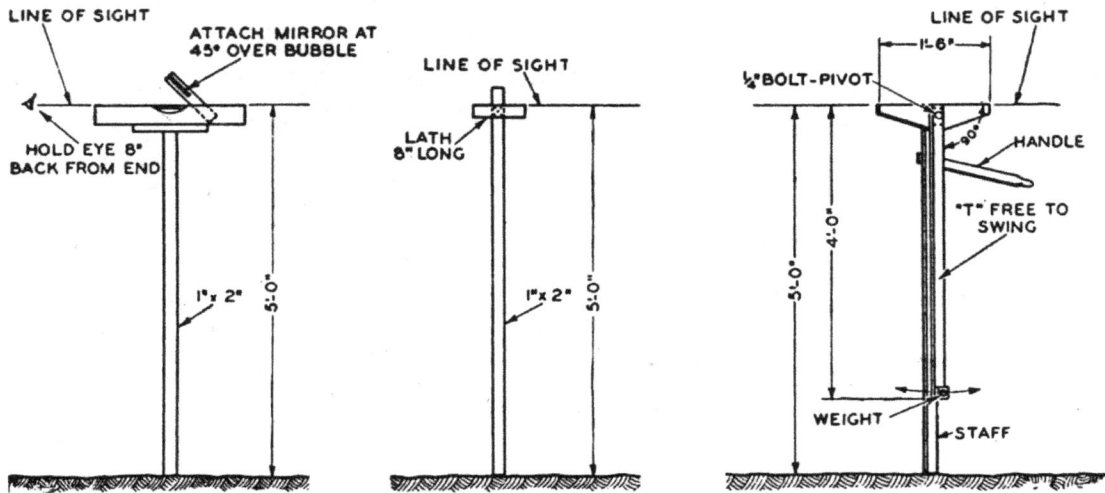

*Fig. 380. Homemade instruments for running contour lines.*

rows together where they can be handled without extra driving. The same system is followed for each successive contour on the slope.

If the turns around a ridge become too sharp to be handled conveniently with regular cultivating and harvesting equipment, these sharply curving rows may be eliminated by cutting across the bend with a well-rounded curve and filling in the cut-off area with point rows. Another solution is to leave a sod strip down the face of the ridge as a turnrow or headland.

**Construction of a Terrace**

The first furrow is turned downhill, with the plow running just above the line of stakes. When this first trip across the field is completed, the stakes are moved straight down the slope, one at a time, for a uniform distance of 8 to 12 feet, depending on the width of terrace ridge desired. This restaking is done by means of a length of light chain or rope, since it is important to keep the width uniform.

On the return trip across the field, the plow is run just below the new line of stakes, throwing the dirt uphill and forming an "island" of uniform width between the two sets of furrows. The tractor should be carefully driven to keep these furrows parallel.

If the terrace is being built with a conventional pull-type, 2-bottom, 14-inch plow, two more rounds are plowed around the "island." If other types of plows are being used, the work is continued until a similar strip of land (seven feet wide) has been plowed on each side of the "island." On the first rounds, the plow is run a uniform depth of about four inches, but on the last round, the plow is adjusted so the last furrow on the upper side of the "island" will be very shallow. This eliminates a sudden step-off from the unplowed land above. On the return half of this

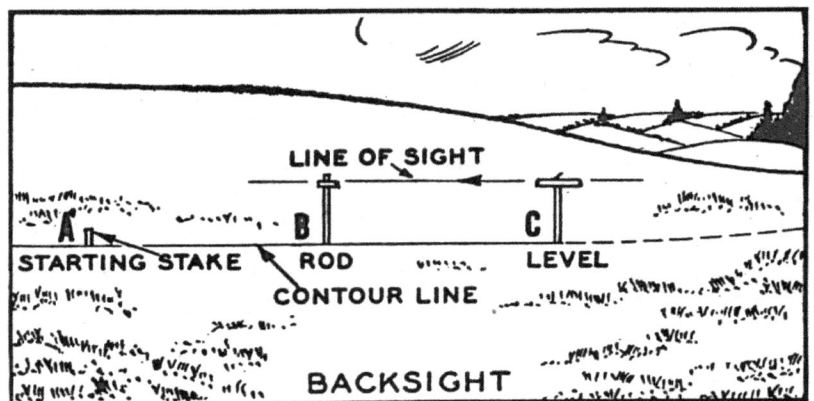

Fig. 381

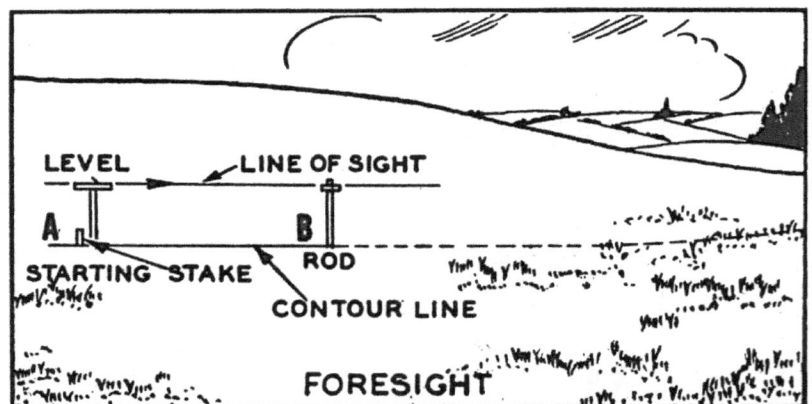

Fig. 382

round, on the lower side of the "island," the plow is again operated at the full depth of four inches.

At the start of the following round (round 4 if a two-bottom 14-inch plow is being used) the operator begins to move the plowed dirt on the upper side of the "island" further toward the center to form the terrace ridge. This is done by starting again at the inside edge of the seven-foot 'wave' of plowed dirt and replowing this same ground, running the plow so that on the first

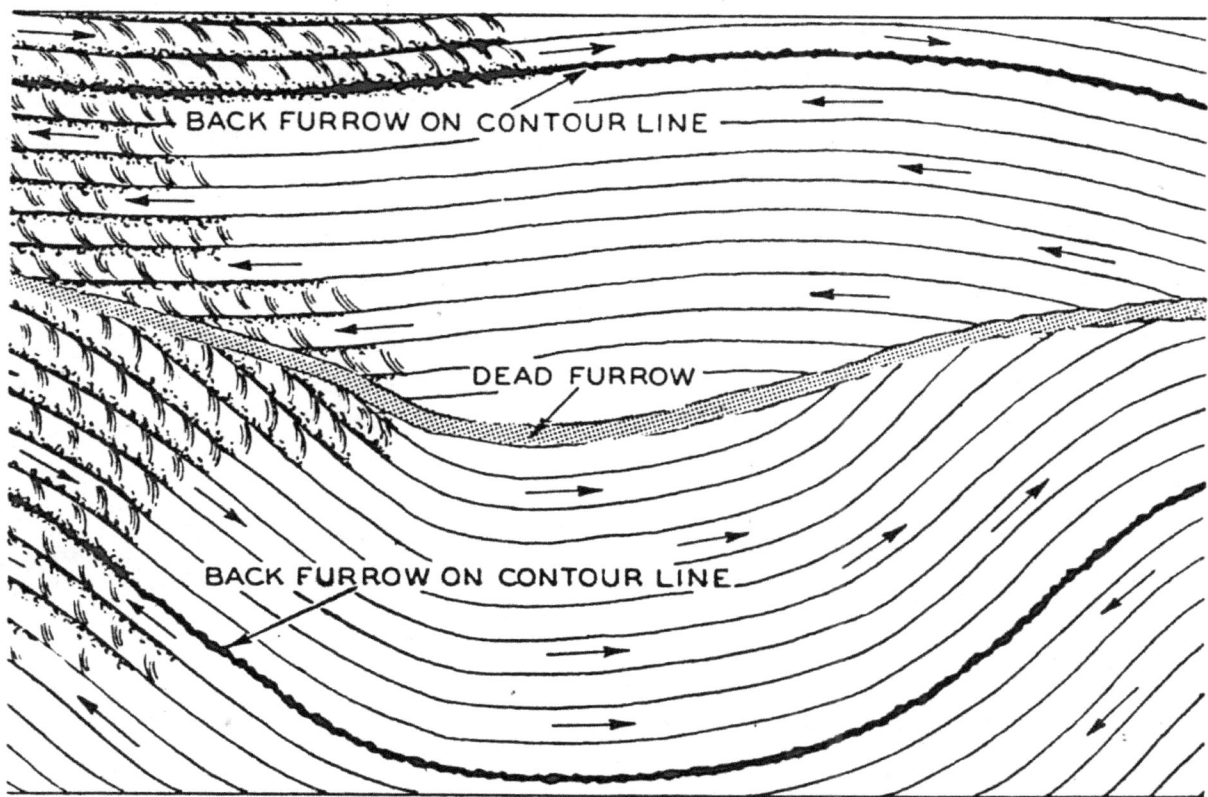

*Fig. 383. This shows contour plowing with proper location of dead furrows.*

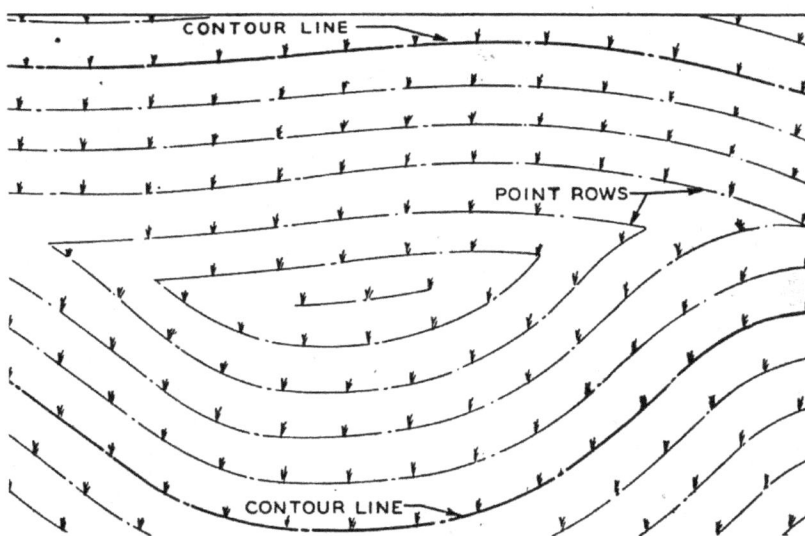

*Fig. 384. This shows location of point rows in the same field.*

to eliminate a ditch below the terrace ridge.

On the next round (number 7 if a two-bottom plow is used) the operator again starts to replow the upper "wave" of soil and at the same time he gains to replow the 14-foot lower "wave," running the plow so the dirt is moved a few inches nearer the center of the "island." No attempt is made to move the lower "wave" any great distance uphill. It is simply replowed to heap the dirt higher along the edge of the "island".

Plowing is continued until the plowed earth from upper and lower sides meet to form a rounded terrace ridge. By this time the water channel will also have the desired dimensions, since the open furrow above the terrace ridge has been widened each time the upper "wave" of dirt has been replowed.

If a nine-foot "island" was left in the beginning, about 27 rounds will be required to complete the terrace with a two-bottom 14-inch plow.

By this time, the upper, 7-foot "wave" of soil will have been replowed eight times (after the initial plowing); and the lower, 14-foot "wave" replowed three times (after the initial plowing). Three return trips across the field are left as extras to be used in smoothing up the lower side of the ridge and in filling any ditch left at the lower edge where the terrace joins the unplowed land.

round the dirt is moved 8 to 12 inches nearer to the center of the "island". The plow is run deep enough so that a thin slice of the firm earth underneath is turned with the loose dirt to make the plow scour.

While the seven-foot "wave" is being replowed on the upper side of the "island", (rounds 4, 5 and 6 if a two-bottom plow is used) the plowed land on the lower side is widened out to 14 feet. On the last trip along the lower edge of this side, the plow is adjusted so the final furrow is very shallow. This is especially important here

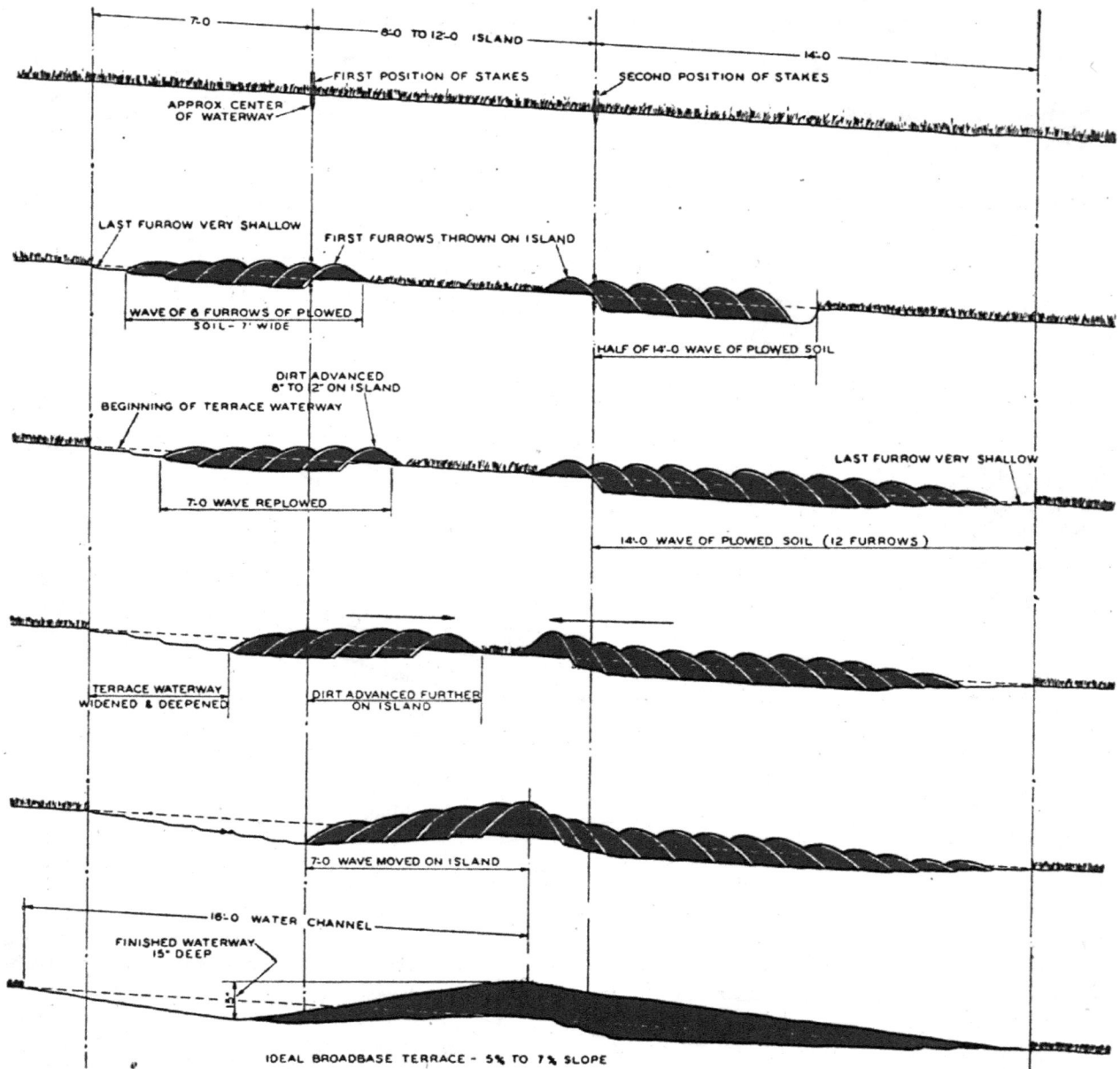

7-0

8-0 TO 12-0 ISLAND

14-0

FIRST POSITION OF STAKES

SECOND POSITION OF STAKES

APPROX CENTER OF WATERWAY

LAST FURROW VERY SHALLOW

FIRST FURROWS THROWN ON ISLAND

WAVE OF 8 FURROWS OF PLOWED SOIL - 7' WIDE

HALF OF 14'-0 WAVE OF PLOWED SOIL

DIRT ADVANCED 8" TO 12" ON ISLAND

BEGINNING OF TERRACE WATERWAY

LAST FURROW VERY SHALLOW

7-0 WAVE REPLOWED

14-0 WAVE OF PLOWED SOIL (12 FURROWS)

TERRACE WATERWAY WIDENED & DEEPENED

DIRT ADVANCED FURTHER ON ISLAND

7-0 WAVE MOVED ON ISLAND

16-0 WATER CHANNEL

FINISHED WATERWAY 15" DEEP

IDEAL BROADBASE TERRACE - 5% TO 7% SLOPE

*Fig. 385. End view of terrace construction.*

If the soil becomes too loose to be moved again before the terrace is completed, it may be packed with a heavily-weighted disc-harrow with the gangs set straight, or with a field packer. Another remedy used where the soil is often very dry, is to partially build the terraces and then wait for a light rain to pack the soil so plowing may be continued without difficulty.

Fig. 387. The best way to learn plowing is to have help. This young man is getting a sense of the walking plow without having to worry about the Shire team. Photo by Heather Erskine.

Chapter Nine

# Teaching Horses and Yourself to Plow

Work animals, those with some rudimentary level of training, learn plowing quite rapidly. Beyond the basics of wearing harness, driving well and acceptance of hitching; what they need to learn, specific to plowing, is not complex. They are asked to pull across a field an implement with a steady resistance. They are asked to pull at a sustained slow pace and to walk straight (or true to the teamster's design). They are expected to accept the rudeness of the plow leaving the soil and causing them to momentarily experience a free-fall forward. They are expected to stand quietly in rest breaks and when adjustments need to be made. And they may be expected to walk in a furrow.

By comparison, the novice horsefarmer has a great deal to learn if he or she is to become an accomplished plowman. Much, if not all, of what a plowman needs to know has been covered in the other chapters of this book. This author believes, born of experience, that understanding technical adjustments constitutes an important half of the challenge. The other half, and certainly for many the 'tricky' half, deals with the balanced marriage of horsemanship to plow performance. Another way to think about these two aspects is to point out that an excellent horseman with years of teamster experience and his or her own animals will still have great difficulty plowing if the technical aspects are not known and understood. Someone who knows and understands all the technical

mechanics of horsedrawn plows can, with quiet well trained horses, take to the actual plowing work very quickly.

We cannot do justice, within the covers of this book, to the expansive subject of horsemanship. That subject which includes not only how to harness and drive horses but how, also, to develop and maintain the essential bond of trust with those working partners. We must insist that the reader invest time into an in-depth inquiry into working horses and mules in harness. Not to do so puts everyone and every thing at jeopardy.

As was noted in a previous chapter, if you proceed inexperienced and without key bits of information, to attempt to hook **dependent** animals to a plow and drag it with accuracy and **SAFETY** through that waiting plot of land, you take enormous risk with the well-being of everything in view, animate and inanimate. See those people with the small child watching? See their new car? See that lovely fence? See that clothesline? See that new outhouse? Look down on yourself and at those fine animals and that lovely plow. You put them all at risk. You owe it to all of them and every thing else to have, at the very least, someone knowledgeable and physically capable close at hand, and preferrably you with head and hands fresh from learning.

**Knowledgeable, physically capable help, while learning to plow, is the very best way to proceed.** Consider a qualified helper to be 100%.

Fig. 388. The author's work horse herd munching hay in the snow.

gentlemen who claim absolute knowledge about all things 'work horse' yet whose advice and counsel has either fallen short or proved destructive. Be intelligent for your own sake, seek evidence of credibility. Your goal should be for a safe education. Don't sell that goal short. And also important, the brief outline presented in this chapter and elsewhere in this book should not be trusted as any guarantor of safety and success.

If the available help is someone who, because of age and/or health, is unable to actually take hold of animals or driving lines or help correct harness or hitch particulars you must consider their value to you in the learning process as 55%. If the available helper is someone who might look too young to know the craft of plowing, someone who admits to having learned recently, someone who has demonstrated an easy success with producing beautiful furrows - this person may rate a 110%. The person to whom the adventure of discovery and recently acquired facility is still fresh may just be the very best teacher. Those of us cranky old men who've done it for a long time take too much of the subtlety for granted. Or enthusiasm, though no less real, may have jelled into a less useful arrogance.

Be careful not to assume that a senior citizen, by virtue of age, is a qualified helper. This author has had too many unfortunate experiences with older

## Teaching Your Horse(s) to Plow.

*It's winter time and you look out at your horses in the pasture, or in the snow. You've poured over this long, intense and sometimes boring book, Horsedrawn Plows and Plowing, and now you are wondering if any of your show horses, your parade horses, your pet horses could actually ever learn to plow. You look at them sedately eating hay, and you try to imagine a rear view of four abreast straining against collars, well-muscled and eager for work...?*

As was previously noted, pulling a plow is amongst the best and most honest work you could ask a horse, mule or ox to do. And, contrary to what the uninitiated might think, if the plow is matched to the soil, sharp and properly set, hitched right and matched for size to the number of animals, the job is not an overly difficult one.

The only aspect of the plowing that a good work animal might need some help getting used to is walking

Fig. 389. We might all wonder if we can put together four such working beauties as these Ohio Amish Belgians.

Fig. 390. Plowing matches, old-time farming demonstrations, and special events such as the Horse Progress Days are all excellent places to meet prospective helpers for the beginning plowman.

out. With some animals it will take one pass, with others it will take several. Eventually it will become second nature. And if your goal is to learn the walking plow we suggest that you walk in the ditch as well. All of this has been done, to this point, with one horse. With some people one horse will be all that is available and necessary. However, if the work is to be done with teams of two or three abreast, there is a second stage to the learning. Referring back to page 61, rig up a team intended for plowing and tie the lines together and put them over one shoulder and under one arm. Speak to the team and attempt ground driving with only a slight leaning pressure back on the lines. Avoid using your hands. When all seems to be going well, have one of the horses step into the furrow or ditch as you do. Practice driving this way. This simple compounded exercise will help both the teamster and the horses (or mules).

Keep in mind that what has just been suggested is for animals which are trained for driving in harness but which may not have any experience actually plowing.

If you are going to attempt **plowing with animals which are not well trained** you MUST have plenty of teamster experience. But know that the repetition of the work and procedures will be quickly

in the furrow. It is surprising how quickly most animals learn this. A simple repeated exercise can help.

Again, an ideal situation would be the availability of an actual furrow to practice in. But barring that, any narrow ditch (preferrably 12 to 16" wide and 6 to 8 inches deep) will work fine. And the exercise is simple. Just ground drive the animal back and forth in the ditch, without being hooked to the plow. Keep doing this, patiently returning him or her to the furrow or ditch whenever they step

Fig. 391. At a Duvall, WA plowing demonstration John Erskine asks his two Shires to pull the two-way plow. John is one of those teamsters who loves to help beginners get a safe start.

Fig. 392. John Erskine of Monroe, WA helps young C.J. Shopbell get the feel of driving six big Shires hooked to a JD Wheatland gang plow. In a very few years C.J. will be helping others get started.

accepted by the new animals.

In the early eighties this author was invited to North Dakota by the state draft horse association to demonstrate the buck-back tie-in system with six head on a gang plow. With lots of good help, six Tweeten family Clydesdales (see Fig.394) were successfully hitched to the John Deere gang plow. What made this occasion somewhat noteworthy is that of those six Clydesdales, one was a young halter broke stallion (later sold to Budweiser), two were brood mares, and only three were trained to work in harness. Two of the

trained horses were put in the lead with the green mare set near (land) lead position. One trained horse was used at the wheel with the other brood mare and stallion. With care and common sense these six were plowing beautifully in a half day's time. Again plowing will be rapidly accepted by good animals of varying levels of training.

It is important however, to understand the implications of just what sort of training animals have had. A particular circumstance which can create difficulties has to do with the animal's experience with

Fig. 392. Dale Hendricks drives the Erskine plow hitch while seated John discusses with the on-deck teamster the subtleties of of life.

actual work, or pulling. If the horse or mule has always pulled a wagon or light implement without steady resistance, it may take some careful persistance to get it to pull the plow. The utmost concern should be taken not to make the animal balky or sullen, refusing to pull. It is best for an experienced horse or mule man to deal with such a circumstance. If no help is available, try working your animals first on a sled for long steady walks before hooking to a plow.

A specific problem of the greatest difficulty arises with animals which have been used in pulling competitions and not much else. These animals will often want to 'jerk' start the plow and pull it very fast and hard for a short distance. This is because they associate a resistance with the contest routine. The only way to correct this is with lots of slow steady repetition using a work sled or stoneboat until the animals have learned a new routine and are ready to take on the plow in a calm and safe manner. A riding plow following a runaway team across rough terrain WILL cause the teamster, horses, and equipment harm. All precautions should be taken to avoid such an occurence.

And this author believes with absolute certainty that it is possible to have a trusting relationship with work animals. A relationship where the teamster can worry about how the plow is functioning and not about the animals.

This book is not about how to work horses or how horses or mules work, it is about plows and plowing. You are encouraged to refer to *The Work Horse Handbook* and *Training Workhorses.*

**Teaching Yourself.** If you've never plowed before, trying to do it on your own with a plow you know nothing about, and with horses or mules you might be unfamiliar with, is asking for frustration at the least. As was hammered on in the previous paragraphs, the best way to learn to plow is with competent help. The way to learn is to start with little introductions. If you know someone who plows with horses, or know of a plowing demonstration or competition, or a work horse workshop which includes plowing, you owe it to yourself to sign up, invite yourself, check in, plead for a taste. The best way to learn to plow, and this is especially true of the walking plow but only slightly less so of the riding versions, is by having someone else worry about the animals while you try your hands on the handles and in the furrow.

If you cannot do that and insist on trying to do it on your own, here are some tips. Assuming you have paid attention to the warnings and have learned about working animals in harness, (again) you should practice ground driving (driving horse[s] which are not hooked to anything - not pulling anything) while the lines, tied or fastened together (see Lines sidebar), are over one shoulder and

Fig. 394. The author driving the Tweeten family Clydesdales at North Dakota Draft Horse Association Meeting in Minot, N.D.  The stallion and two mares were halter broke and had not been driven before this weekend. They are hitched to a two-bottom John Deere plow.

under the other. Practice maintaining a perfect light tension on the lines  without touching them with your hands, only by slightly leaning back and/or slowing yourself down. You'll find that it is tricky at first and you'll be reaching, with both hands, for the lines. Try not to. The horse(s) should start out smoothly on a quiet voice command, slow when you apply pressure to the lines, and stop when you say "whoa".

When you think you are ready to actually plow ***do not do it on your own the first time!*** Have someone at the very least physically capable of helping you with fetching tools, holding animals, etc. At the risk of sounding alarmist, we cannot stress enough how important it is that you exercise caution and make safety a prime concern. Plowing with horses or mules should be immensely enjoyable. It can be relatively risk free. It does not require brute force. It can be as frustrating as the first time with ice skating.

The beginner's first sensation might be that the plow, regardless of what speed it's going, is going too fast and is out of control. Or the beginner's first sensation might be one of complete awe and exhiliration. Most people report years later that that first time plowing feeling is still with them. We sincerely hope that this book and these precautions do not damper a beginner's interest but rather offer a helpful outline for structuring a safe beginning.

*(For information specific to starting out with a walking plow please refer to page 61.)*

Fig. 395, These three handsome mules look ready for the plow.

Chapter Ten

# Hitch Gear

Fig. 396. The Harvey Racine Company sold this doubletree as "Favorite Plow Set" in 1920. Singletrees were 28 inch and double tree 34 inch. Its Northwestern Plow Set featured a 38 inch doubletree.

Fig. 397. This was called the "Harvey Heavy" Plow Set, 36" x 30".

Fig. 398. The neckyoke below features a design still common today and available in 2" increments to match the length of the doubletree. (i.e. the Harvey Heavy would take a 36" neckyoke to match.)

In this chapter we provide some engravings from old equipment catalogs illustrating eveners as well as diagrams of various tandem hitches. The criteria we used for selecting engravings and diagrams was somewhat arbitrary, as we were looking for good illustrations and valuable information. All the major plow makers offered their own line of eveners.

Next in importance to the problem of setting up a plow that will work under average conditions is the problem of hitching the plow to the prime mover, or the power that is to draw it. The hitching of horses requires different arrangements from that of tractors. The same principles, however, are involved, but they must be handled differently. The problem is to get all the pulling forces of the power and the resistance forces of the load in equilibrium, both vertically and horizontally.

**Hitch.** The hitch is composed of the parts connecting the plow with the power. It may be simple, consisting of only one or two parts, or it may consist of a multiplicity of bars, braces, angles, and levers arranged to absorb certain vertical and horizontal forces.

**Center of Power.** The center of power is often described as the true point of hitch, or center of pull. Whatever the term used, the point referred to is the center of the power, which is mostly horizontal, but the vertical forces must also be considered.

When horses are used, the center of power with one horse is midway between his shoulders or hame tugs. If two horses are used, the center of power will be halfway between the two animals.

**Center of Load or Resistance.** The center of load is often termed center of draft. As shown in Fig 75, page 37, this is the point within the plow about which all the forces acting on the plow are balanced.

**Line of Hitch.** The line of hitch or line of draft is an imaginary straight line passing from the center of load, or resistance, through the clevis or hitch to the center of power where the hitch is attached to the power. This definition applies to both the vertical and horizontal adjustment.

**Side Draft.** Side draft is produced when the center of load or resistance is not directly behind the center of power. The center of load is out of line or to one side of the true line of hitch or draft. When side draft is present, there may be a pull sidewise on either the power or the load, depending upon the hitch.

**Vertical Adjustment of Hitches.** With the horse as a source of power, the proper arrangement for the hitch is that there should be a straight line extending from the center of resistance through the clevis to the point where the tugs are fastened to the hames. (See Figs. 133 & 134.) This should be the proper adjustment vertically. If the hitch at the clevis is too high the tendency will be to throw the plow deeper into the soil because of the fact that the line of hitch is seeking the straight line just mentioned. The reverse action will be true if the hitch is lowered below that of a straight line. This principle applies to all horse-drawn plows from walking to gang, and also to tractor-drawn plows, both moldboard and disk.

**Horizontal Adjustment of Hitches**. To hitch the plow to make it take the proper furrow width, the center of load or resistance must also be considered. A straight line must pass from the center of load through the clevis to the center of power between the tugs. If the plow bottom is in perfect condition and the hitch properly adjusted, the ordinary walking plow should operate with very little assistance from the operator. When three or more horses are used, the hitch problem

Fig. 399. A strap end
cultivator or plow singletree.

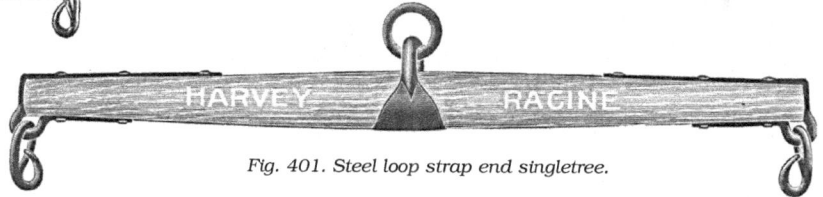

Fig. 401. Steel loop strap end singletree.

Fig. 400. Full ironed back strap singletree

Fig. 402. Wagon doubletree with clevises

Fig. 403. Plow doubletrees with clevises.

is greatly increased because the right hand horse walks in the furrow throwing the other two upon the unplowed land so that the center of power does not coincide with the center of load, thus creating side draft. Some operators attempt to remedy this on walking plows by the use of short eveners. The effect upon the team can be seen readily because the pull to the side causes additional trouble. A good length of singletree is 26 inches. For the good of the horse, nothing shorter than this should ever be used. It is even better to use a 28 or 30 inch singletree, which will give more clearance for the horse.

**Balancing Teams.** When working two horses to a load, it is often found that they are not well matched. It is not an uncommon sight to see a heavy and a light horse being worked together. Naturally, the lighter horse is at a disadvantage. There are two ways of constructing the evener to correct this.

1. By adjusting the pin holes at the ends of the evener.

2. By adjusting the draft hole at the center of the evener.

Adjusting the pin holes at the ends of the evener is nothing more than making one end of the evener shorter than the other by having another hole closer in. The method used to determine the distance the hole is moved as shown by the following example: Assume that a team consists of a 1,500 pound and a 1,000 pound horse. The load each should pull is 150 and 100 pounds, respectively. If the pin hole for the lighter horse is 24 inches from the draft hole, the length of the evener arm for the other horse is found as follows:

Take the length of the evener arm for the lighter horse and multiply that by the pounds it should pull. Divide the result obtained by the pounds the heavier horse should pull. The result will be the length of evener are for the heavier horse.

Hence:
$$24 \times 100 = 150 \times X$$
$$150X = 2,400$$
$$X = 2,400/150$$
$$X = 16$$

When the draft hole is used it is moved only half the amount the pin hole was moved, because 1 inch at the draft hole is equivalent to 2 inches, since it means 1 inch on each end of the evener. The length of the ends is obtained by the same method as with the pin holes.

This rule will work equally well for a three, four, or five horse evener when two or three horses are considered as one.

Fig. 404. The Heider" brand three horse plow evener. Though this style, in new-old-stock, is sometimes still found around, the author does not recommend it as the old wooden singletrees are prone to break.

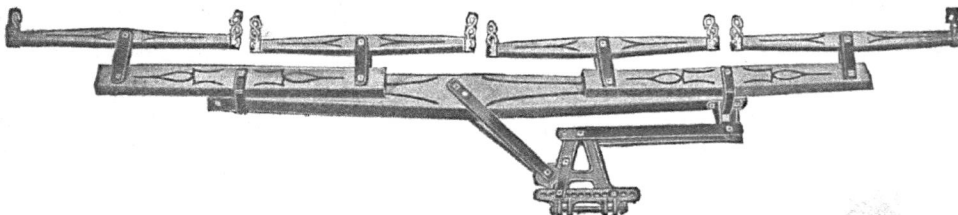

Fig. 405. This Heider four abreast is built with an eccentric equalizer that permits one horse in the furrow and three on the land. It is similar to the new Pioneer four abreast which is by far the better instrument.

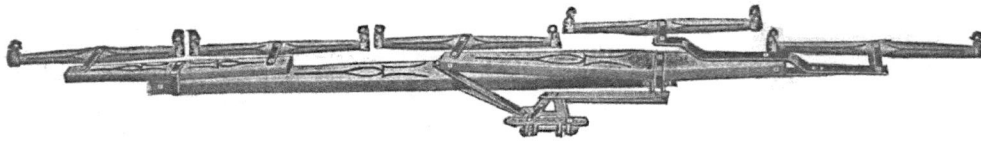

Fig. 406. Heider brand five abreast gang plow evener.

Fig. 407. Six Percherons in 1930 pull a gang plow and harrow with a hitch diagramed in Fig. 411.

Fig. 408. These four depression era Belgian colts walk out easily hooked to the four up tandem diagramed as the swing (mid) and lead of Fig. 410.

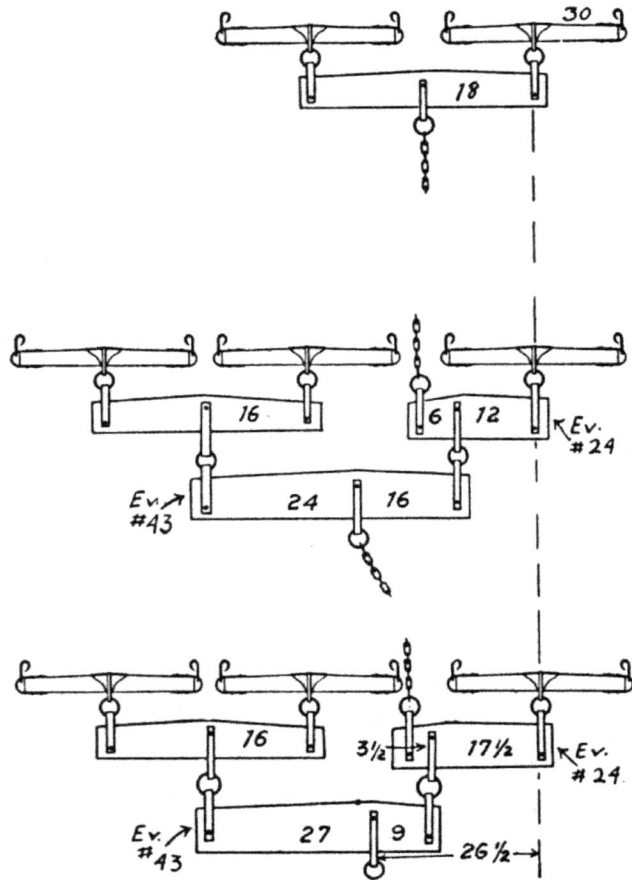

Fig. 409. A Talkington style eight-up hitch.

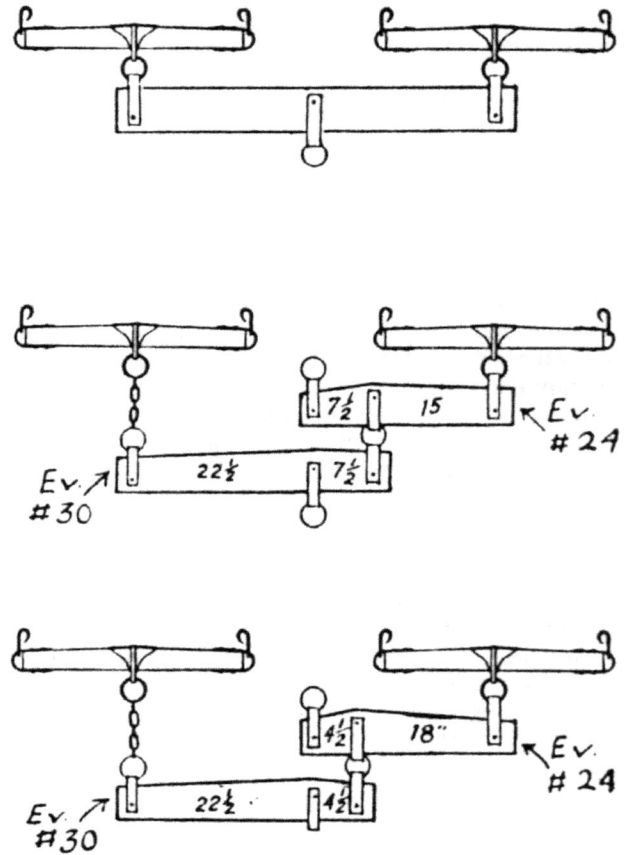

Fig. 410. The lead and swing of this 6 horse hitch form a four horse tandem hitch for gang plows.

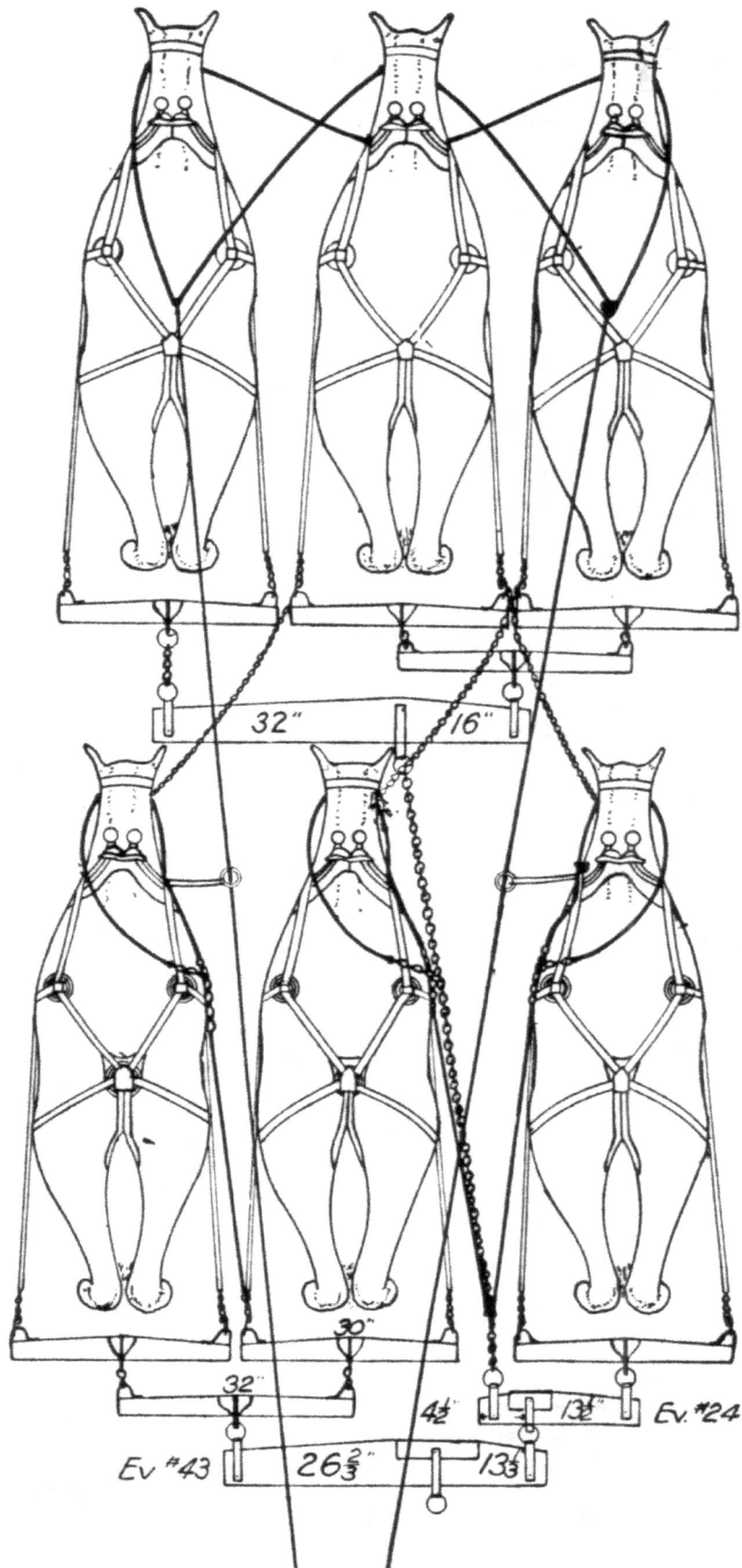

32"    16"

30"

32"    4½"    13½"    Ev. #24

Ev. #43    26⅔"    13½"

*Fig. 411. A six-up hitch detailed.*

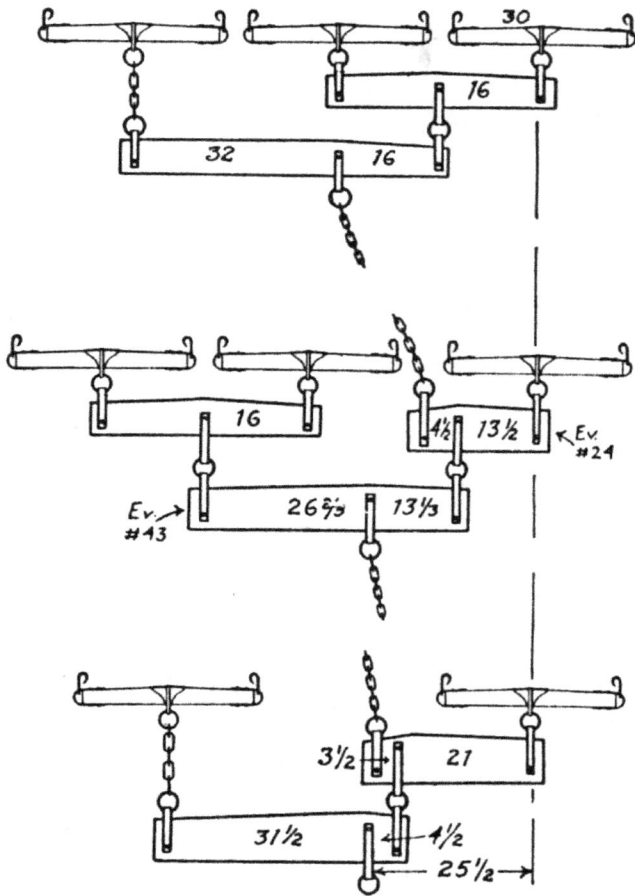

Fig. 412. Eight horse hitch with two at the wheel allowing a spread for improved teamster visibility.

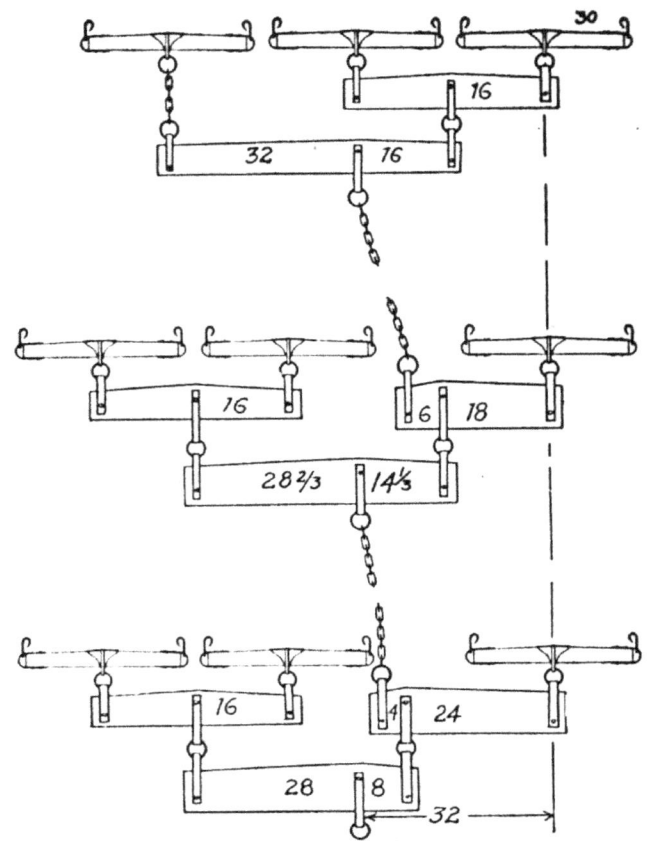

Fig. 413. A good 9 horse hitch for a 3 bottom plow (14") in heavy soil.

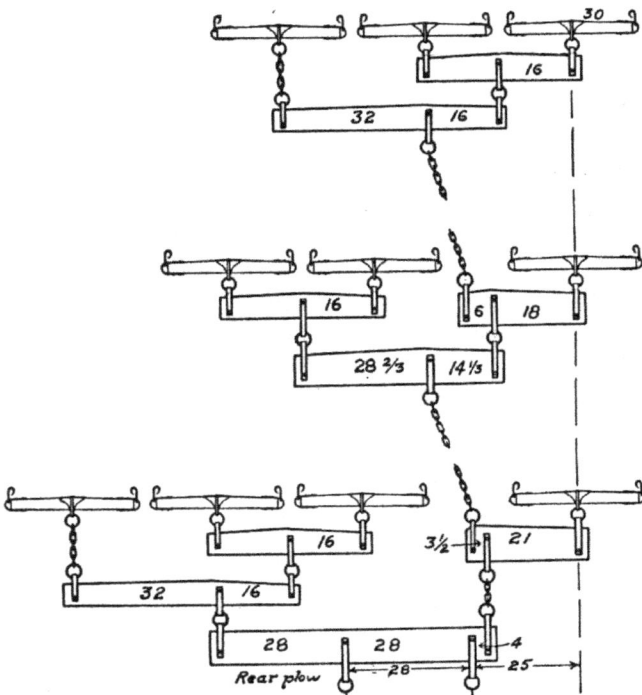

Fig. 414, A Ten horse hitch for three or four bottoms depending on soil.

Fig. 415. A 1930's five horse Percheron plow hitch utilizing a Talkington evener setup with the classic homemade plank horizontal equalizers..

Fig. 416. An Illinois six horse hitch on a gang plow.

Fig. 417 & 418 are front and back views beautifully illustrating how six horses may work with comfort in a six horse hitch on a two bottom gang plowing extra hard summer ground.

Fig. 419 & 420. Two different hitches (one right, with eight mules and one left with six horses) showing tag harrows.

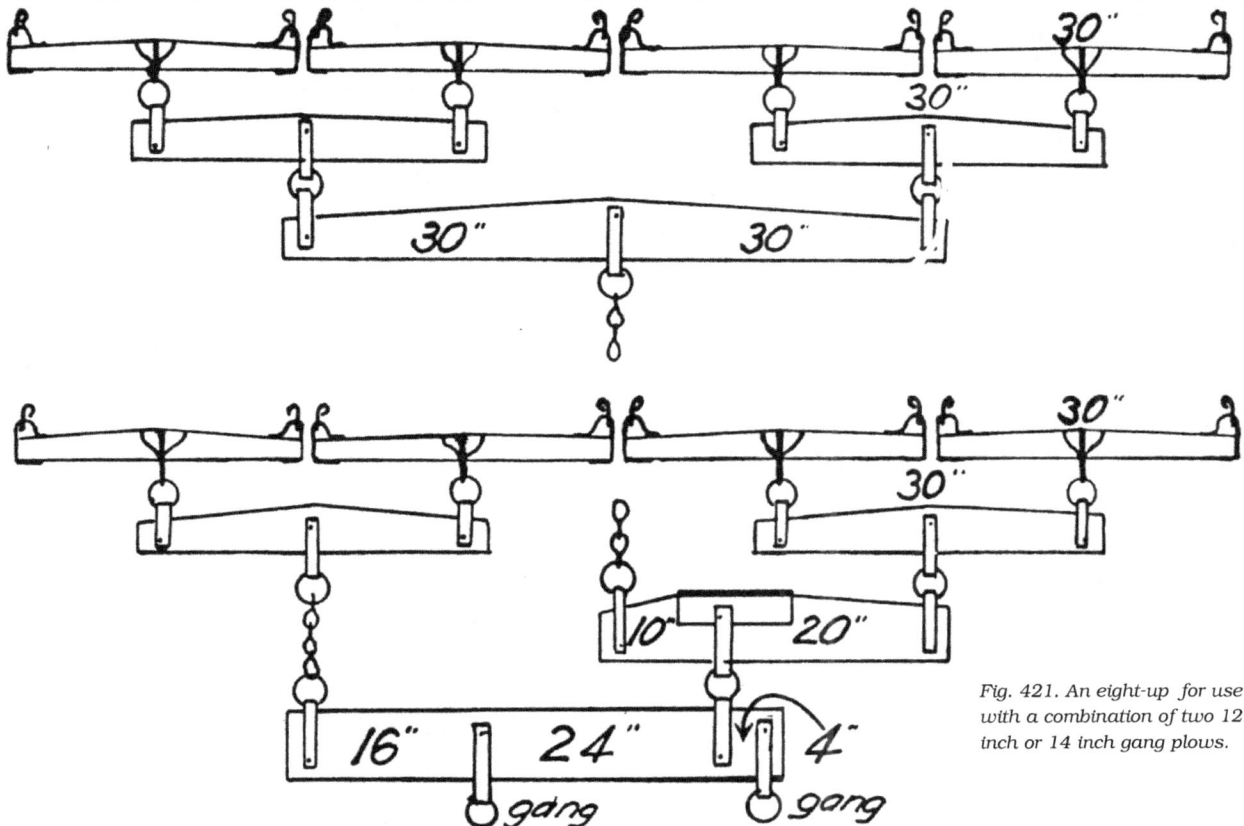

30"
30"
30"
30"
30"

30"
30"
10"
20"
16"
24"
4"
gang
gang

Fig. 421. An eight-up for use with a combination of two 12 inch or 14 inch gang plows.

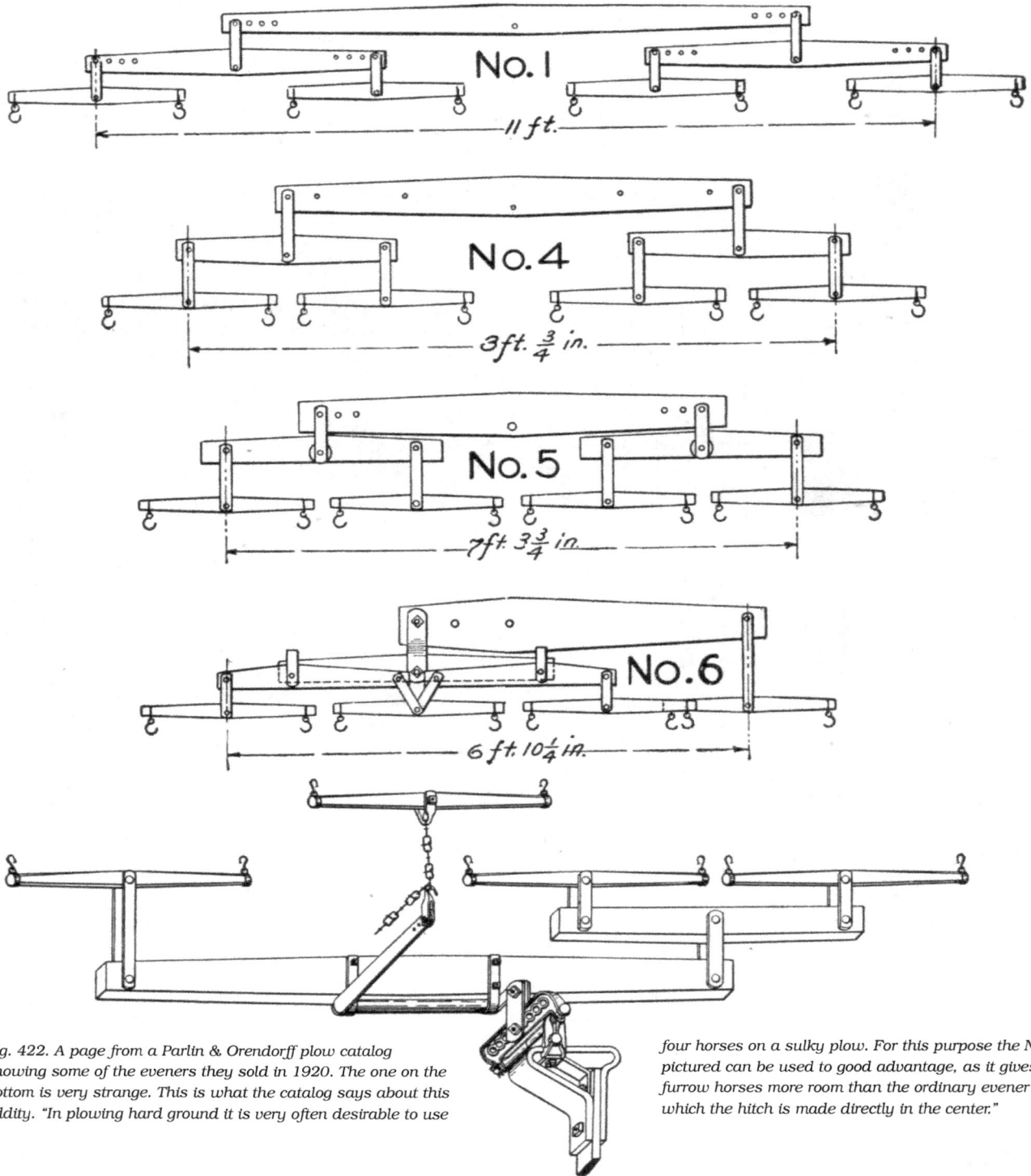

No. 1

*11 ft.*

No. 4

*3 ft. ¾ in.*

No. 5

*7 ft. 3¾ in.*

No. 6

*6 ft. 10¼ in.*

*Fig. 422. A page from a Parlin & Orendorff plow catalog showing some of the eveners they sold in 1920. The one on the bottom is very strange. This is what the catalog says about this oddity. "In plowing hard ground it is very often desirable to use four horses on a sulky plow. For this purpose the No. 7 pictured can be used to good advantage, as it gives the furrow horses more room than the ordinary evener on which the hitch is made directly in the center."*

The factory eveners sold by such companies as P&O, John Deere, Moline, Oliver, McCormick Deering etc. were almost all of metal from circa 1914 on. A rolled flat steel design was most prevalent and in fact so many millions were manufactured that thousands still circulate through antique farm equipment auctions all across the country. This author cautions especially beginners against using the old wooden eveners. Though they may look to be in fantastic

No. 22.

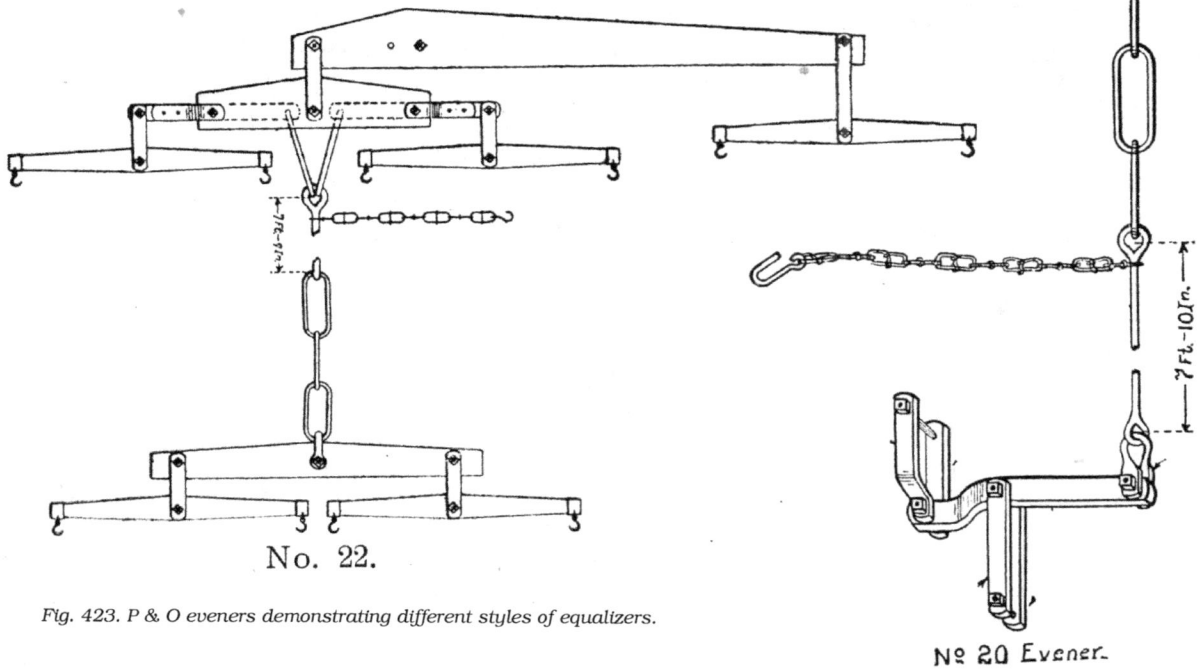

Fig. 423. P & O eveners demonstrating different styles of equalizers.

Nº 20 Evener.

Fig. 424. P & O six horse evener with lead bar.

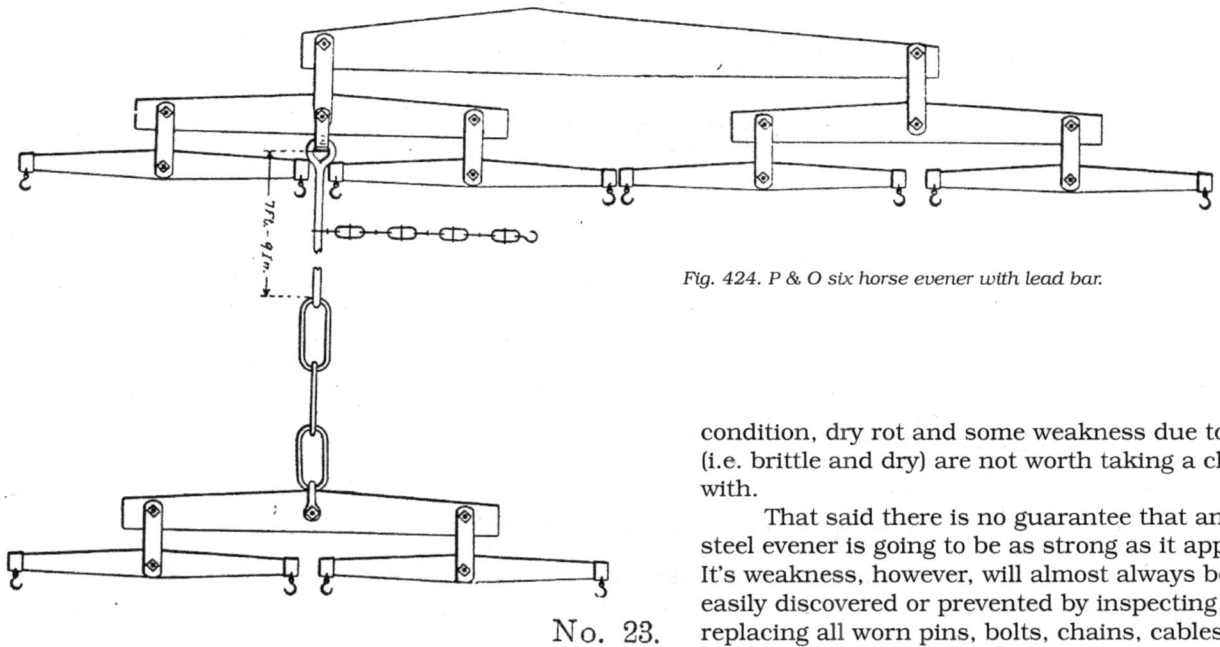

No. 23.

condition, dry rot and some weakness due to age (i.e. brittle and dry) are not worth taking a chance with.

That said there is no guarantee that an old steel evener is going to be as strong as it appears. It's weakness, however, will almost always be easily discovered or prevented by inspecting and replacing all worn pins, bolts, chains, cables, keys clevises. One aspect of the rolled metal eveners that this author appreciates is that for the relative strength the big hitch steel eveners weigh a fraction of the equivalent strength wood eveners. A wheel 4 abreast of the Talkington style is extremely difficult to pack around.

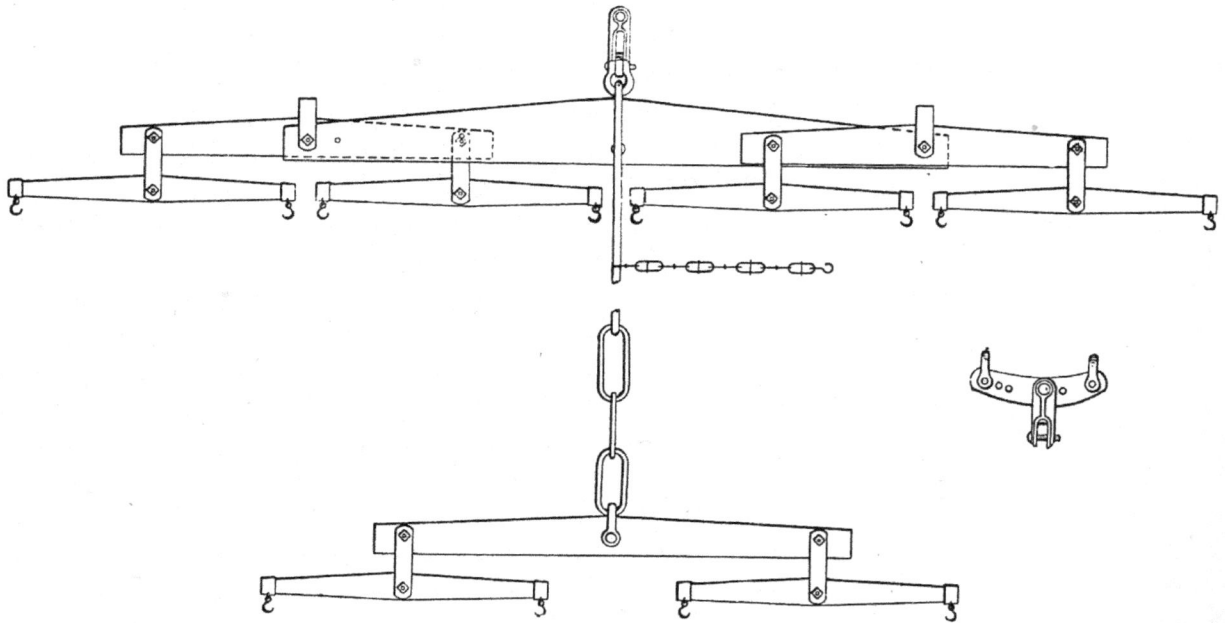

No. 24.

Fig. 425. P & O center-fire six-up.

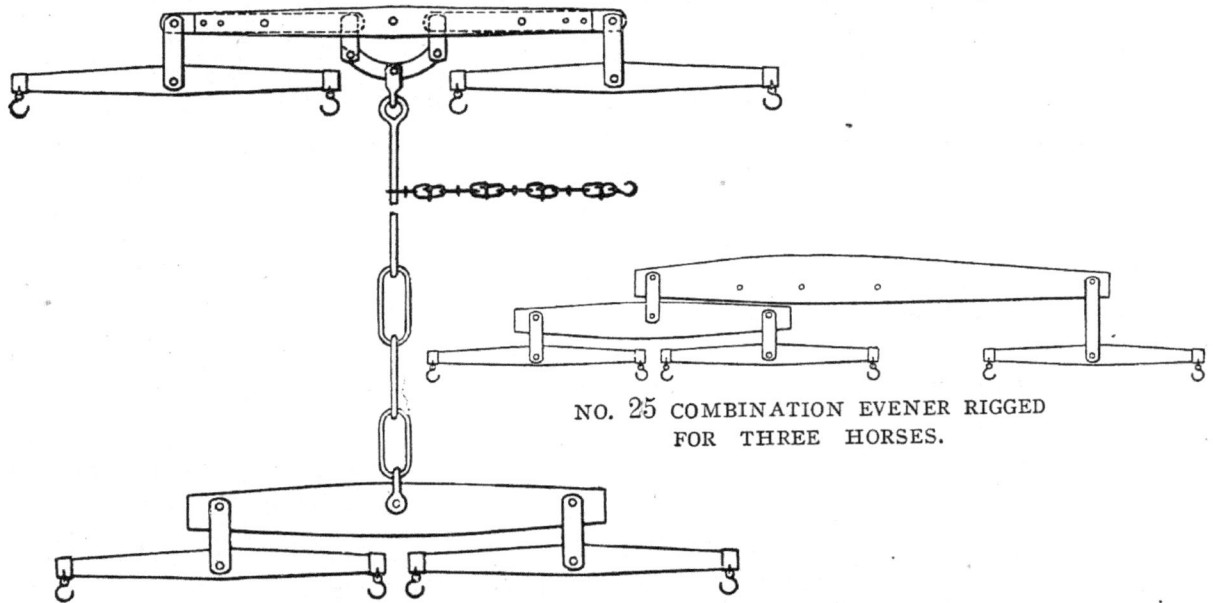

NO. 25 COMBINATION EVENER RIGGED
FOR THREE HORSES.

NO. 25 COMBINATION EVENER RIGGED FOR FOUR HORSES TANDEM.

Fig. 426. P & O eveners including the most conventional style of factory steel four-up.

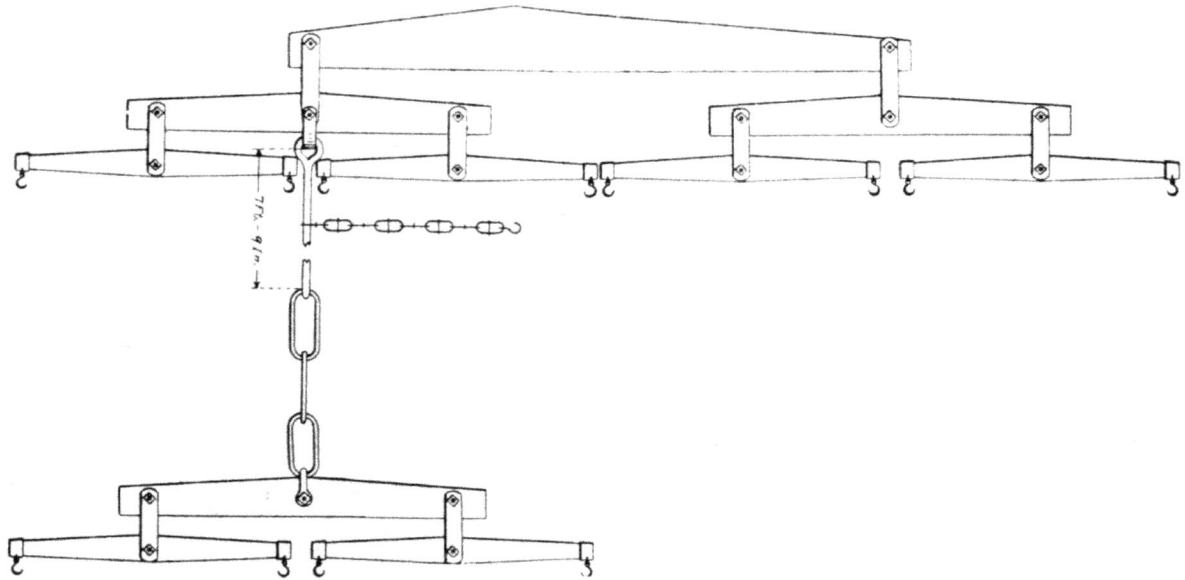

No. 23.

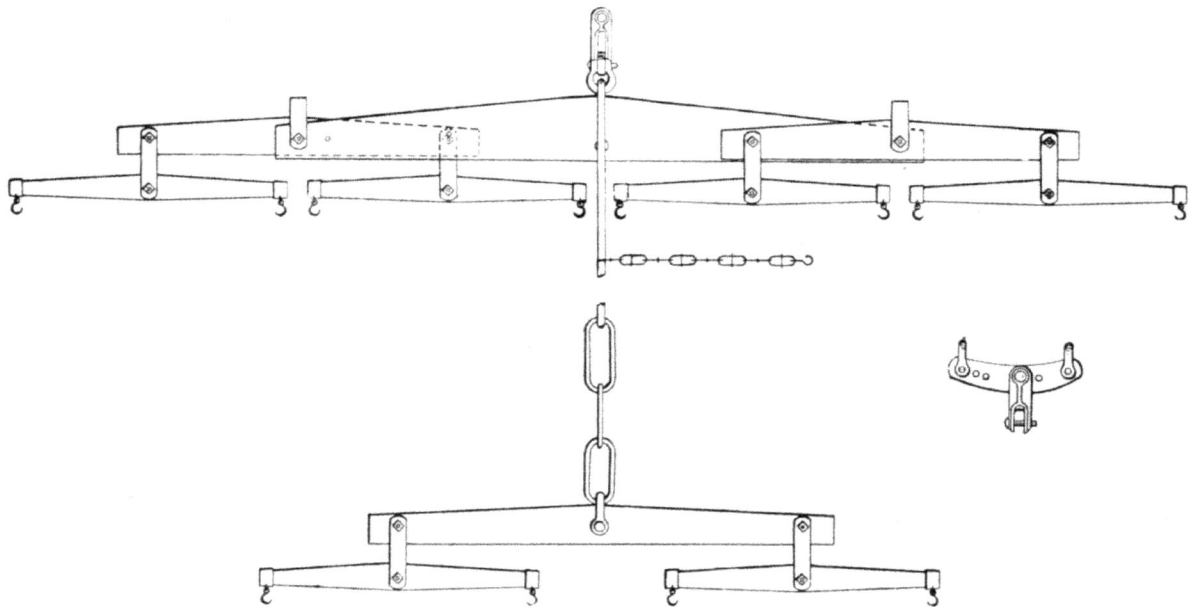

No. 24.

Fig. 427. Additional P & O evener variations on six-ups.

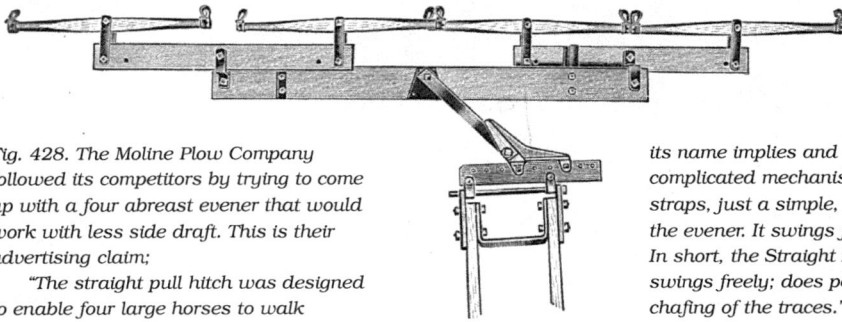

Fig. 428. The Moline Plow Company followed its competitors by trying to come up with a four abreast evener that would work with less side draft. This is their advertising claim;

"The straight pull hitch was designed to enable four large horses to walk straight, ahead of a 12 inch gang plow (when working abreast) with the furrow horse walking in the furrow. It is just what its name implies and does the work with but few parts - no complicated mechanism, no intricate arrangement of levers and straps, just a simple, neat, strong connection between the clevis and the evener. It swings freely and works to the best possible advantage. In short, the Straight Pull Hitch is easily understood quickly attached; swings freely; does perfect work with no crowding of horses or chafing of the traces."
All other unbiased information of the time seemed to indicate otherwise.

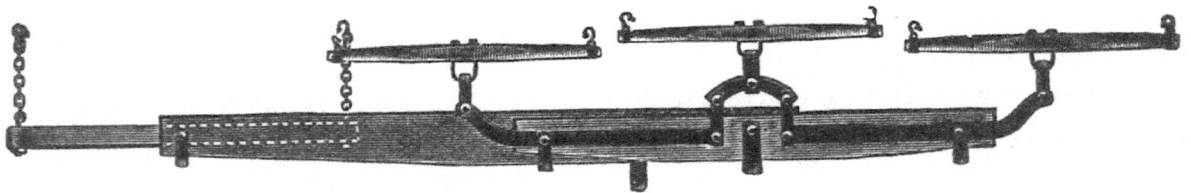

Fig. 429. The Moline Patented four abreast equalizer was unusual as it provided chains for the near horse. The company also offered an extra heavy version which featured springs instead of chains, which absorb all shocks, protecting the fourth horse should the other three suddenly stop.

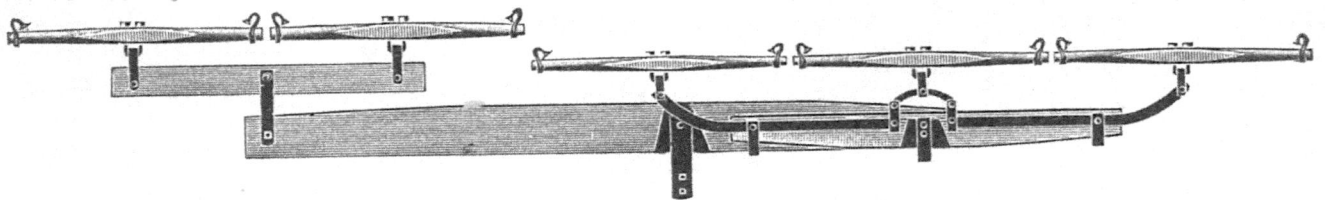

Fig. 430 Moline's convertible evener allowed the farmer to use team, three abreast, four abreast and five abreast all with the equipment provided.

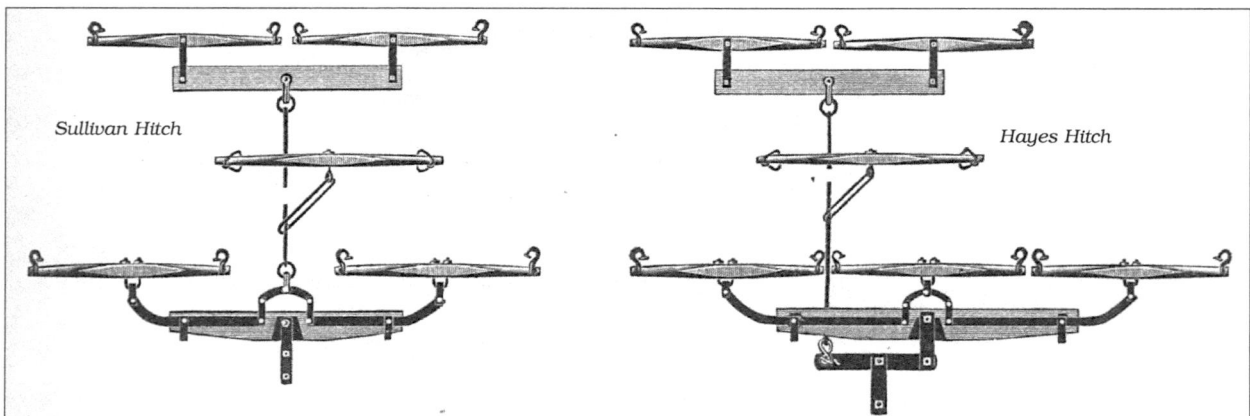

Sullivan Hitch

Hayes Hitch

Fig. 451. Moline Plow Company offered these two "strung-out" hitch sets complete with neckyoke.

3-Horse.

4-Horse.

5-Horse.

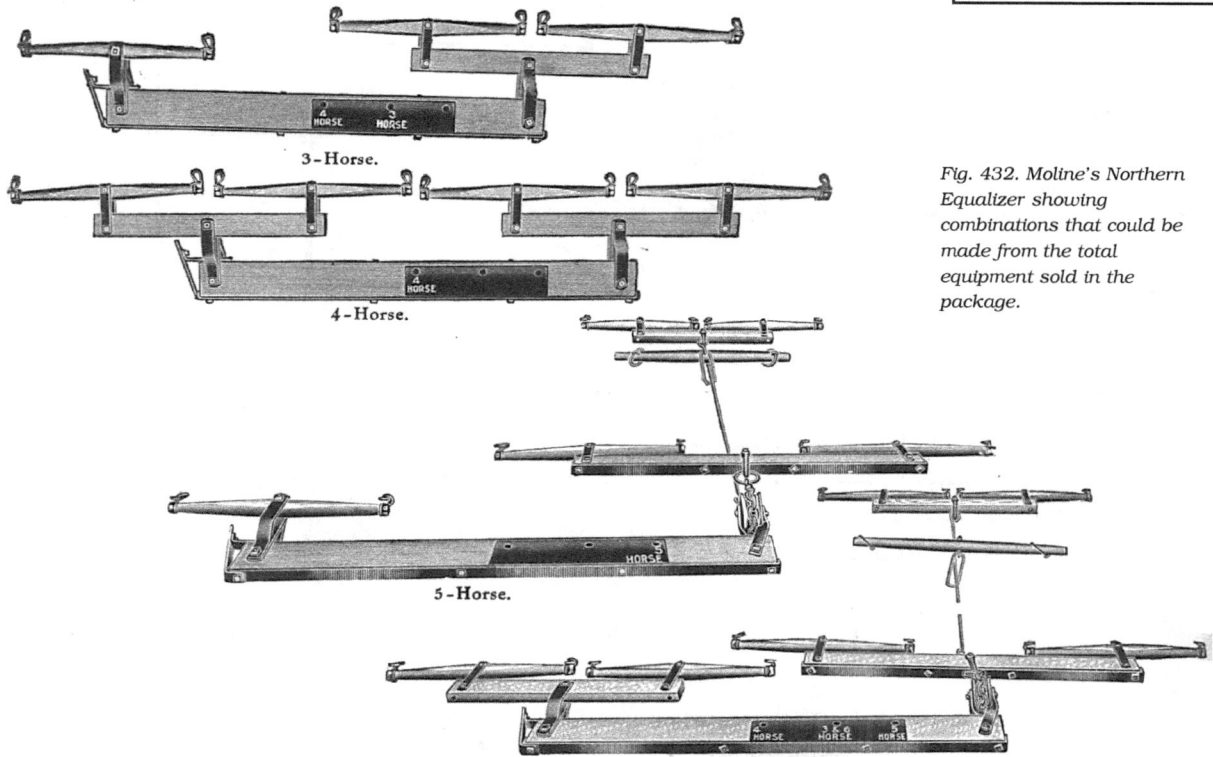

*Fig. 432. Moline's Northern Equalizer showing combinations that could be made from the total equipment sold in the package.*

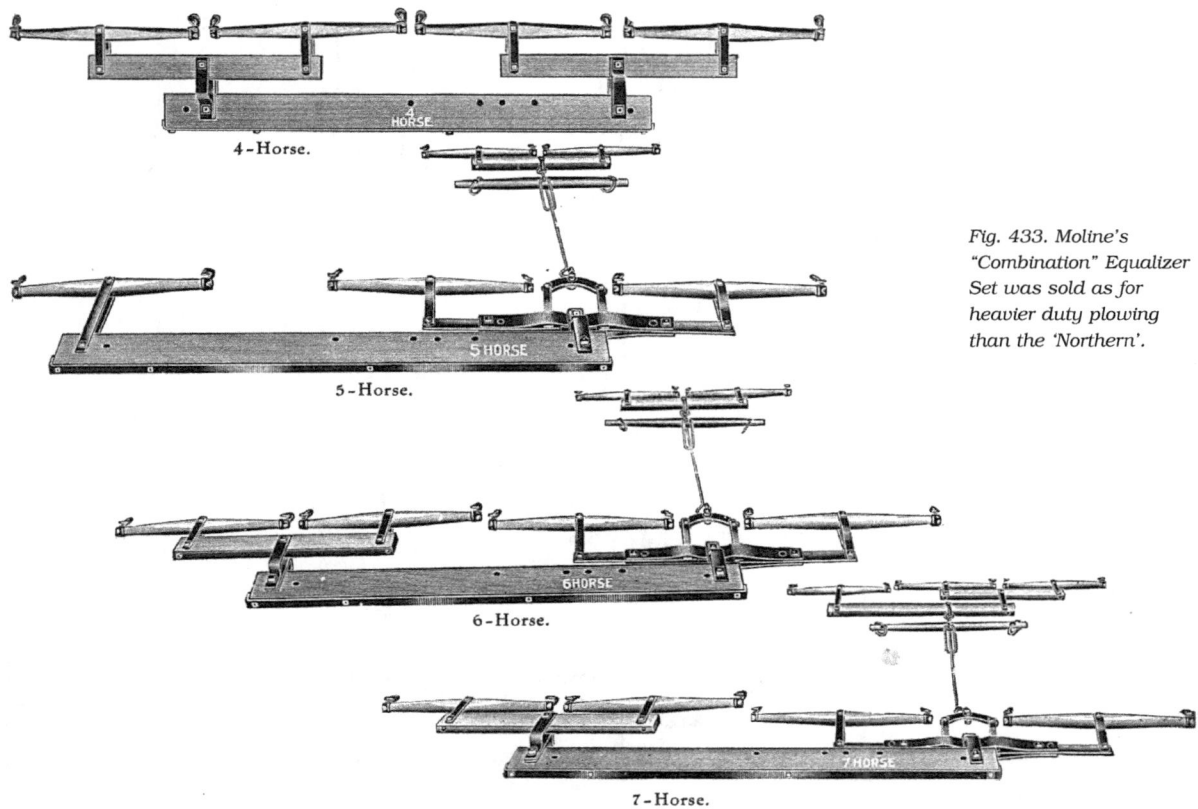

4-Horse.

5-Horse.

6-Horse.

7-Horse.

*Fig. 433. Moline's "Combination" Equalizer Set was sold as for heavier duty plowing than the 'Northern'.*

Fig. 434. The Moline combination evener as furnished for Moline 'Rotary' Plows can be converted into either a 2, 3, 4, 5 or 6 horse evener. When used with six horses it is attached at C and used as shown in cut. It can be converted into a 5 horse by moving hitch to B and equipping the large evener with a singletree. To convert it into a 4 horse, it is hitched at D and the lead rod and bar A are taken off, replacing with doubletree E. For a 3 horse the main bar H is used and is coupled at C, putting on the singletree at J and leaving the doubletree at L. Any of the doubletrees may be used for 2 horses.

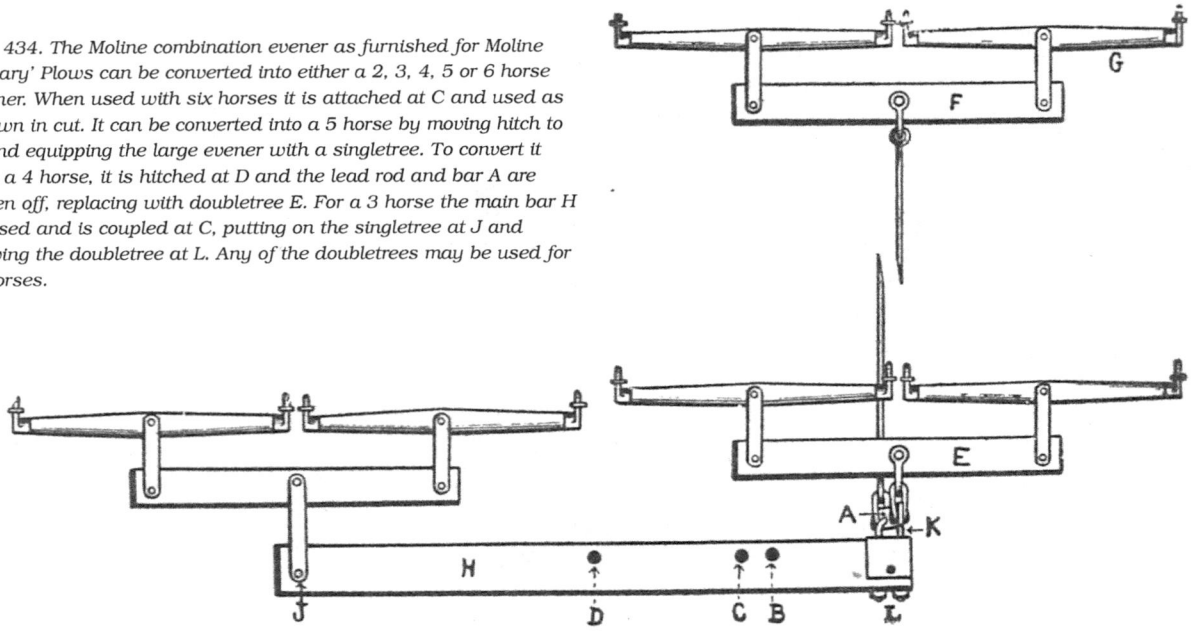

Fig. 435. John Deere, in 1910, offered a rope and pulley hitch much the same as the one being offered by White Horse Machine (see page 112).

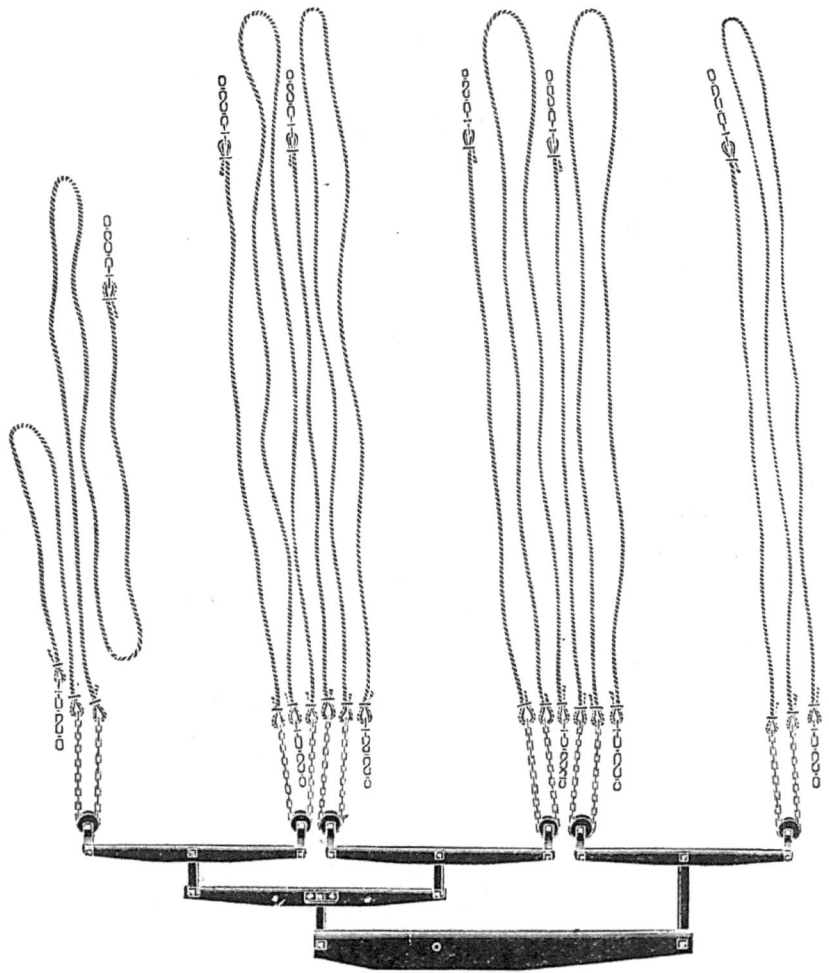

No. 81—Six-Horse Rope Evener

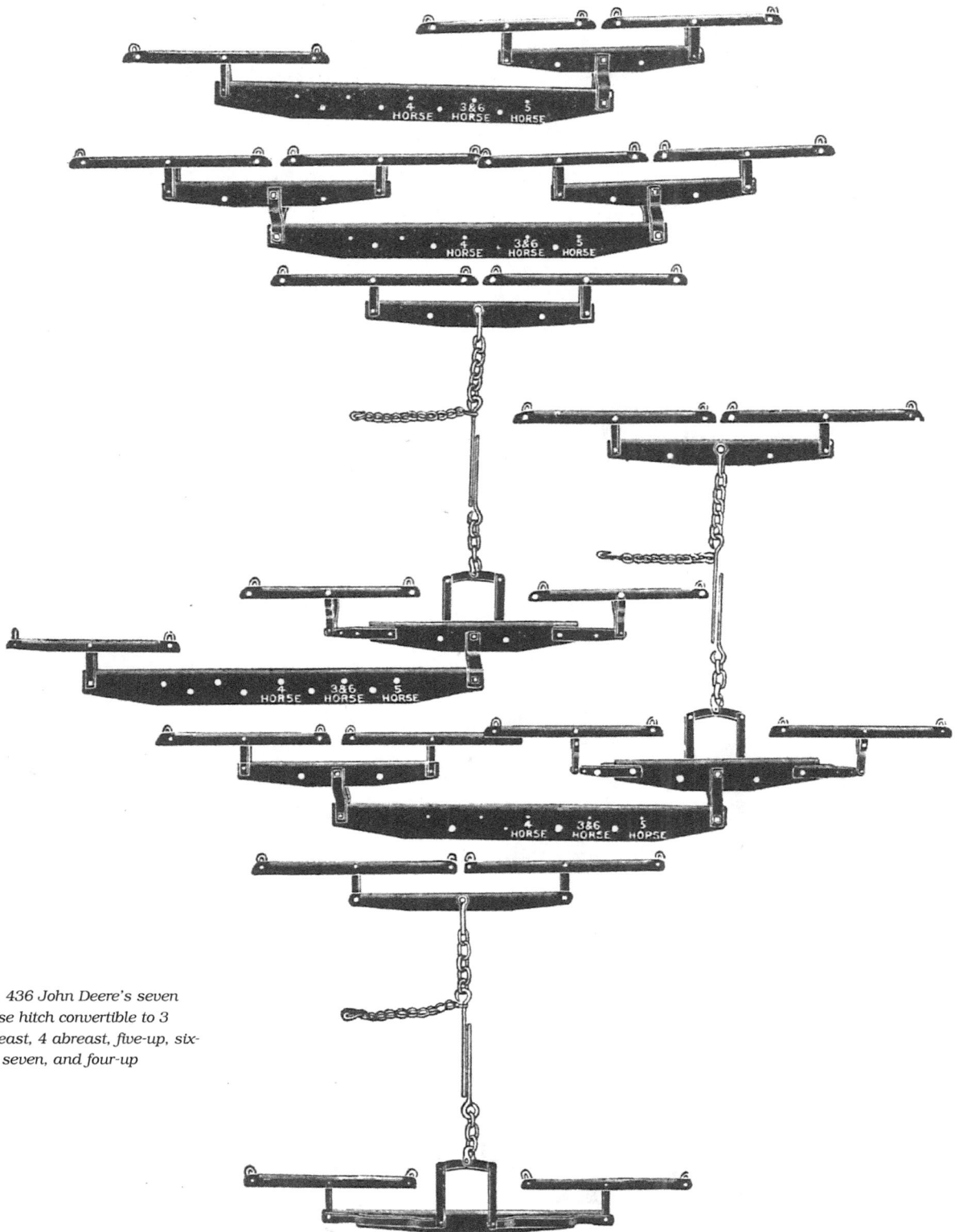

*Fig. 436 John Deere's seven
horse hitch convertible to 3
abreast, 4 abreast, five-up, six-
up, seven, and four-up*

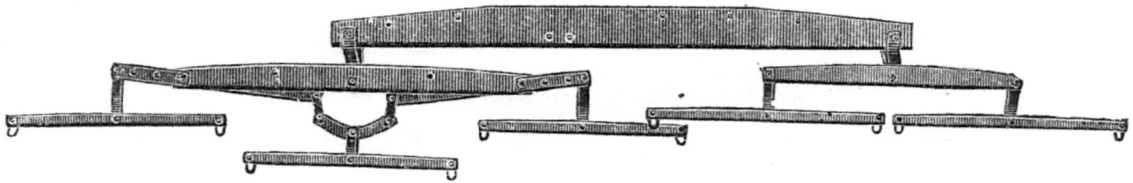

No. 65—Five-Horse-Abreast—Steel—Export

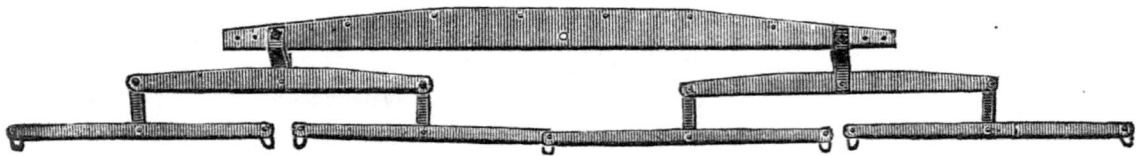

No. 66—Four-Horse-Abreast—Steel

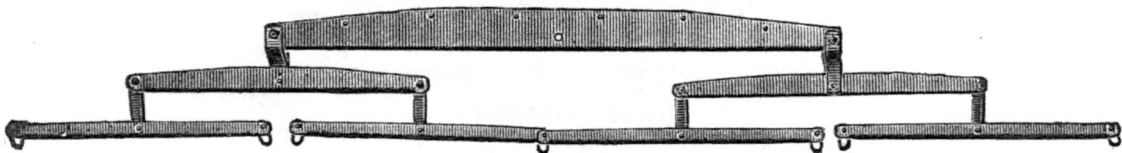

No. 67—Four-Horse-Abreast

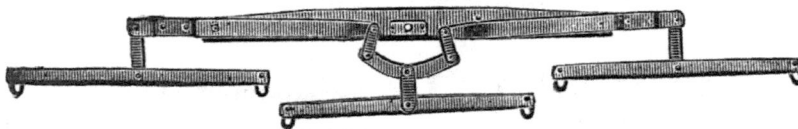

No. 68—Triple Tree—Steel

*Fig. 437. John Deere factory eveners from 1910*

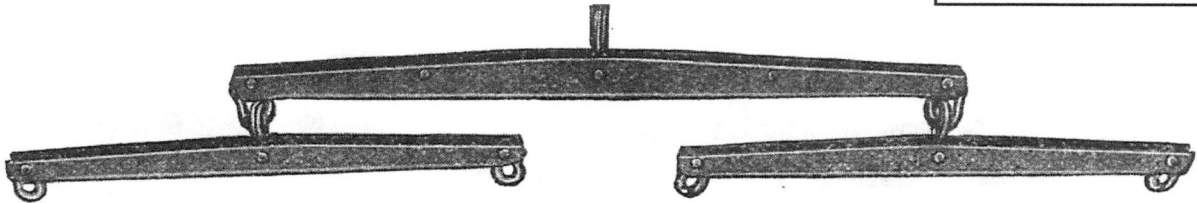

**No. 71—Steel Doubletree with No. 10 Singletrees—(Special St. Louis)**

**No. 72—Steel Doubletrees**

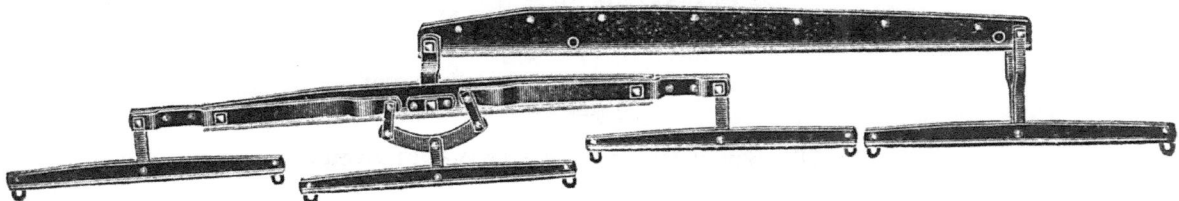

**No. 74—Four-Horse-Abreast Steel Evener—Export**

*Fig. 438. John Deere eveners from 1910*

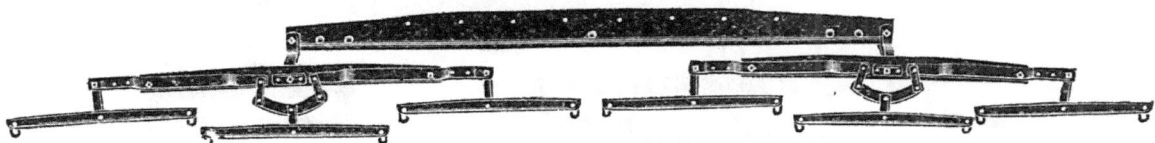

**No. 76—Six-Horse-Abreast Steel Evener—Export**

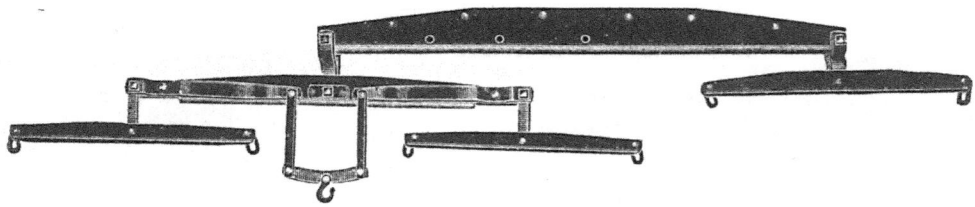

No. 77—Five-Horse Steel Strung-Out Evener—Less Lead Doubletree and Rod

*Fig . 439. John Deere eveners from 1910.*

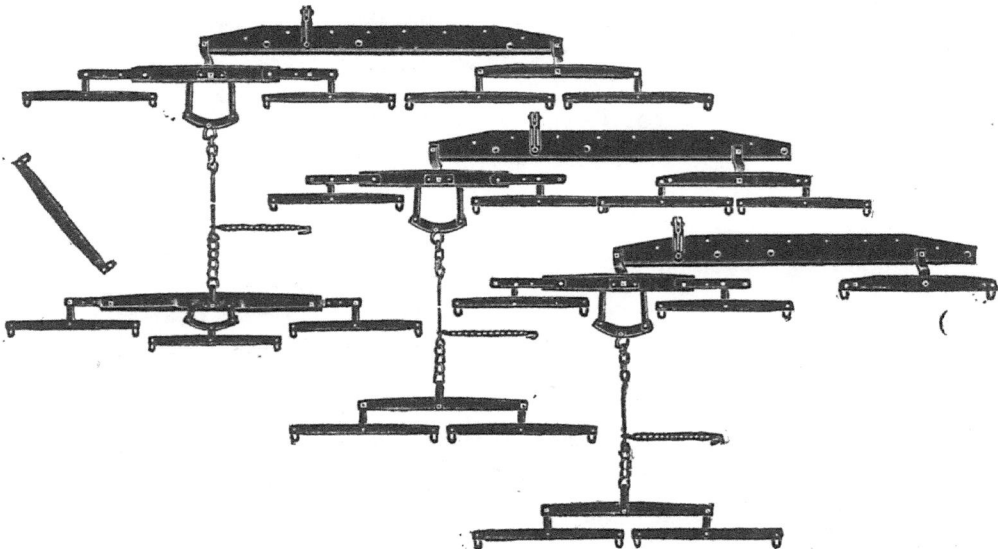

No  78—Seven-Horse Steel Strung-Out Evener

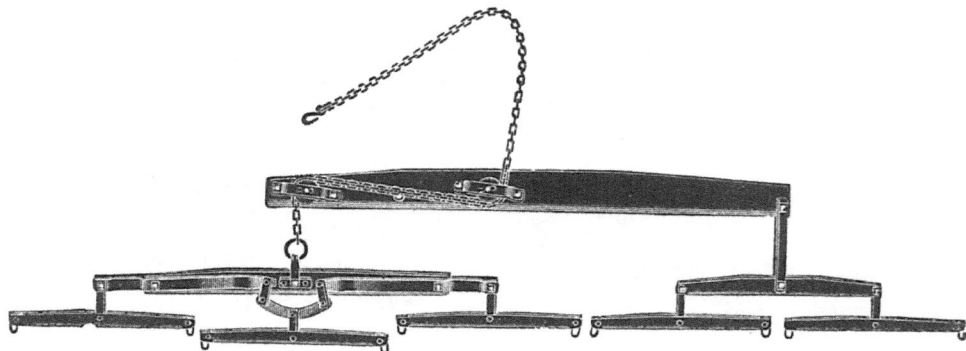

No. 80—Five-Horse-Abreast Chain Evener—Export

## Pioneer Eveners

Pioneer Equipment of Dalton, Ohio, makes a full range of new steel and wood eveners and neckyokes. These items, which are sold all over North America, receive extensive field testing from the neighboring Amish farm community.

Fig. 440. (Top left) A hitched view of a heavy duty Pioneer four abreast evener being used on Pioneer Gang plow.

Fig. 441 & 442. Two working views of the Pioneer 4 abreast offset plow evener for the Pioneer sulky plow.

## White Horse Machine Shop - rope and pulley systems.

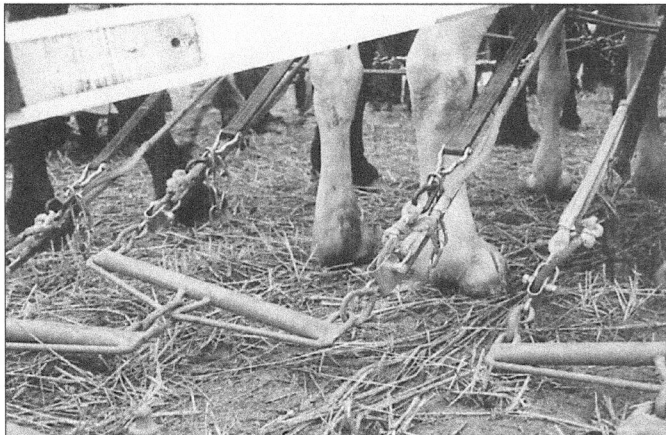

White Horse Machine of Gap, PA, makes a wide assortment of eveners for sale. At recent Horse Progress Days their rope and pulley systems have drawn quite a bit of attention. With this system, see page 112, only one evener is used at the wheel position with all other animals hooked by ropes and pulleys in line. It is so dramatically different from what most modern horsefarmers are accustomed to that, in spite of its obvious strong points, it is slow to catch on in certain circles.

Fig. 443 & 444. These closeups illustrate how the wheel tug or trace is fastened to the nylon rope which passes through the pulley, fastened to the evener, and then forward to the horse or horses directly in front. Hooks are braided or tied into the ends of the ropes to facilitate a quick hookup. The geometry of this system has the in-line animals working as integral power units which are equalized at the evener.

Fig. 445 & 446. These side by side photos at Horse Progress Days Ohio show an eight horse hitch, 4 x 4, using the White Horse rope & pulley rig.

Fig. 447. Tugs tight and horses plowing the pulleys are making sure everyone is equal.

Fig. 448. A closeup view of the rope and pulley rigging required between wheelers and leaders with a eight horse hitch.

Chapter Eleven
# Makes and Models

*Rolland Chilled Plow*

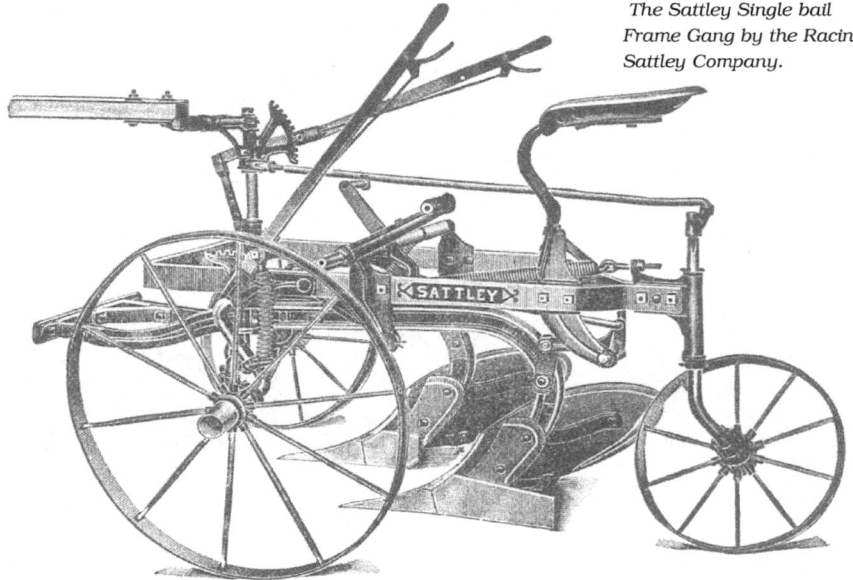
*The Sattley Single bail Frame Gang by the Racine-Sattley Company.*

When we first conceived of this book there existed, momentarily, the crazy notion that it would include an illustration of every horsedrawn plow ever made! If we had access to that information, and we don't, it would have required a book in excess of a thousand pages. What we have done is present the largest collection of plow illustrations ever gathered between the covers of one book. We have identified over 65 different plow makers, most from the period of 1890 to 1925, and those companies had (and have) models and attachment options which number in the several thousands. We hope you find, here, the information you might seek.

*On this page are a smattering of lesser known makers. This chapter is organized alphabetically except for those misc. makes and the new companies which appear at the end. Keep in mind that the bigger outfits, i.e. International and John Deere, bought out many others. You may discover your John Deere plow with a different maker's name on it. This is NOT a complete set of horsedrawn plows.*

**"LITTLE YANKEE" THREE WHEEL PLOW**
MANUFACTURED BY THE
**GRAND DETOUR PLOW CO.**
DIXON, ILLINOIS.

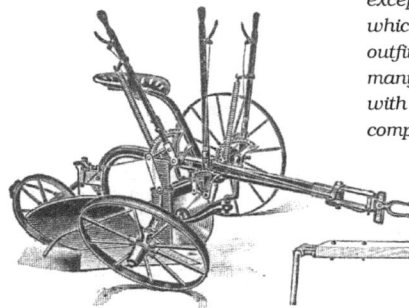
*The Sattley Royal Blue frameless.*

NEW ECLIPSE SPRING LOCK HIGH FOOT LIFT
**PLOW WITH FULLER & JOHNSON NEW ECLIPSE GANG**

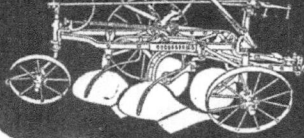
**SOLID COMFORT**
TONGUELESS.
THE **"WONDER ON WHEELS"**
**Self Guiding.** Uses a wheel landside. Two horses instead of three. A ten year old boy instead of a plowman. No pole (except among stumps). No side draft. No neck weight. No lifting at corners. Easier driving, straighter **LIGHTER DRAFT THAN ANY** furrows, and **PLOW** on or off wheels. Will plow any ground a mower can cut over. No equal in hard, stony ground, or on hillsides. Our book, "FUN ON THE FARM," sent Free to all who mention this paper.
**ECONOMIST PLOW CO.** SOUTH BEND INDIANA.
☞ Special prices and time for trial given on first orders from points where we have no agents

B.F. Avery & Sons, founded in 1825, were located in Louisville, Kentucky and manufactured an extensive line of excellent horsedrawn plows, tillage implements, and harvesting machinery early in the twentieth century. The following illustrations were originally published in a 1925 catalog.

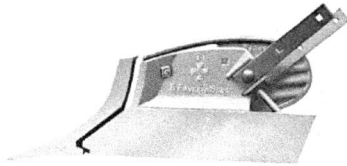

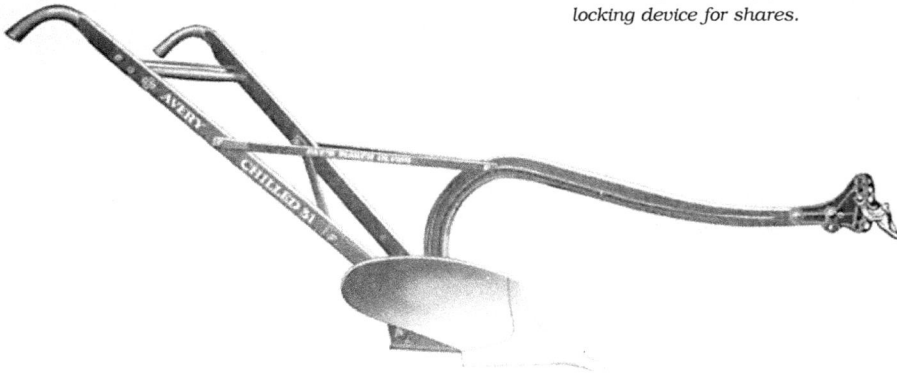

Illustrating the Avery patented locking device for shares.

Showing extension moldboard attached

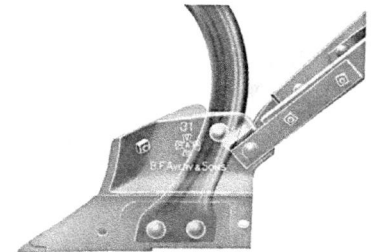

Model 31 Chilled.

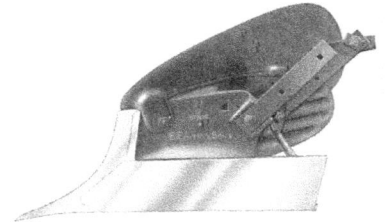

X-ray view showing Avery method of attaching beam foot to standard.

Common share

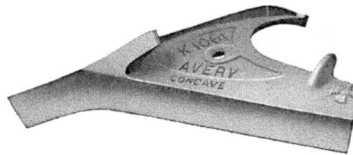

Concave share

Drop-forged steel share

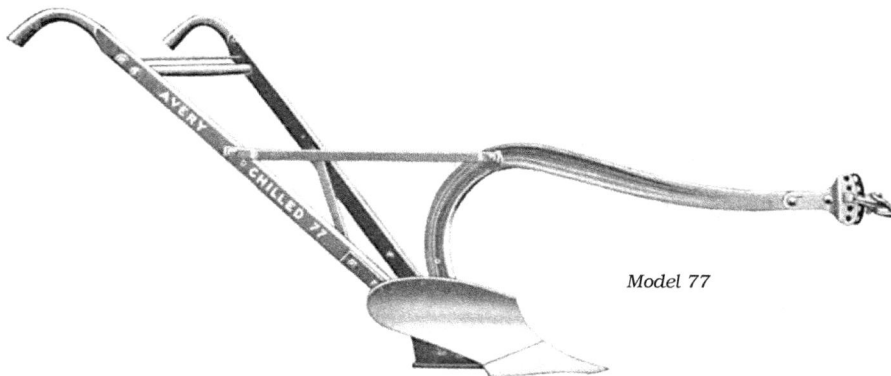

Model 77

Cricket one-horse

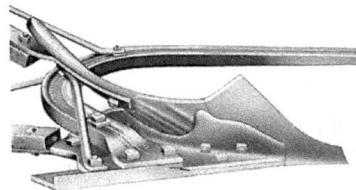

Model 77 frog shown

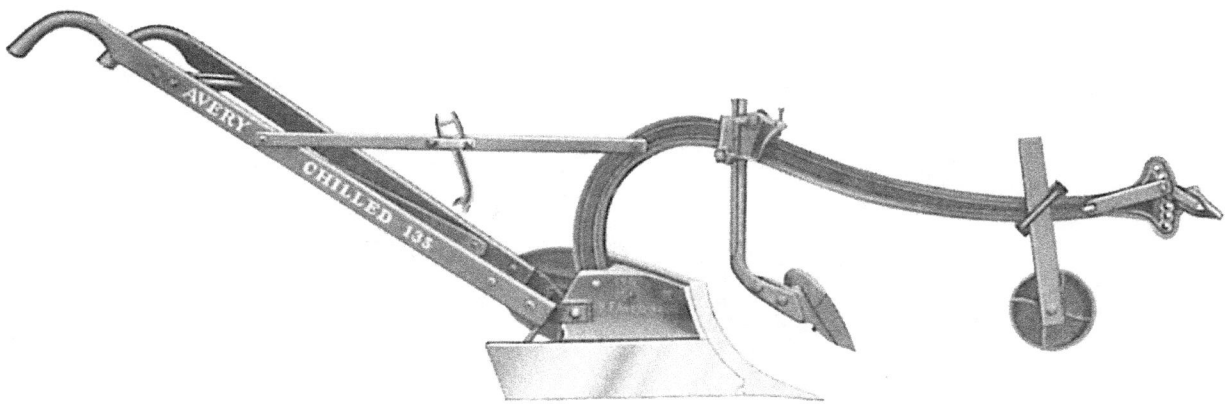

*Model 135 with gauge wheel and jointer.*

*Slatboard backside*

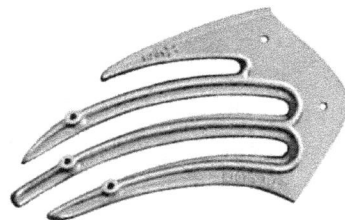

*Slatboard*

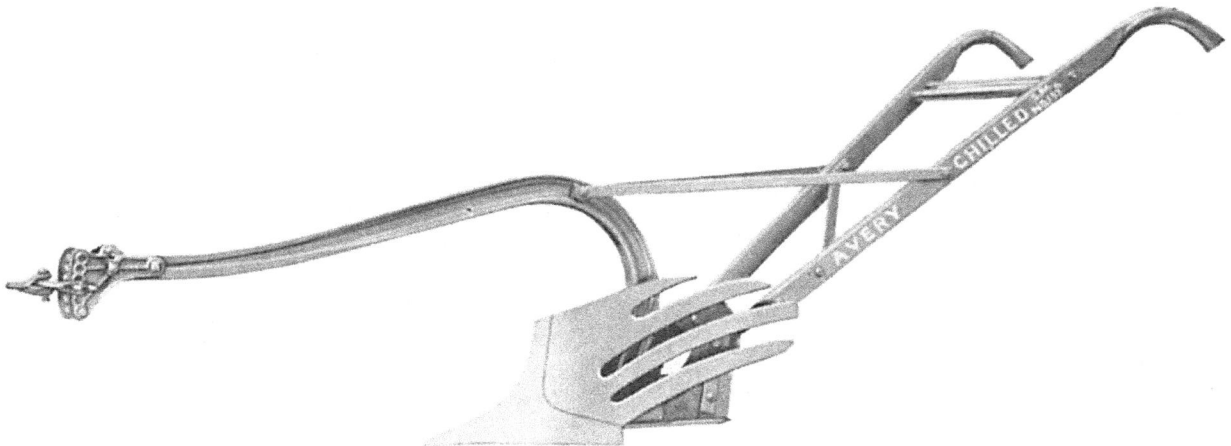

*Slat Moldboard Walking Plow*

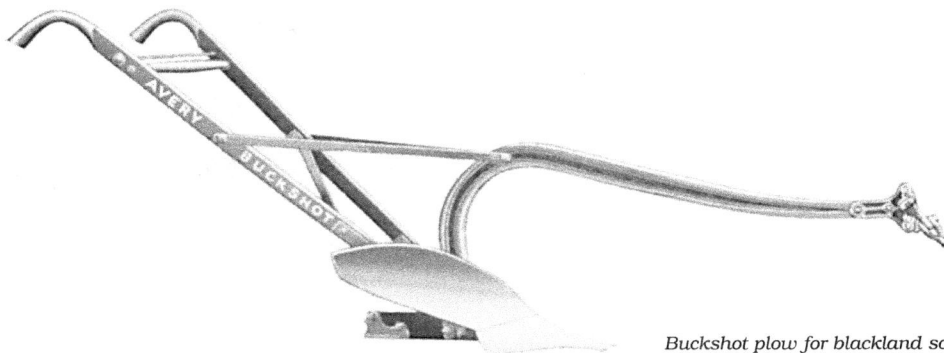

*Buckshot plow for blackland soil*

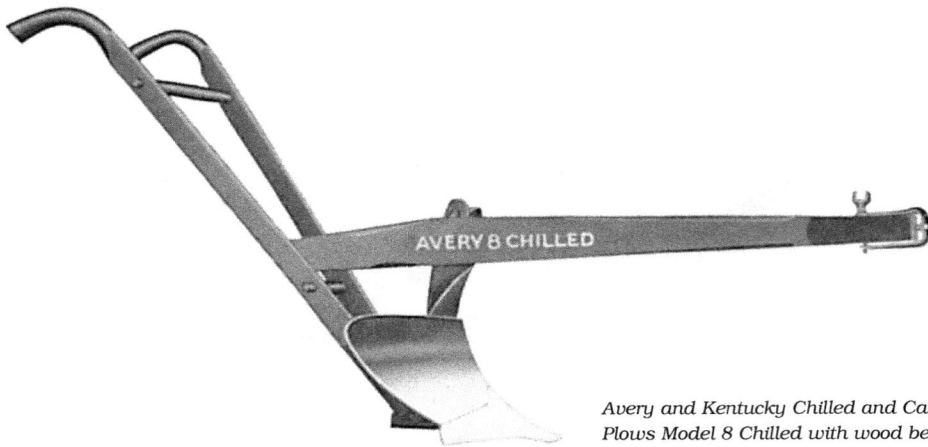

Avery and Kentucky Chilled and Cast
Plows Model 8 Chilled with wood beam.

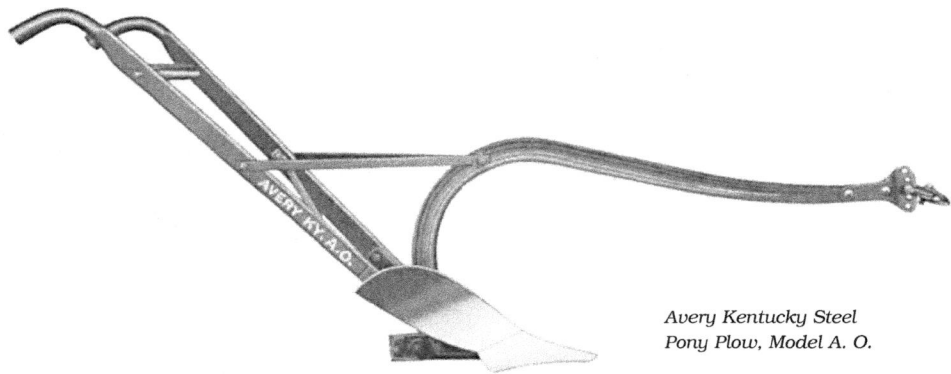

Avery Kentucky Steel
Pony Plow, Model A. O.

Avery Kentucky wooden beam Pony walking
plow sold with from 7 inch to 11 inch shares
and weighing from 39 to 46 lbs.

Oriole Blackland Plow
for sticky, waxy lands.

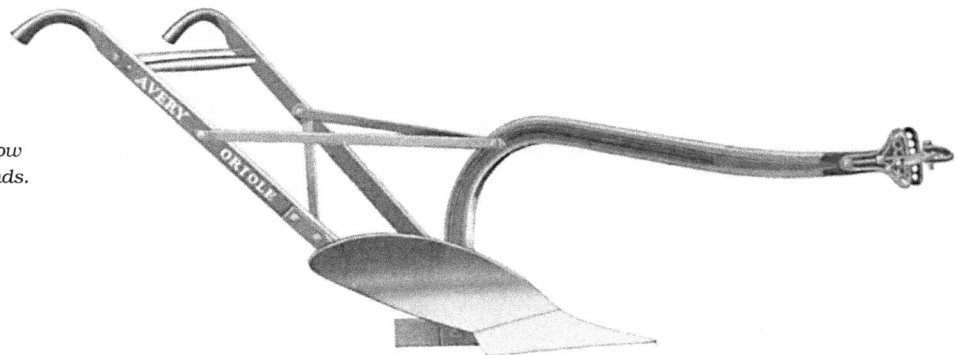

# Avery Plows

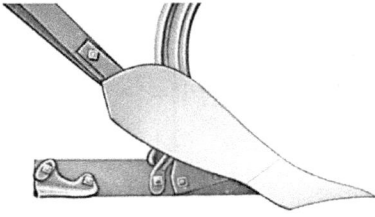

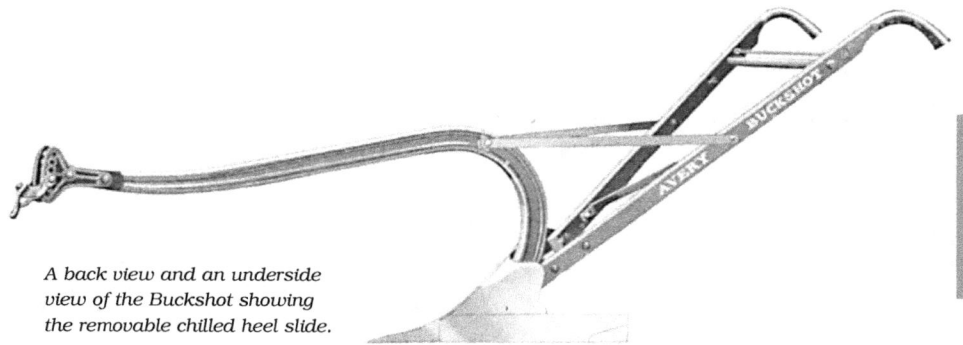

*A back view and an underside view of the Buckshot showing the removable chilled heel slide.*

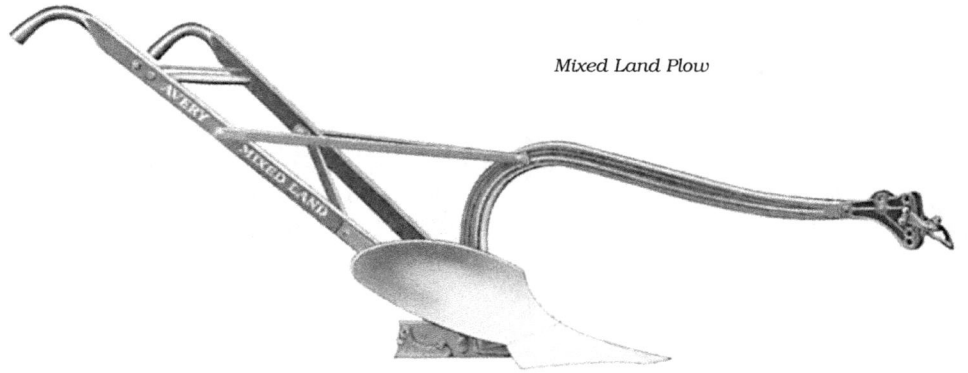

*Mixed Land Plow*

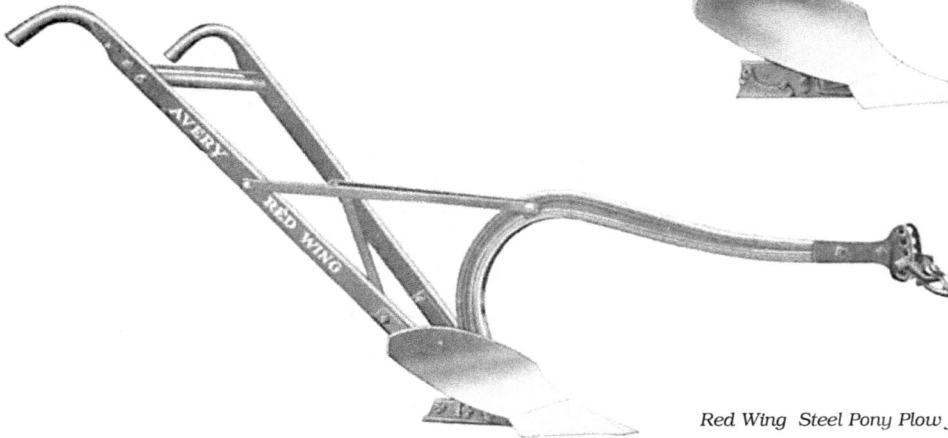

*Red Wing  Steel Pony Plow for mixed lands.*

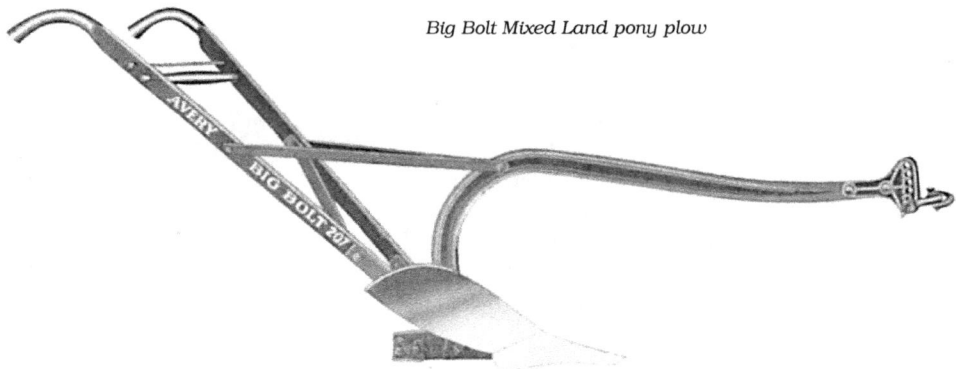

*Big Bolt Mixed Land pony plow*

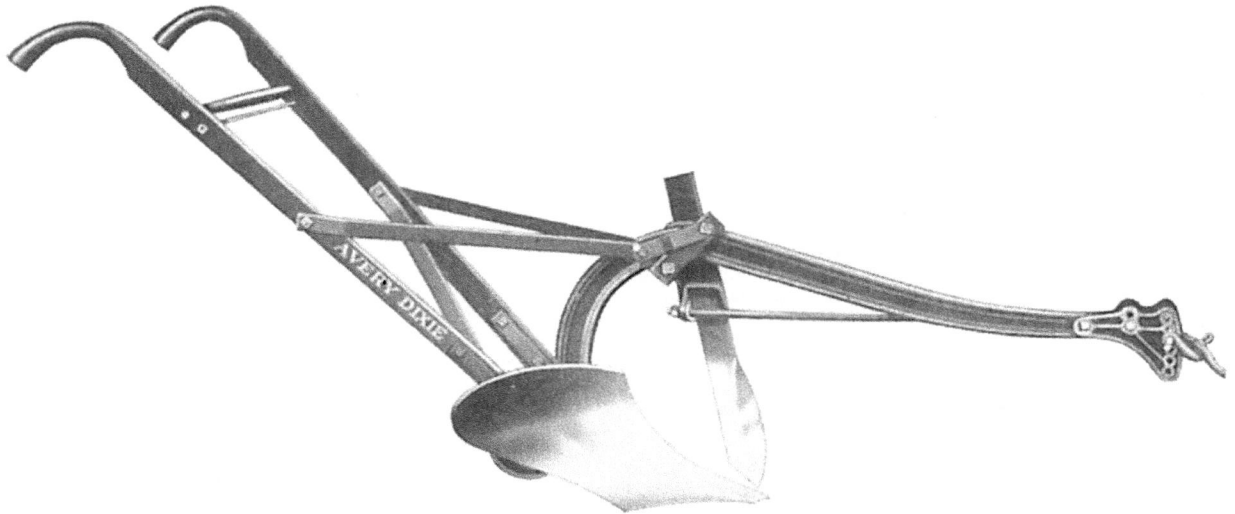

Dixie New Ground with rolling landside and root knife.

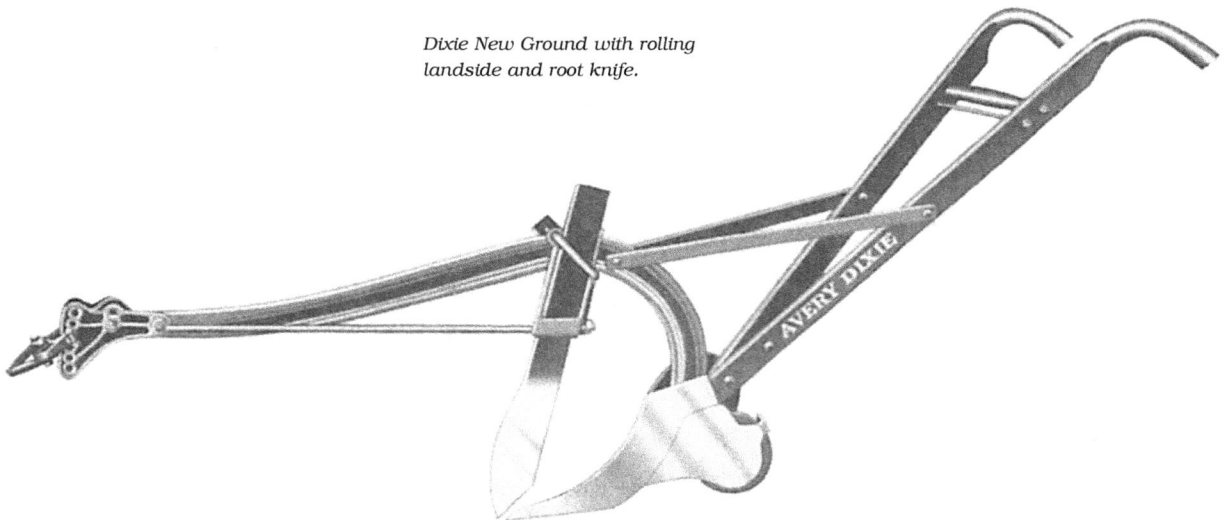

Samson New Ground. Built for plowing rough and stony new ground. It came with a 6 and 7 inch share and weighed 67 lbs.

Makes
& Models

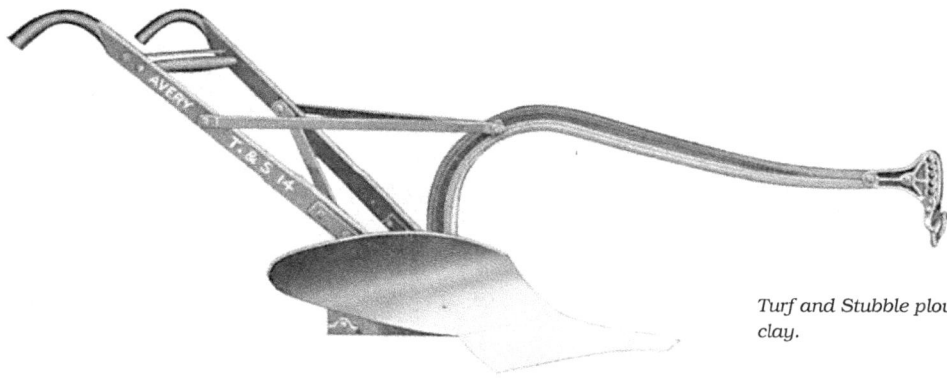

Turf and Stubble plow. For sod and heavy clay.

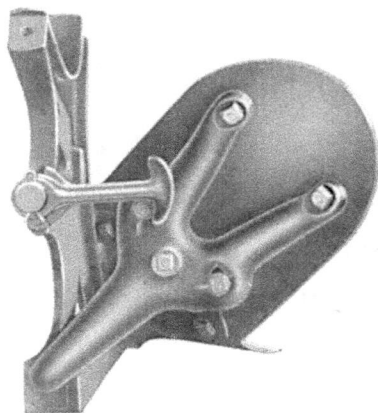

Lock mechanism

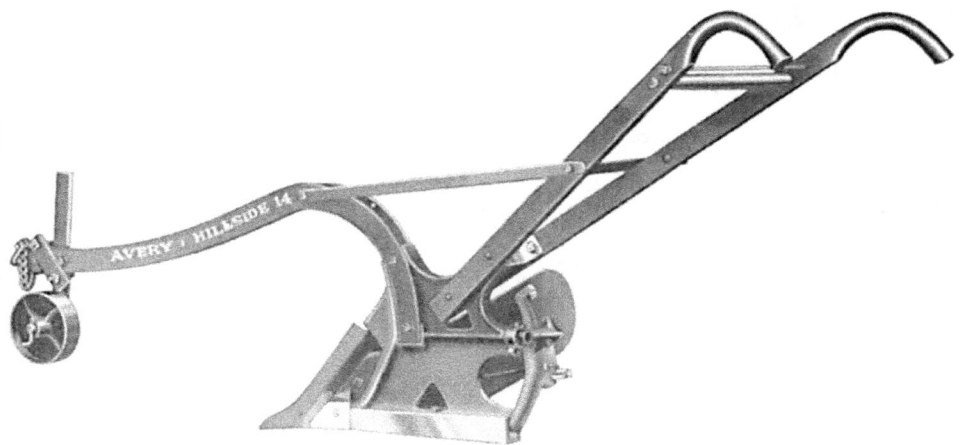

Avery New Series Hillside Plow Model 14 with a 10 inch share

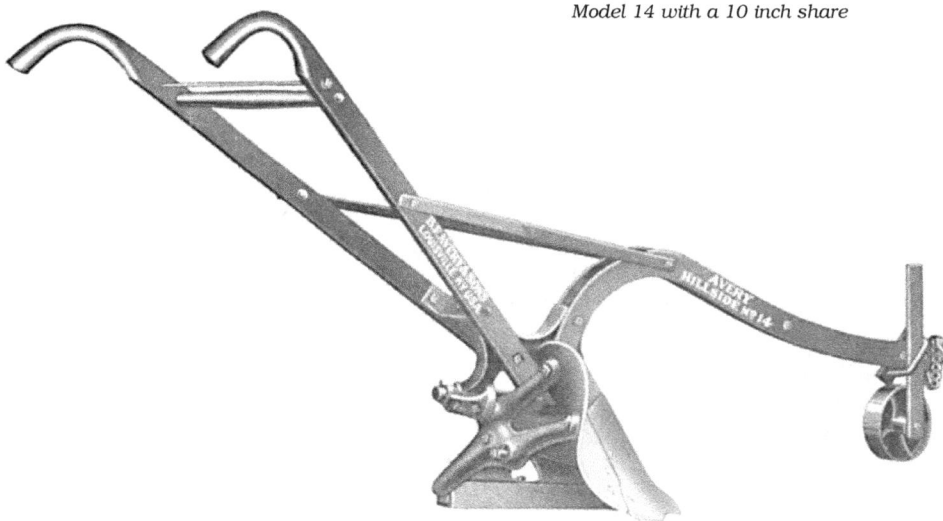

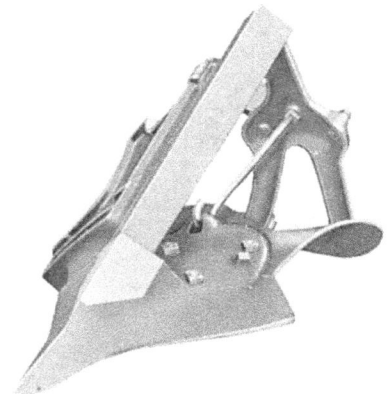

Detail of frog shield

*Bee Line Middle Bursters, Yellow Jacket model*

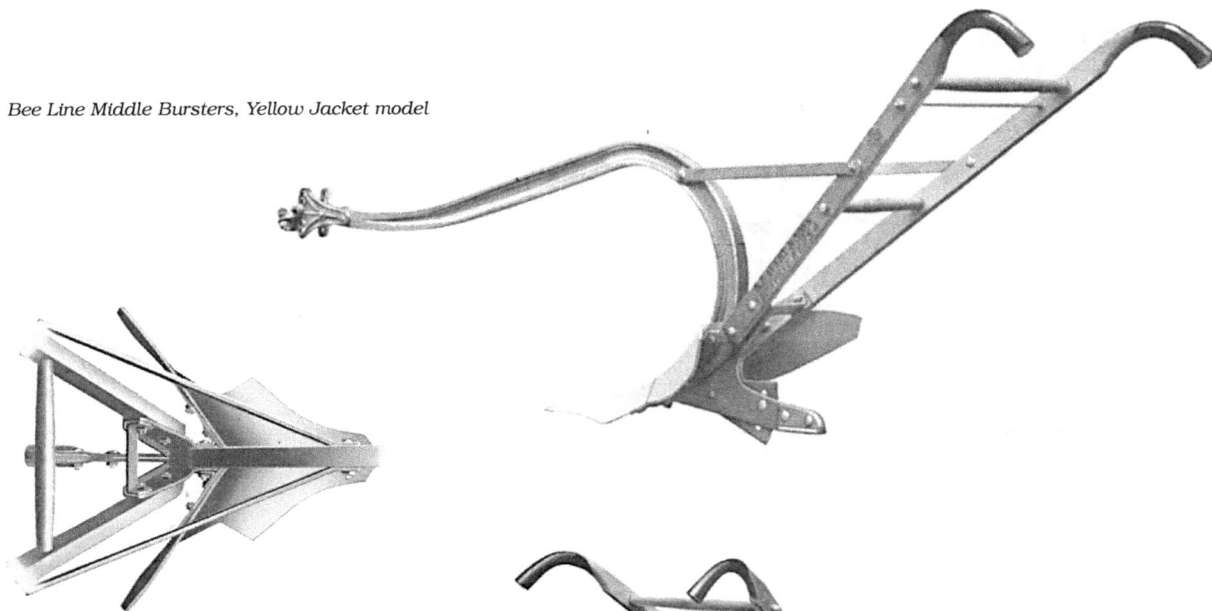

*Detail of bottom construction*

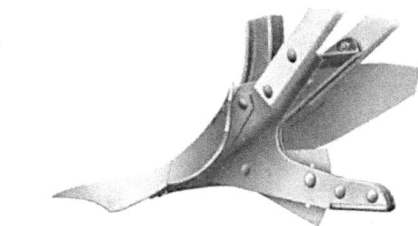

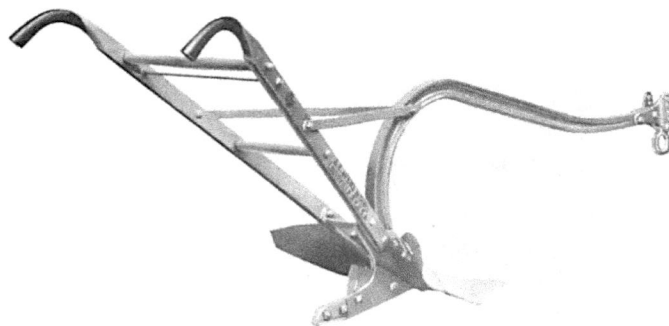

*Bee Line Middle Bursters, Sand Fly model*

*Bee Line Middle Bursters, Yellow Jacket H series with 16 inch share and 15 inch rolling coulter.*

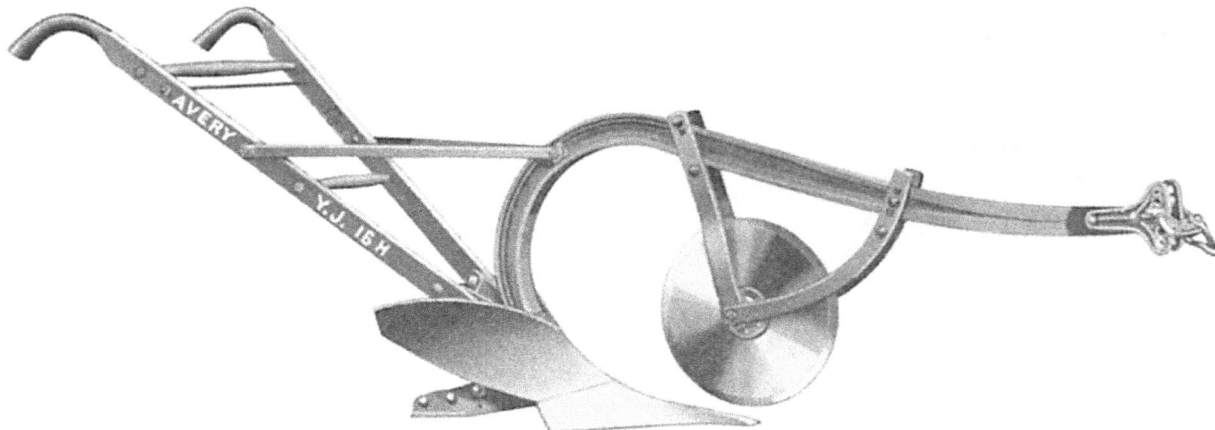

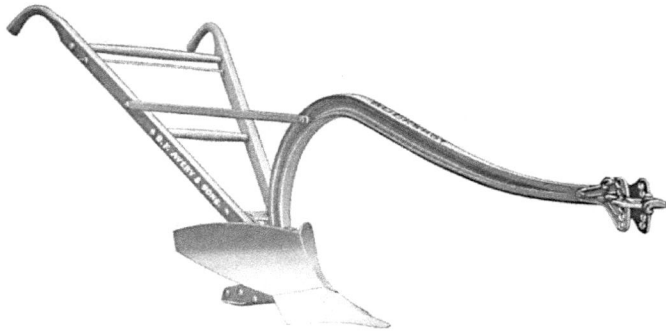

*Bee Line Middle Burster,*
*Buckshot model*

*Hard Pan Subsoiler*

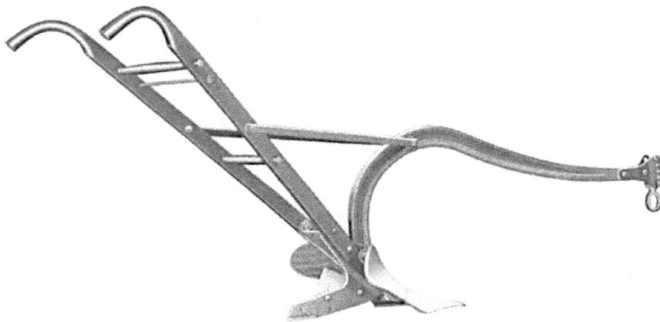

*Bee Line Middle Burster, Hornet Chilled.*

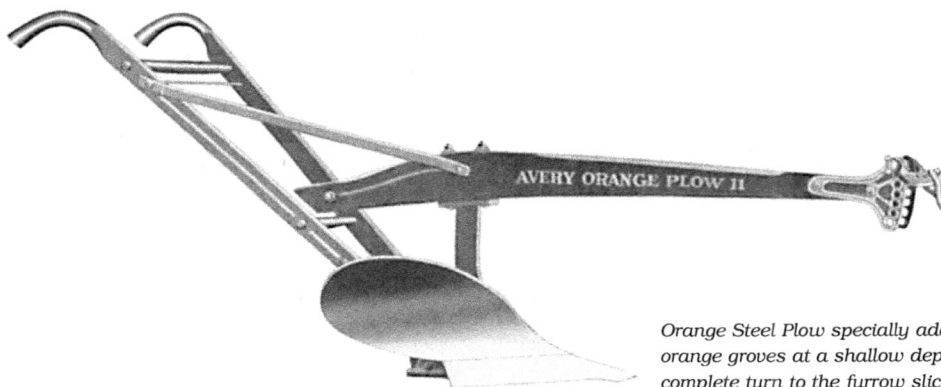

*Orange Steel Plow specially adapted to plow*
*orange groves at a shallow depth but with a*
*complete turn to the furrow slice.*

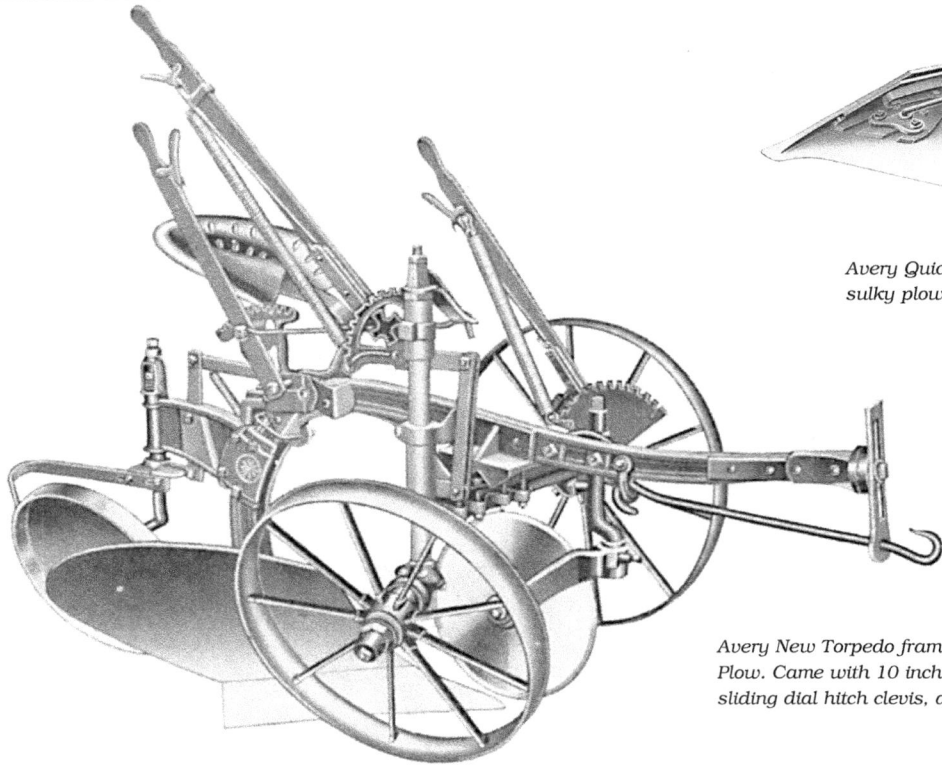

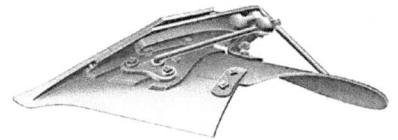

*Avery Quick detachable
sulky plow shares.*

*Avery New Torpedo frameless heavy duty Sulky
Plow. Came with 10 inch to 14 inch shares, a
sliding dial hitch clevis, and it weighed 458 lbs.*

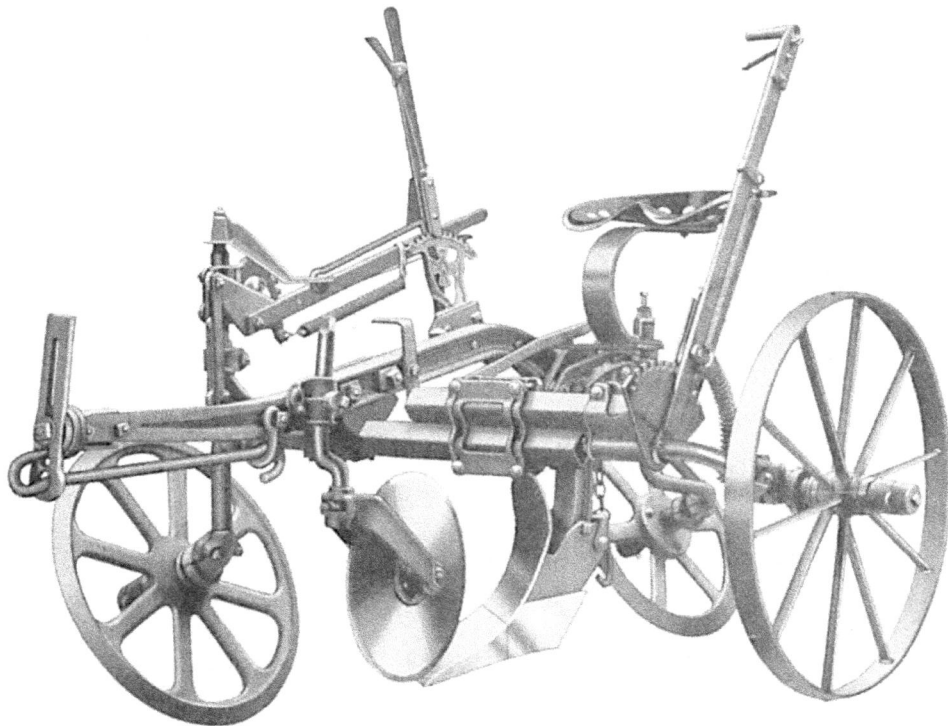

*Avery Blue Ribbon Sulky Plow. This frameless beauty was built extra
heavy for the Southern blackland cotton farmer. The wheels were
adjustable from 40 to 62 inches wide.*

*The Avery Blue Ribbon Sulky Plow setup with interchangeable middlebreaker (burster).*

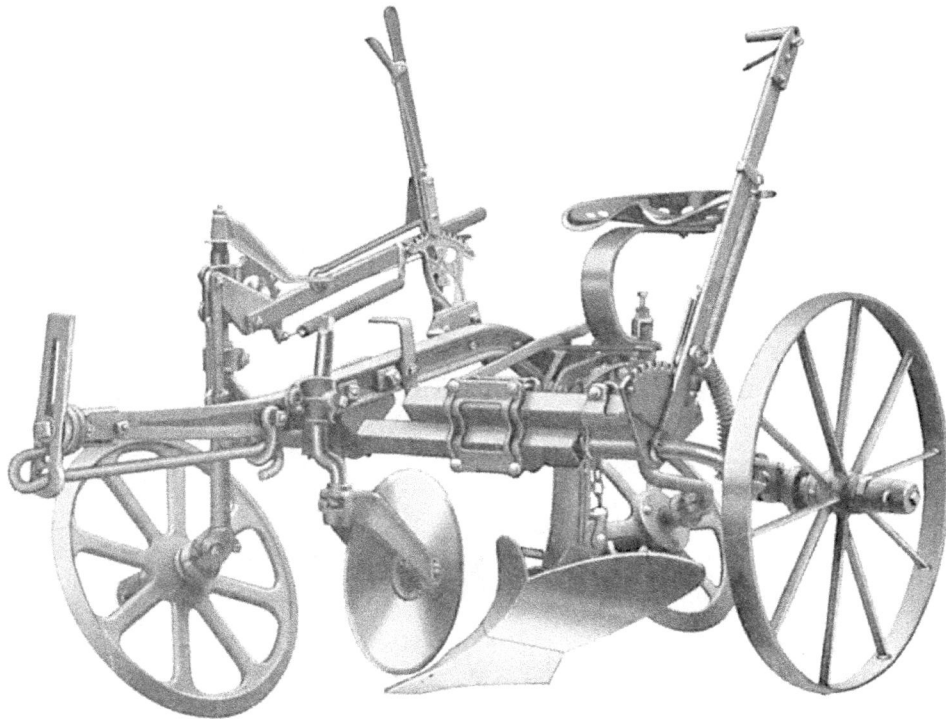

*No. 2 Foot-lift frame-style sulky plow*

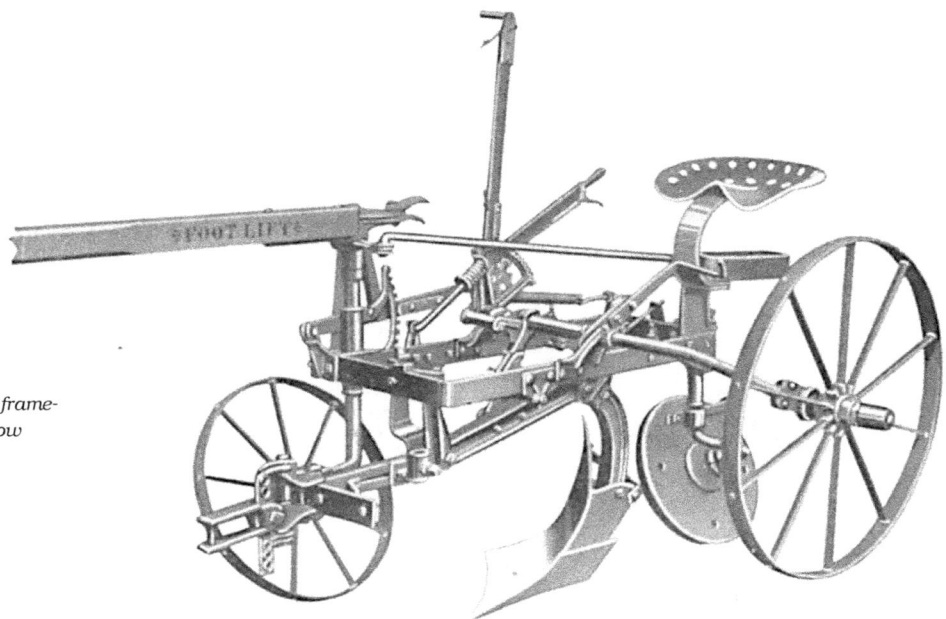

*Two-way Sulky Plow with steel
or chilled bottoms, came
equipped with 12 to 16 shares.*

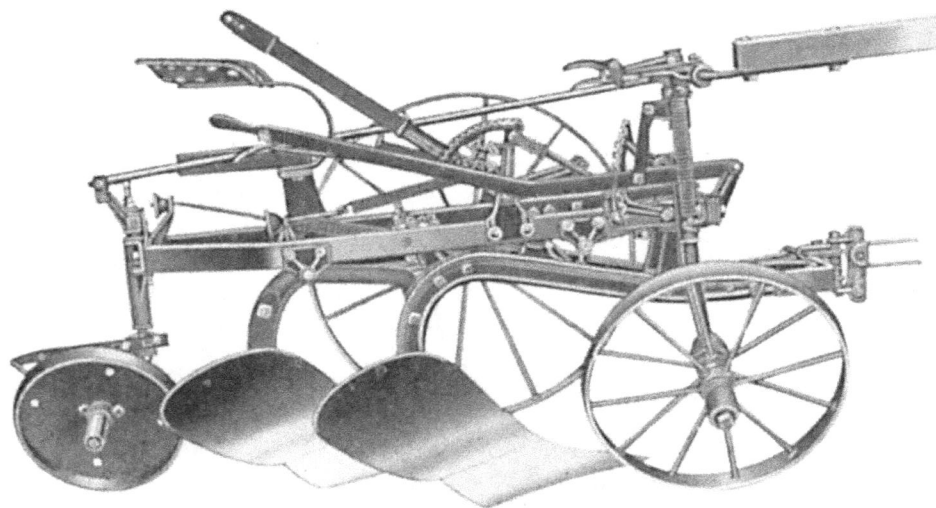

*Foot-lift double-bail gang plow
came with 10, 12, or 14 inch
bottoms and weighed 777 lbs.*

# Avery Plows

*Avery Bob Cat Disk Plow*

*Avery Red Lion three bottom Disk Plow*

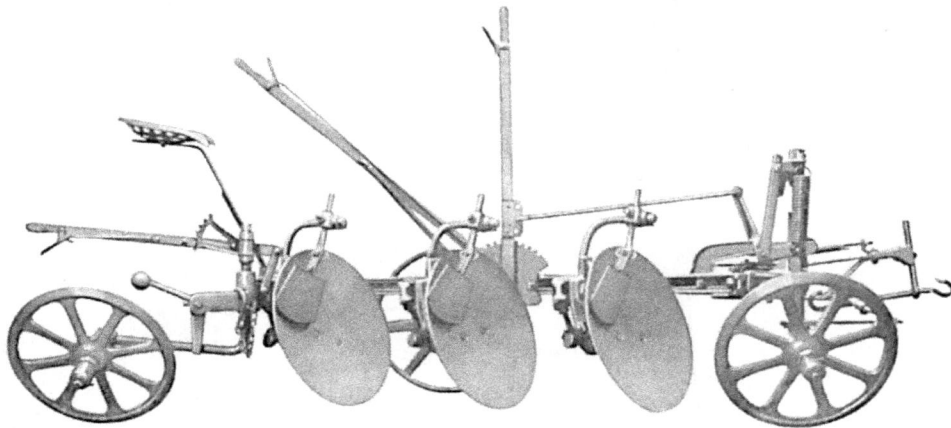

## Benecia Hancock Disc Plows

*Benecia Reversible Disc Plow. Recommended even for steep hillsides.*

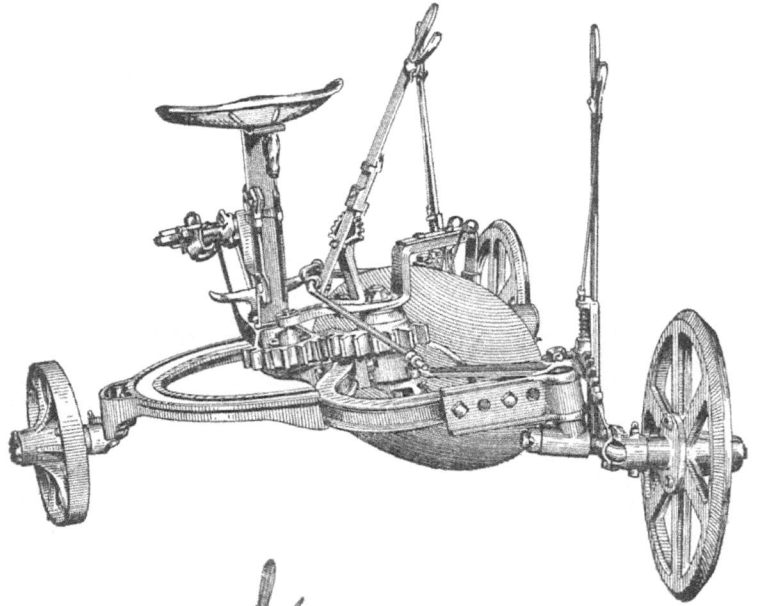

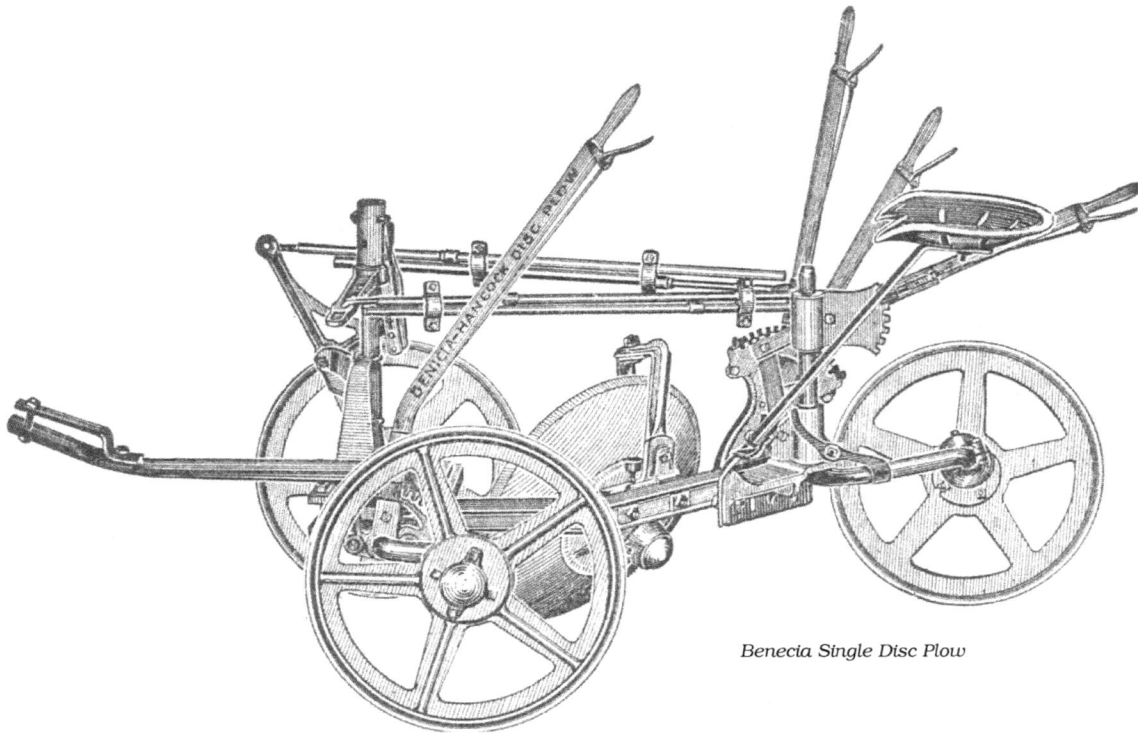

*Benecia Single Disc Plow*

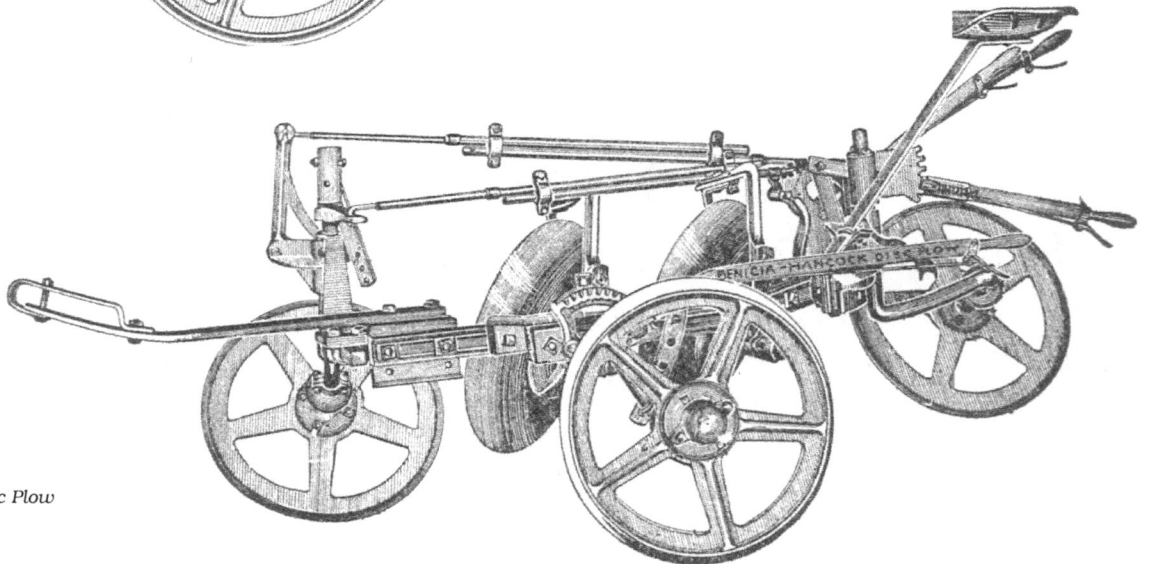

*Benecia Double Disc Plow*

# J.I. Case Plows

*The J. I .Case Plow Works were located in Racine, Wisconsin.*

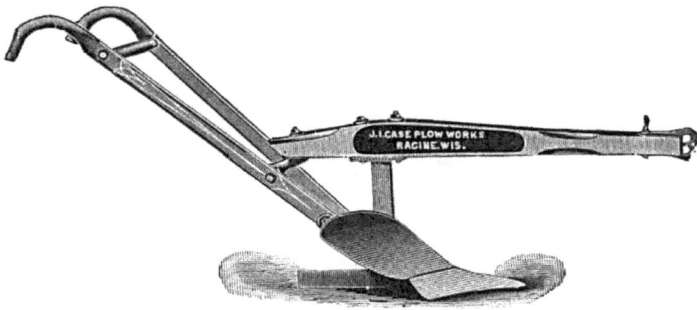

*Eight inch Corn or Vineyard Plow also suitable for gardens.*

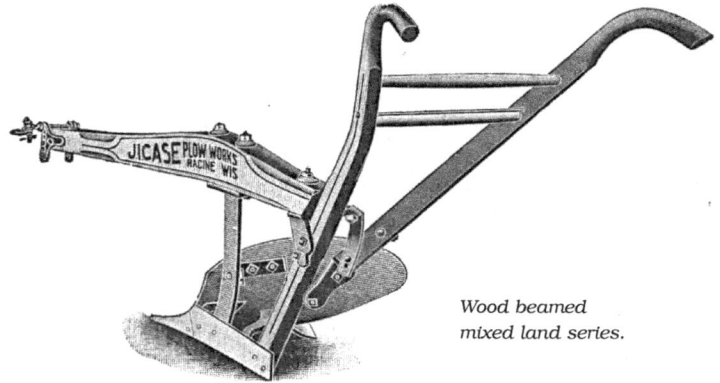

*Wood beamed mixed land series.*

*Steel beam Mixed Land series.*

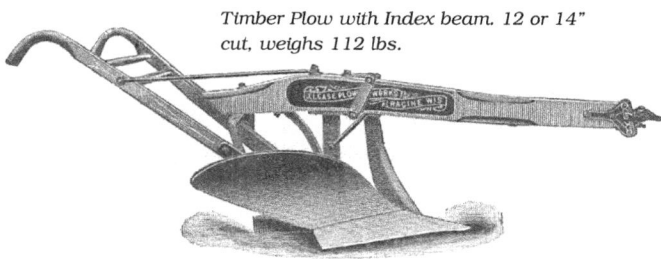

*Timber Plow with Index beam. 12 or 14" cut, weighs 112 lbs.*

*New Vineyard Plow with side chain attachment allowing the operator to hitch seven inches to either side of the center of the beam. This helps when plowing up against berries or small trees. The handles swivel either way.*

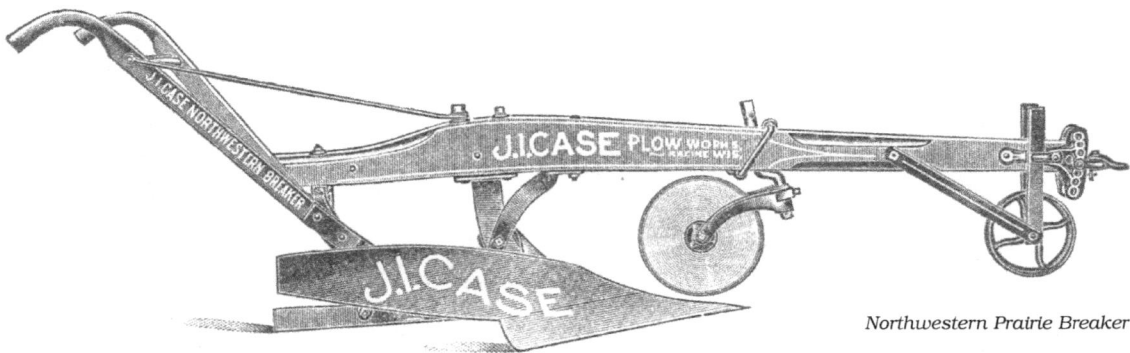

*Northwestern Prairie Breaker*

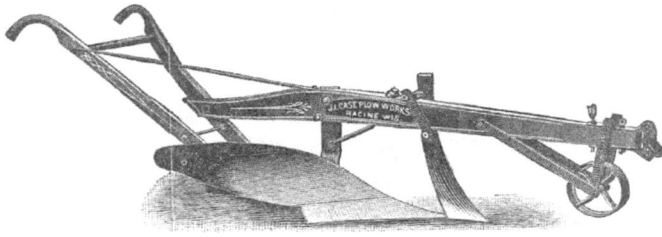

Brush and grub breaker

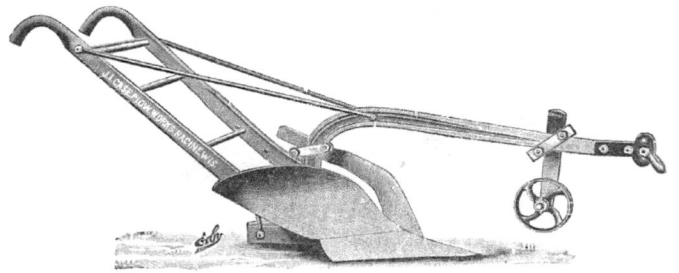

Road or Township plow

Wood standard plow

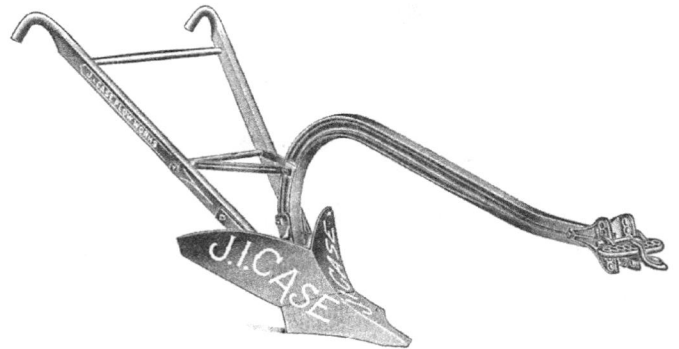

Middlebursters T.M. series

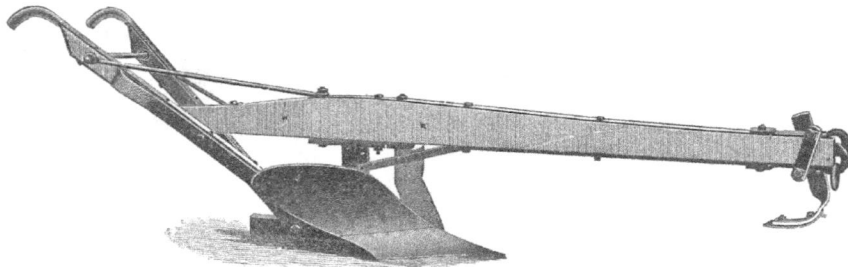

Railroad and township grading plow

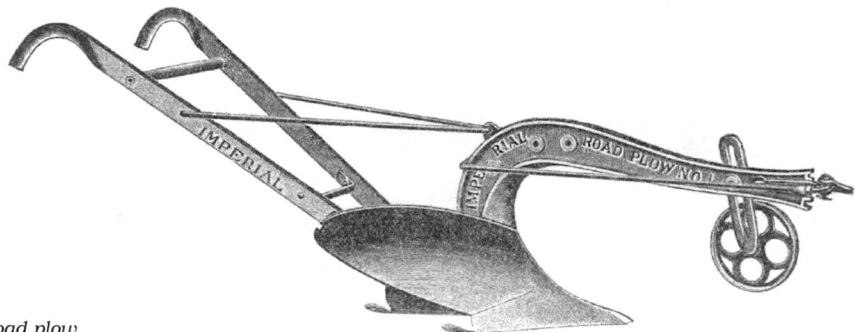

Imperial road plow

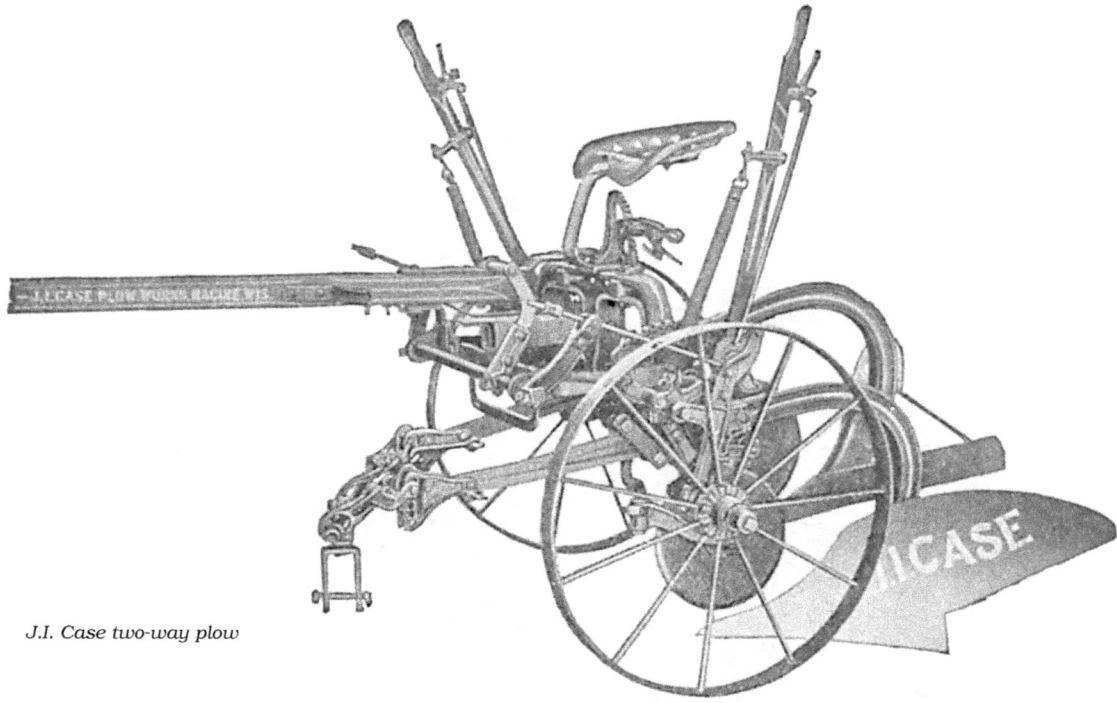

*J.I. Case two-way plow*

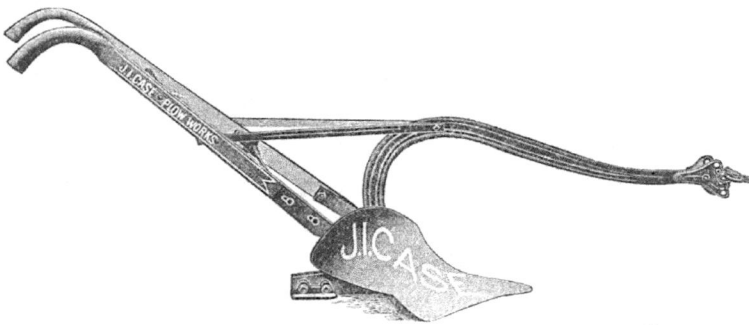

*Stubble Plow*

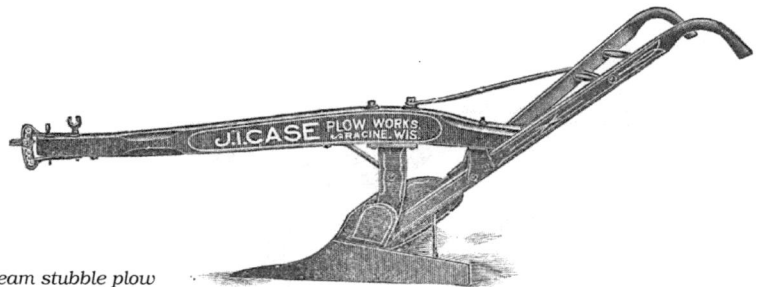

*Wood beam stubble plow*

*Sod and stubble plow, (wood beam
above steel beam right).*

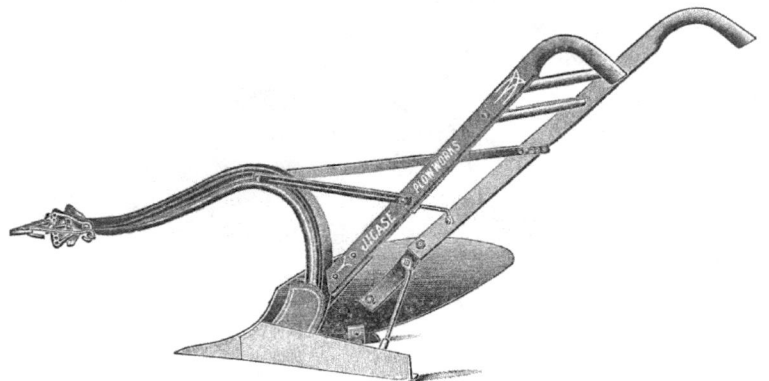

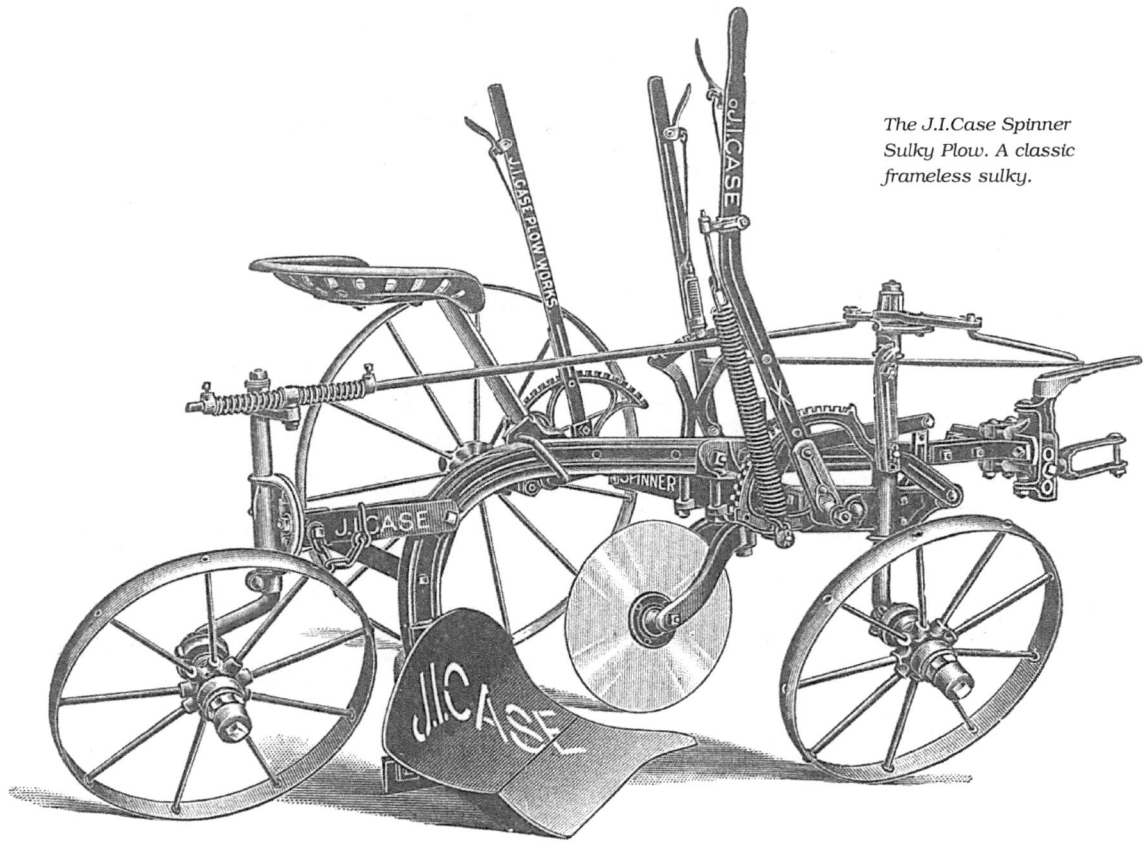

The J.I.Case Spinner
Sulky Plow. A classic
frameless sulky.

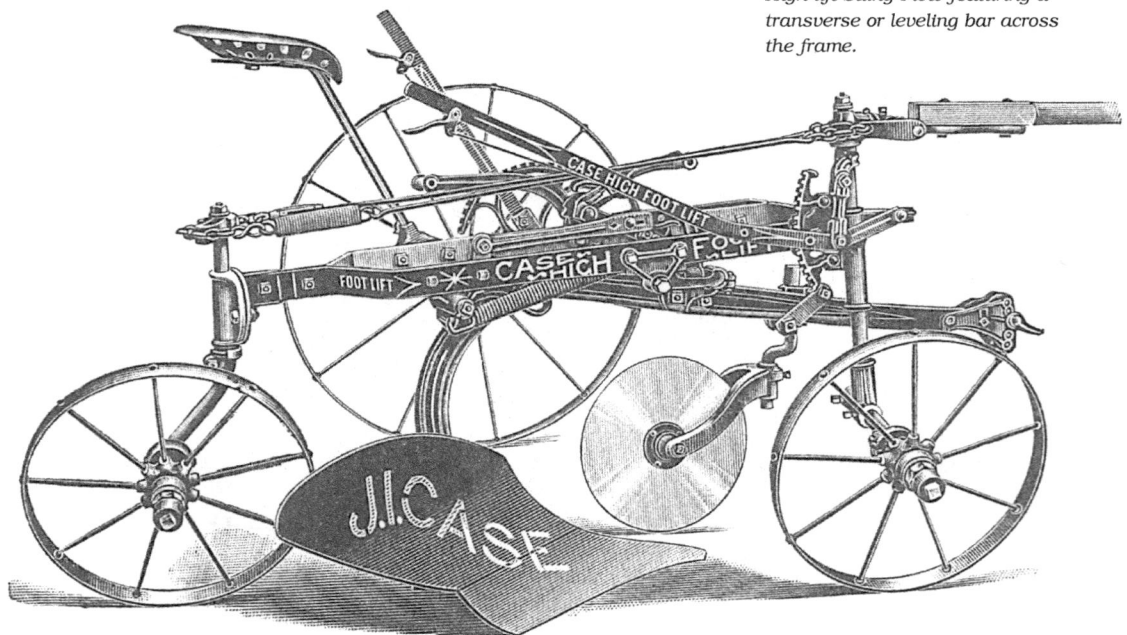

High-lift Sulky Plow featuring a
transverse or leveling bar across
the frame.

*J.I. Case Farmer-boy Low Lift
(frameless) sulky plow.*

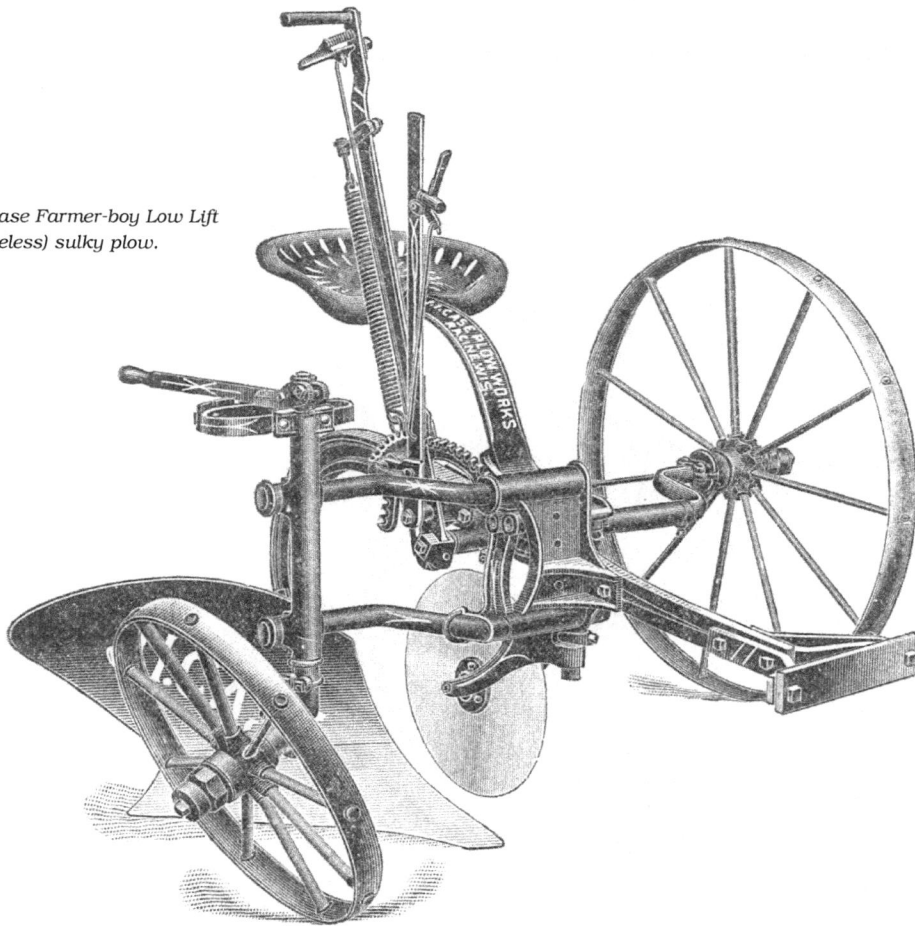

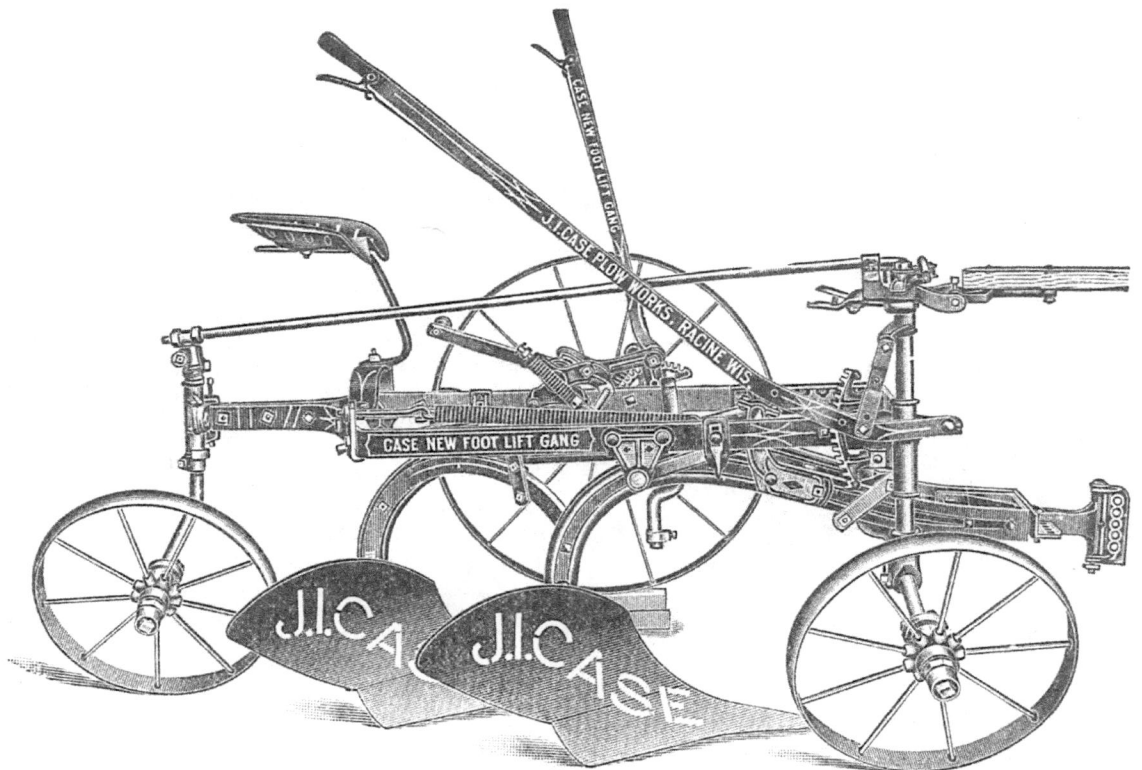

*New Foot-lift gang plow*

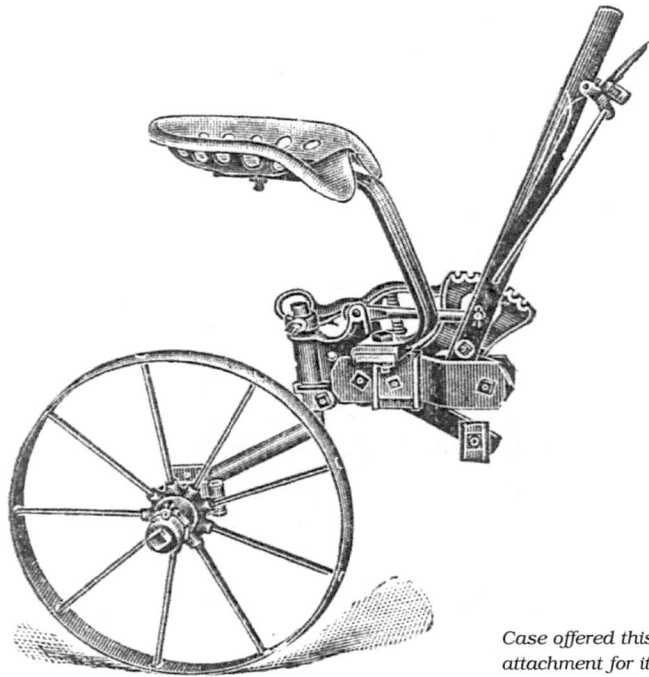

*Case offered this seat attachment for its walking gang plows.*

*New Foot-lift sulky plow featuring a single bail.*

# Emerson Brantingham

Emerson Brantingham was a much respected equipment line.

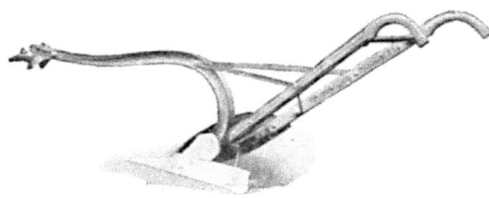

Model OS-2 Steel beam 12 or 14" old ground plow.

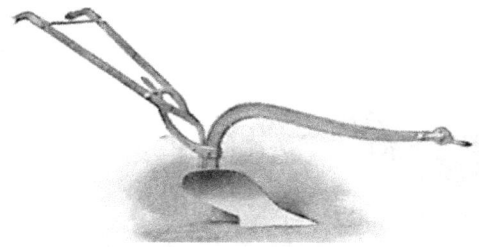

Small Vineyard plow.

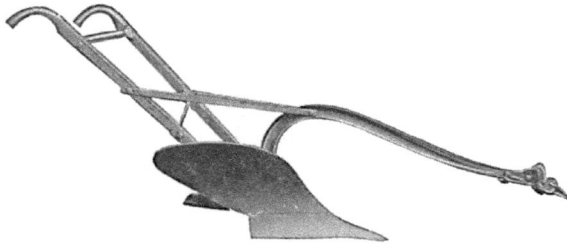

C.S.2 Steel Beam and Double Shin Turf & Stubble Plow

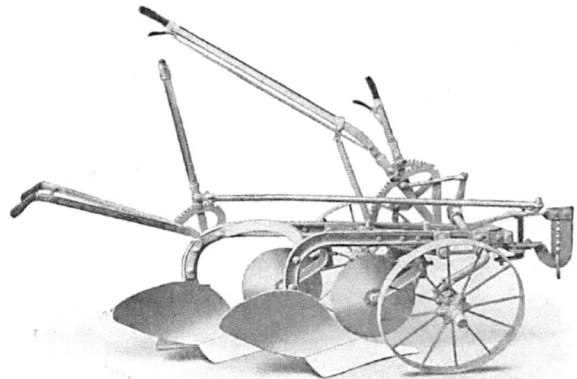

E. B. No. 5 Walking Gang Plow

E.B. Footlift Sulky Plow.

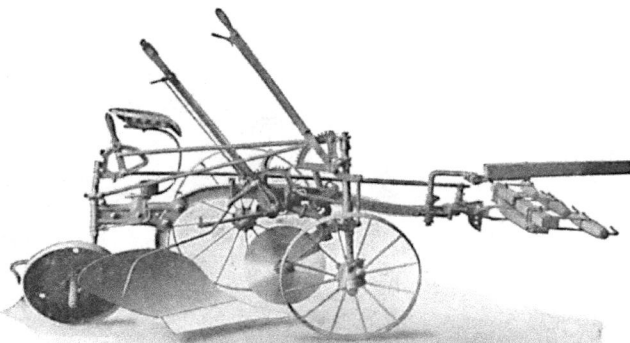

E.B. Frameless Foot Lift Plow. Unusual as almost all foot-lift were 'frame-style".

E.B. Foot lift gang plow.

*E. B. Power Lift two way plow*

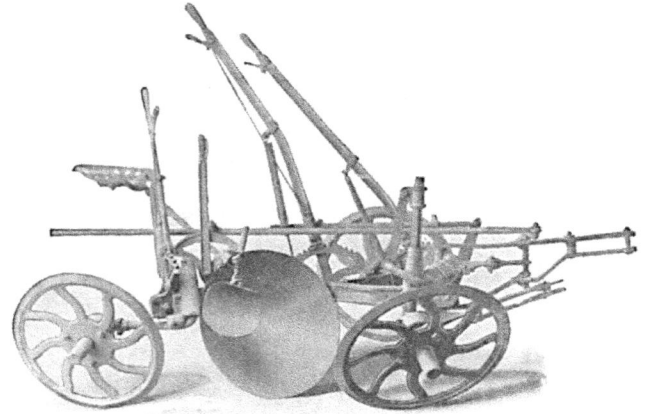

*E.B. Horse Disk Plow*

*Terry Beyer and Bob Champion from Elizabeth, Ill. with their 12 mules on a big-wheeled forecart and three bottom plow. Photo by Bob Mischka*

# John Deere

*Deere & Company manufactures John Deere plows. These illustrations came from a 1915 catalog from their Moline, Illinois offices.*

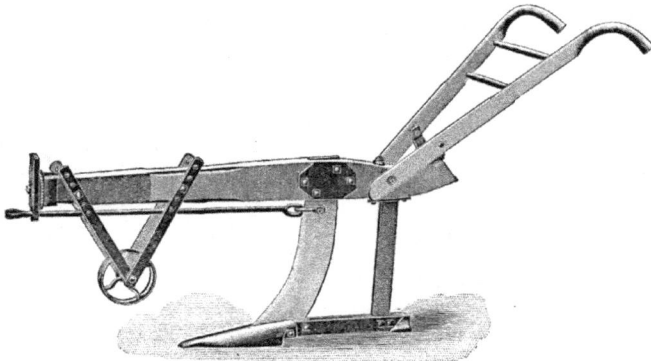

Makes & Models

*Deere wood beam subsoil plow*

*Deere steel beam subsoil plow*

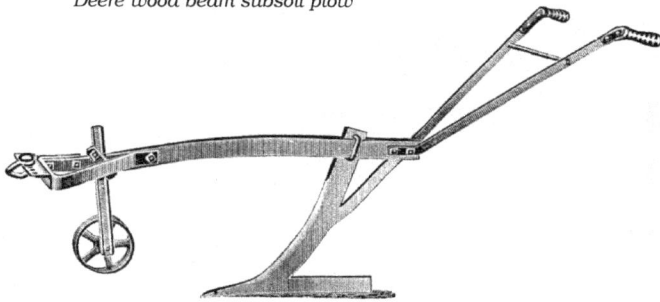

*John Deere Taylor subsoiler*

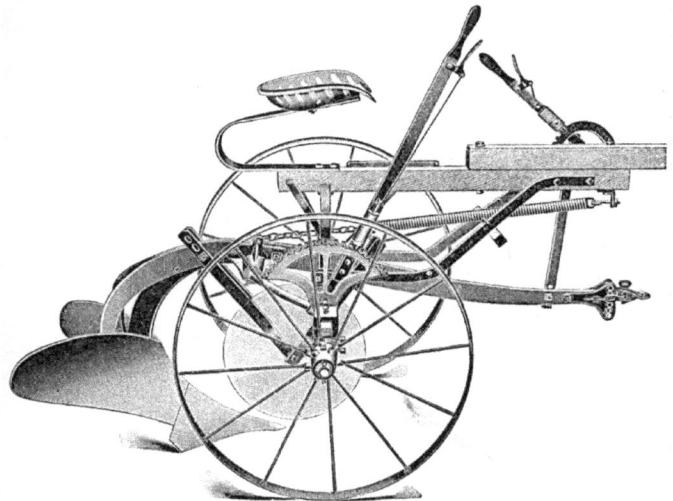

*Deere convertible sulky  middlebreaker*

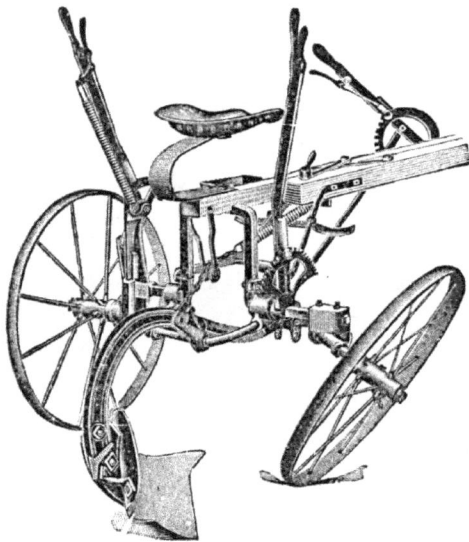

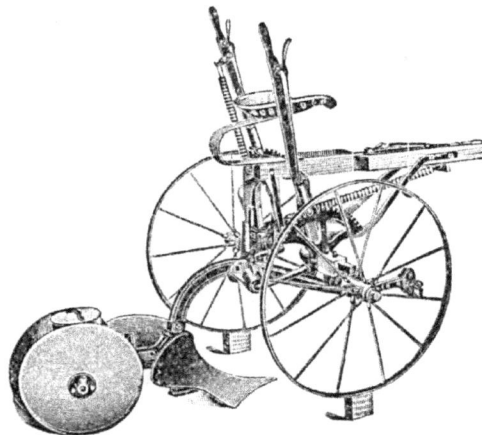

*Deere Common Sense sulky plow and sulky lister*

*Two-way sulky plow pre 1908*

*Two-way sulky after 1908*

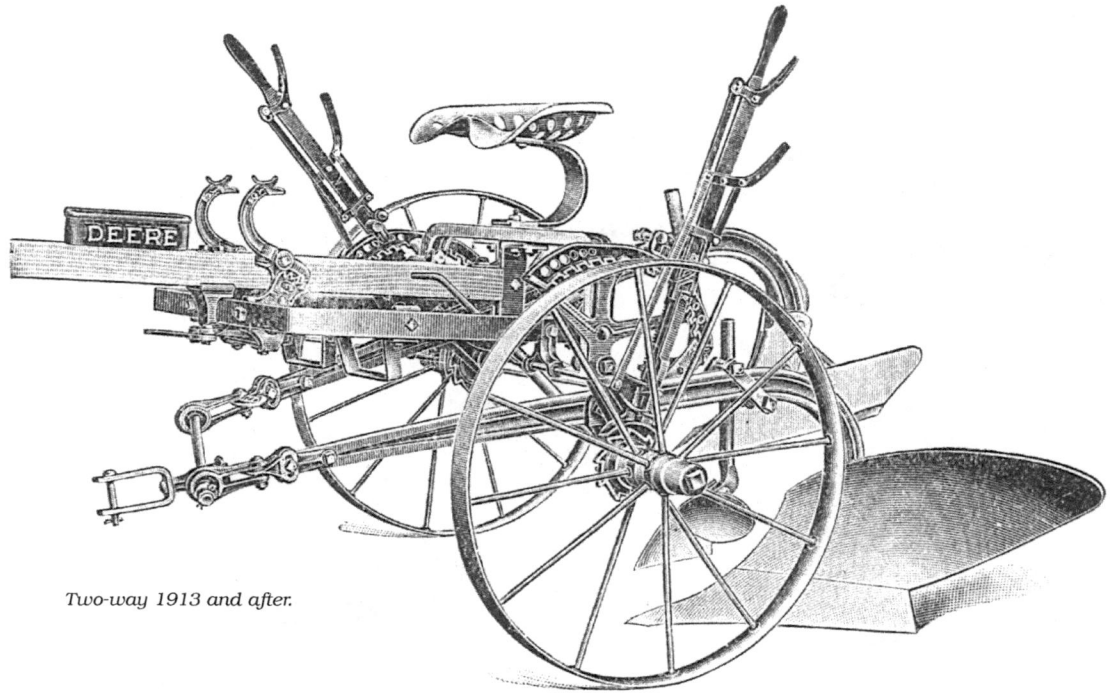

*Two-way 1913 and after.*

*Middlebreaker carriage with plow attached.*

*Two-way sulky No. 2.*

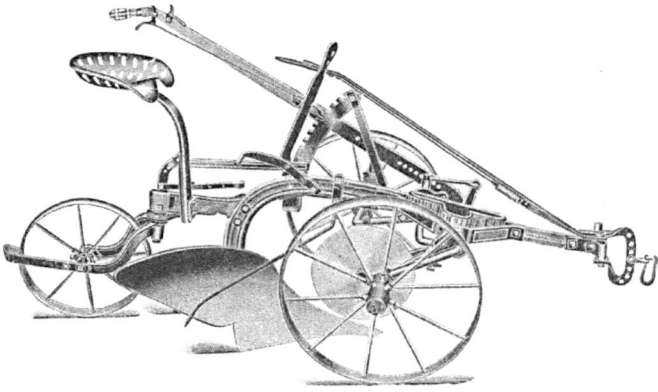

*New Deal single frameless*

*New Deal steel walking gang,*

*John Deere walking gang*

*Three-bottom steel gang, Portland model.*

*Deere Ranger sulky, pivotal.*

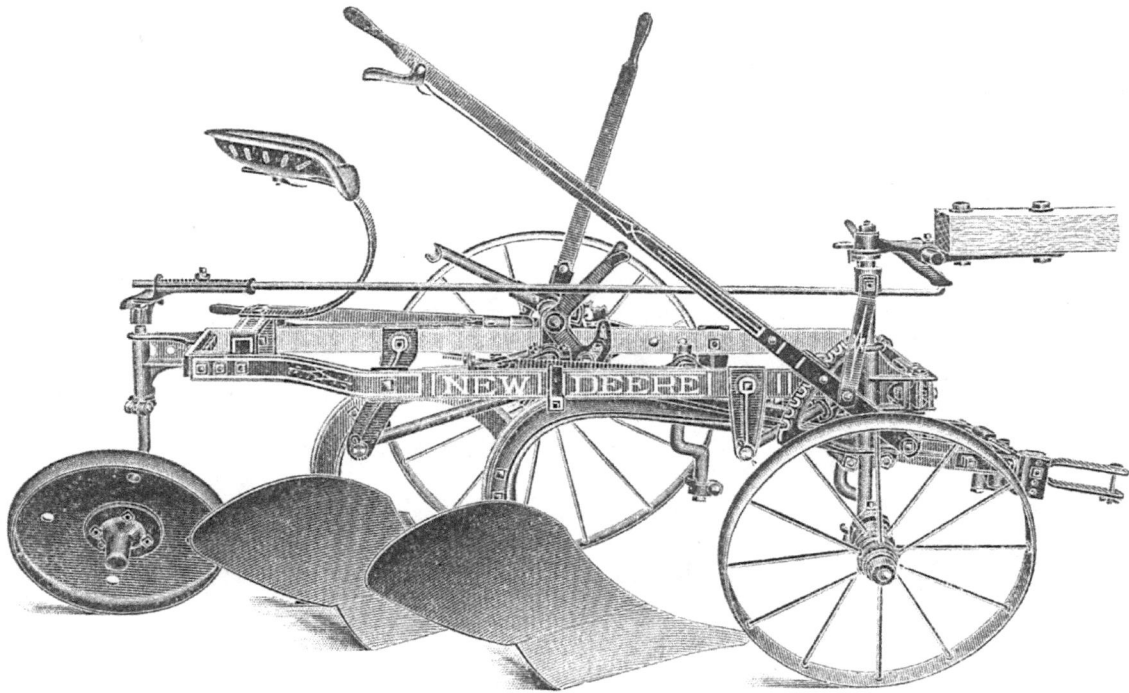

*New Deere Foot-lift gang plow prior to 1915*

*New Deere Foot lift suky*

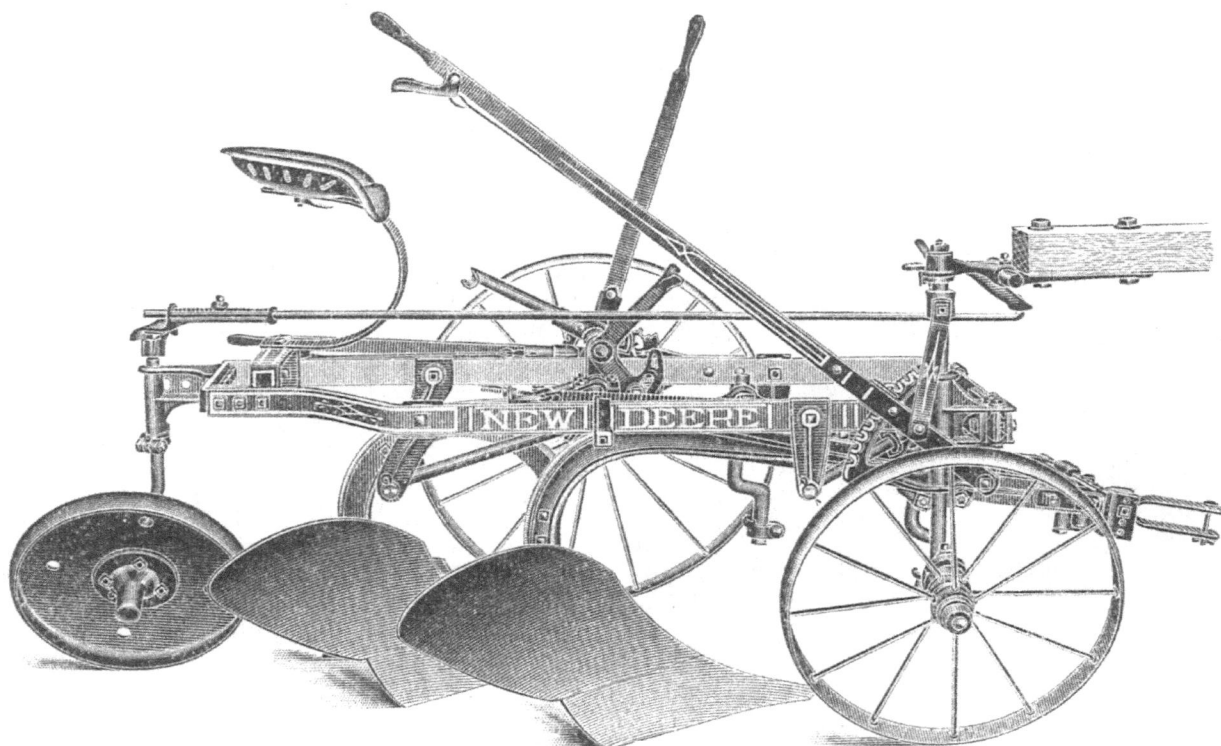

*New Deere gang, Minneapolis and Winnipeg style*

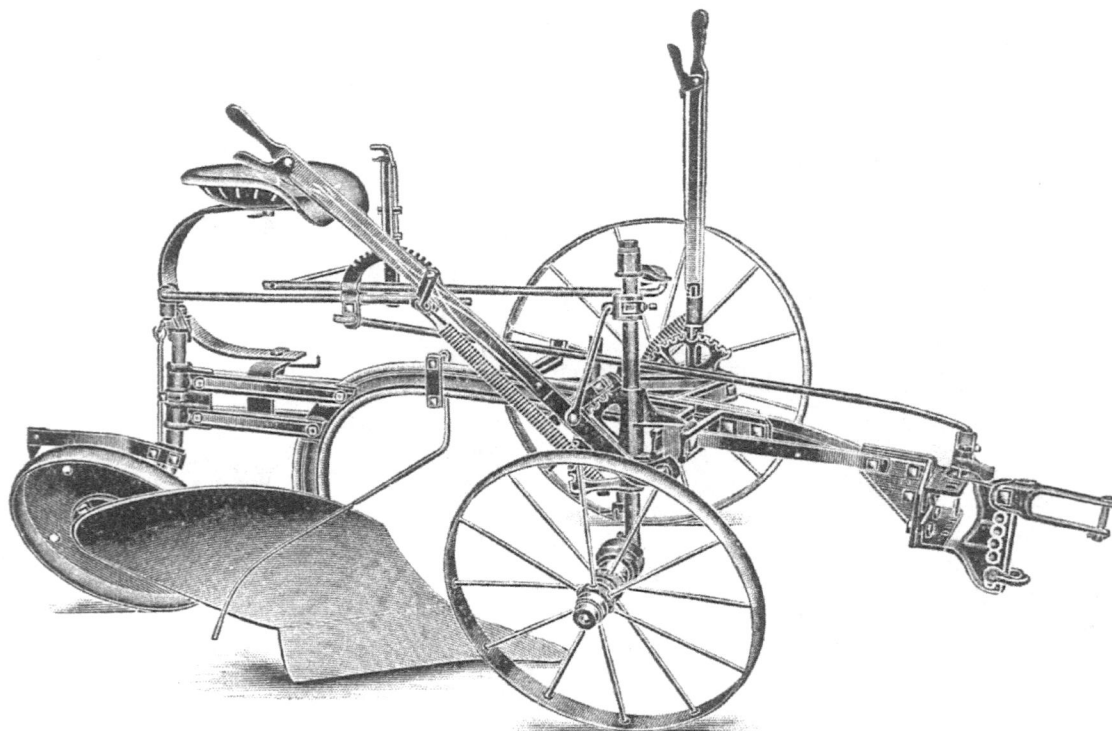

*California Stag sulky*

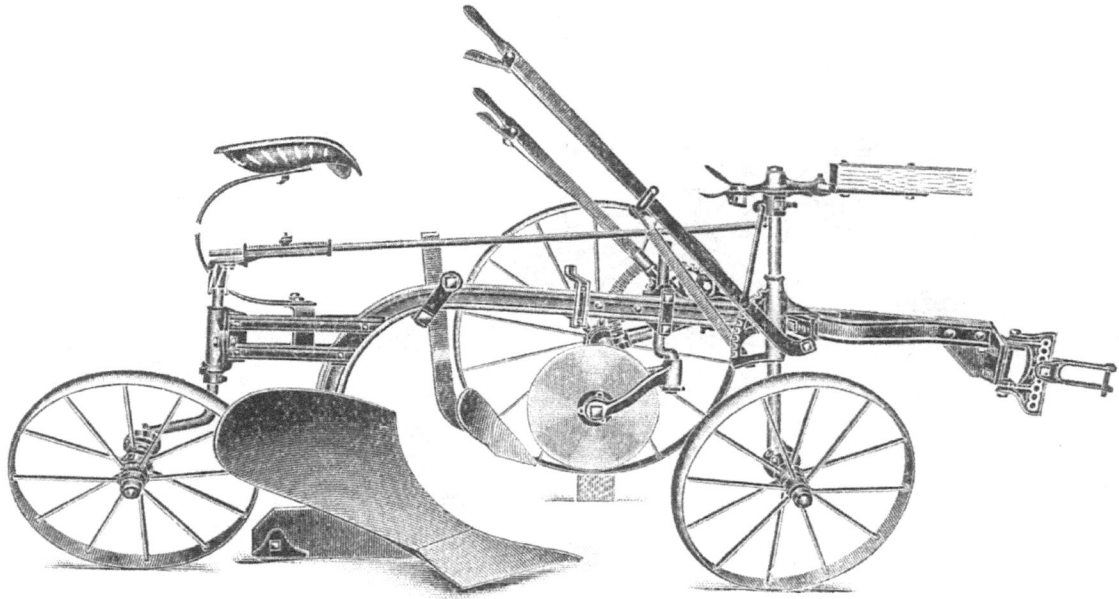

*No. 10 Stag sulky, deep tilling with No. 18 beam (pole plow only)*

*Stag sulky No. 5 and No. 7*

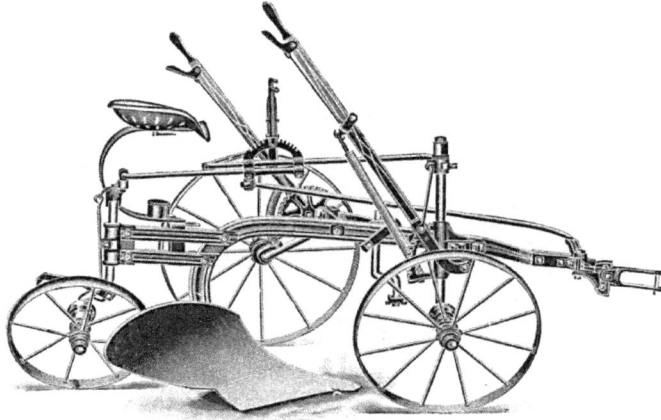

*Stag sulky, frameless.*

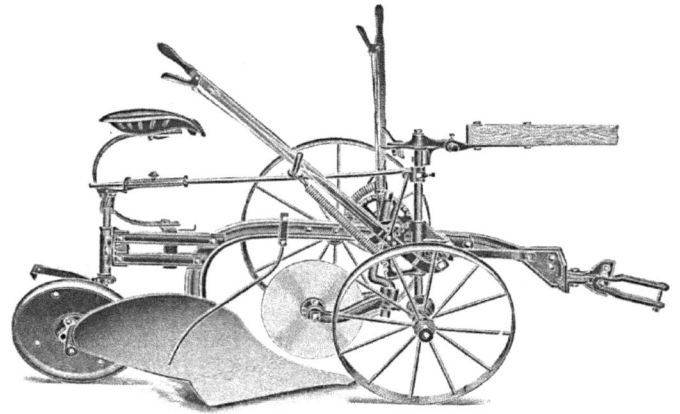

*John Deere Stag No. 6*

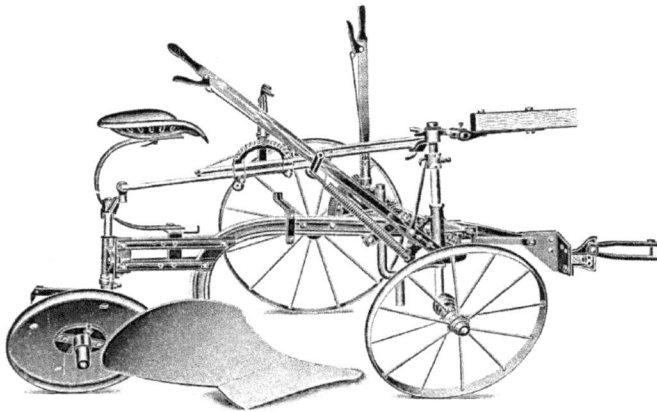

*John Deere Stag sulky No. 8*

*Stag gang 1913*

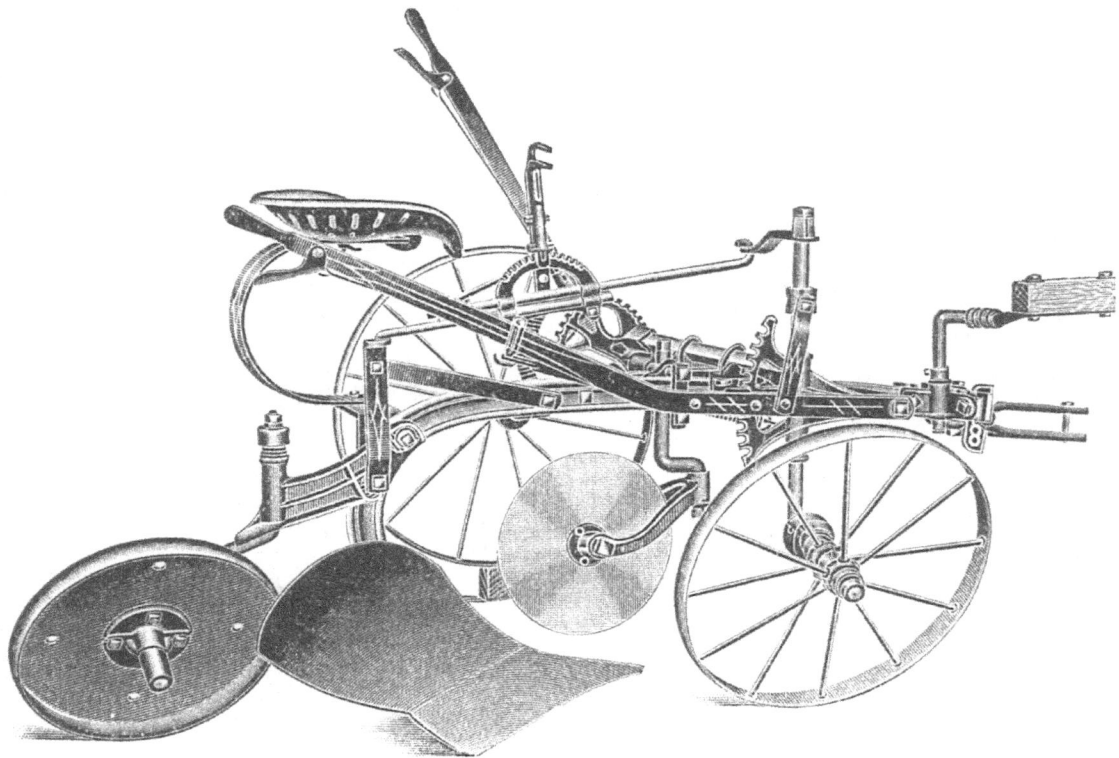

*"Triunfo" sulky Nos. 13 & 14.*

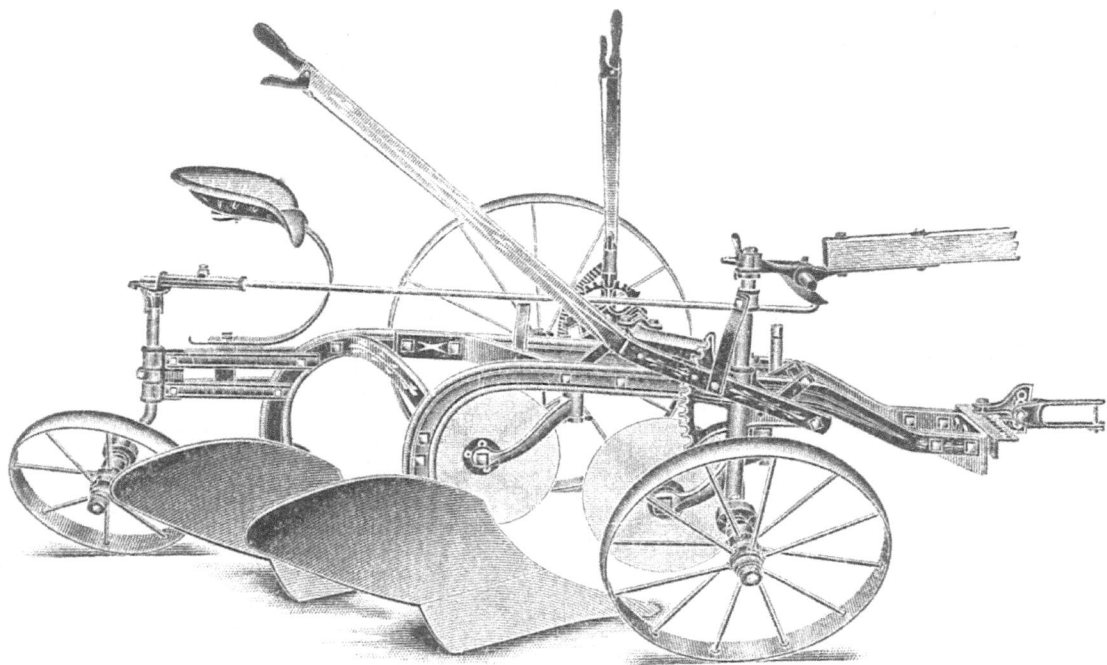

*Stag gang No. 3 Winnipeg (with pole)*

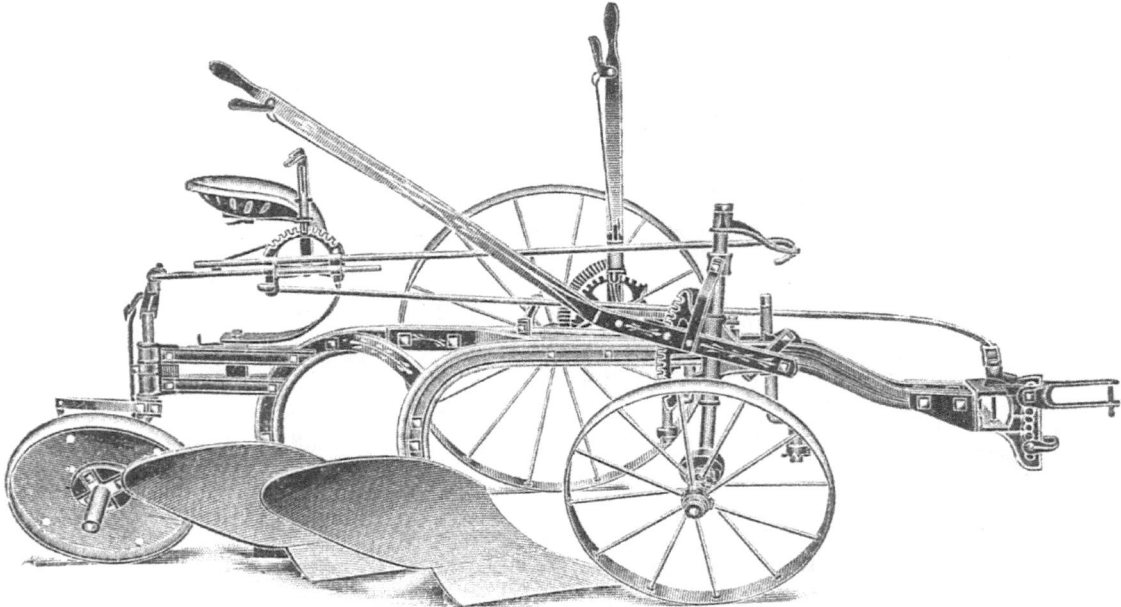

*Stag gang No. 4 California*

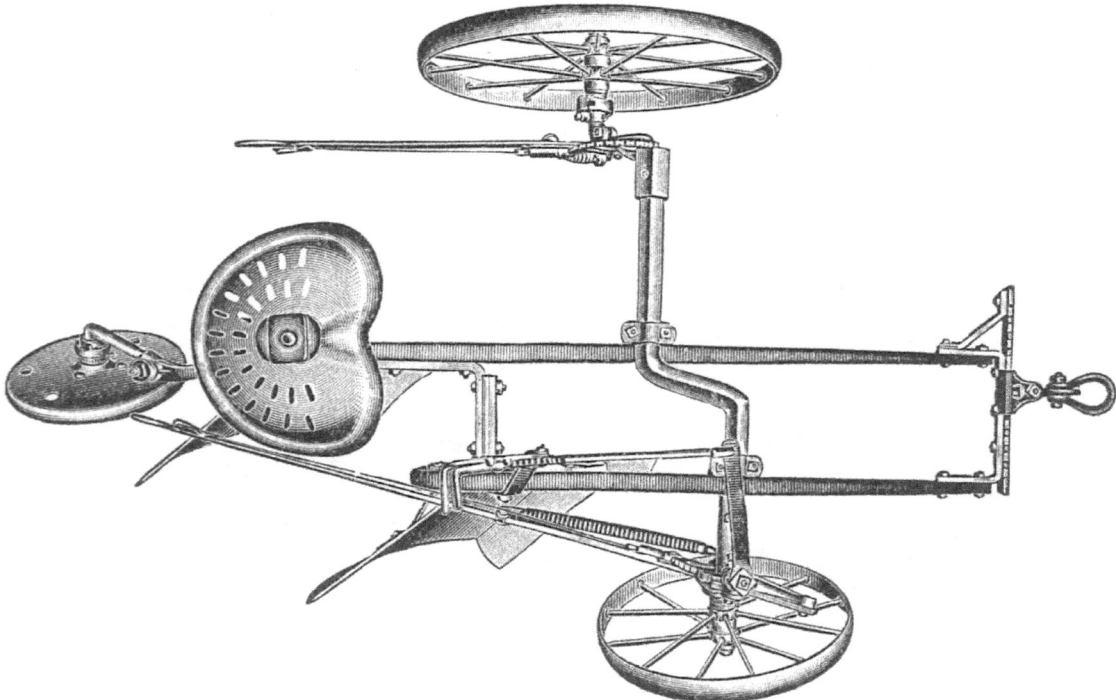

*Stag gang No. 5 Russia*

# John Deere

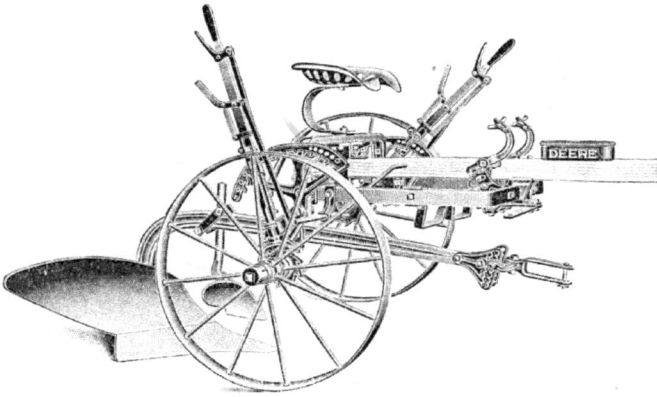

*Two-way No. 2, 1913 and after*

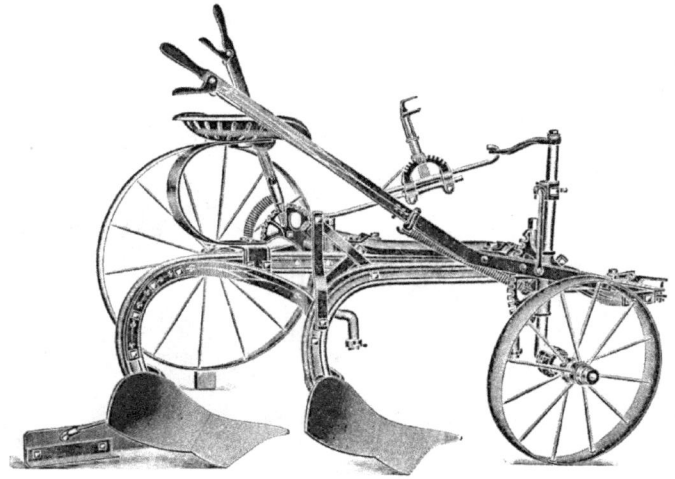

*No. 8 Rice stag gang*

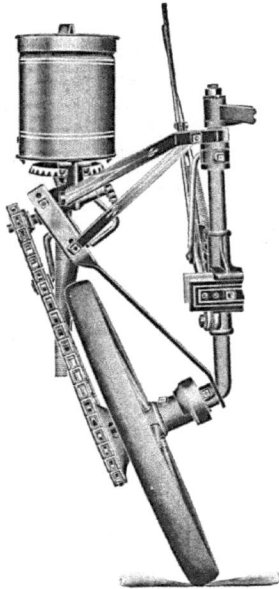

*Stag sulky planter attachment*

*Deere Pony gang with 2 levers*

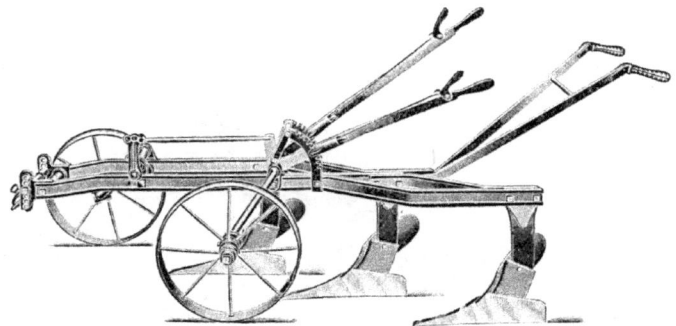

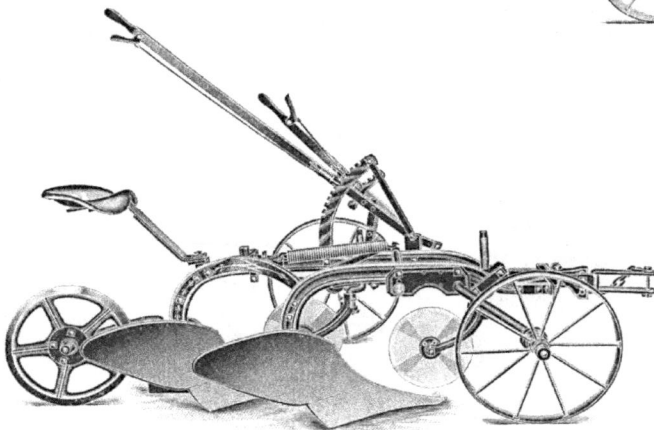

*John Deere Light walking gang*

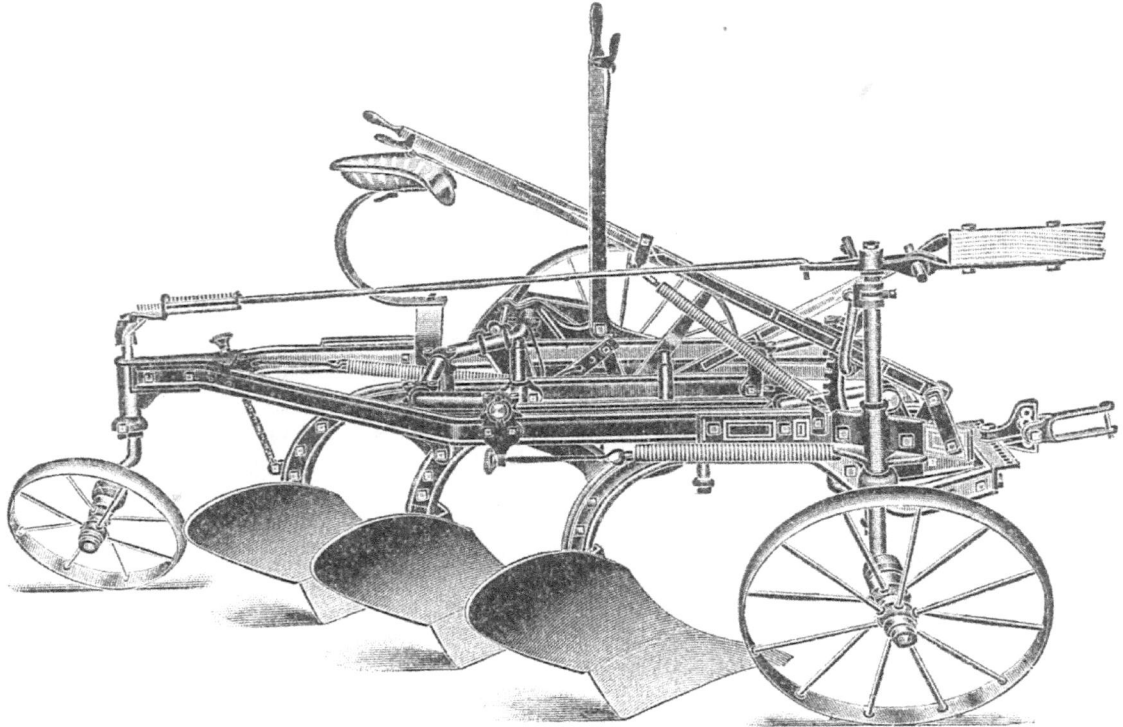

John Deere Three furrow gang

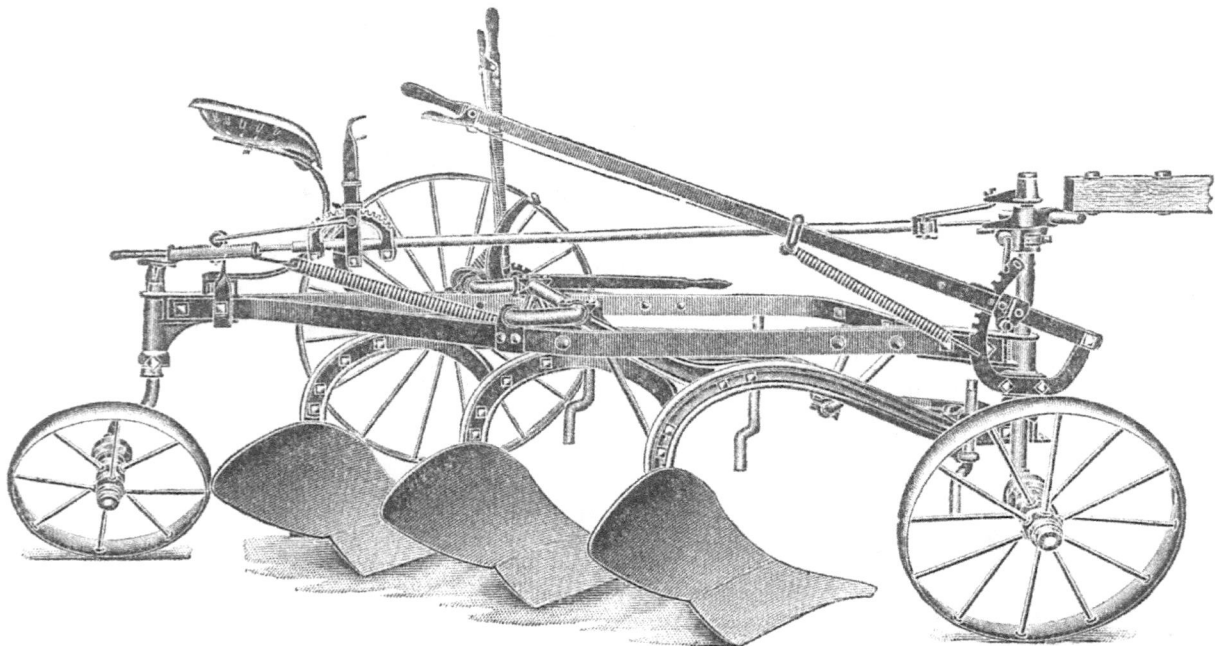

John Deere Triple gang No. 2

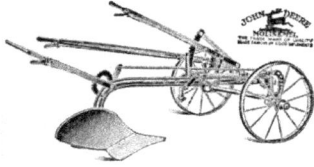

Koodoo Power lift sulky

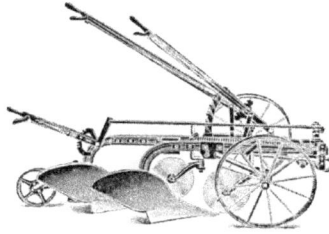

Koodoo power lift gang

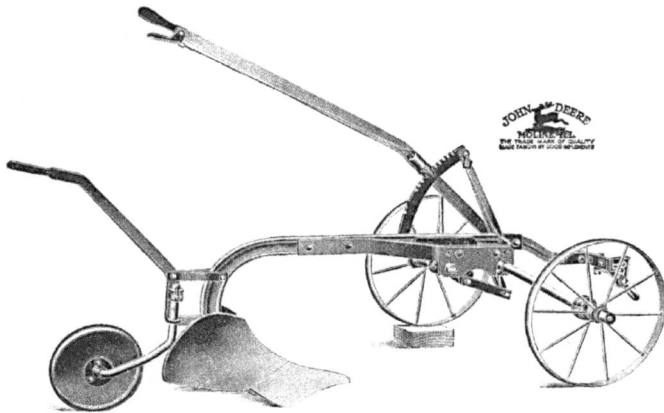

Chilean gang

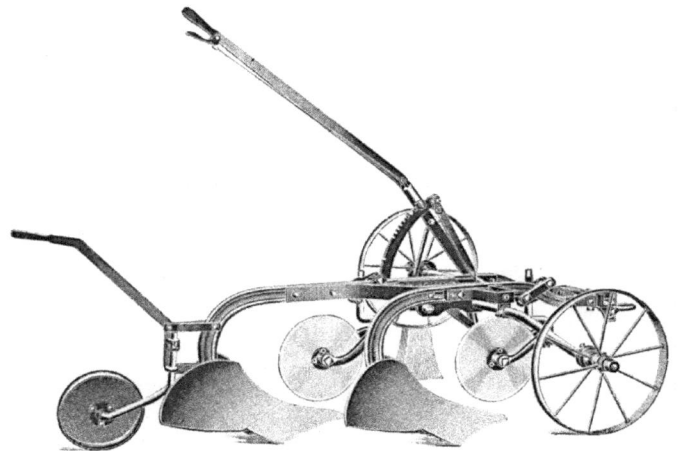

Chilean Sulky

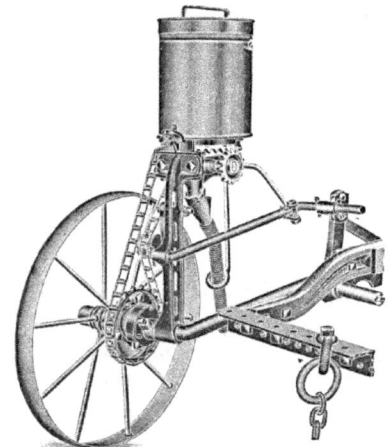

Planter attachment for  Ciclon

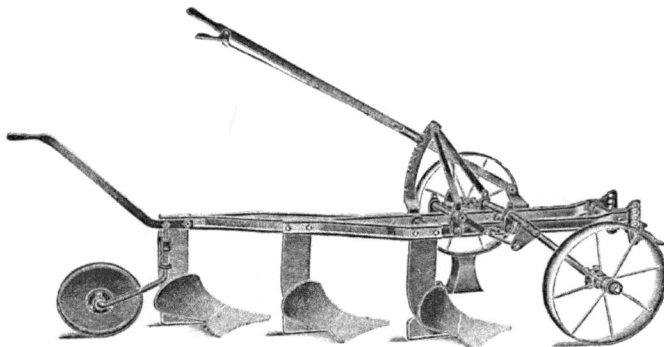

John Deere pony gang for Chile

*Jumbo Grub breaker*

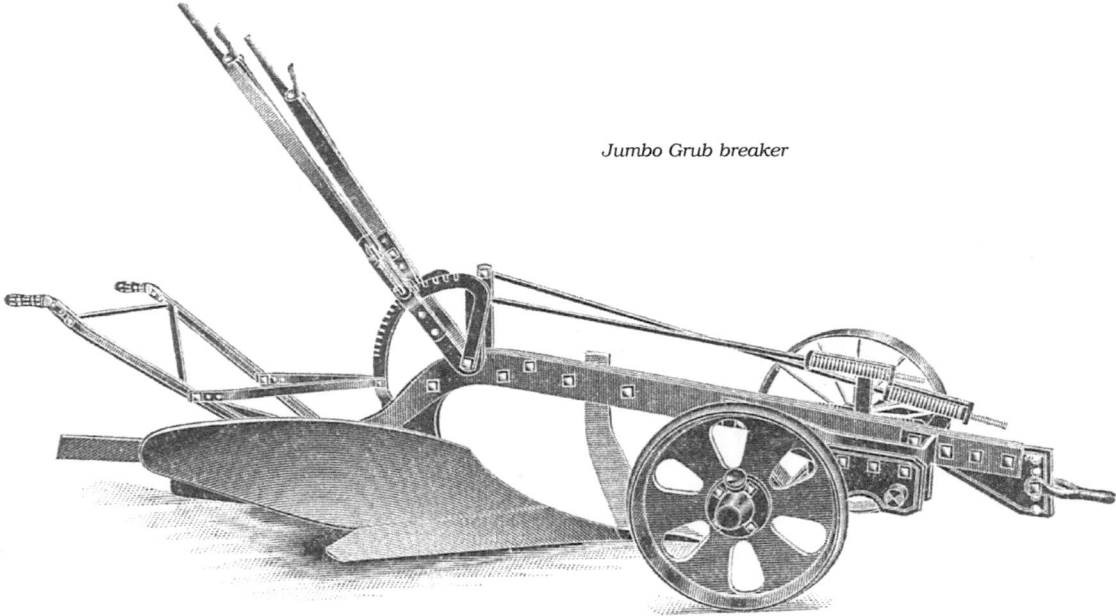

*Reversible Disc plow, 20 inch*

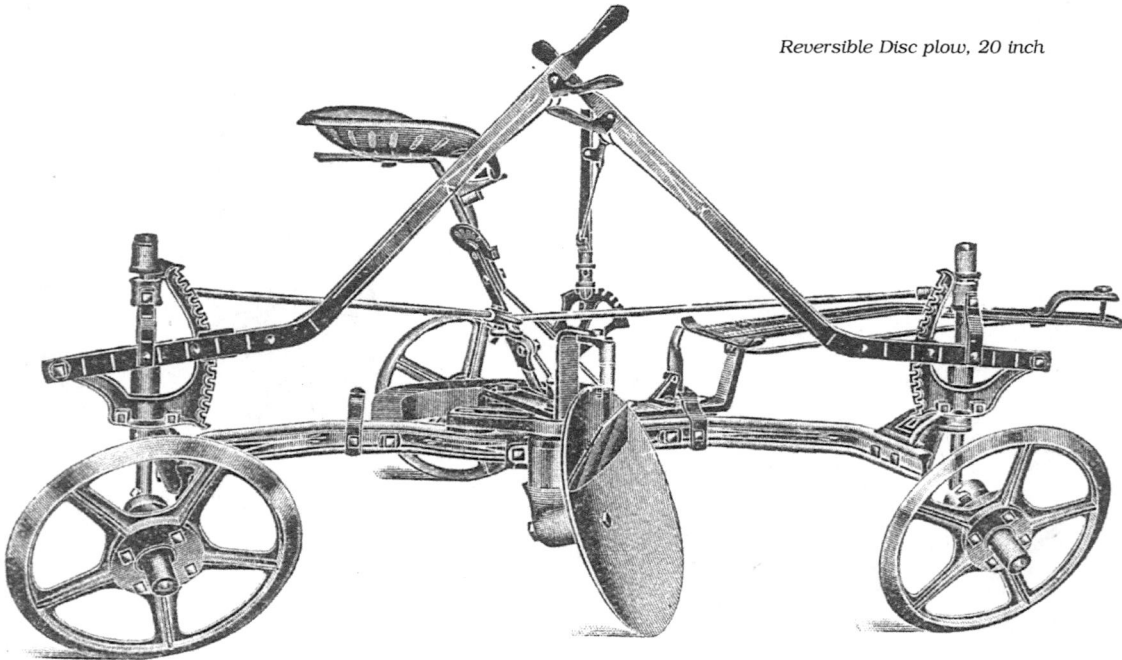

Ciclon No. 4

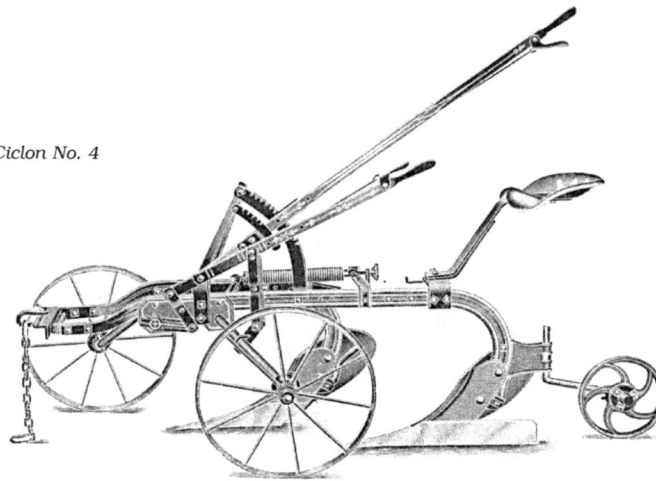

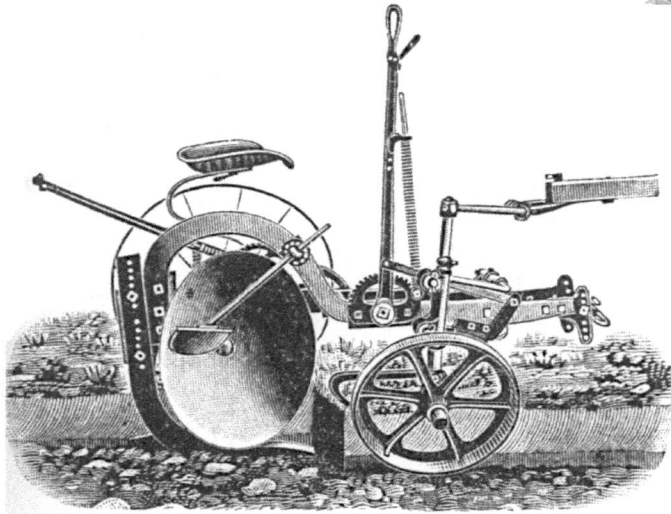

Secretary Disc plow, single

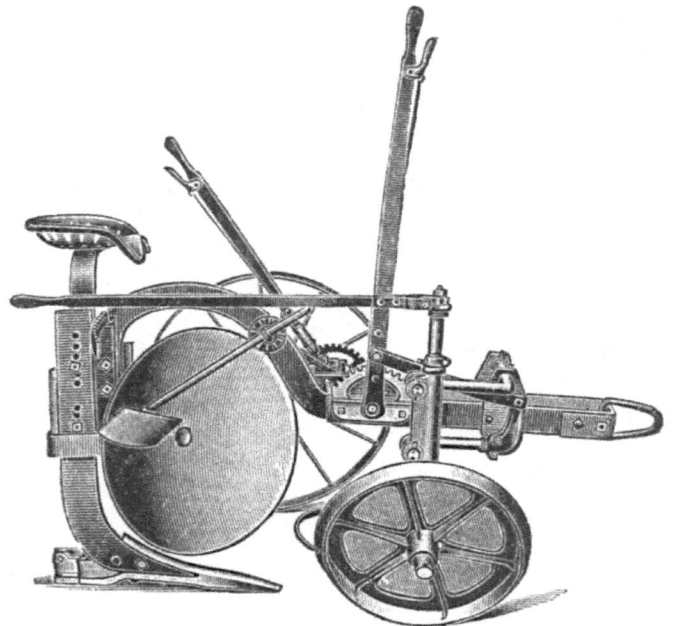

Secretary Disc plow, Calif.

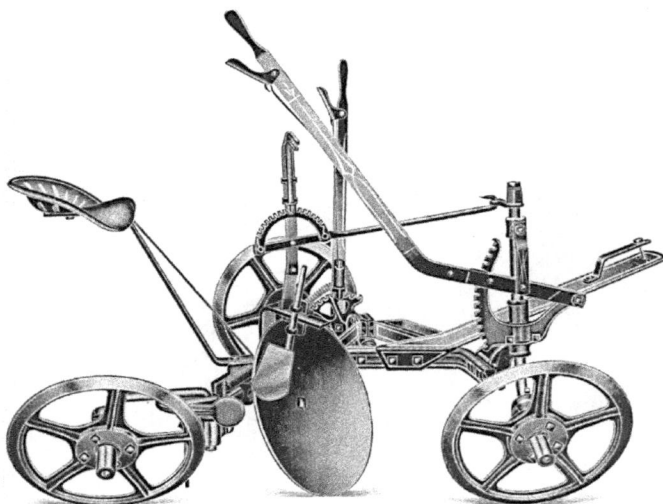

Pony Disc plow

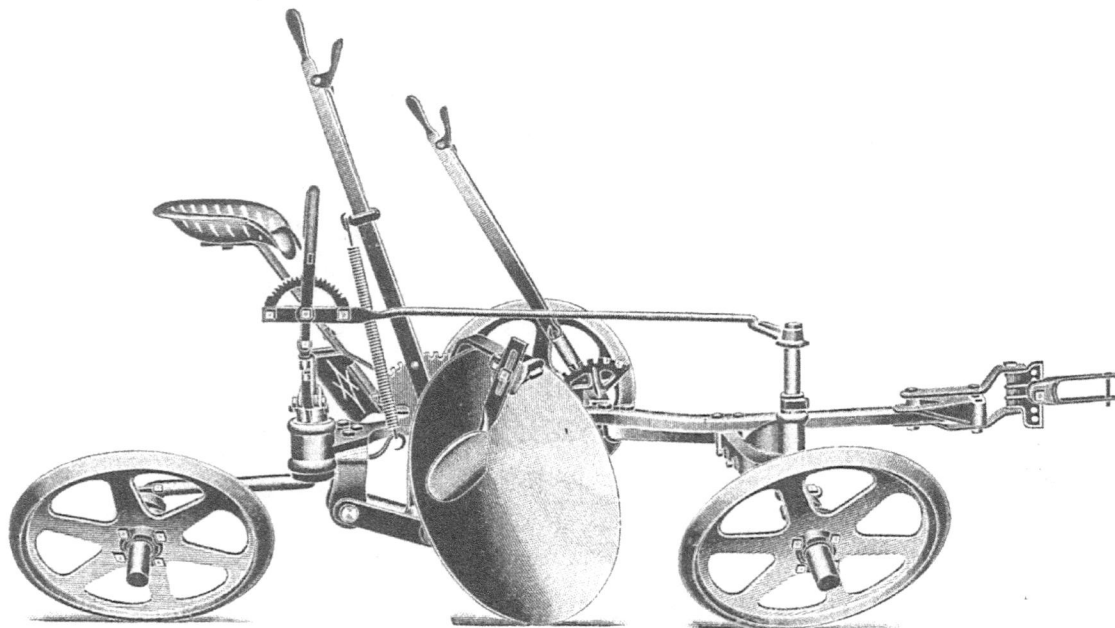

*Deere single disc plow*

*Deere double disc plow*

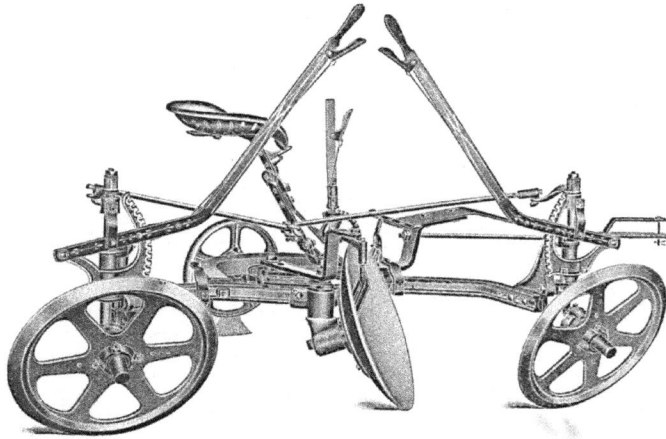

Reversible Disc Plow, 24 inch disc

Planter attachment for
disc plows

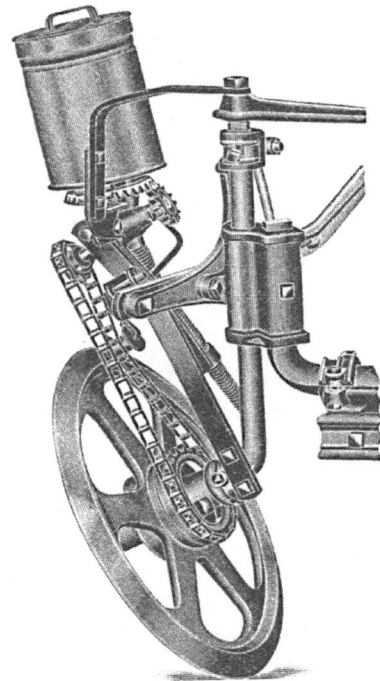

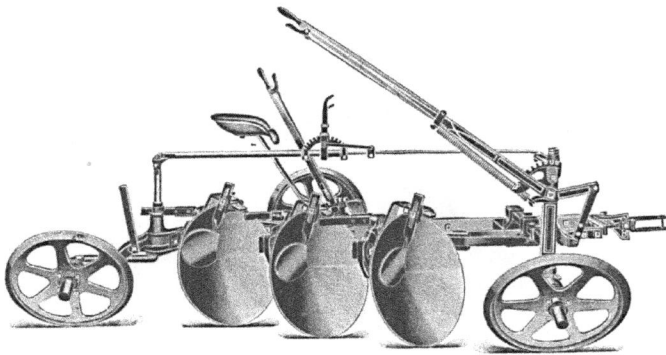

Deere triple disc plow

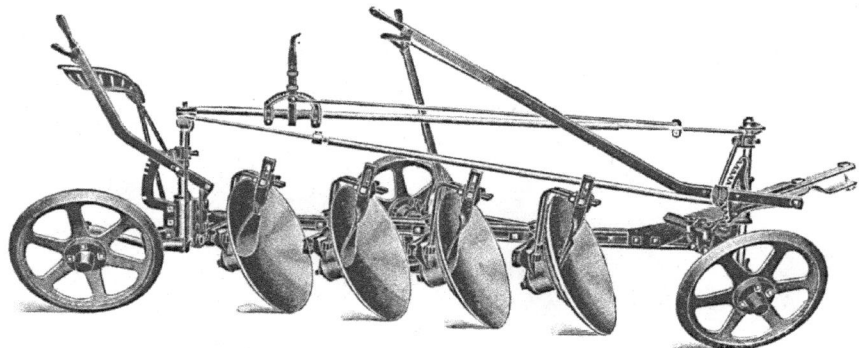

Quad Disc plow

McCormick Deering/ International Harvester Company of Chicago, Illinois acquired the P & O Plow Company and Chattanooga Plow Company (amongst others). In this book we've attempted not to duplicate the representations of P & O plows in their second incarnation as International plows. The plows in this segment are from a 1941 Parts Catalog.

Subsoil plow

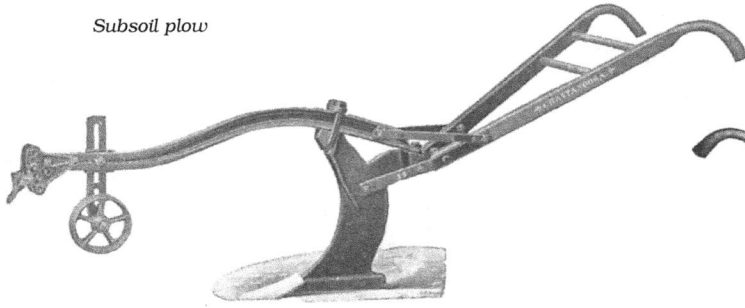

One horse Middlebreaker

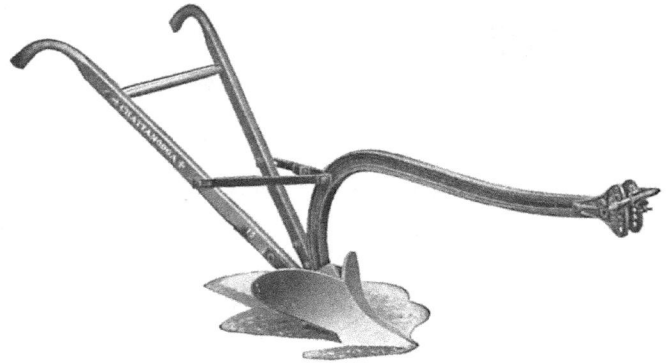

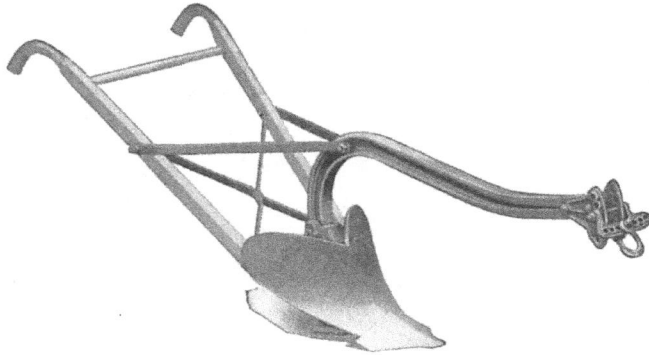

No. 18 Middlebreaker

No. 20 Middlebreaker

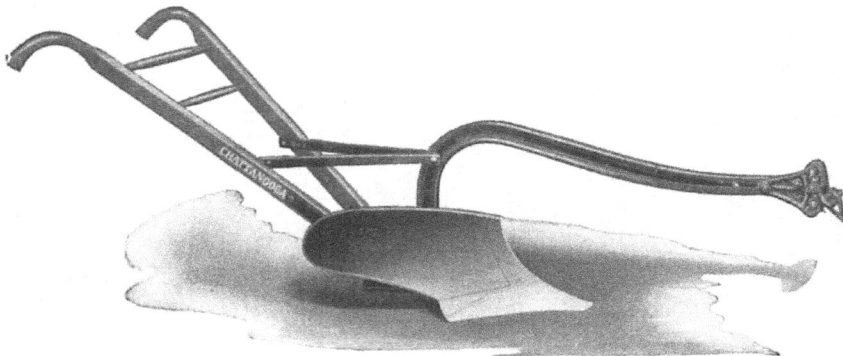

No. 45 Chilled plow

# McCormick Deering / International

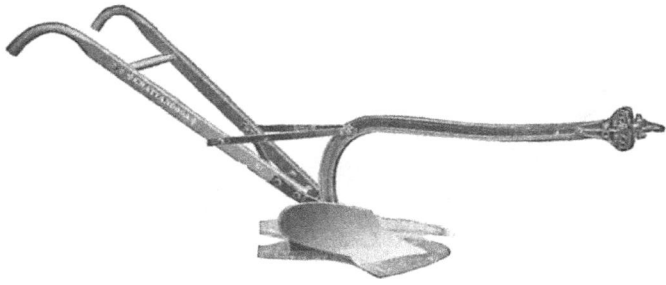

*No. 61 One horse plow*

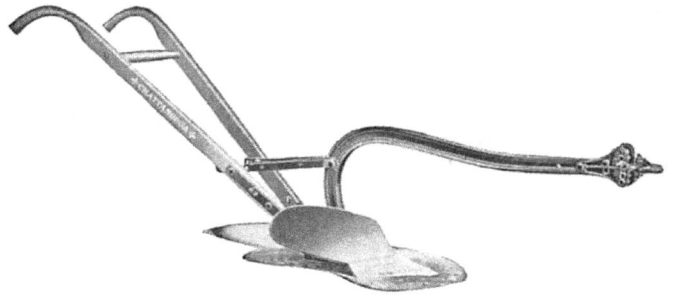

*No. 62 One horse*

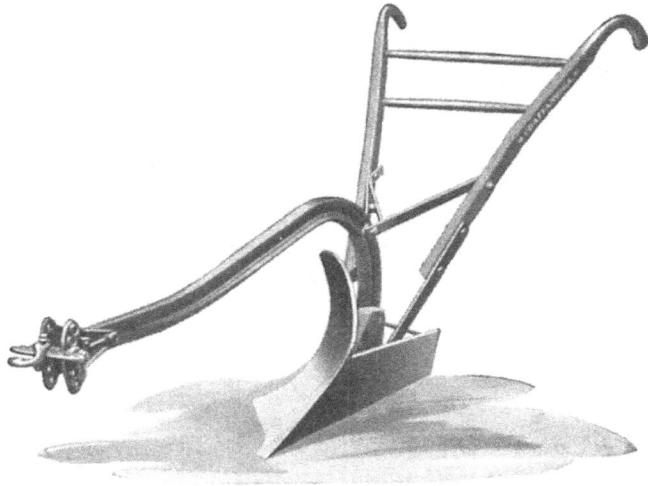

*No. 63 Walking plow*

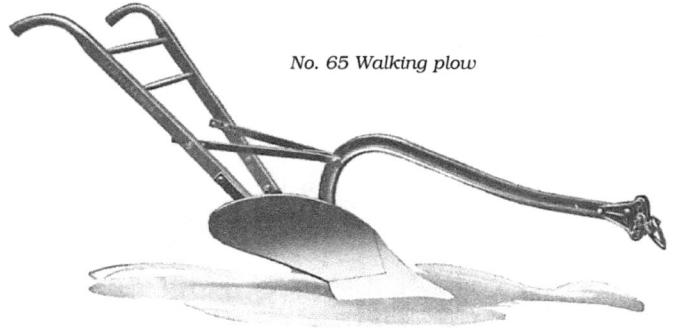

*No. 65 Walking plow*

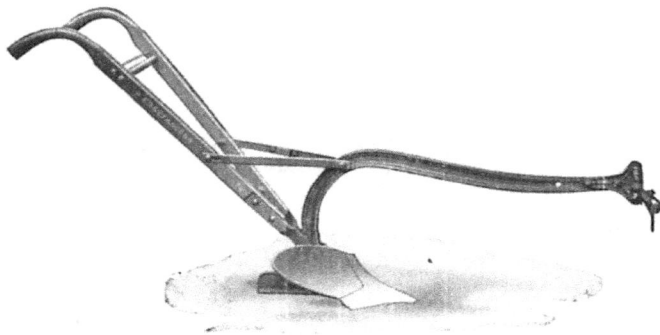

*No. 70 One horse walking plow*

*No. 72 One horse walking plow*

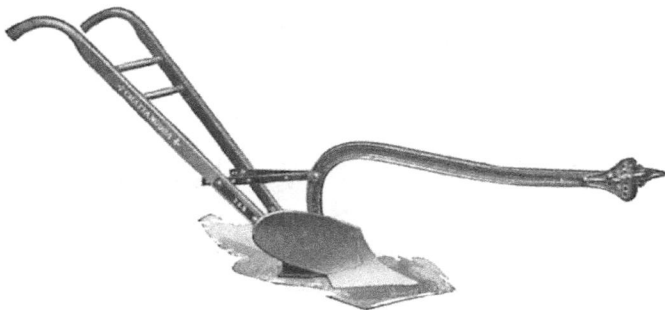

*No. 72 $^{1/2}$ B  One horse walking plow*

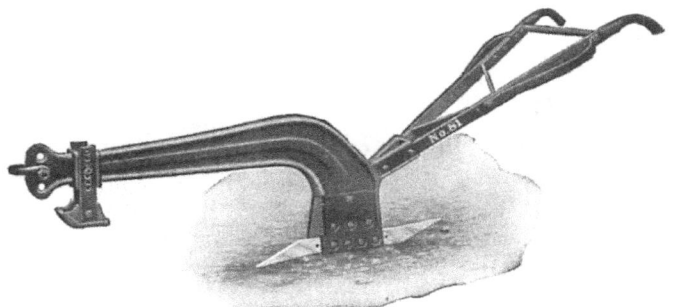

*No. 81 Cast Beam Rooter*

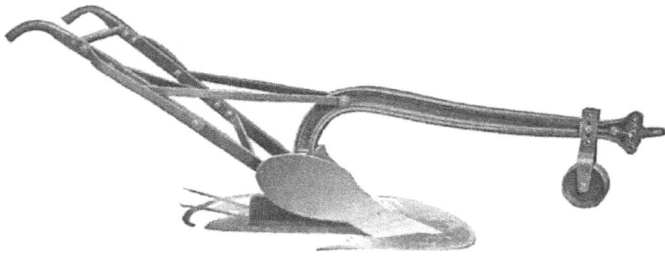

No. 84 Steel Beam Road plow

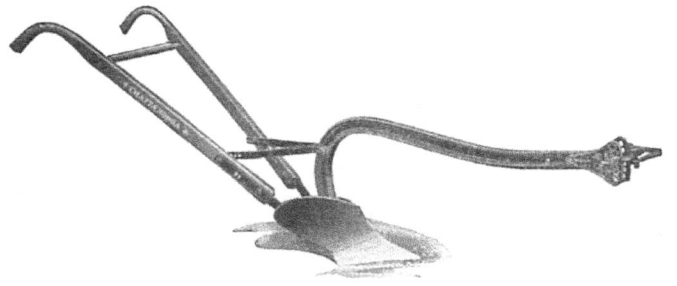

No. 91 One horse

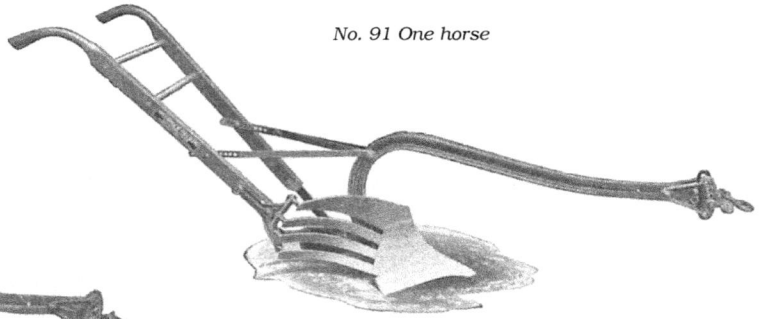

No. 95 Two horse

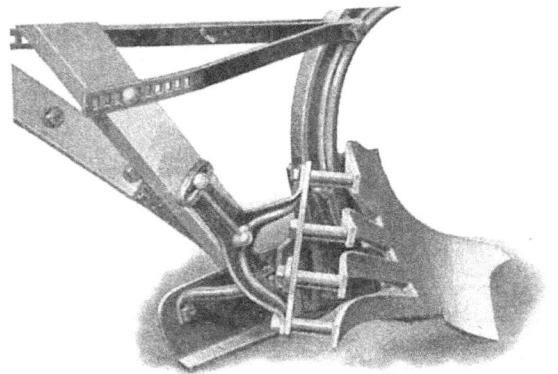

No. 95 Slat Mold

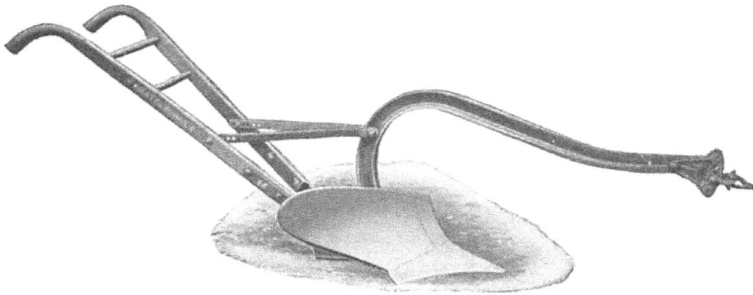

Slat Mold

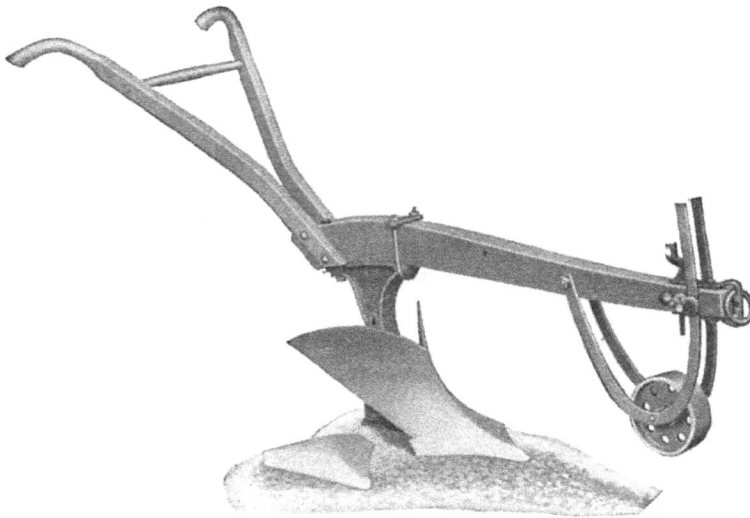

No. 251 or No. 252 Middlebreaker

No. 262A Chilled

No. 272 Chilled

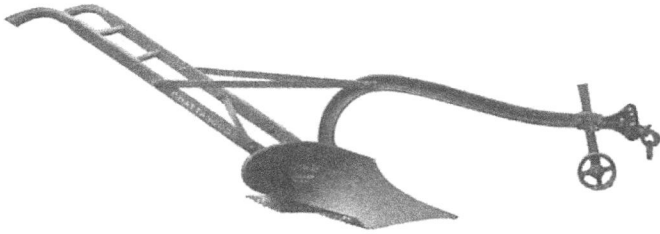

No. 145 Chilled

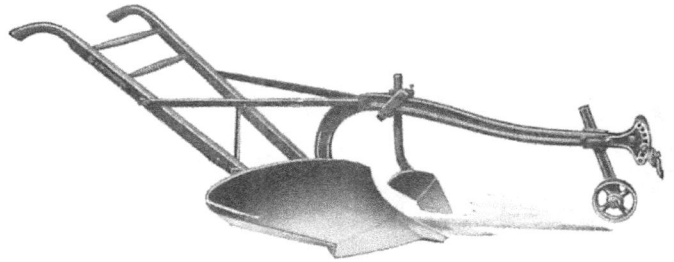

No. 545

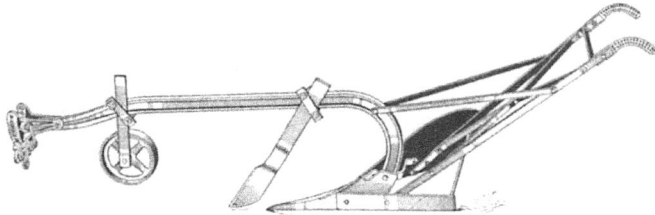

Sugar Cane plow

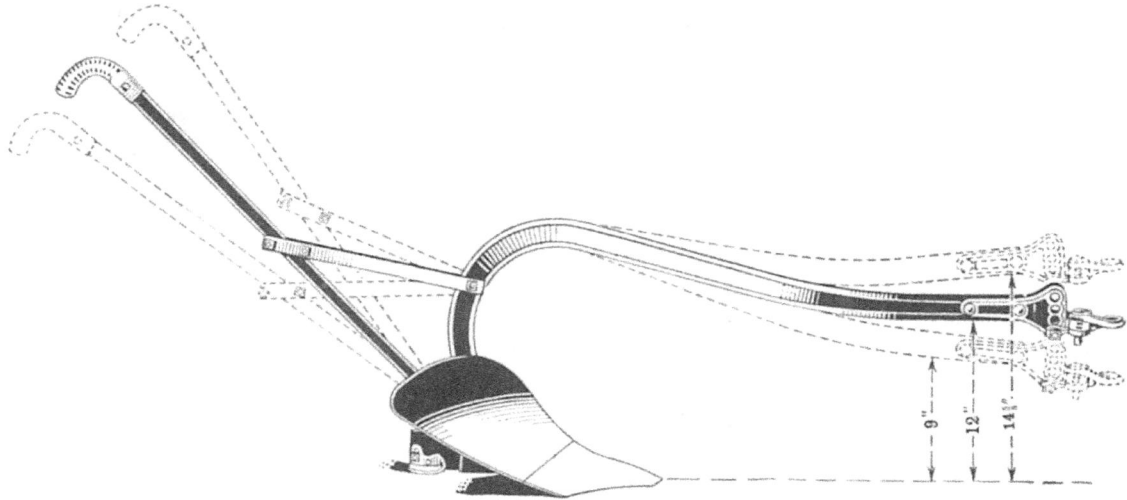

J-7A adjustable built since 1925

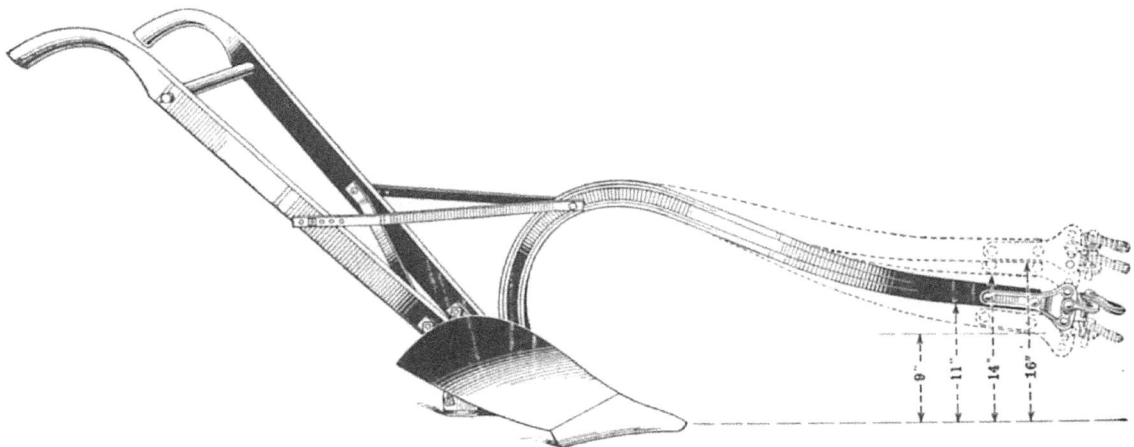

JA-7C, JA-7E-AO or JA-7F-BO adjustable

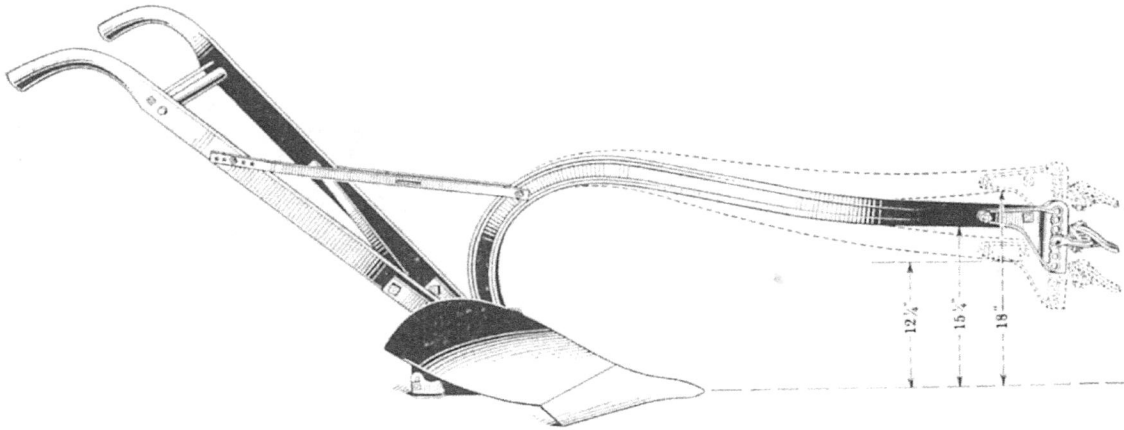

*J-7-G and J-7-I adjustable*

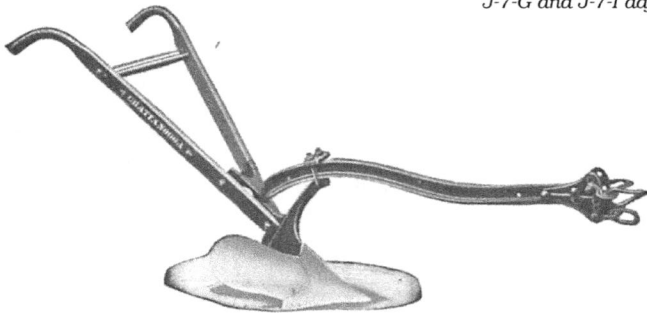

*No. 51 One horse Hillside*

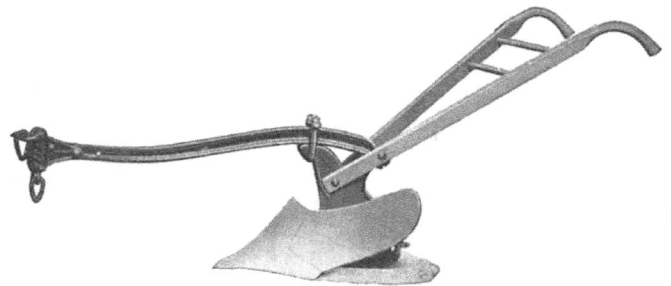

*No. 55 Two horse hillside*

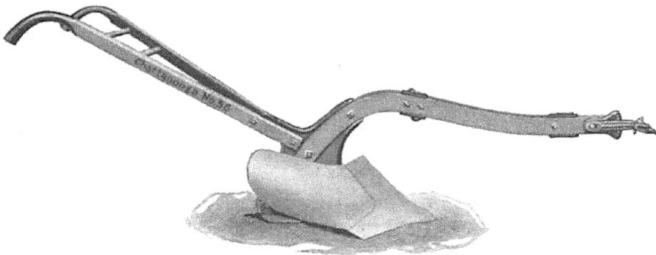

*No. 56 Three horse hillside*

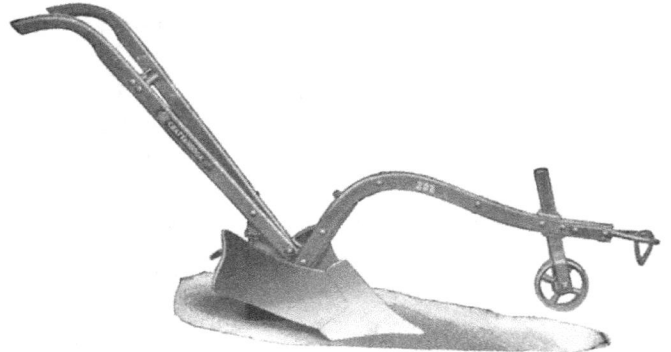

*No. 208 One horse hillside*

*No. 209 Hillside*

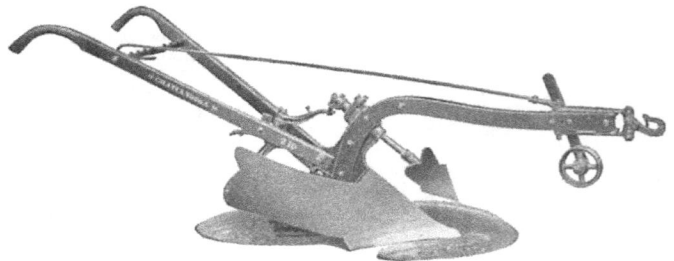

*No. 210 Hillside*

*Pond Disk Gang plow*

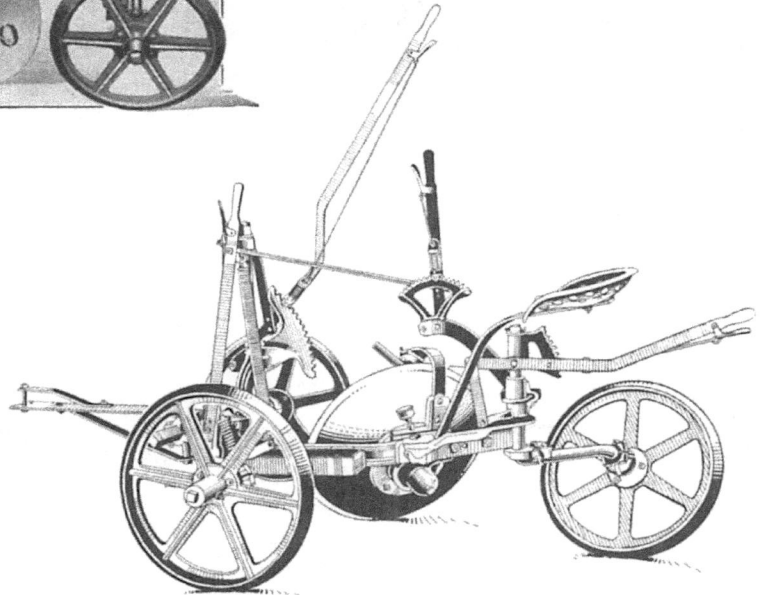

*No. 4 Disk Sulky plow*

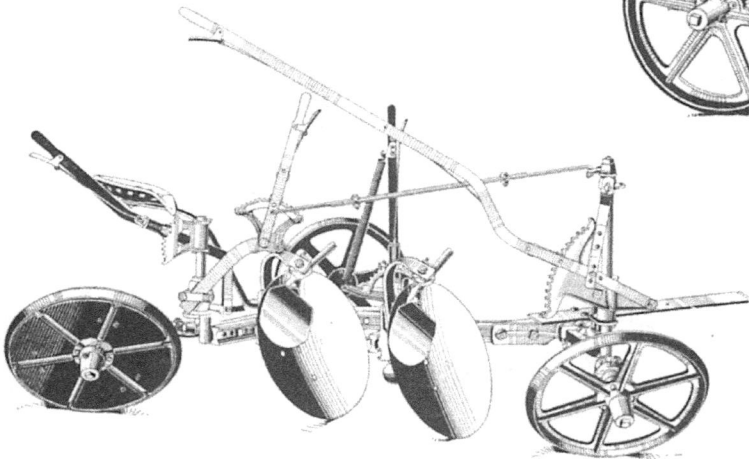

*No. 4 1/2 Disk Gang plow*

*Chattanooga Reversible Disk Sulky plow*

*Top view Chattanooga Reversible Disk Sulky*

*Three Horse Twin Disk plow*

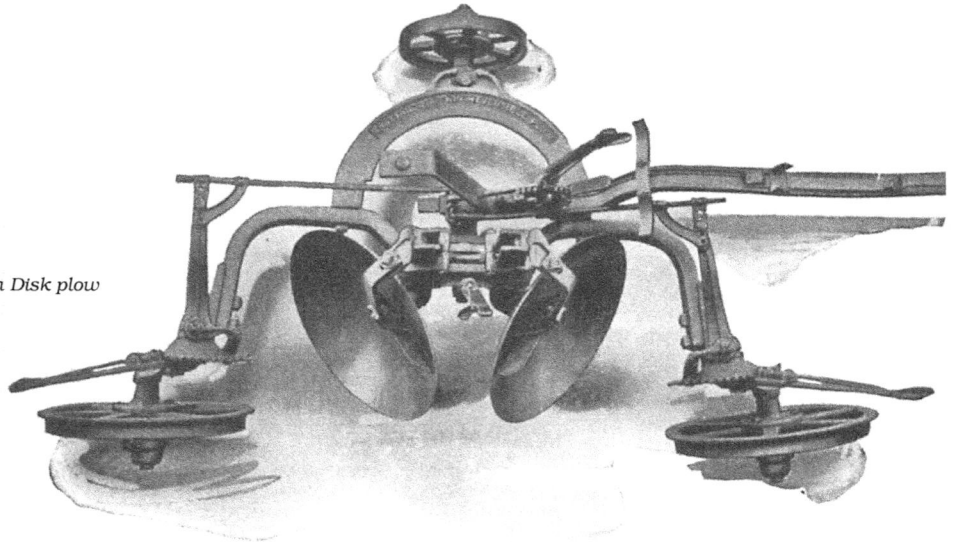

*Chattanooga "Hancock" Disk plow*

*Little Chief Sulky Plow*

*Riding Ditcher or Middlebreaker*

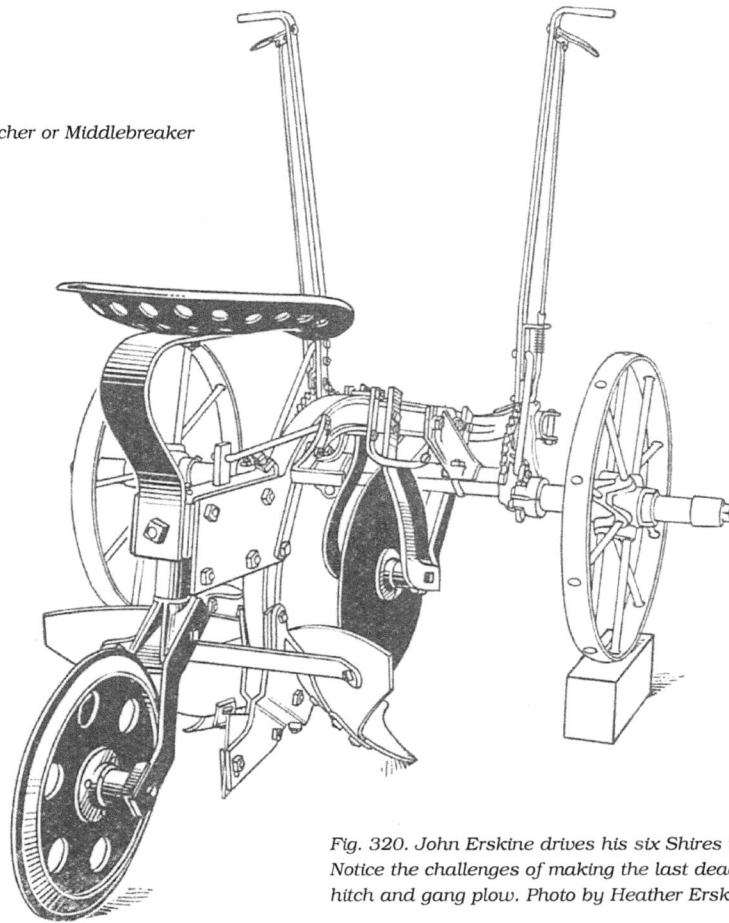

Fig. 320. John Erskine drives his six Shires while Marianne Frank rides. Notice the challenges of making the last dead furrow passes with the big hitch and gang plow. Photo by Heather Erskine.

Their catalog opened with these words; 'Moline Plows and other Flying Dutchman Farm Tools'. The illustration on the left is the trade mark 'Dutchman'. Moline Implement Co. was a massive enterprise with manufacturing plants and offices in fifteen states in the 1920's. Moline was an industrial umbrella for Adriance-Platt, Mandt Wagon, Freeport Carriage, Henney Buggy, and Monitor Drill.

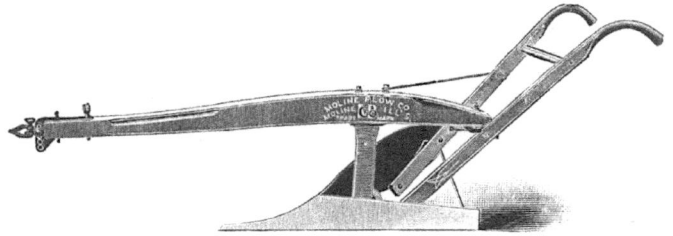

C or Scotch Clipper Series in both wood beam (right) and steel beam (left) came in 12 to 16 inch bottoms

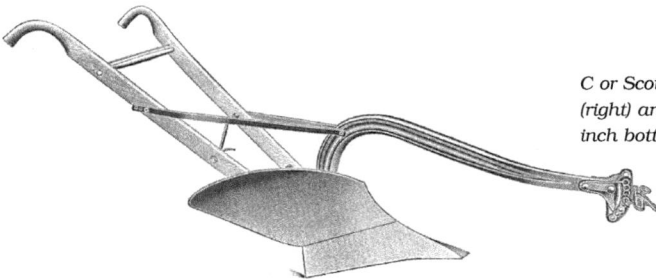

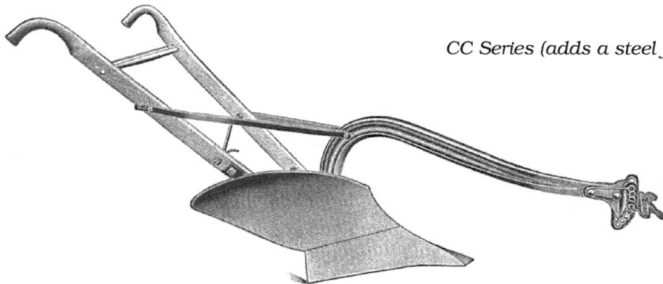

CC Series (adds a steel frog to the C series)

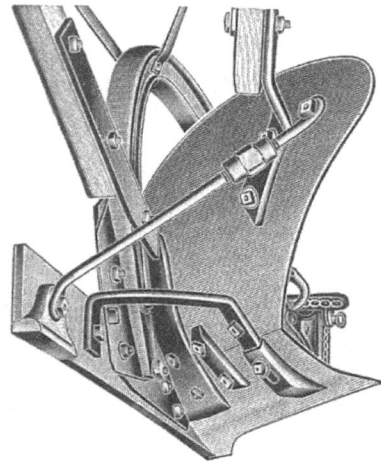

C Special Series

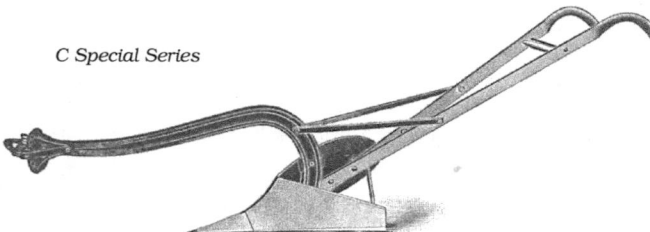

D Series with stubble bottom

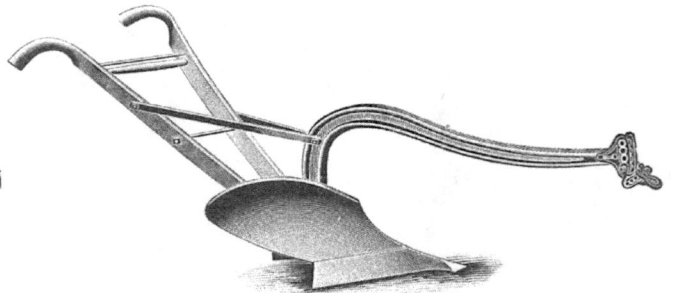

*D Series with wood beam*

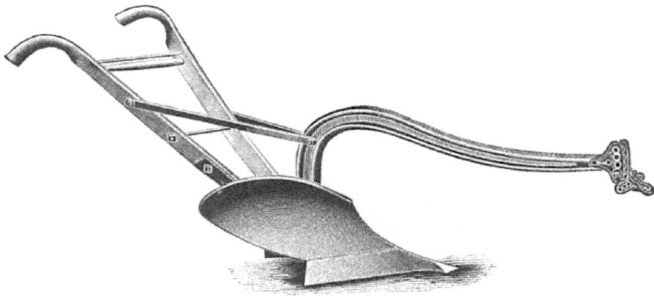

*D D Series for old ground and mellow soils*

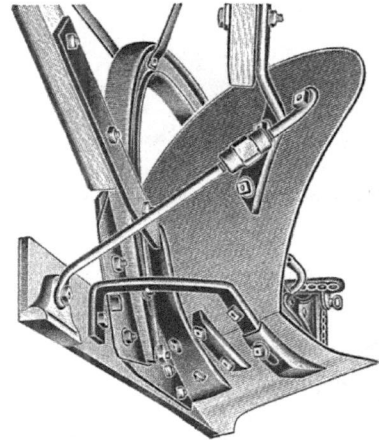

*D D Series plows featured a removeable frog wedge which would convert plow from two to three horse use.*

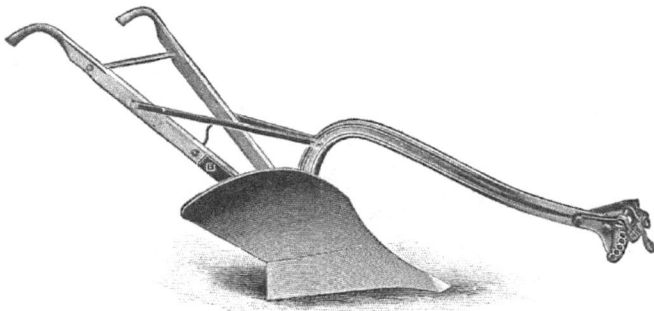

*Nebraska Clipper No. 2 for light fluffy difficult to scour soils in Nebraska and Northern Iowa.*

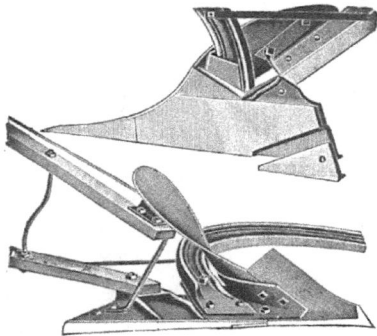

*Blue Bird bottom with high landside and replaceable cast landside shoe*

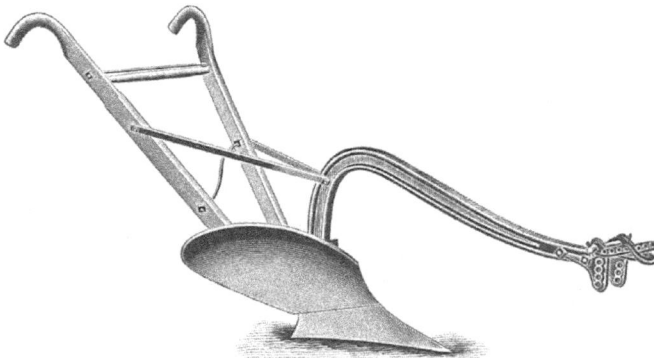

*Blue Bird general purpose plow*

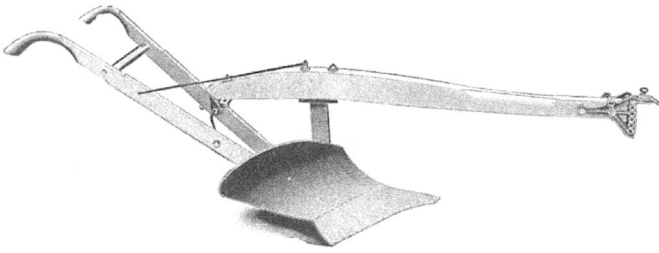

Pacific Coast Special Blue Bird with adjustable wood beam

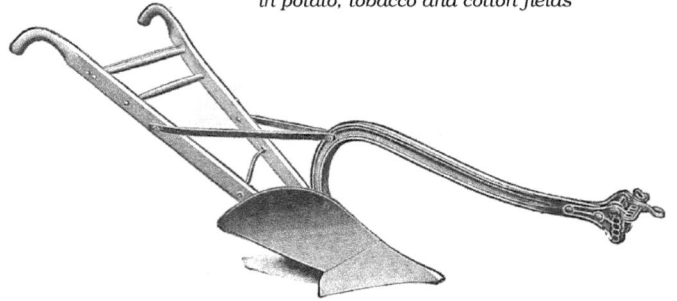

Famous plows shaped for cultivating in potato, tobacco and cotton fields

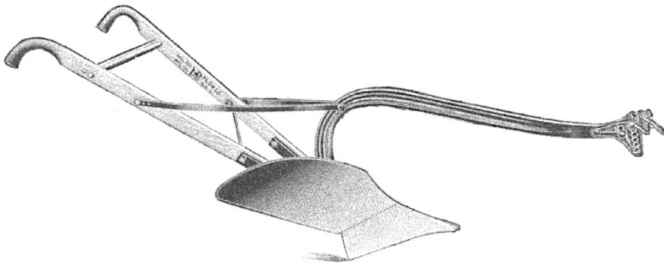

AA Blackland for southern black, waxy, sticky soils

Lousiana Blackland differs from AA in clevis and hanging cutter.

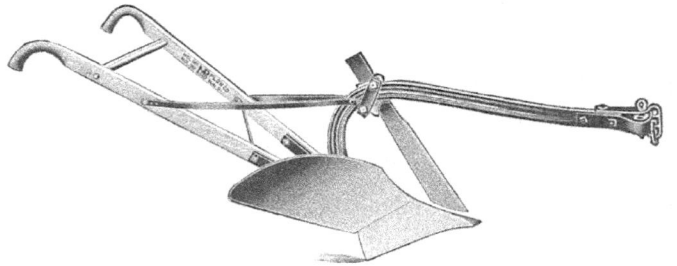

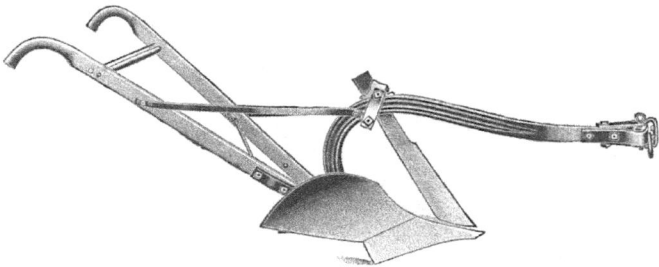

LX Series for cane growers, clevis adjustments are designed for "wrapping the center" and "barring off."

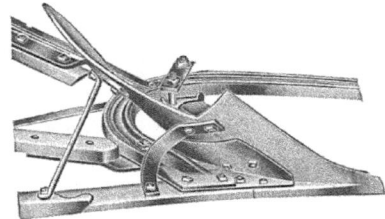

Showing the LX Series steel frog

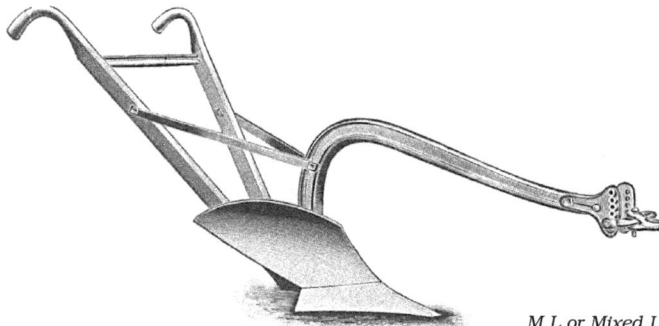

M L or Mixed Land

New Vineyard plow No. 2. Beam is
adjustable at the foot, handles swing
right or left.

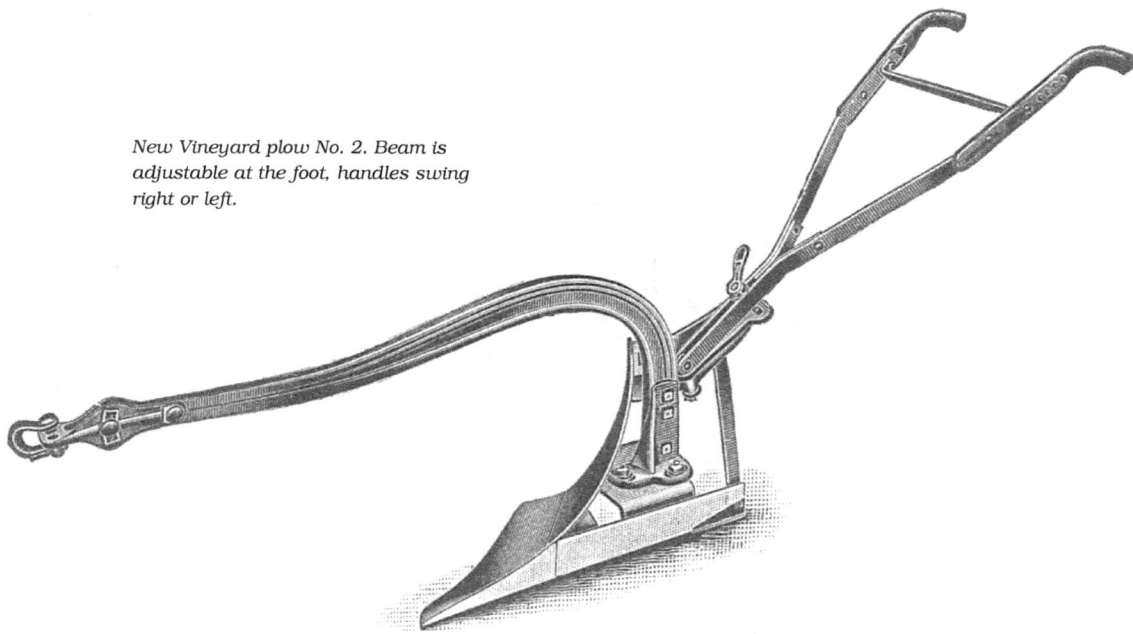

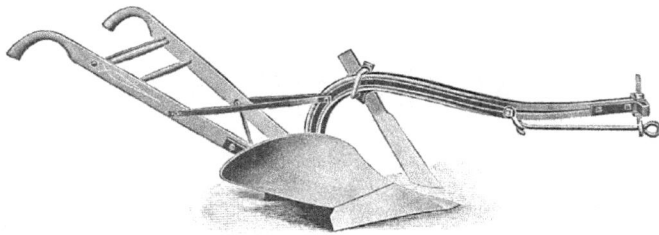

Lousiana Four Mule plow

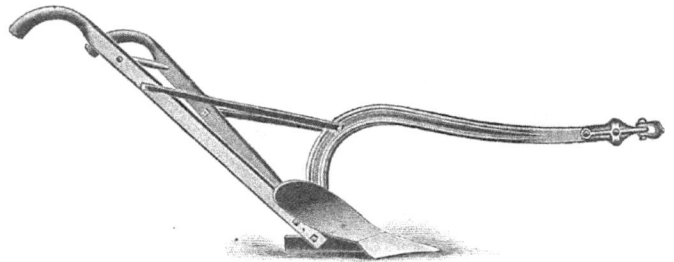

S plow was available in 6, 7 and 8 inch bottoms.

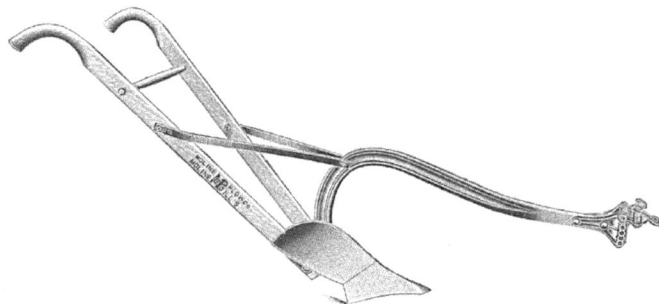

Famous Pony plows adapted for plowing weedy and
trashy land; for cultivating small trees, shrubs, and for
laying by corn

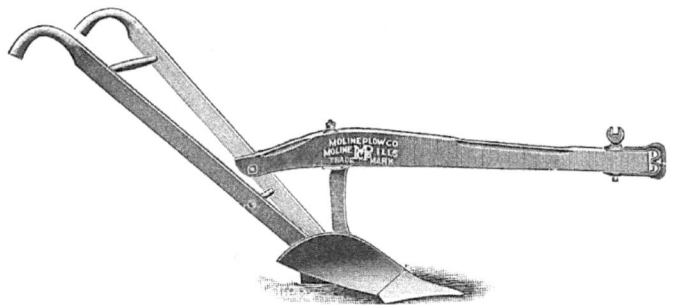

Wood beam Famous Pony plow

*North Texas No. 2 Middlebreaker. Bottom equipped with root cutter.*

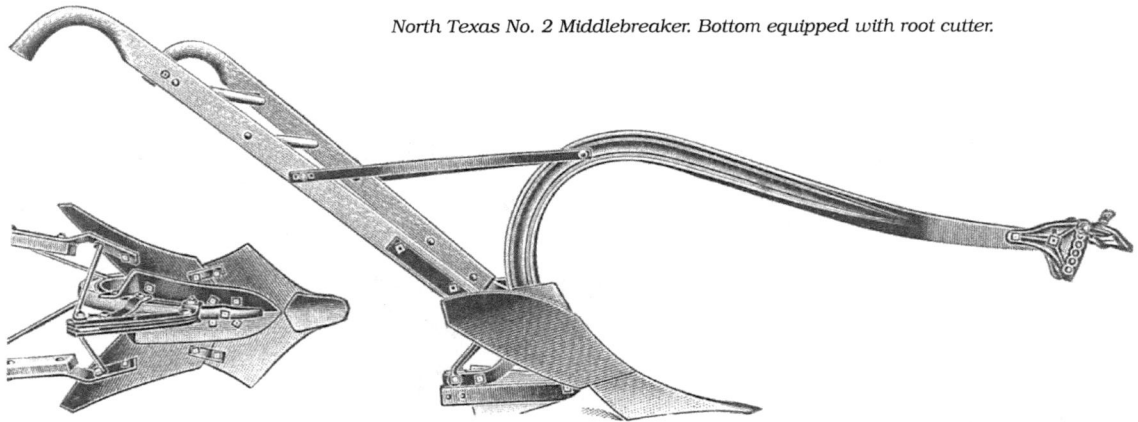

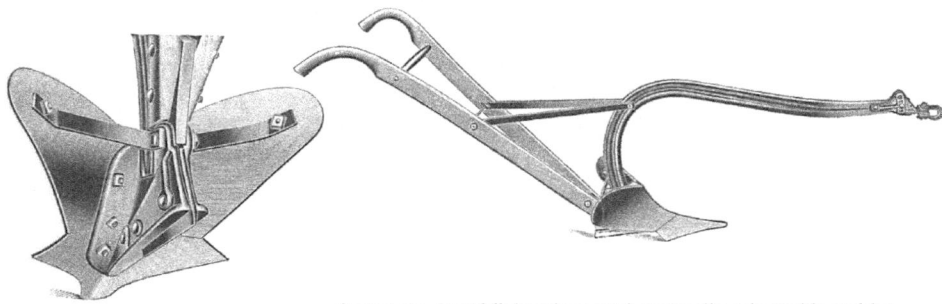

*Comet No. 2 Middlebreakers with vertically adjustable rudder.*

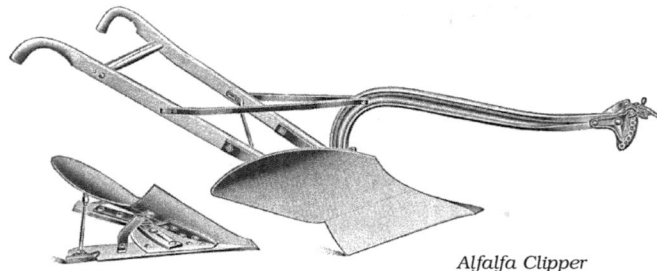

*Alfalfa Clipper*

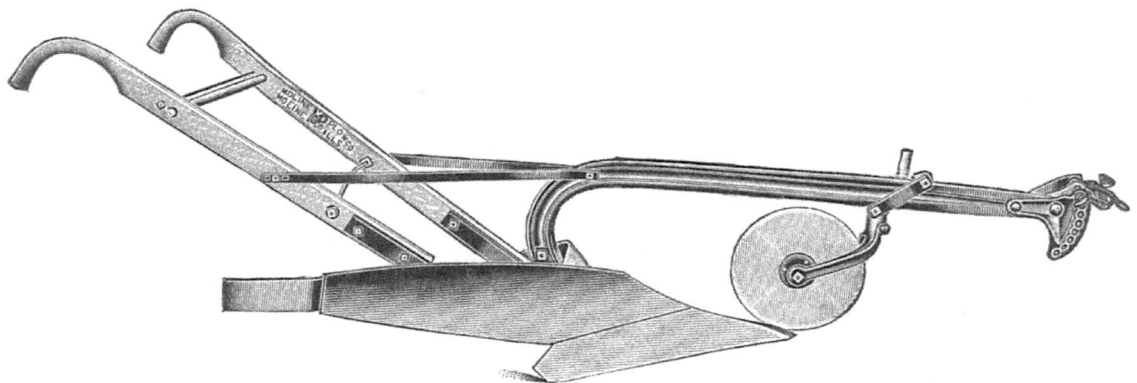

*Western Queen Steel Beam Breaker*

Western Queen Common Breaker

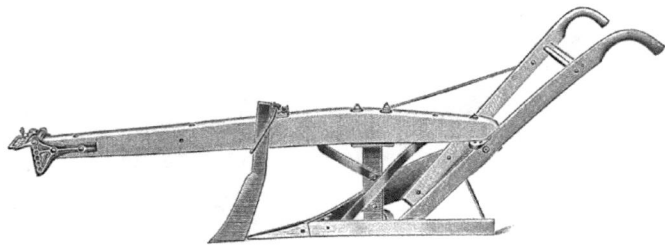

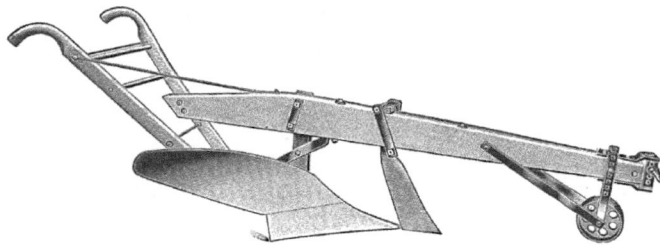

Northern Queen Breaker

Dakota Queen Rod Breaker

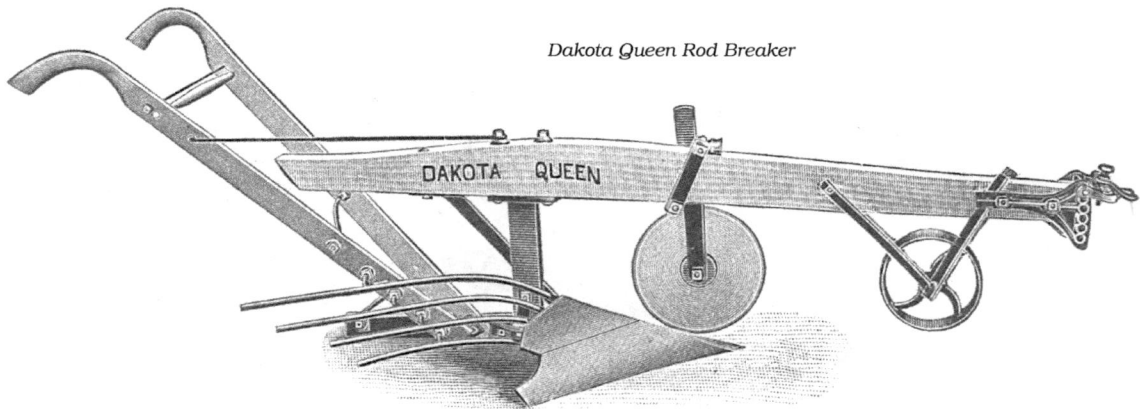

DAKOTA QUEEN

Kansas King Rod breaker

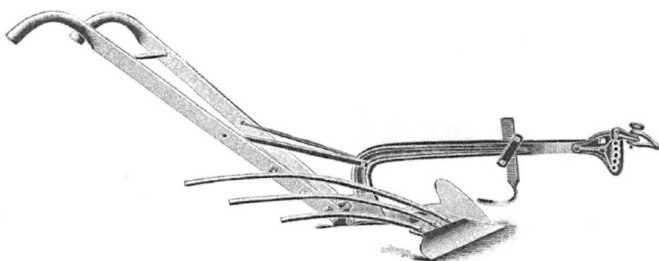

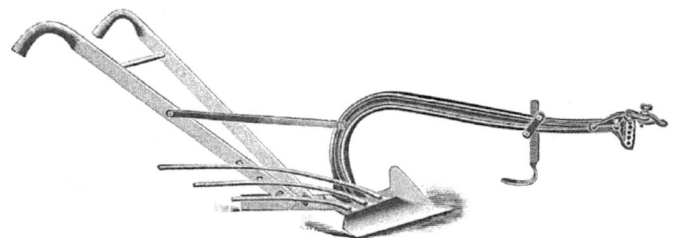

Cricket Rod breaker

*Western Queen Extra Breakers*

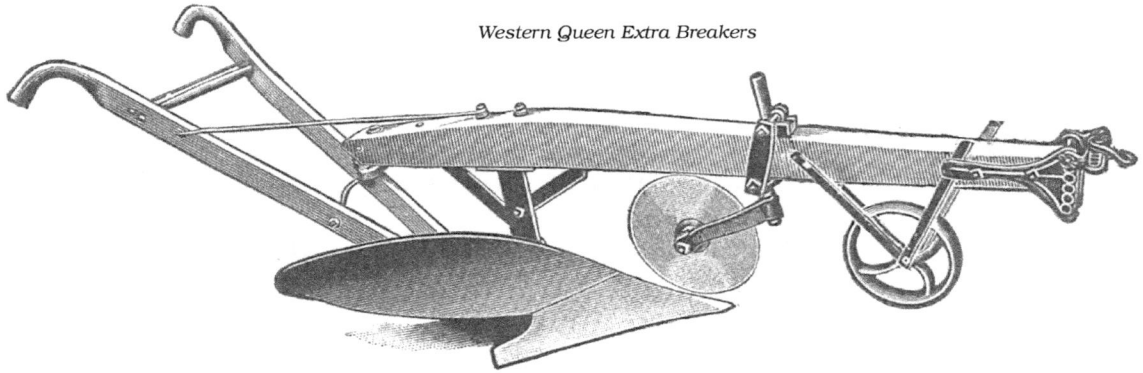

*Hazel Plow built for timber land, equipped with Quincy reversible cutter, sharpened both ends*

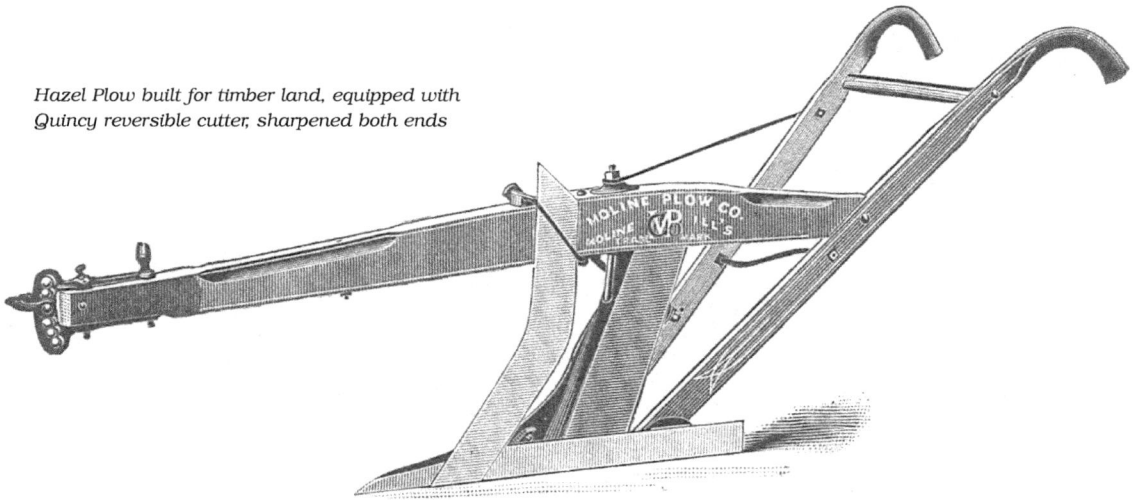

*County Road plow, Quincy Cutter.*

*Forestry Series plow, Robin Hood. For timber land
and heavy soils, here equipped with knee cutter.*

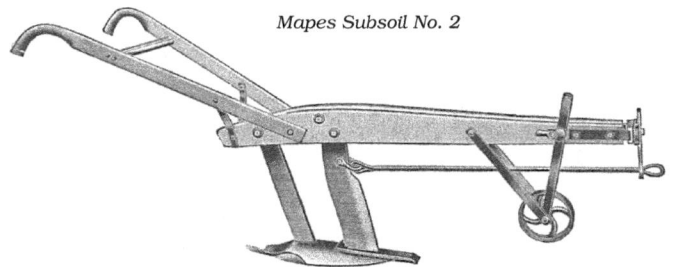

*Mapes Subsoil No. 2*

*Railroad plow. Steel grips on the handles, Heavy
steel runner on the right handle protects it when
the plow is dragged on the side.*

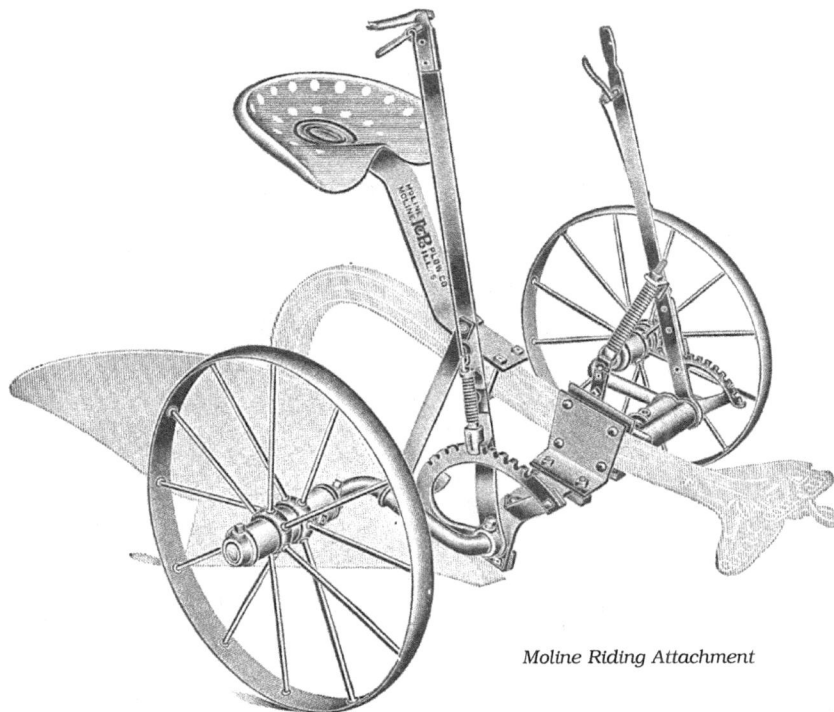

*Moline Riding Attachment*

*Little Dutch Sulky*

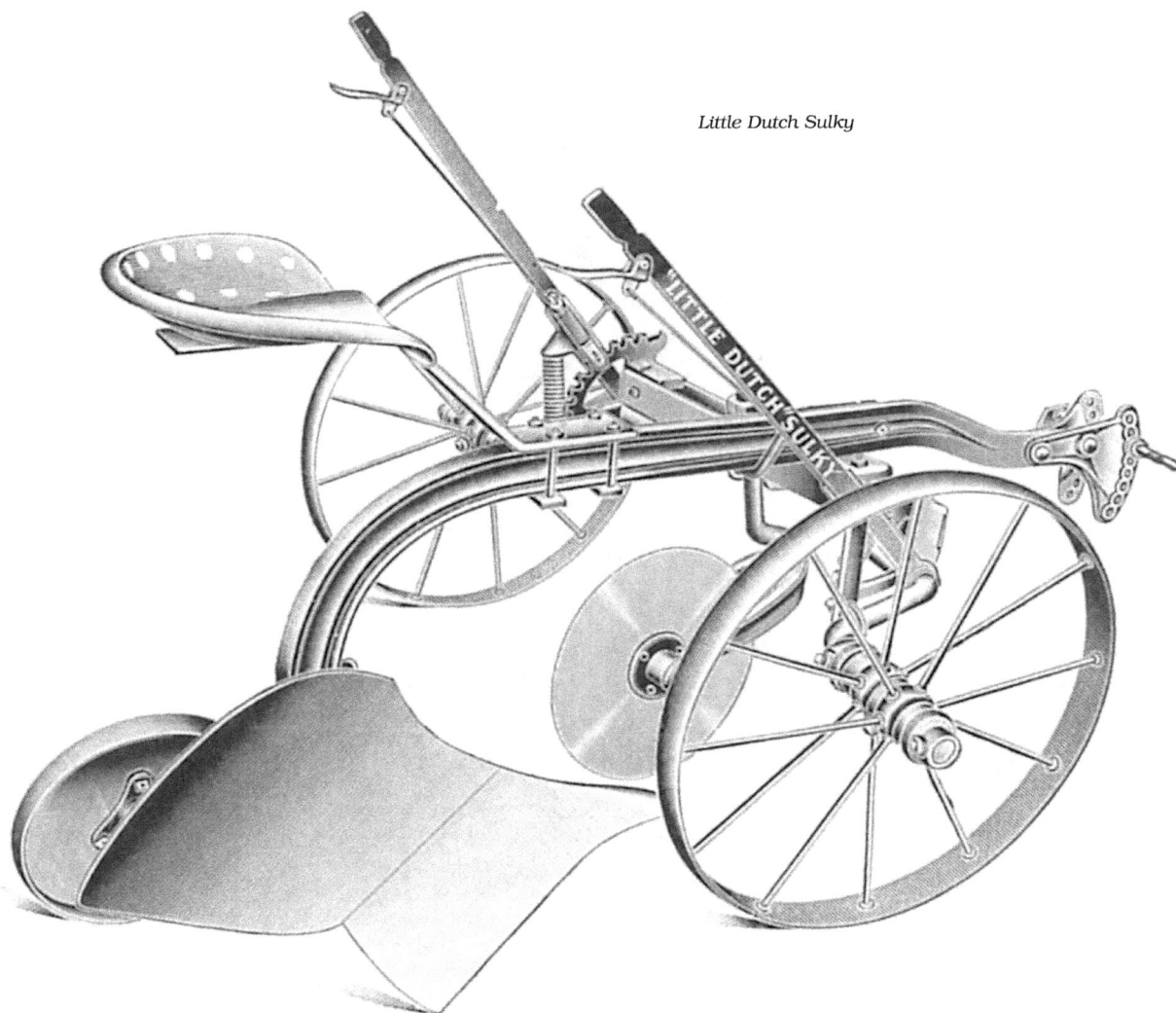

*Good Enough Sulky No. 1*

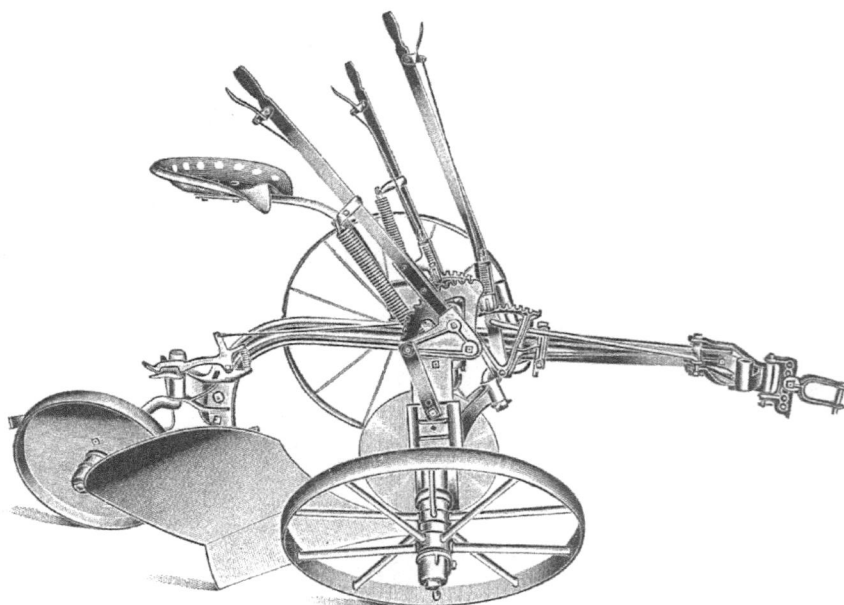

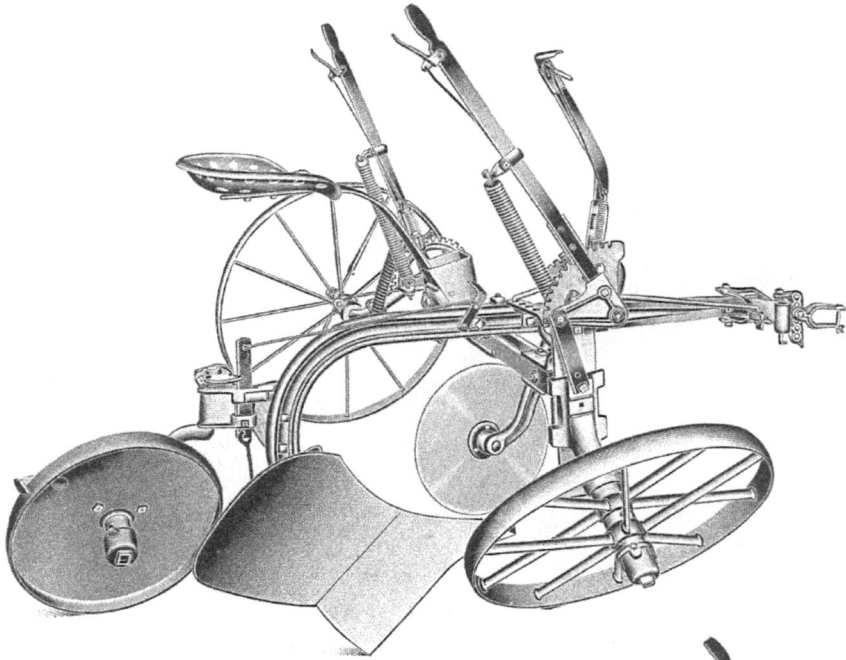

*Good Enough No. 3 with landing lever controlling swinging clevis.*

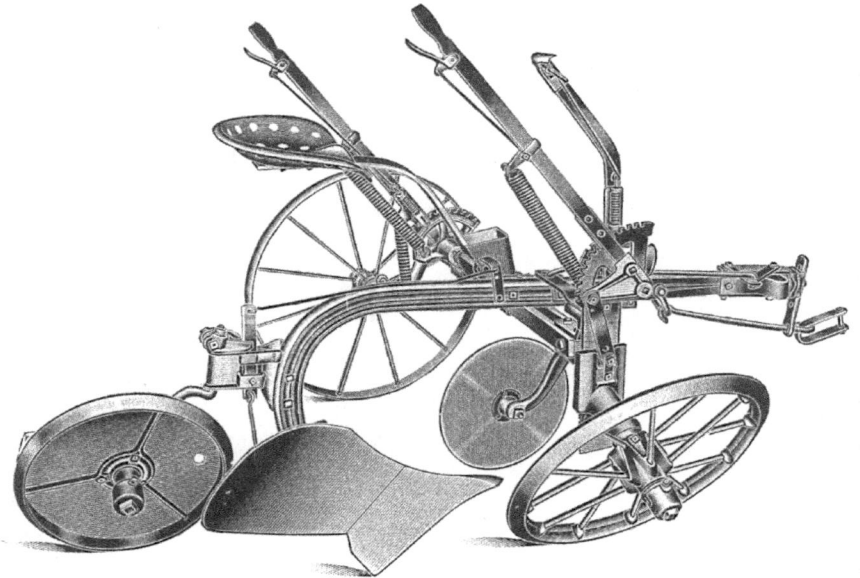

*Texas Good Enough No. 2*

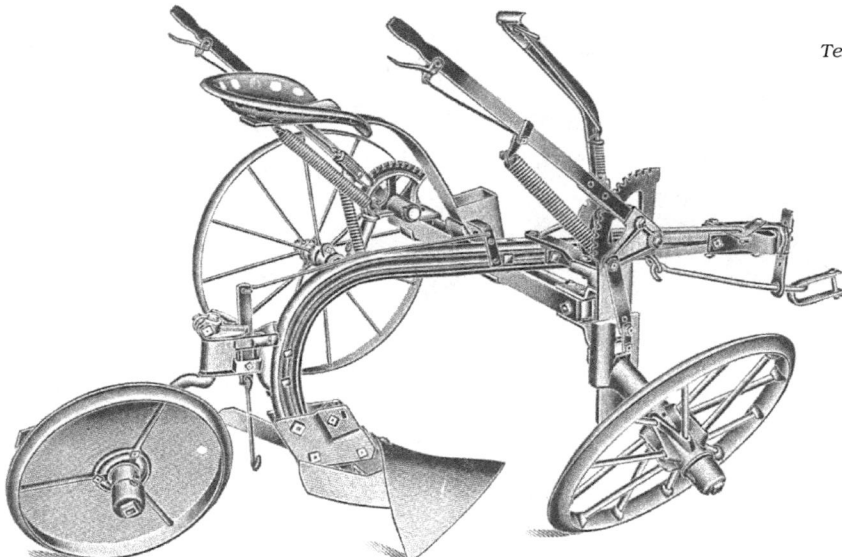

*Texas Good Enough No. 2 with Sweep Attachment*

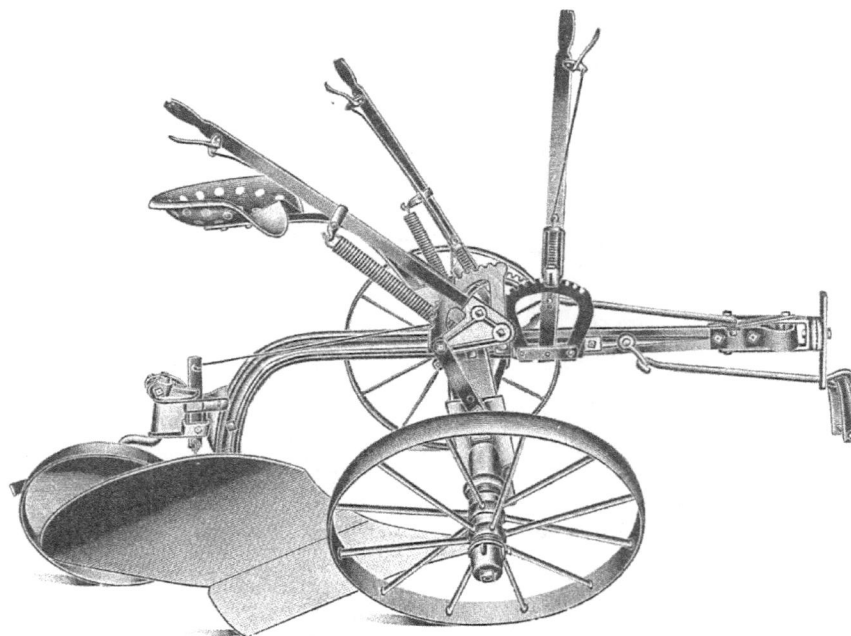

*California Good Enough Sulky. Recommended that teamster use tongue attachment when hitching colts or using plow on steep hills.*

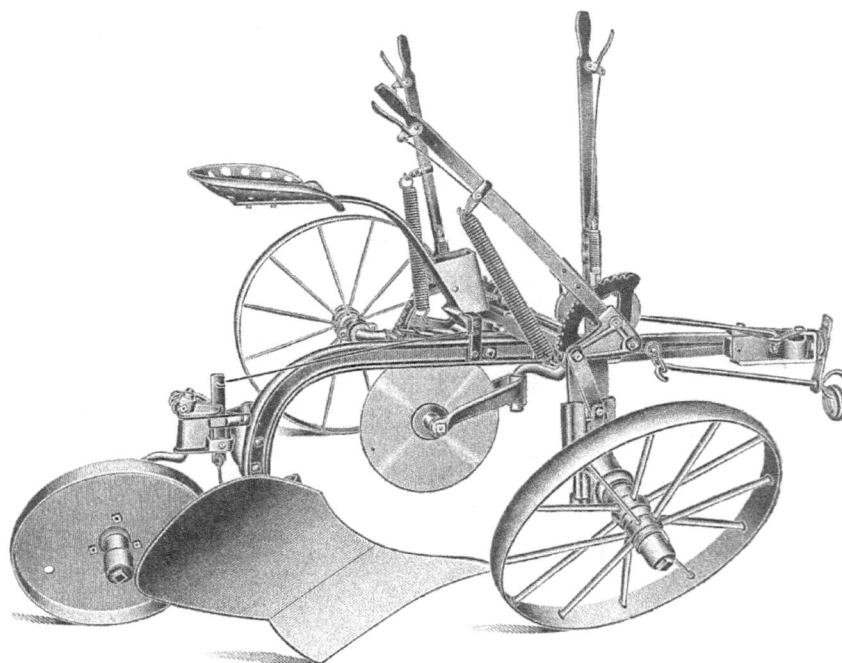

*California Good Enough Sulky with high frame, will plow to twelve inches deep.*

# Moline Plows

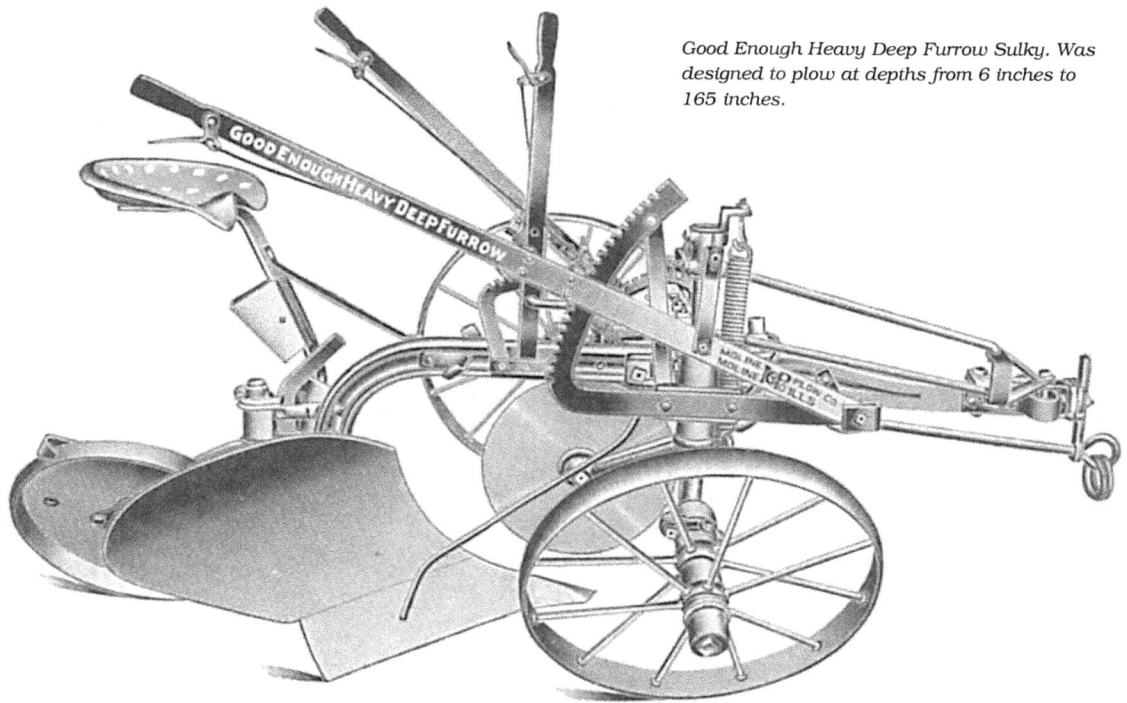

*Good Enough Heavy Deep Furrow Sulky. Was designed to plow at depths from 6 inches to 165 inches.*

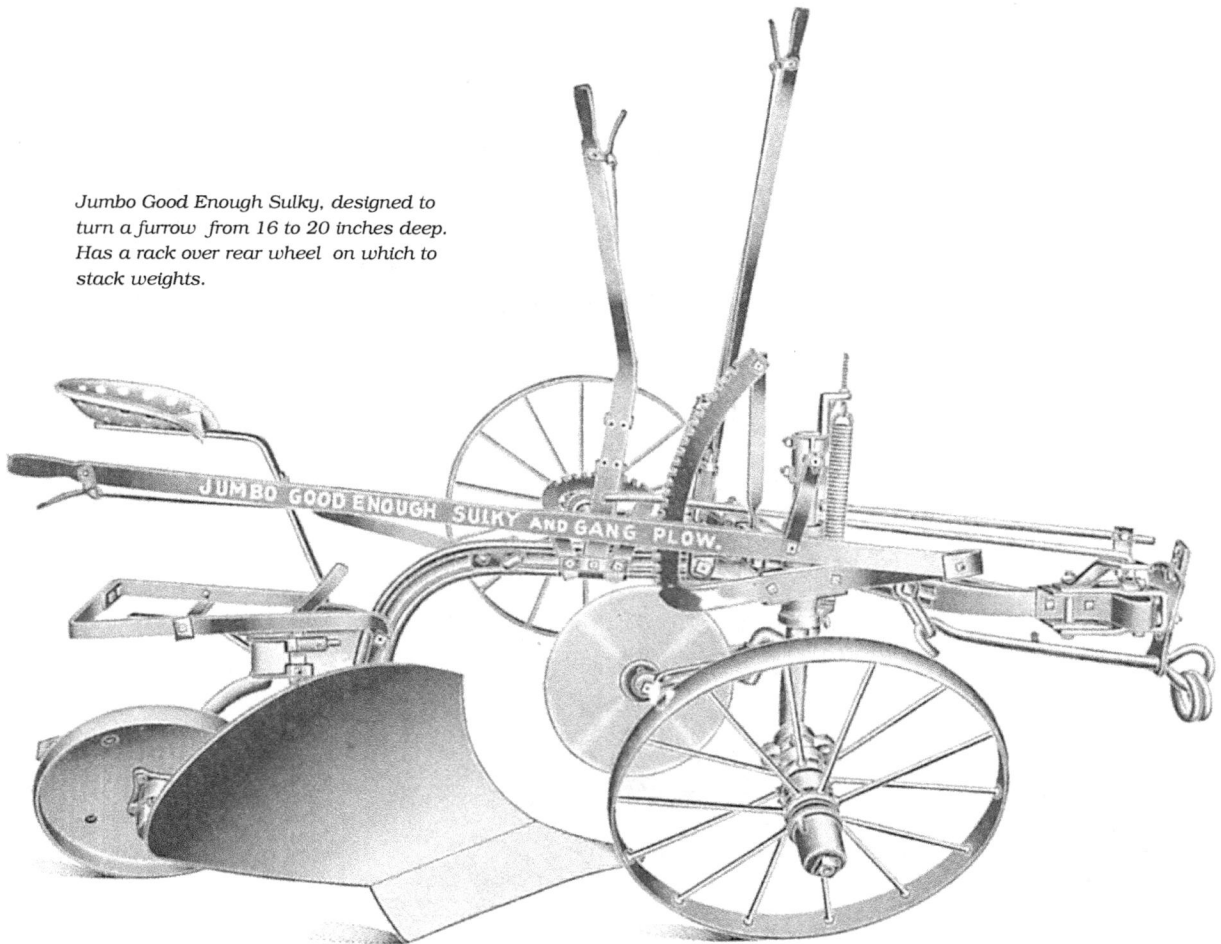

*Jumbo Good Enough Sulky, designed to turn a furrow from 16 to 20 inches deep. Has a rack over rear wheel on which to stack weights.*

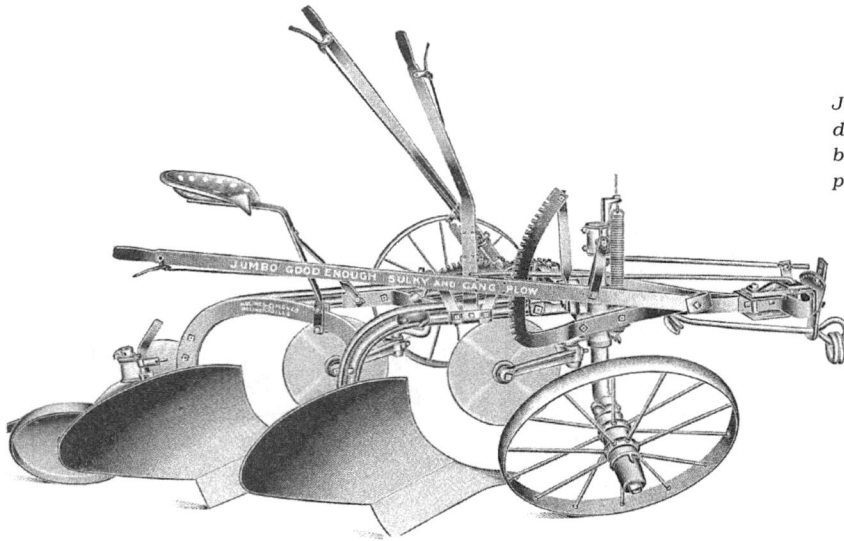

*Jumbo Good Enough as a gang plow. By detaching the second bottom this plow becomes the same as the one on the previous page.*

*Good Enough Gang plow*

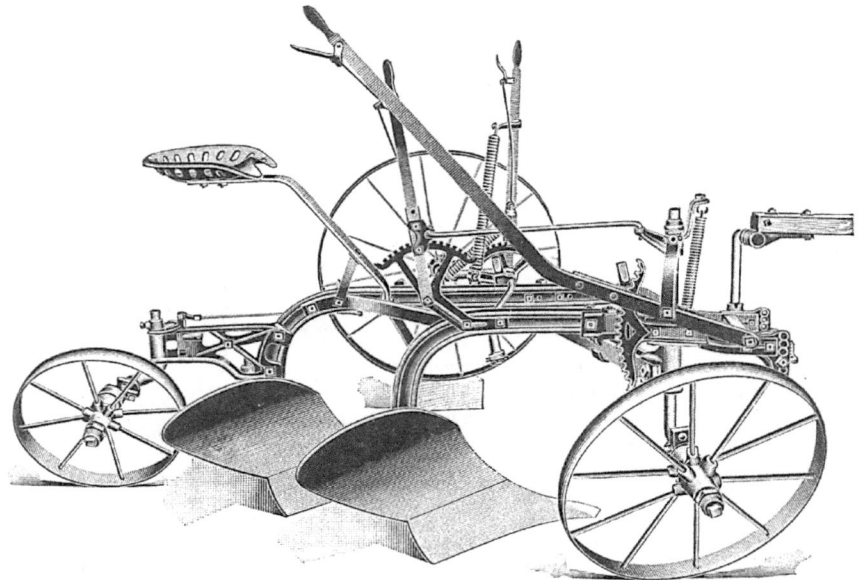

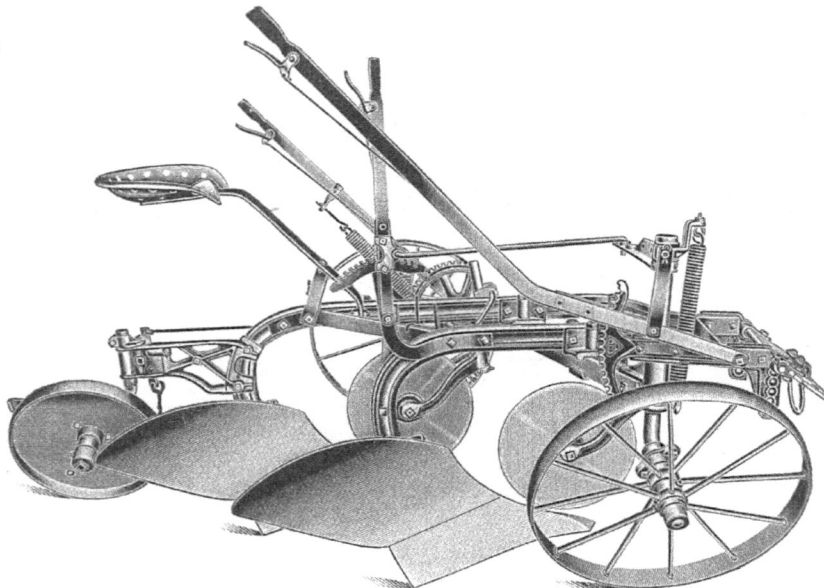

*California Good Enough Gang plow with low frame*

*California High Frame Good Enough Giant of the 'Open and Shut Type' (meaning the width of the cut may be changed from 12 to 14 inches by moving the beams closer to together or farther apart.)*

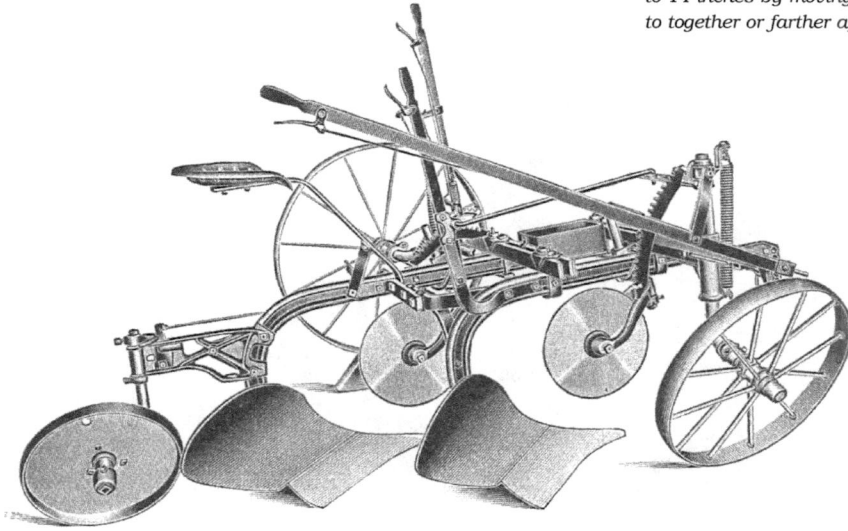

*Best Ever Gang plow*

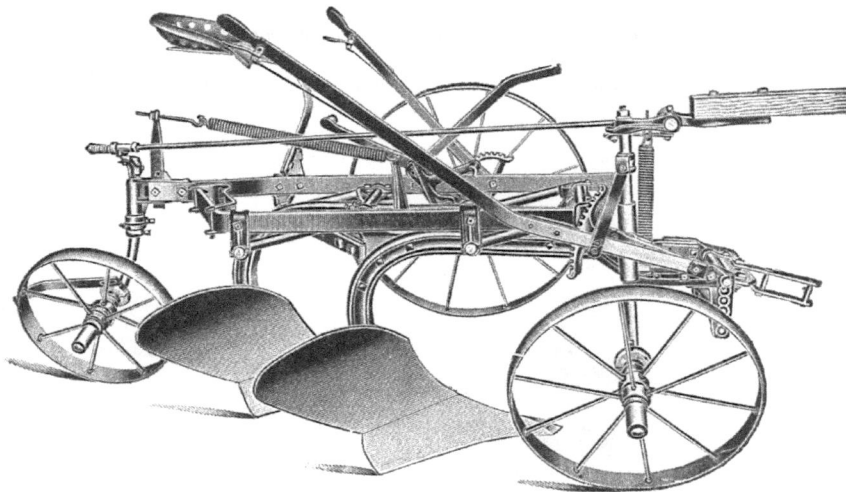

*Best Ever No. 2 Sulky*

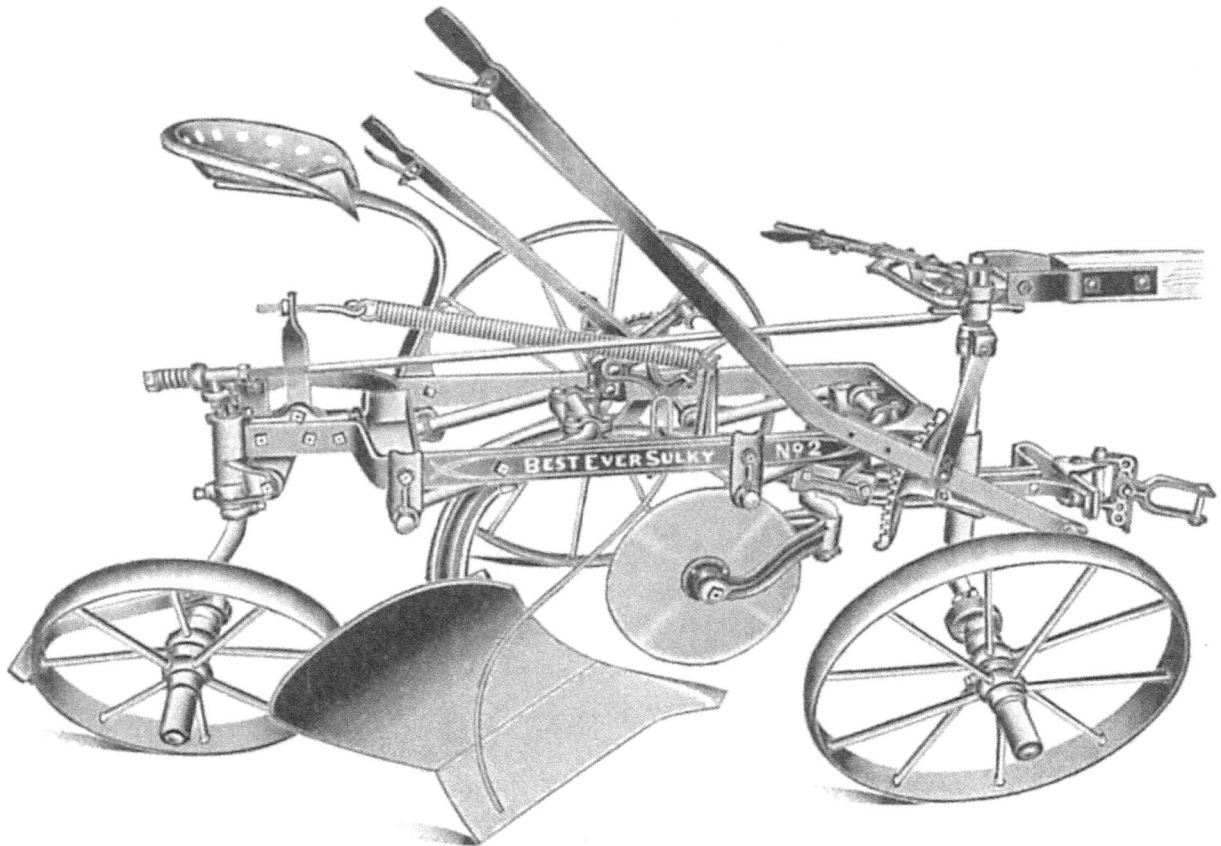

The connecting rod device between the wheels prevents the rear wheel from pushing forward on the rod and causing the pole to crowd the furrow horse, and it allows no motion of the rear wheel to affect the position of the tongue and no movement of the tongue to change the position of the rear wheel. This is accomplished by means of a slotted connection. When the plow is running straight, the front end of the rod is locked directly back of the axle stem. Thus the pressure against the moldboard is carried by the rear furrow wheel and the connecting rod does not in any way affect the front furrow wheel. When making a turn either to the right or left, the rear wheel acts as a castor wheel, pushing the rod to the opposite end of the slot in the front connection. When the team straightens up, the rear wheel is also straightened.

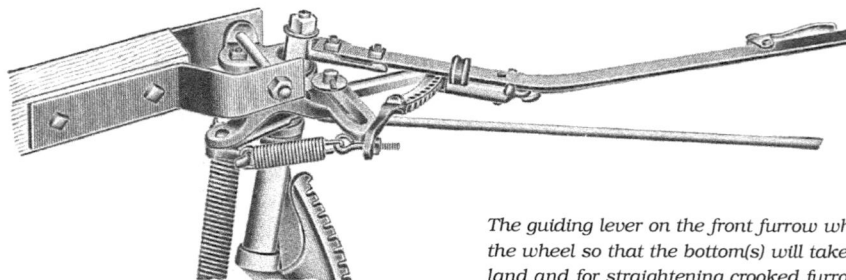

The guiding lever on the front furrow wheel adjusts the wheel so that the bottom(s) will take more or less land and for straightening crooked furrows.

Makes &
Models

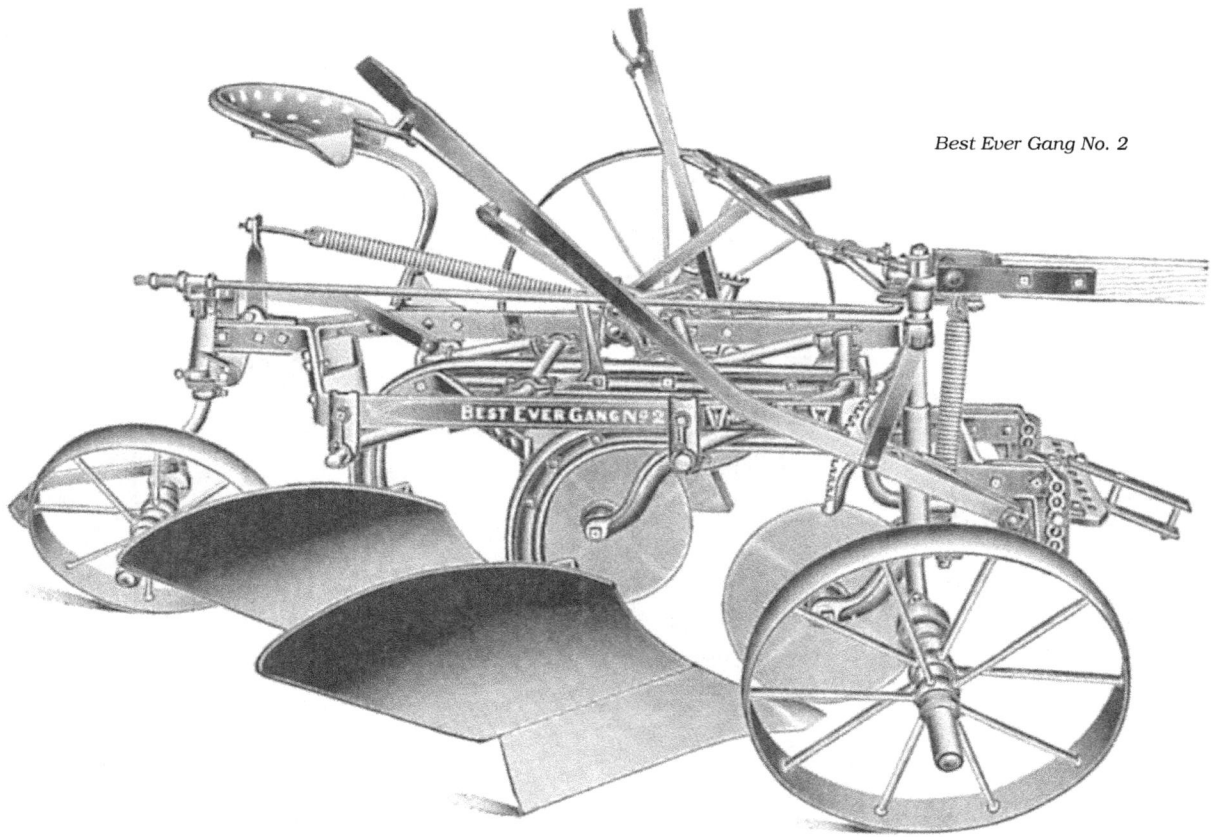

*Best Ever Gang No. 2*

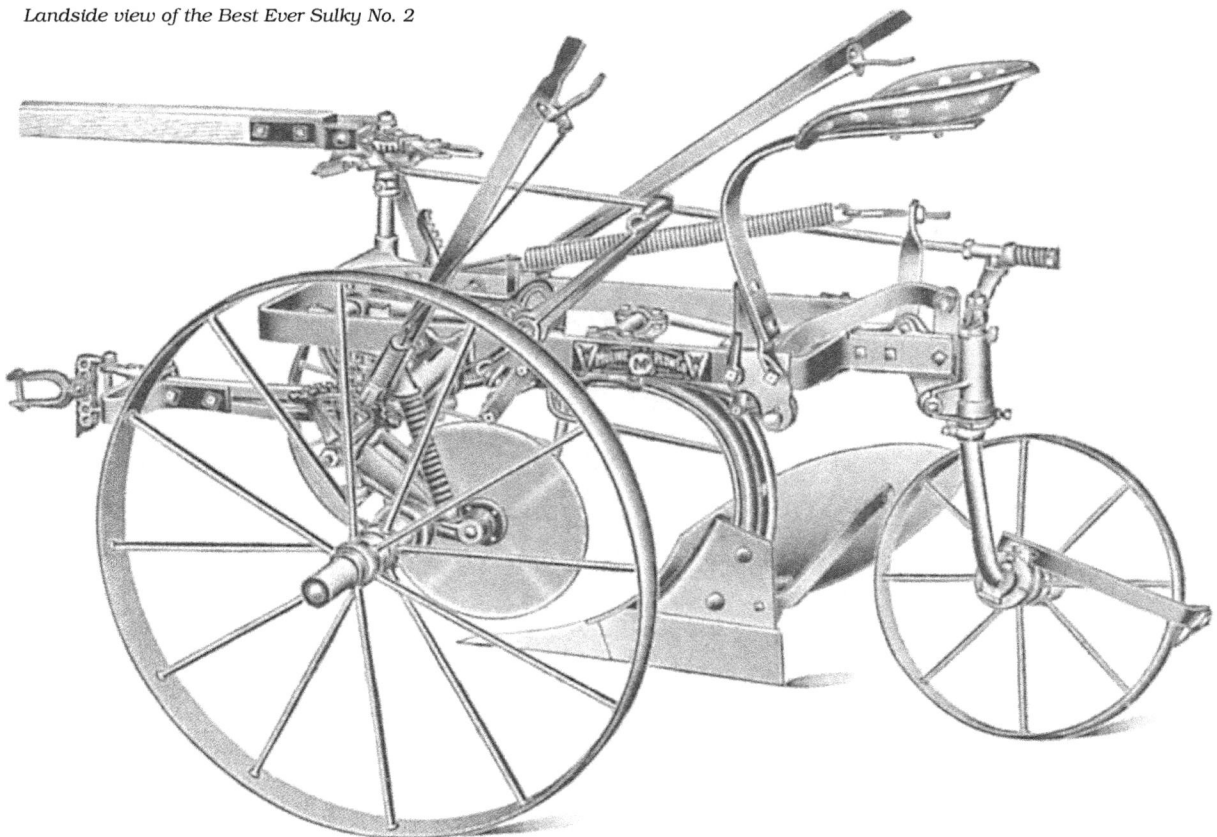

*Landside view of the Best Ever Sulky No. 2*

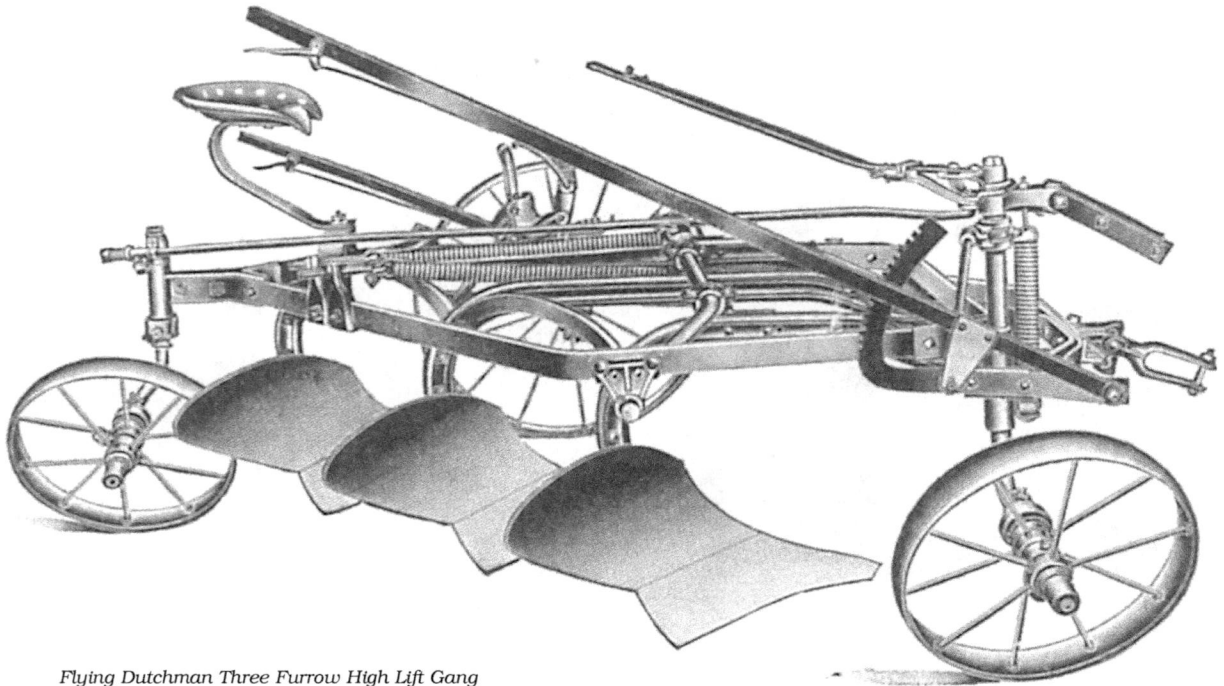

*Flying Dutchman Three Furrow High Lift Gang*

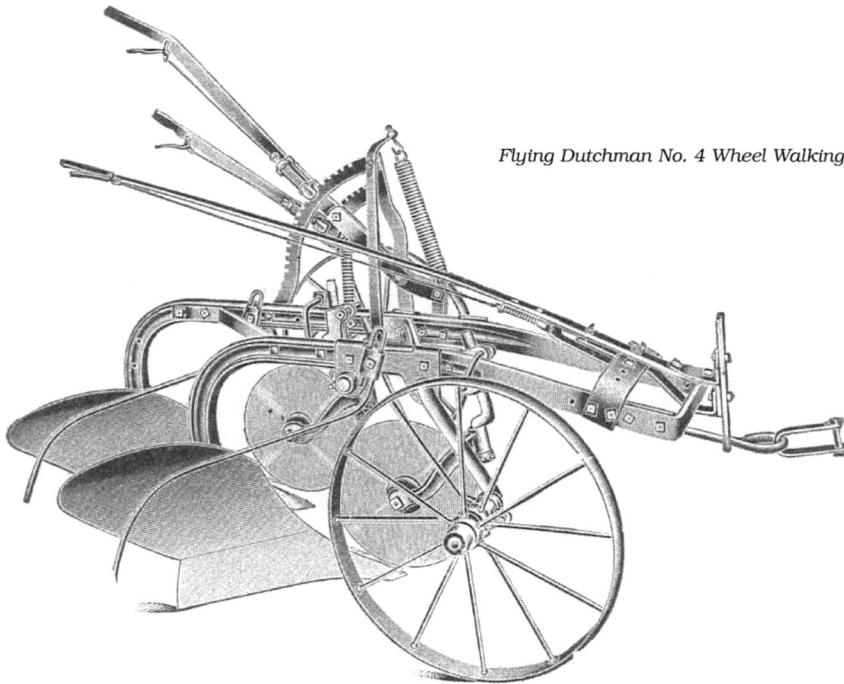

*Flying Dutchman No. 4 Wheel Walking Gang*

*Fylying Dutchman No. 4 Wheel Walking Gang with Riding Attachment.*

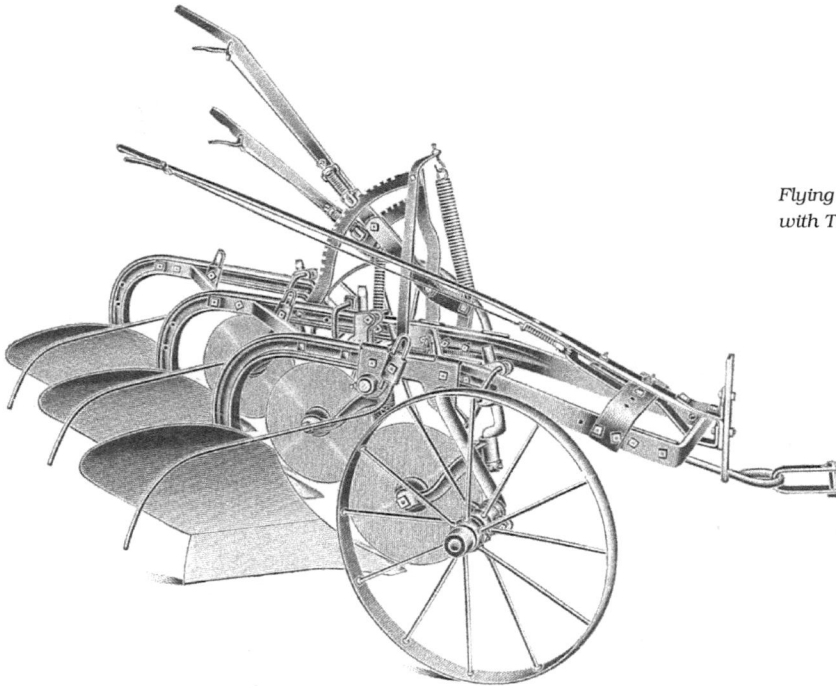

*Flying Dutchman No. 4 Wheel Walking Gang with Third Bottom Attachment*

*Power Lift Wheel Guide Single*

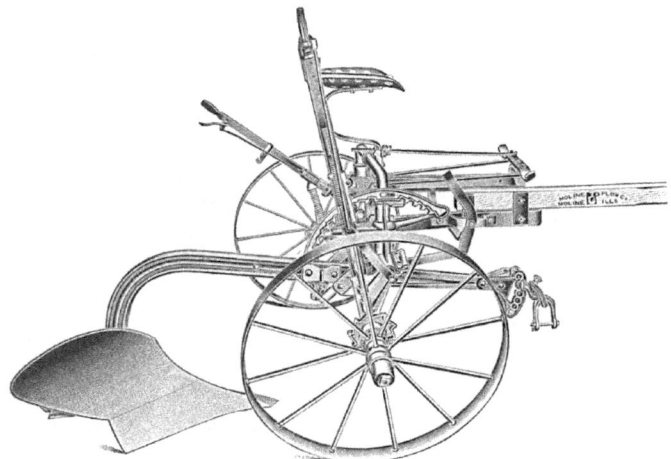

*Wheel Guide Two Way Dutchman No. 2*

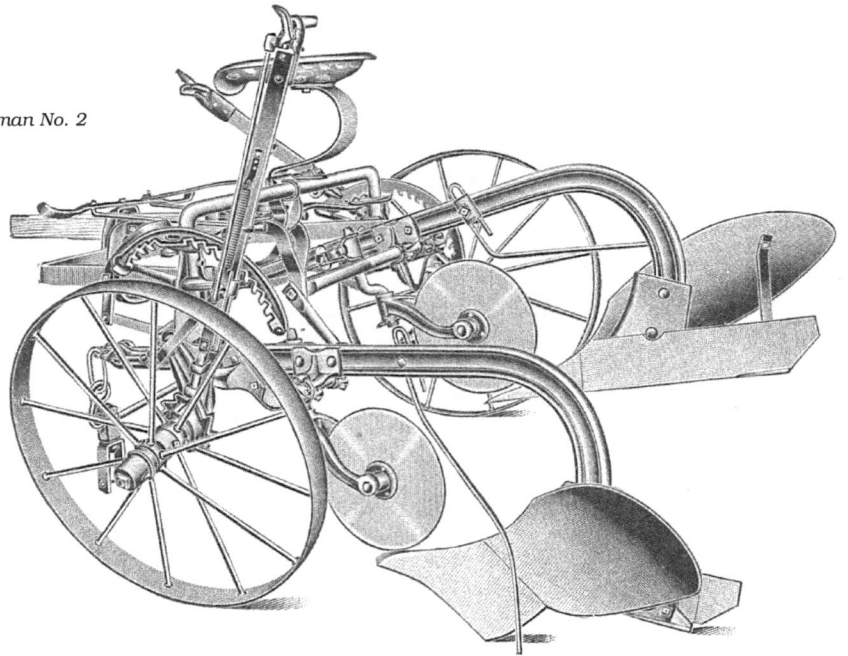

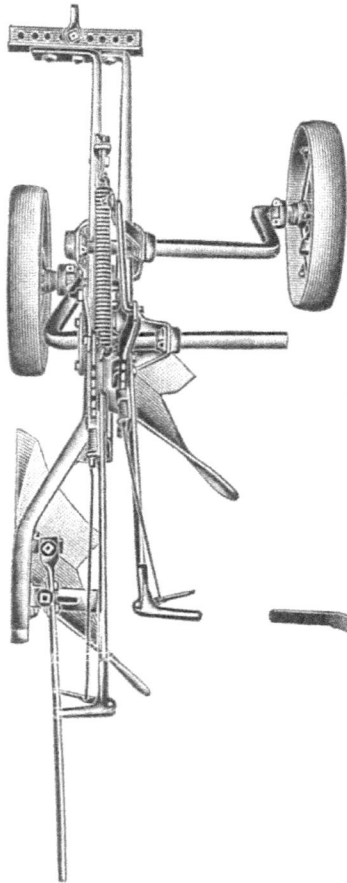

*Top view of Blue Bird No. 3 Vineyard
illustrating adjustable axles*

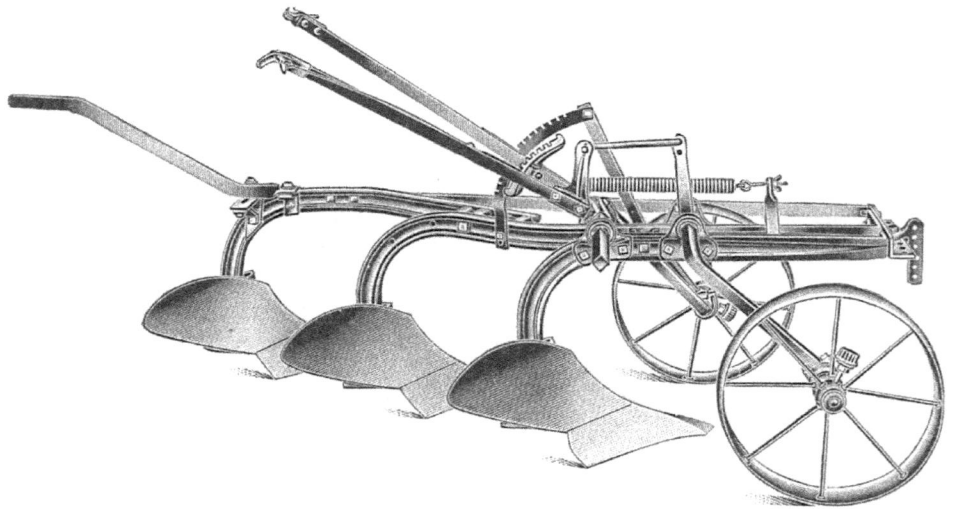

*Blue Bird No. 3 Vineyard Gang*

# Moline Plows

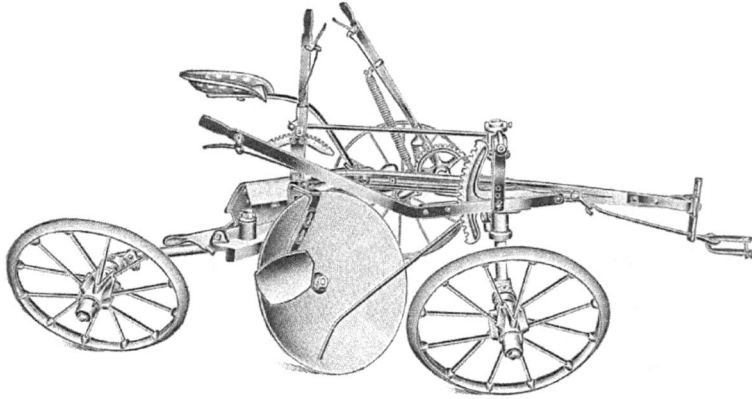

Rotary Good Enough Disc plow

*Adjustable Disc plow hitch*

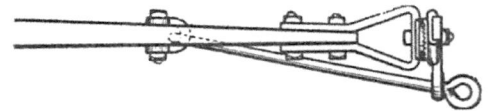

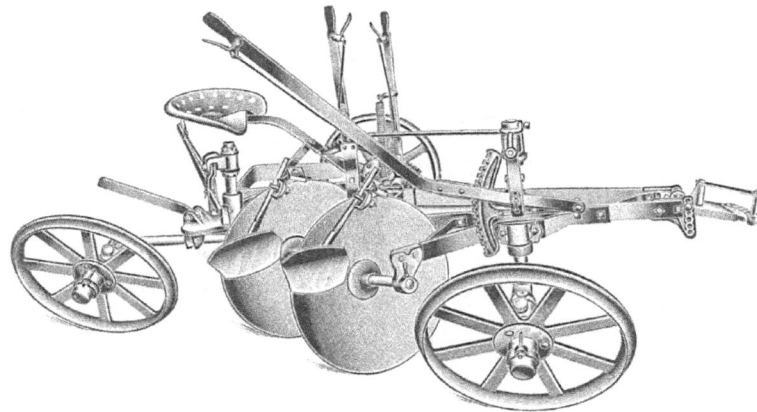

Rotary Dutchman Disc plow (double)

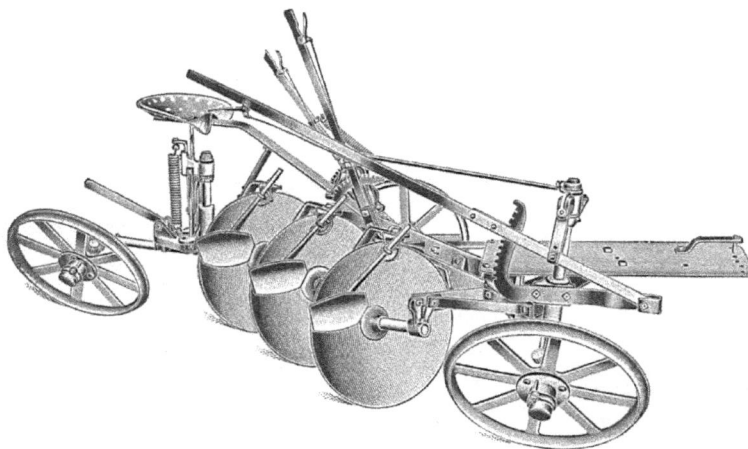

*Rotary Dutchman Three-Two (readily converted into double disc)*

Rotary Dutchman Four-Three (readily converted into triple disc)

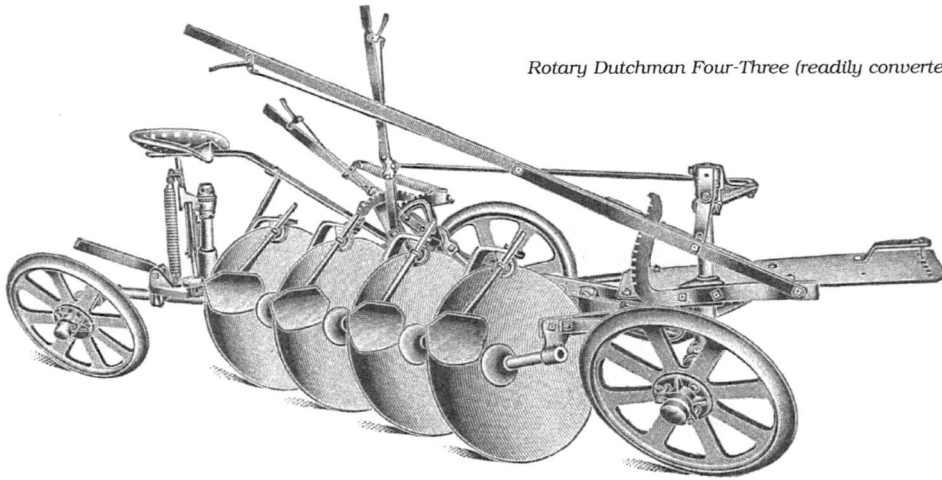

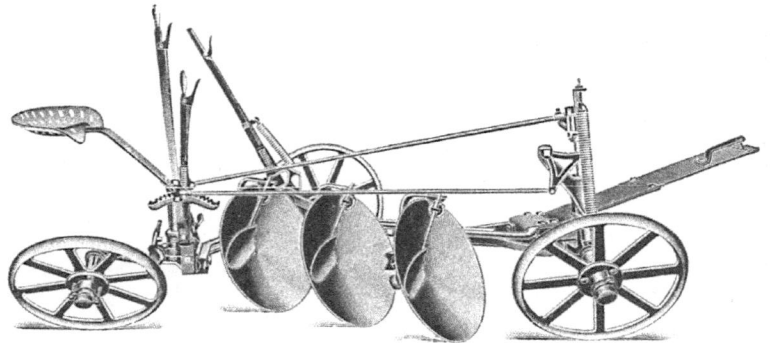

Southern Chief No. 1 Disc (Three-Two)

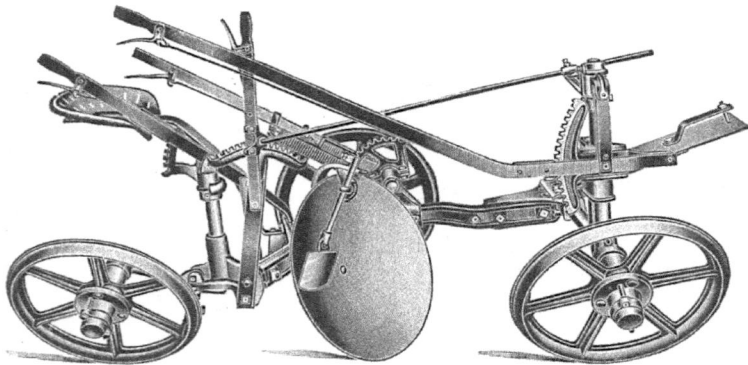

Southern Chief No. 2 Single Disc

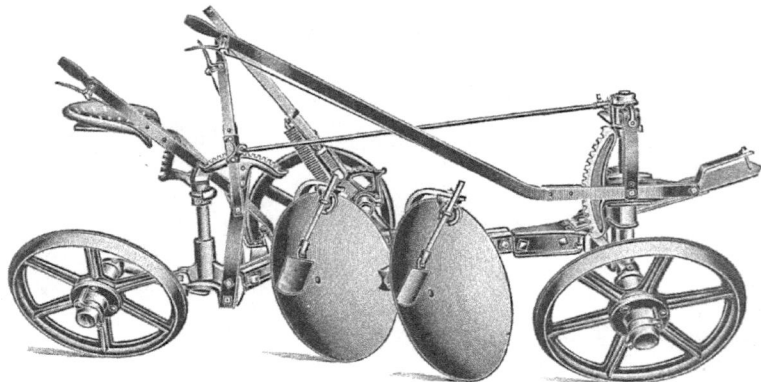

Southern Chief No. 2 Double Disc

# Moline Plows

*Southern Chief No. 2 (Triple)*

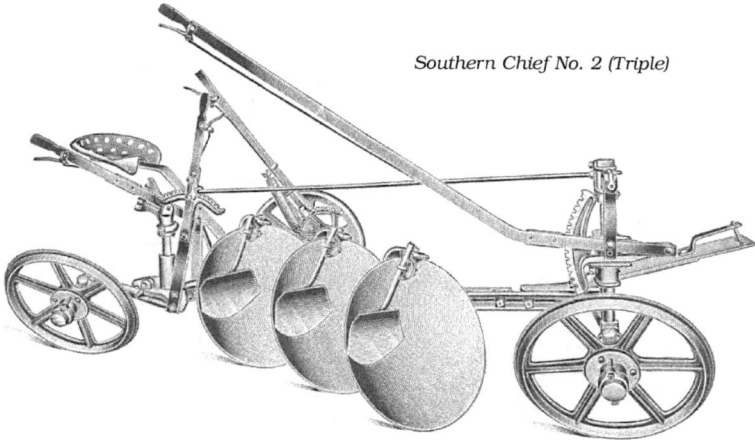

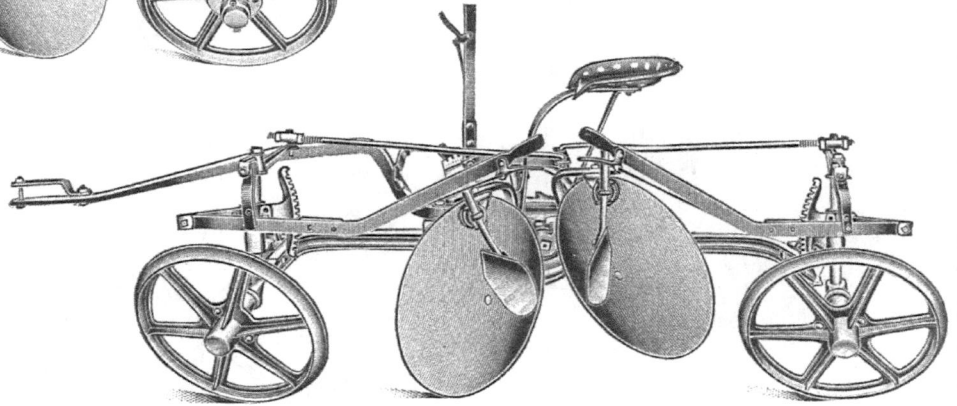

*No. 2 Reversible Disc. Beam and seat
pivot. One disc goes down one goes up.*

*Moline Riding Attachment*

*Rear Wheel Attachment*

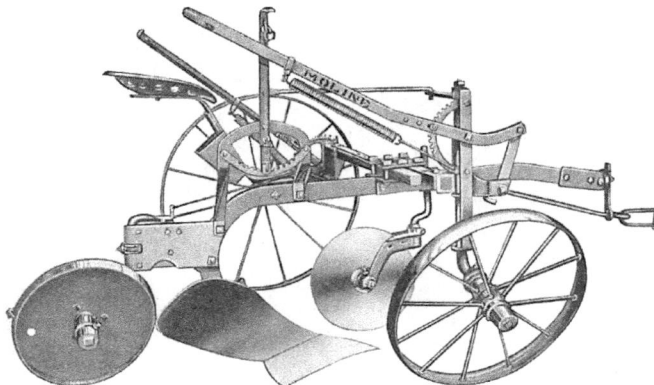

*TSA Northern Type sulky*

*Moline HL 21 Sulky (High Lift)*

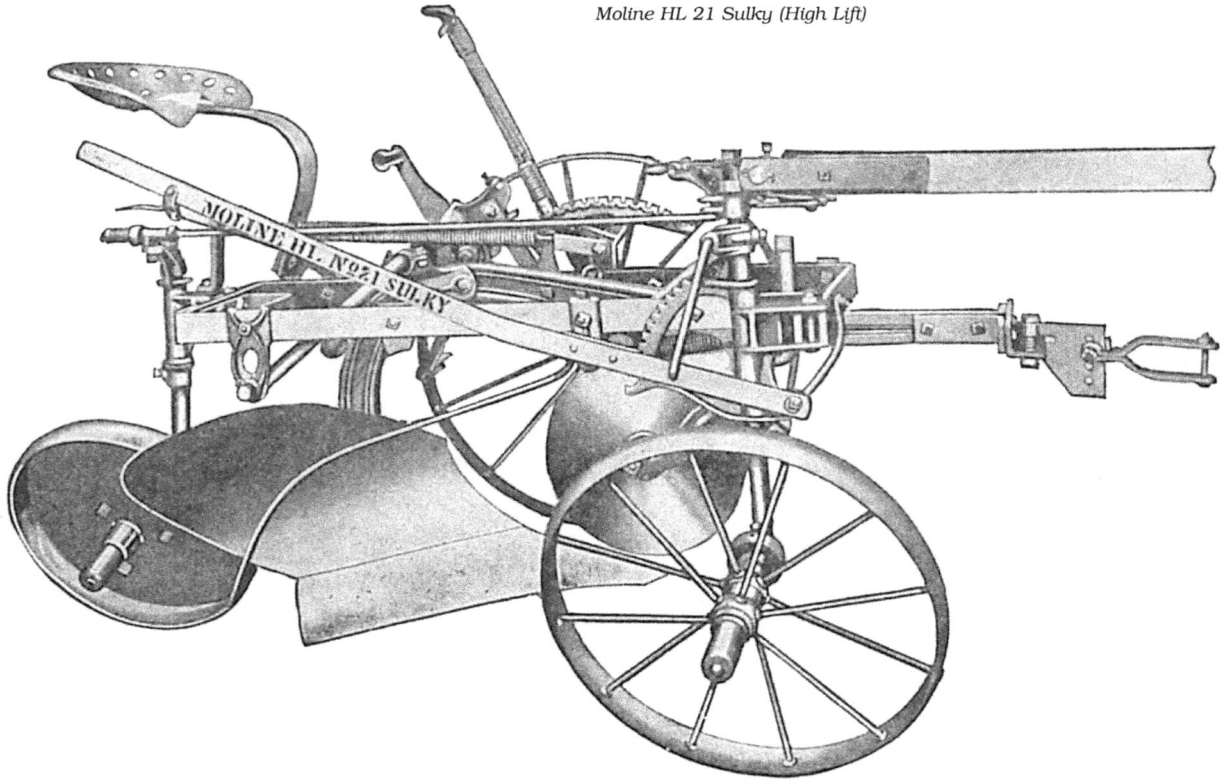

*Moline HM Gang*

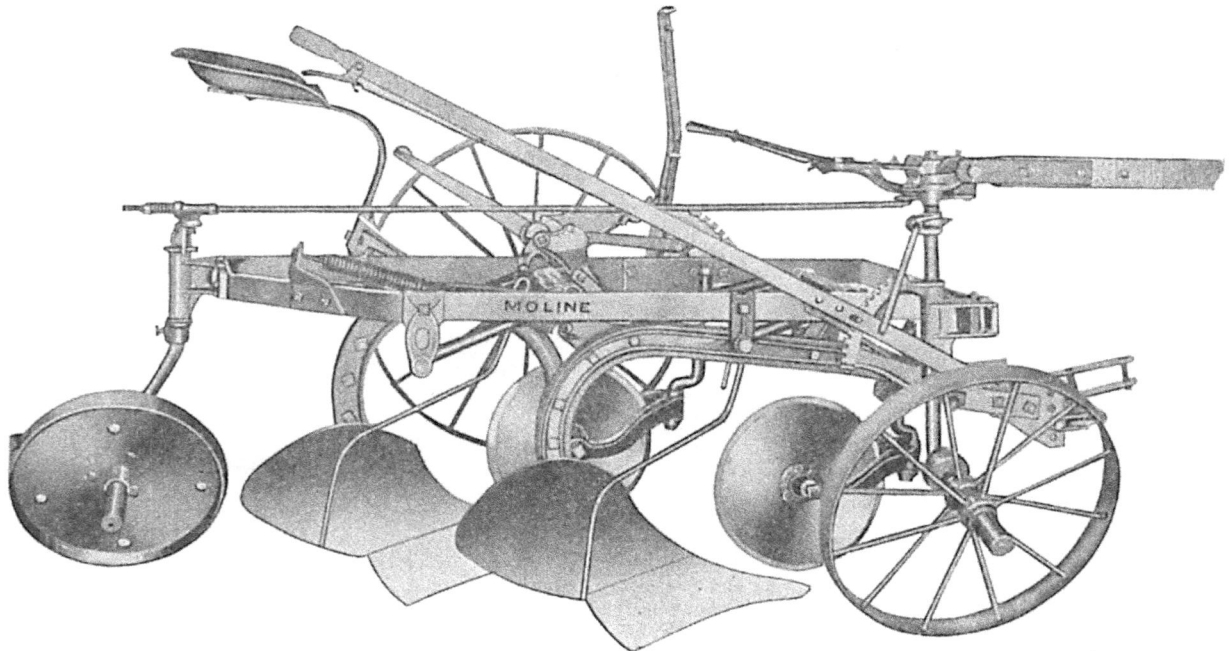

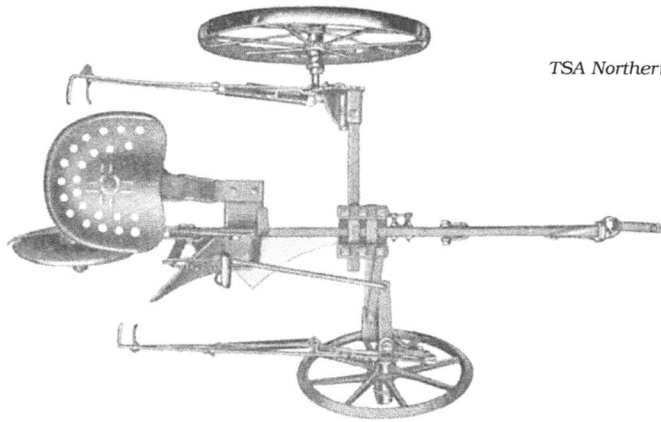

*TSA Northern Type, Overhead view.*

*GE 11 Gang*

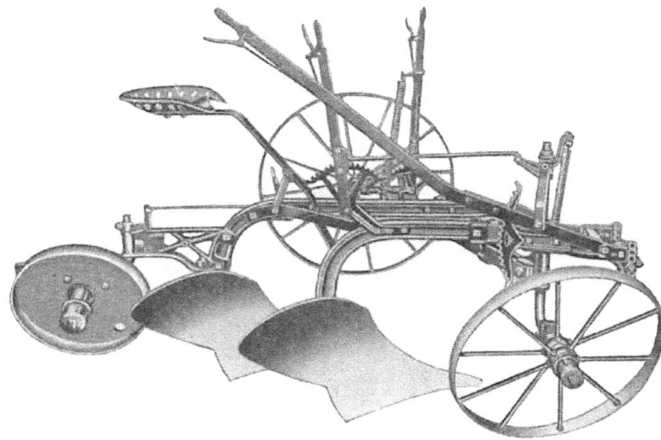

*HL 23 Gang*

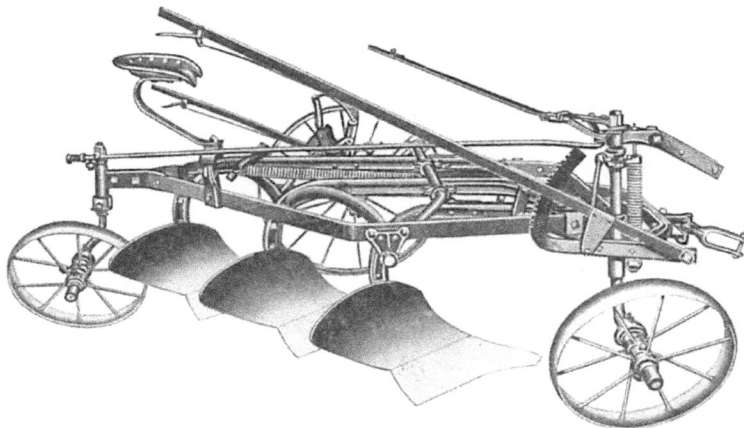

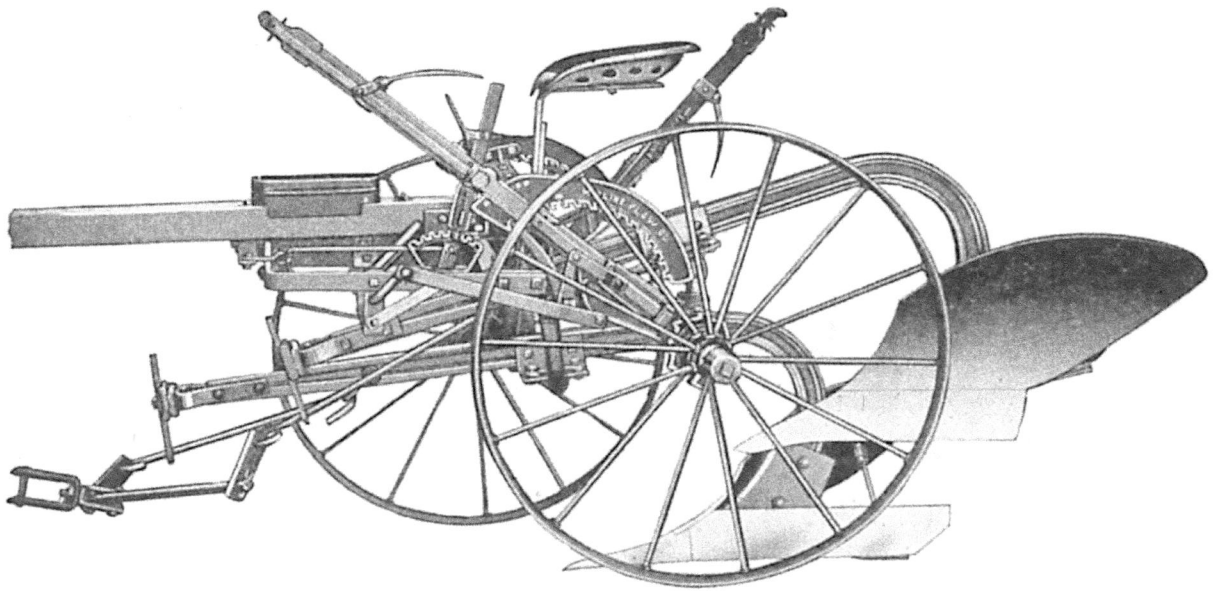

*Moline TW 3 Sulky (Two Way)*

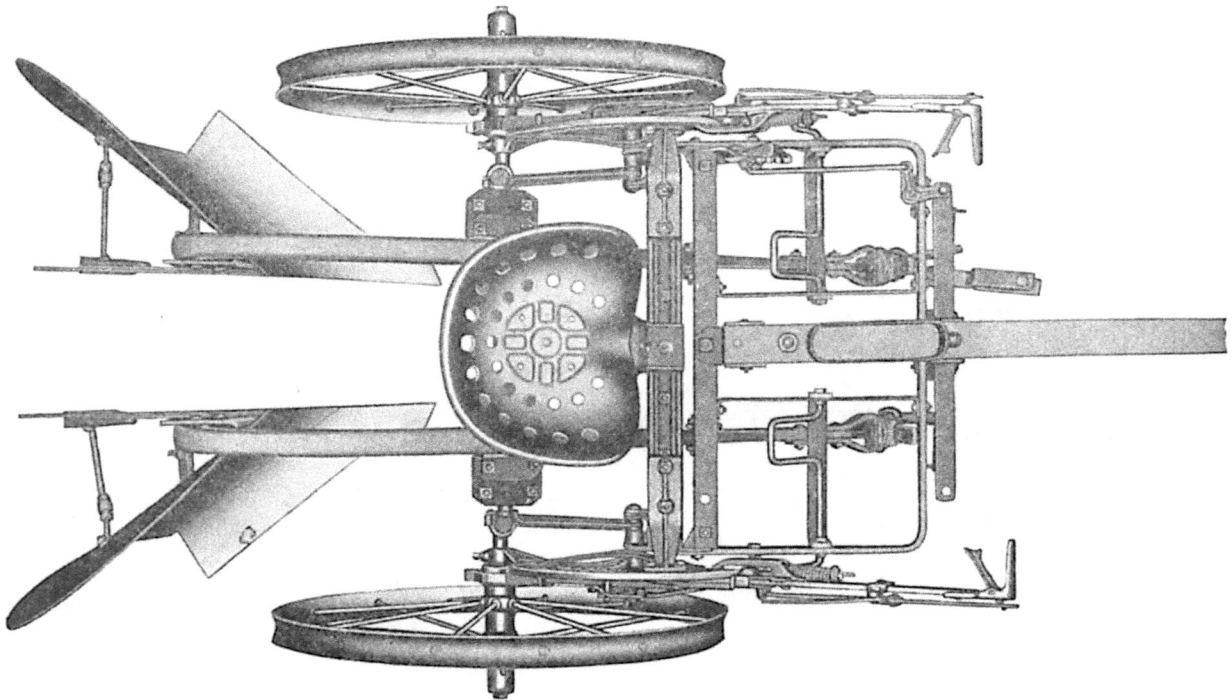

*Top View of Moline TW 3*

# Oliver

The GENUINE OLIVER CHILLED TRADE MARK *which is cast in the metal on the under-side of* EVERY *genuine Oliver chilled mold-board, share, landside and standard, and is also stenciled on the surface of every new walking plow base.*

In 1928 Oliver claimed over 70 years of plow making. Located at that time in South Bend, Indiana, Oliver was truly one of the premier farm implement companies with a strong history of innovation and quality.

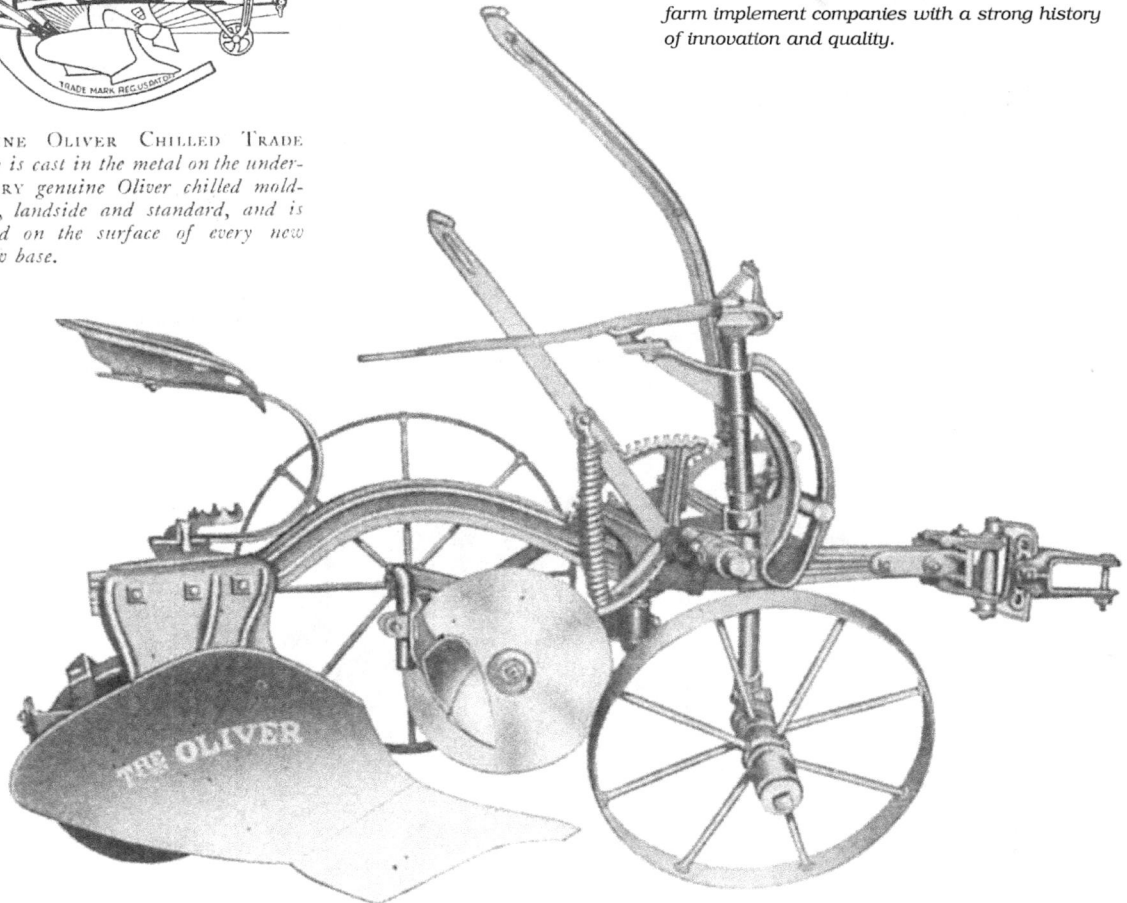

*The Famous James Oliver No. 11 Improved Sulky. In 1928 Oliver laid claim to this being the most famous and best built sulky in North America. It incorporated design features which made for easy square corners with the bottom still in the ground while traveling. Also trademark was the landside wheel with three scrapers.*

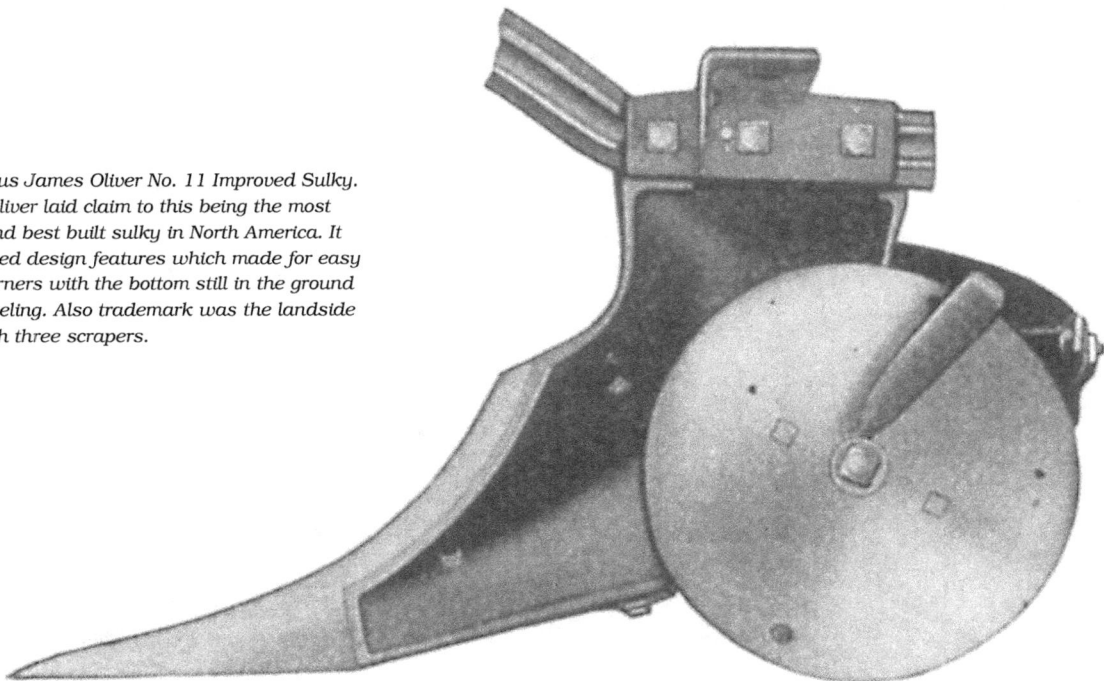

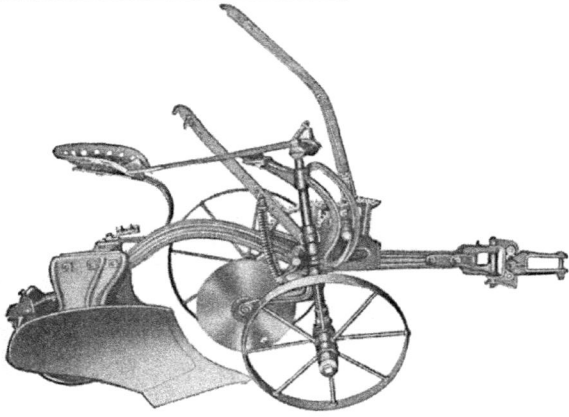

James Oliver No. 11 with plain
rolling coulter and No. 40 chilled
base

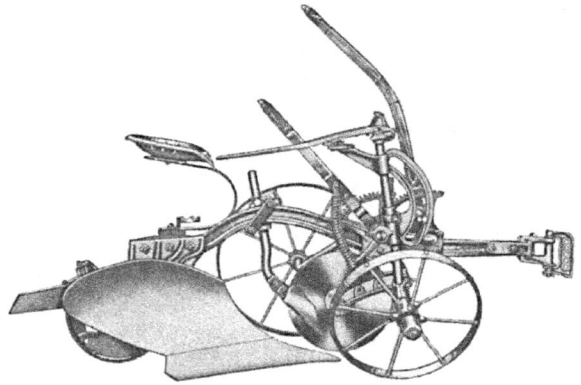

James Oliver No. 11 with big base, jointer
and rolling coulter.

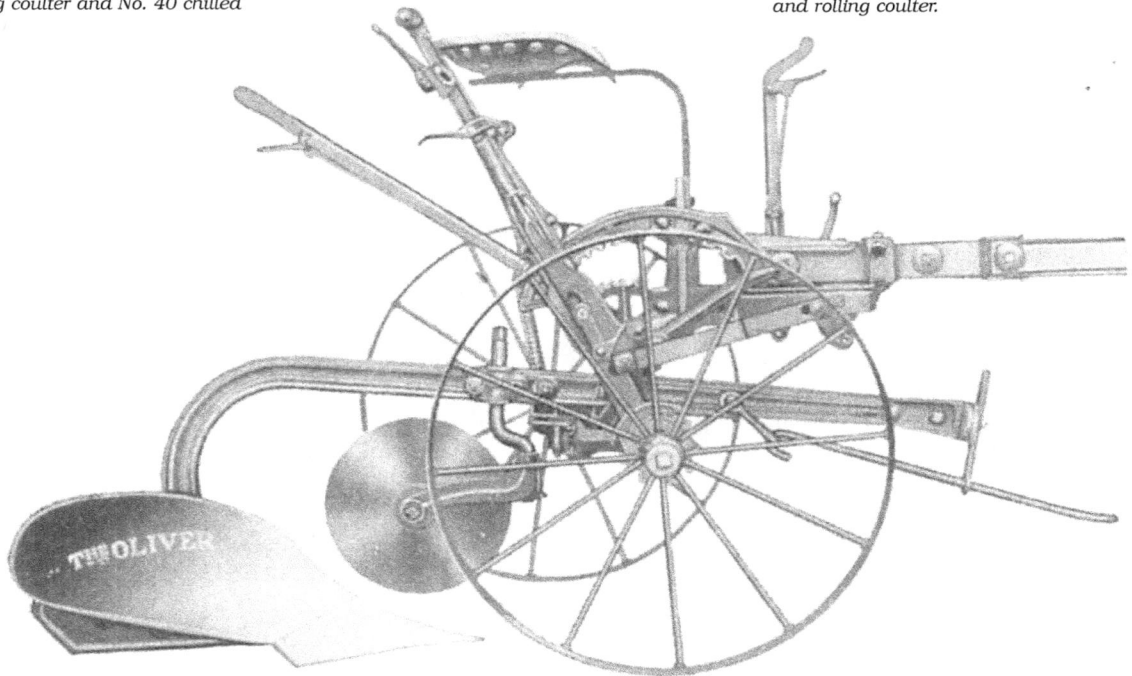

No. 26B with a rolling coulter and draft rod.

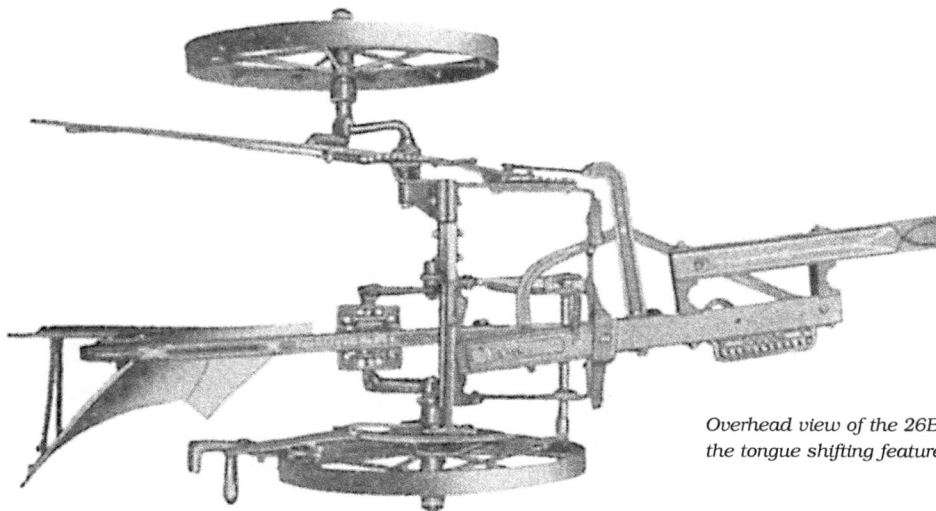

Overhead view of the 26B illustrating
the tongue shifting feature.

23 BXX heavy western type two way sulky came with 12, 14, or 16 bottoms

23A light eastern type two way sulky  has a sliding draft bar
clevis. Came with 10, 12, and 14 bottoms

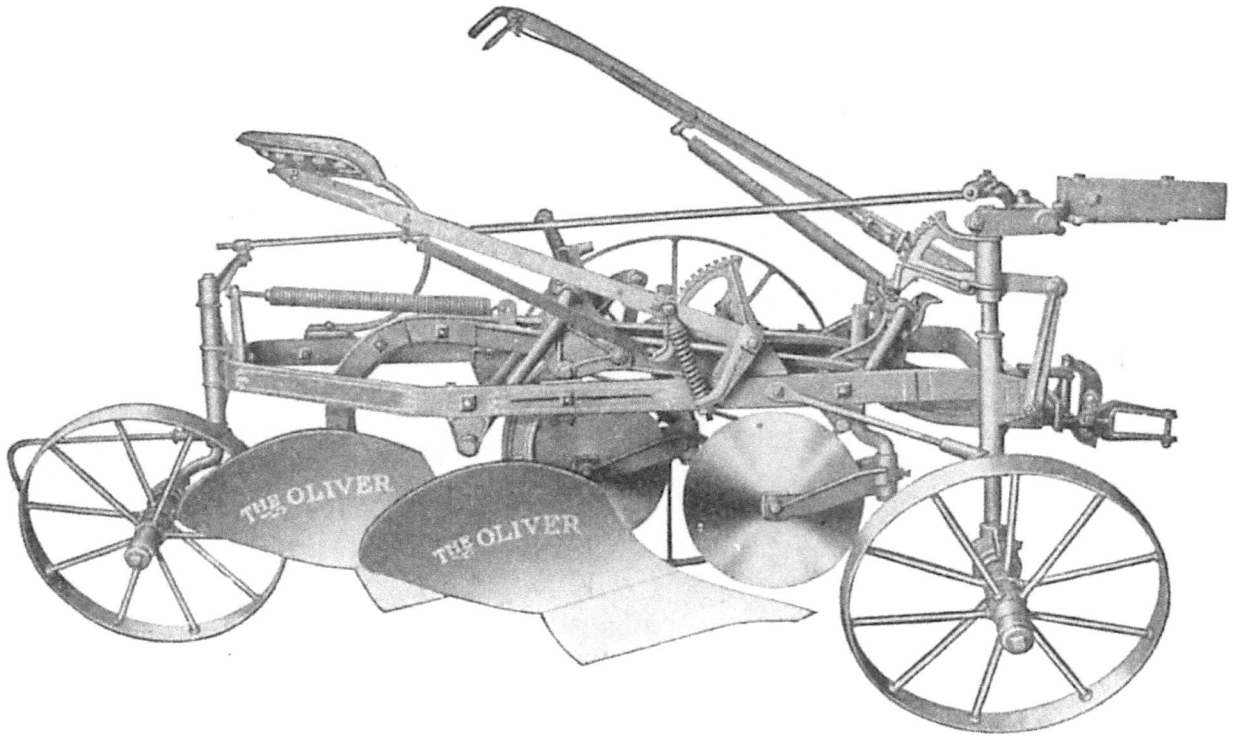

*Model 52 high lift gang*

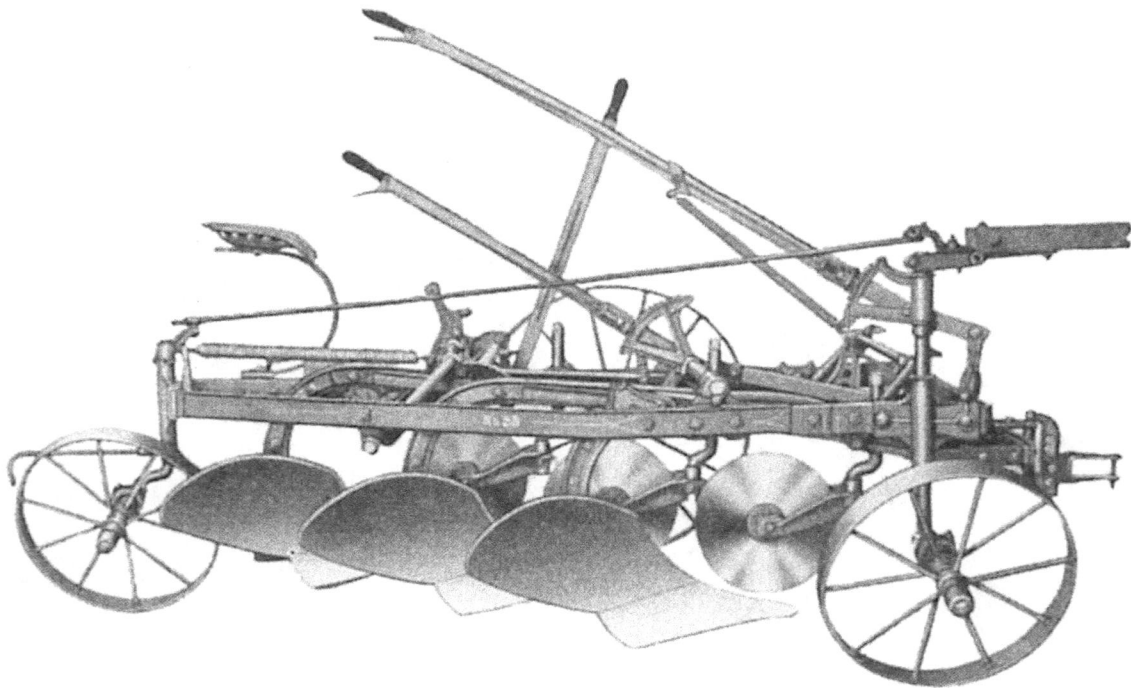

*Model 53 high lift gang*

# Oliver

*51 XX High Lift*

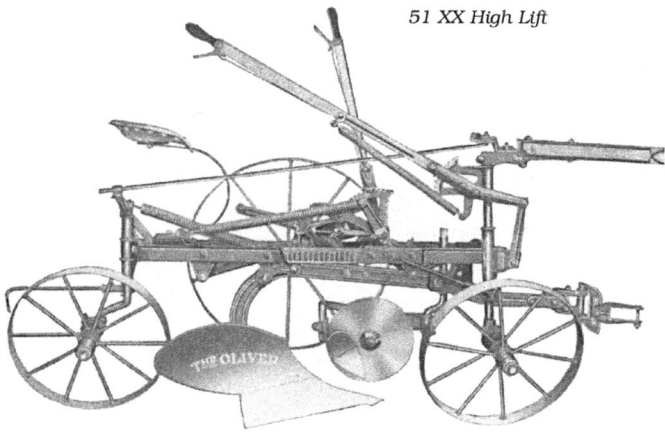

*The foot lift is assisted by a powerful spring. It was adjustable to the weight of the plowman*

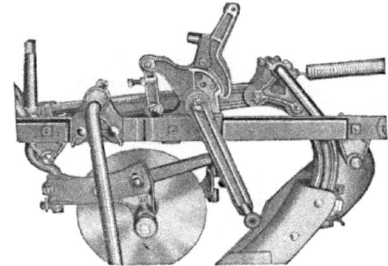

*No. 82 Two bottom gang also available as three bottom*

*No. 81 Texas Sulky is heavier than the regular 81. It features a guiding lever which allows the plow to turn in and out corners without raising the bottom out of the ground.*

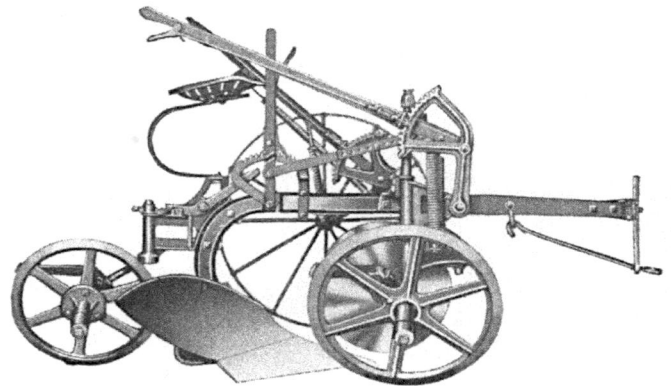

*No.81 Sulky with adjustable frame to receive 14, 16, or 18 inch bottoms.*

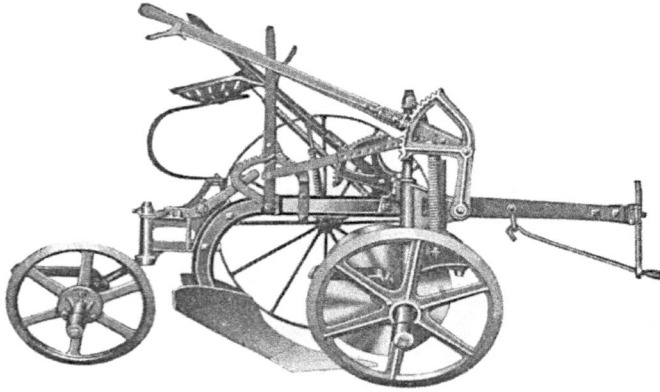

No. 81 Texas Sulky with middlebreaker base (or bottom). May also have been equipped with sweep attachment.

No. 1 Improved High Lift Gang. Featured landed beams to help offset side draft with four abreast. Beams are hung in double bails with the front bail shorter than the rear one to facilitate the bottoms entering the ground point first.

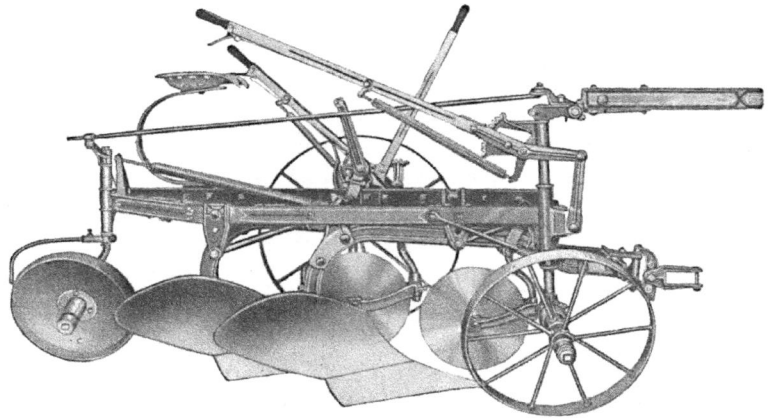

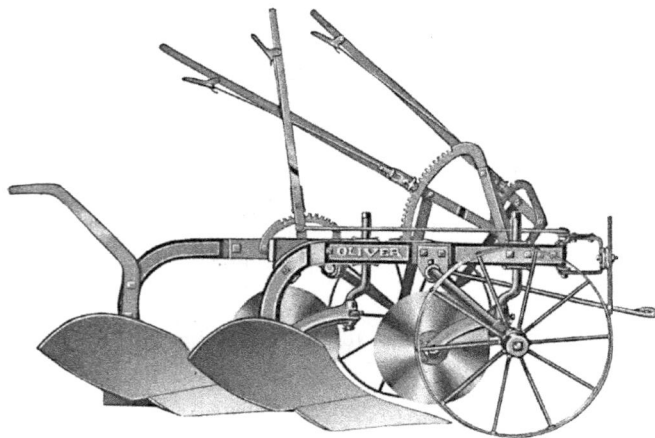

No. 12 Adjustable walking gang with an unusual amount of beam clearance for use in trashy fields.

# Oliver

22-A Adjustable walking gang, originally offered
with either 10 or 12 inch bottoms. Notice third
wheel on castor swivel.

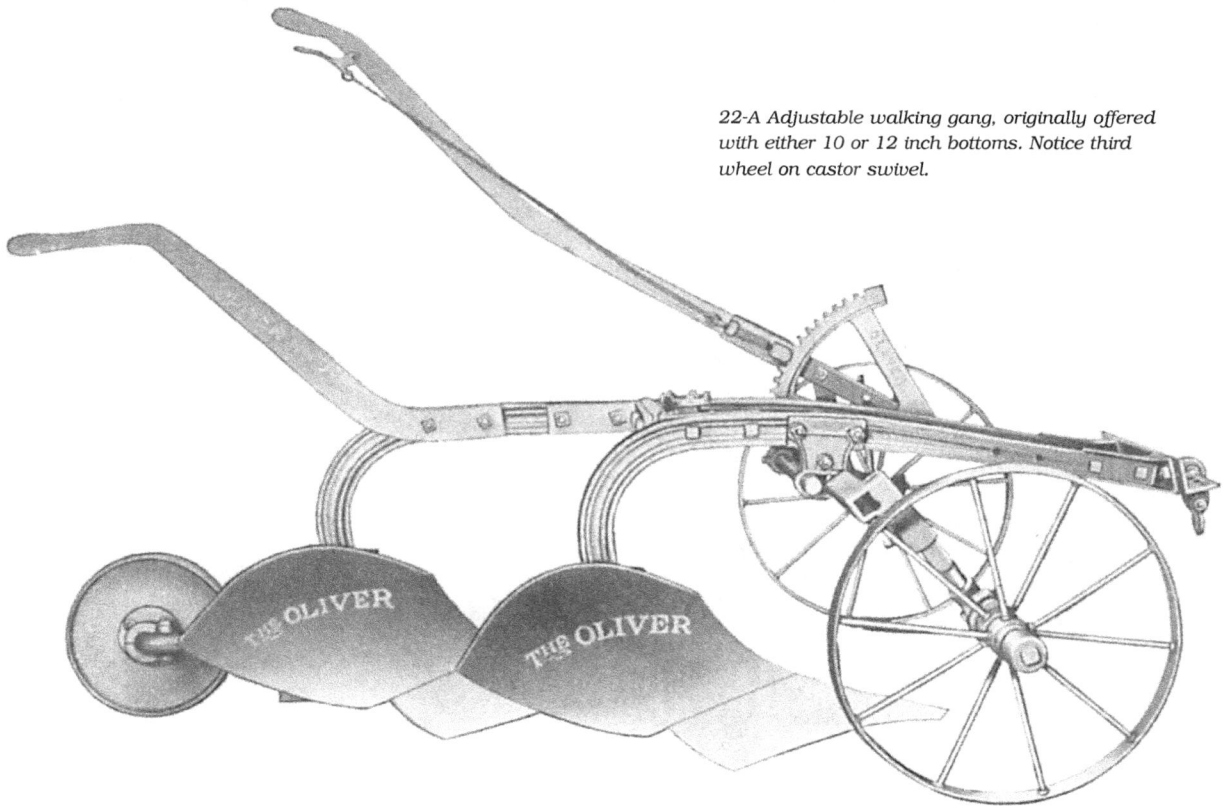

No. 95 Two Bottom gang would receive third
bottom. Horizontal hitch shift. Handles designed
for either walking or riding. Came with either 14
or 16 inch bottoms.

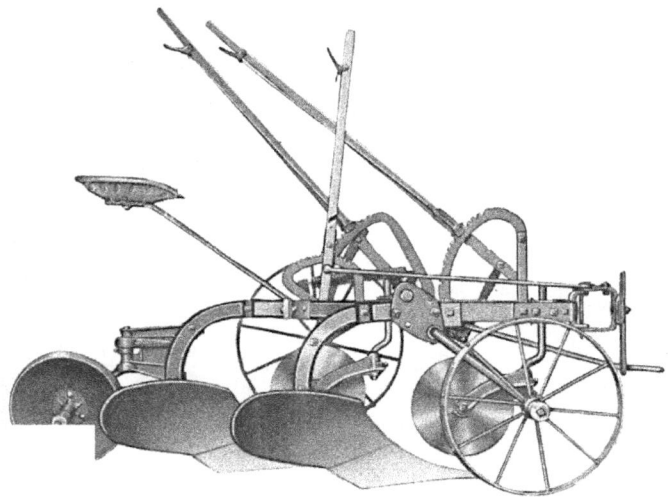

No. 95 Three bottom. By removing the third bottom
and beam this plow, with addition of seat,
duplicates one above.

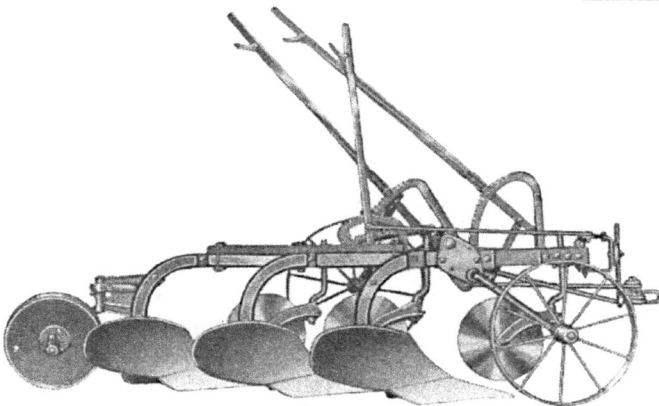

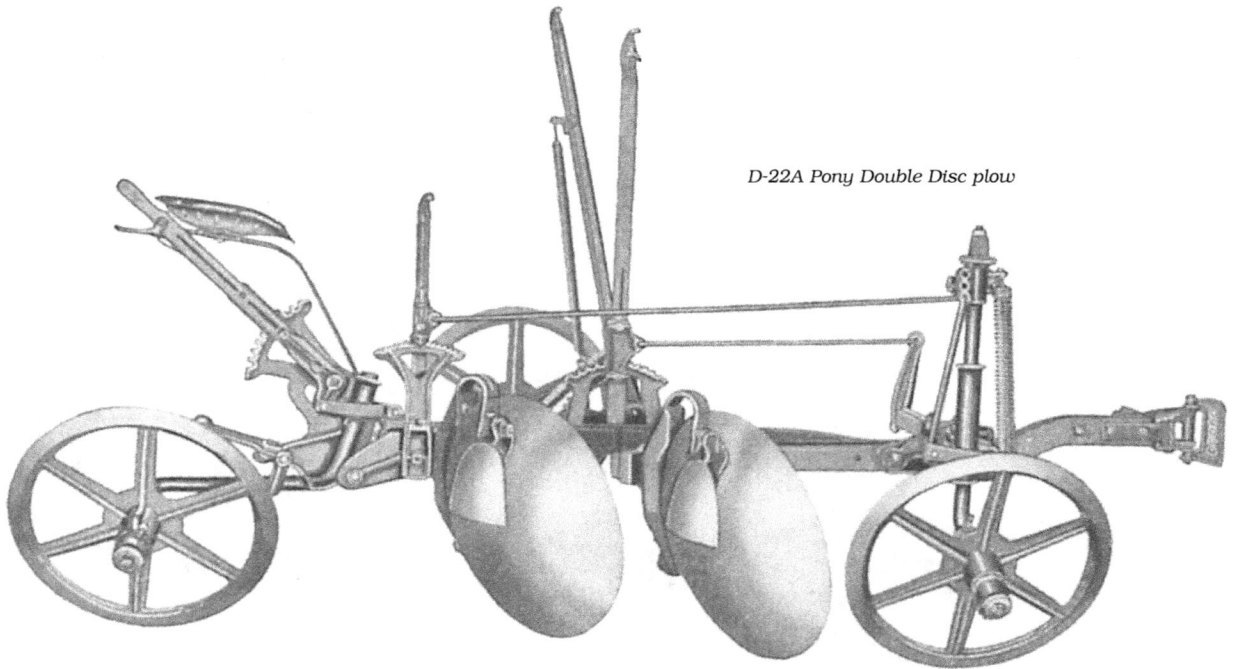

*D-22A Pony Double Disc plow*

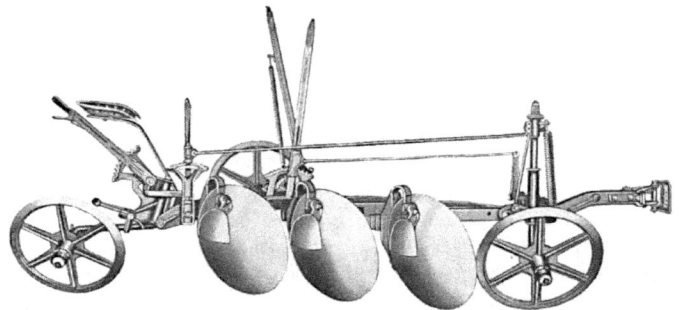

*D-33A Horse Triple Disc plow. This implement weighed 1195 lbs.*

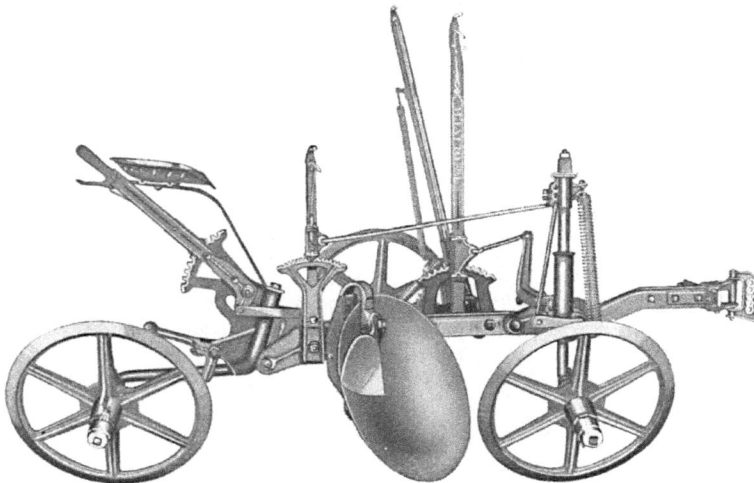

*D-21A Pony Disc plow. This plow weighed 500 lbs. A model D-31A, similar to this one but  heavier, was also offered.*

Makes &
Models

No 40N, largest of the No. 40 series, is a two or three horse plow. First released right after the Civil War, this proved to be one of the most popular walking plow models ever built. Equipped all the way to 16 inch.

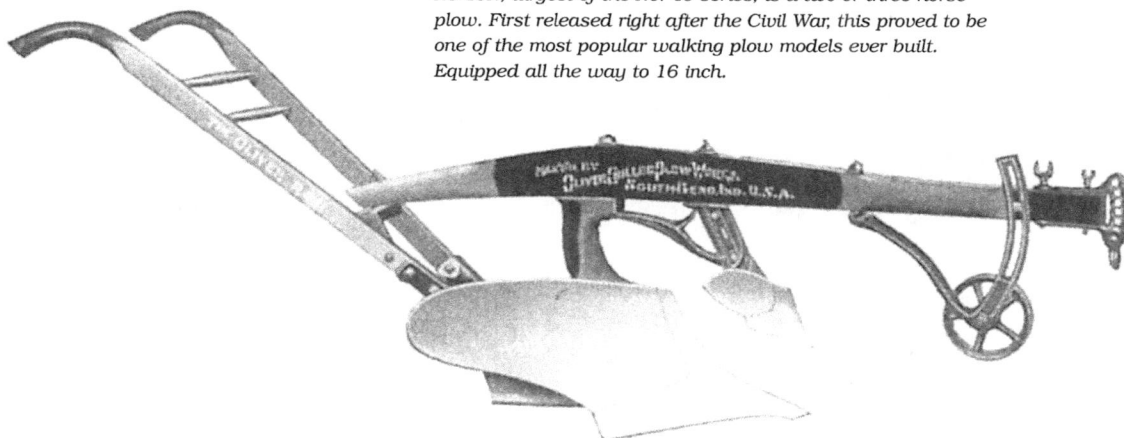

No. 10N, the smallest of the No. 40 series, is either a one horse or two horse plow. It came with an 11 inch bottom.

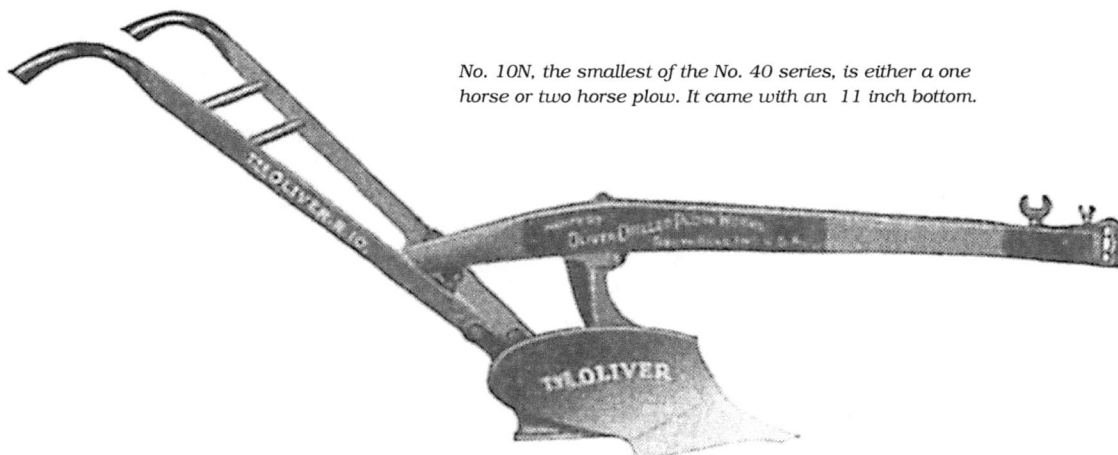

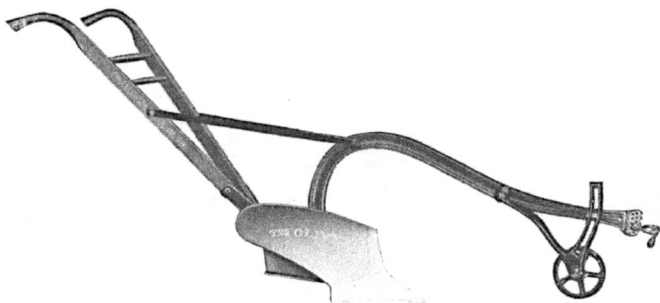

No. 13N, Light two horse plow. The steel beam came right or left hand, wood beam was only offered in right hand.

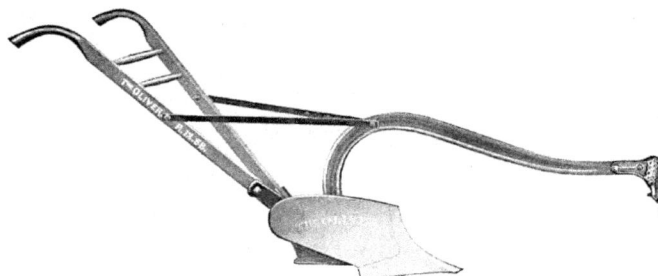

No. 19N, a two horse plow of medium weight and capacity. Here setup with depth wheel.

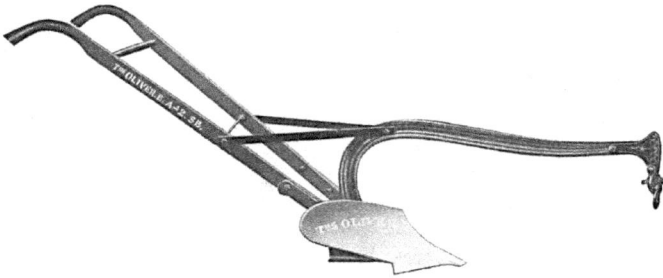

A-2 steel beam plow with 9 inch share

BC-N wood beam chilled plow came with 10 inch share

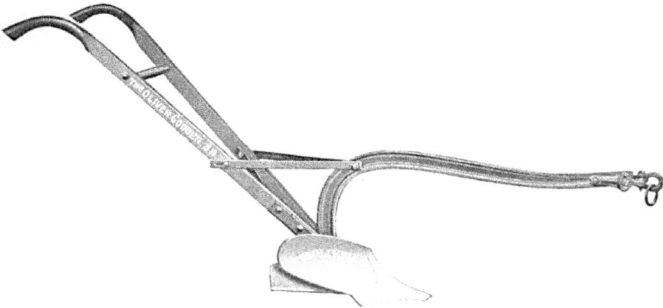

Oliver Goober 8" especially suited for work in sandy or gritty soils.

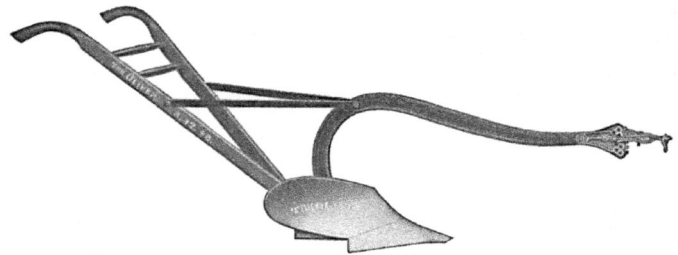

Oliver No. 12 with 11 inch share

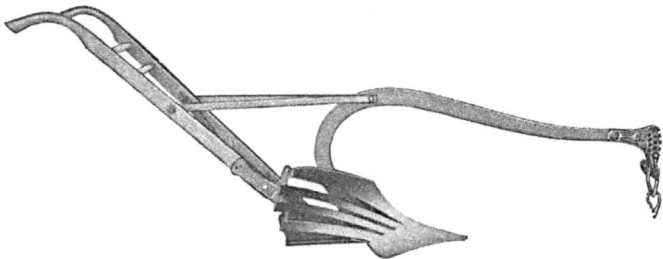

Slat plow for difficult to scour soils

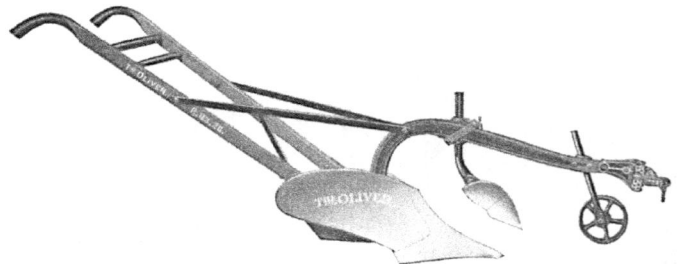

No. 83N general purpose plow

# Oliver

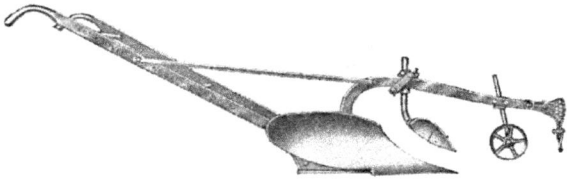

No. 221 Scotch type plow with NC 221 tractor plow bottom and long handles.

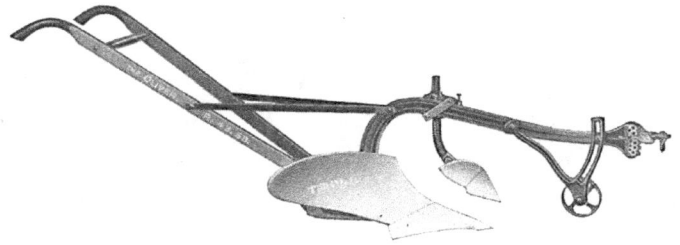

No. 42. with 13 inch bottom

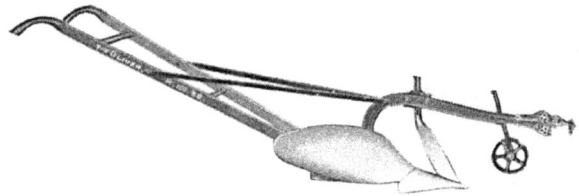

No. 101N with long handles and 10 inch bottom

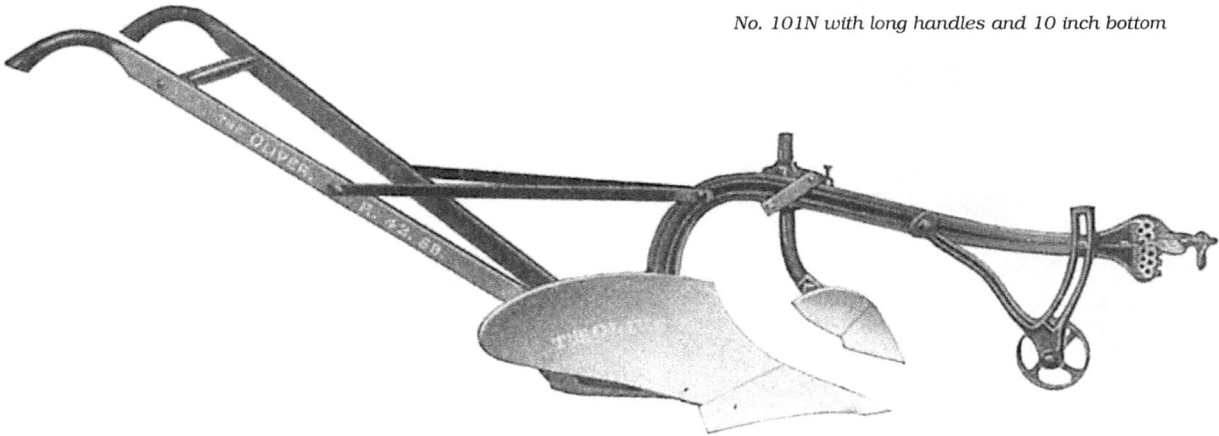

Oliver No. 42 with 13 inch share.

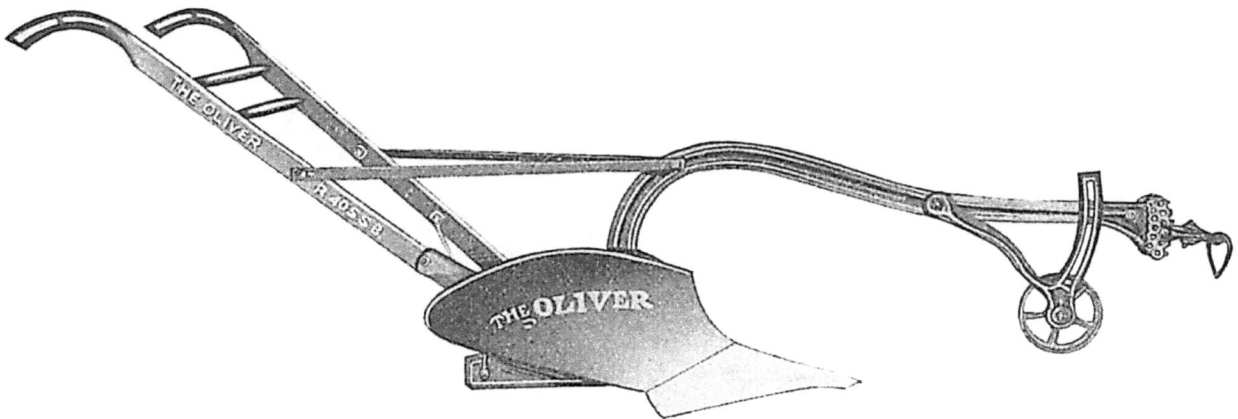

No. 405 with from 8 inch to 14 inch bottom available.

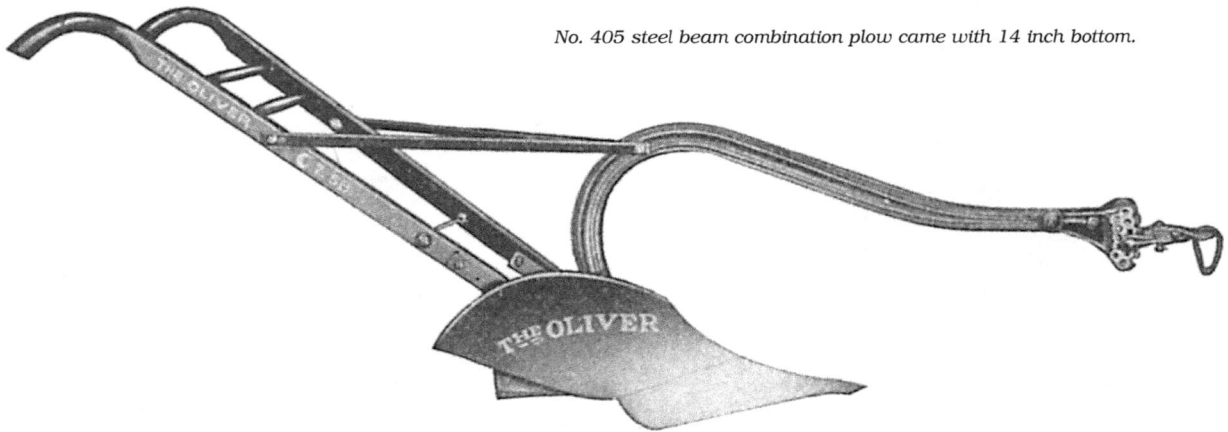

No. 405 steel beam combination plow came with 14 inch bottom.

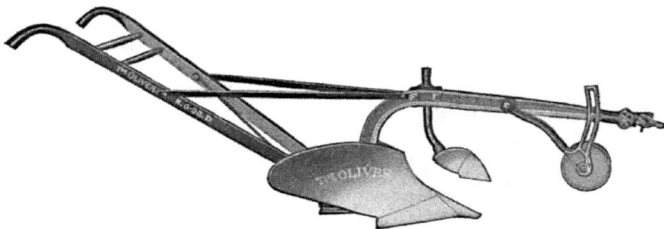

O-98-B with 13 inch bottom

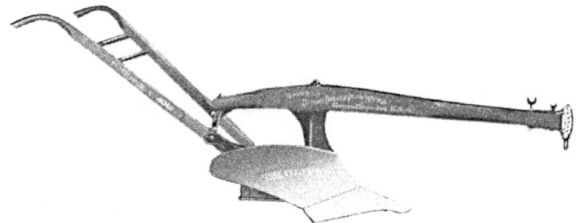

No. 404X wood beam with rigid standard.

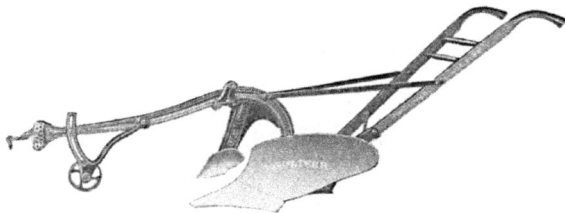

Oliver No. 404-X left hand with 12 inch bottom.

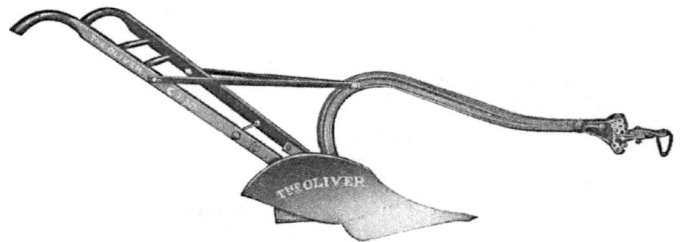

Oliver Crescent 7 with high beam clearance for trashy conditions.

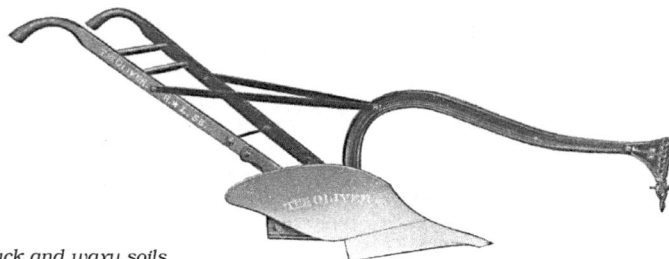

Star 3 for black and waxy soils

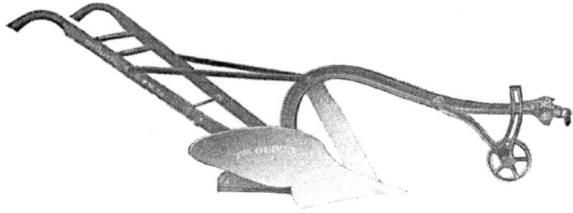

*Oliver Star W with hanging cutter*

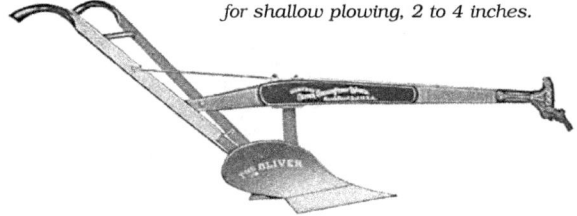

*No. 92 Orange for loose loam soils. Built for shallow plowing, 2 to 4 inches.*

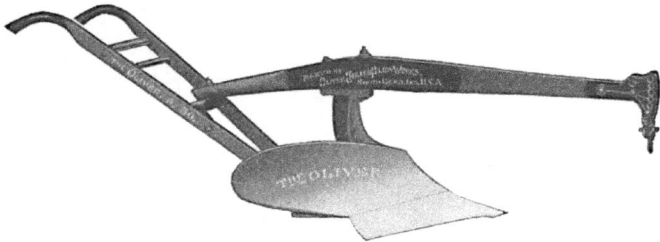

*No. 50 N For hard jobs.*

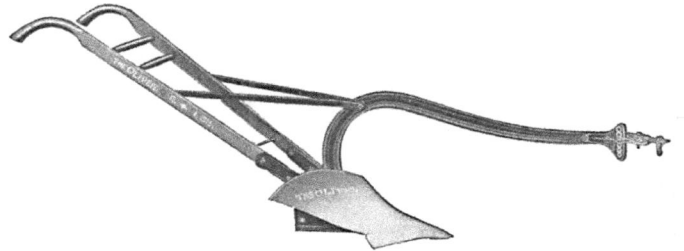

*Diamond No. 1*

*Diamond No. 1 Special, a steel beam version of the 92 Orange*

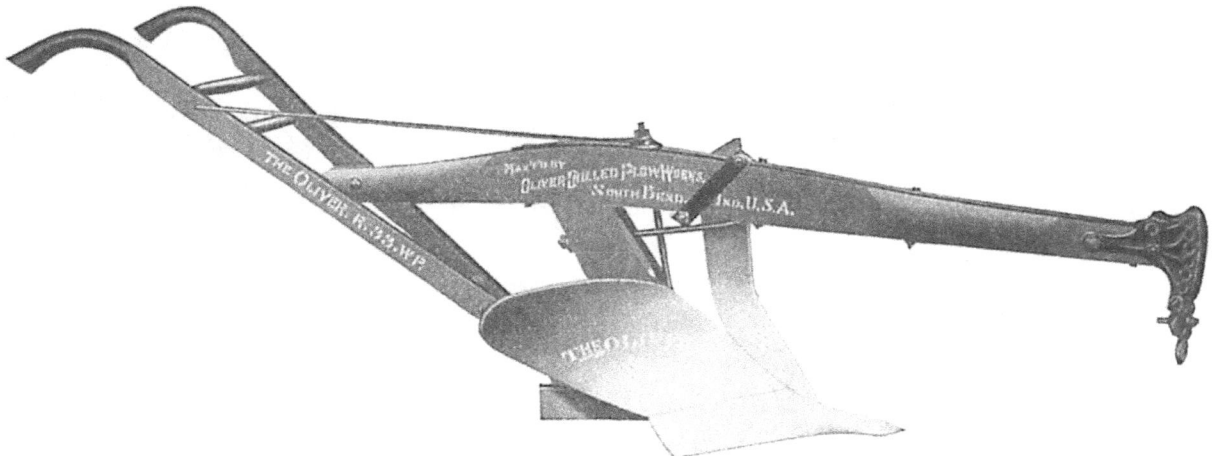

*No. 33 Wood Post plow.  For brush work and light road work. Beam can be landed for either two or three horses.*

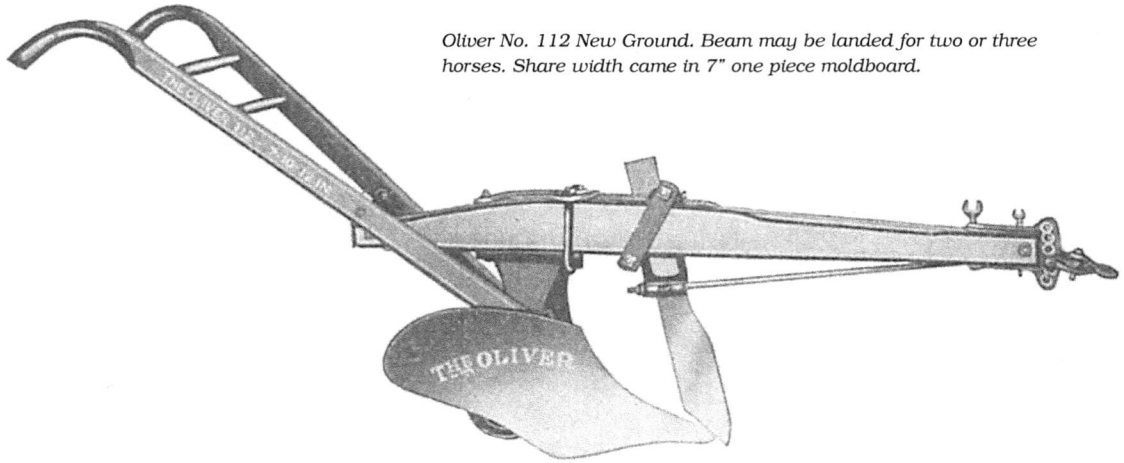

Oliver No. 112 New Ground. Beam may be landed for two or three horses. Share width came in 7" one piece moldboard.

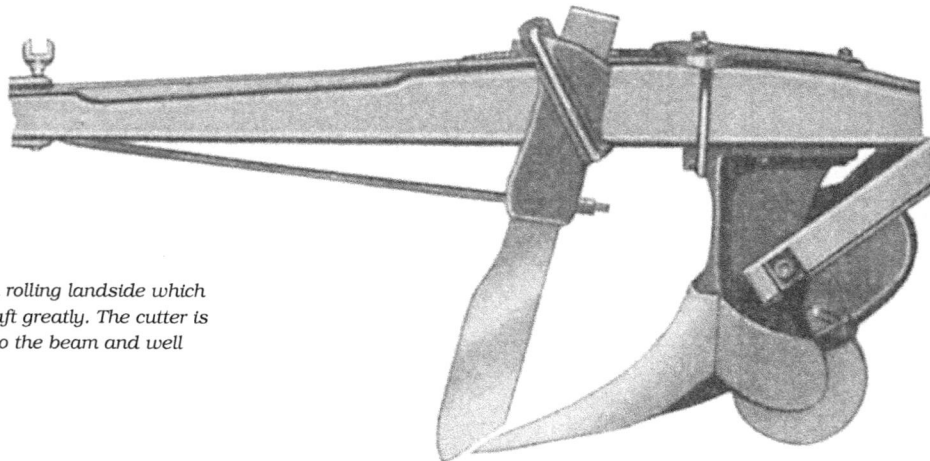

The 112 has a rolling landside which lightens its draft greatly. The cutter is rigidly bolted to the beam and well placed.

# Oliver

Makes & Models

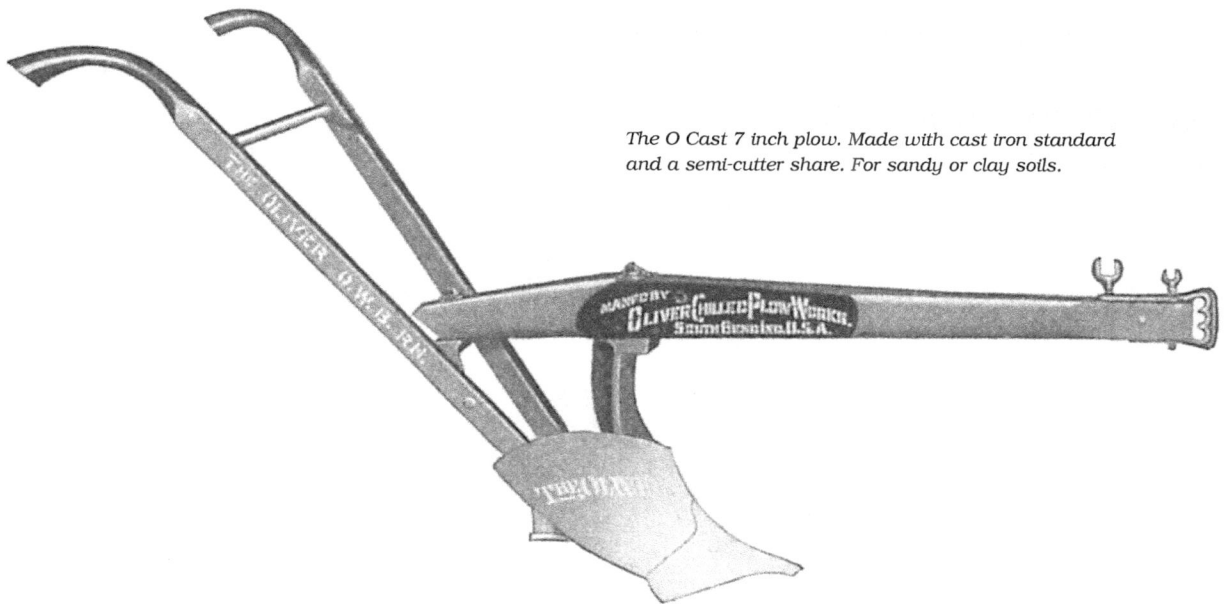

The O Cast 7 inch plow. Made with cast iron standard and a semi-cutter share. For sandy or clay soils.

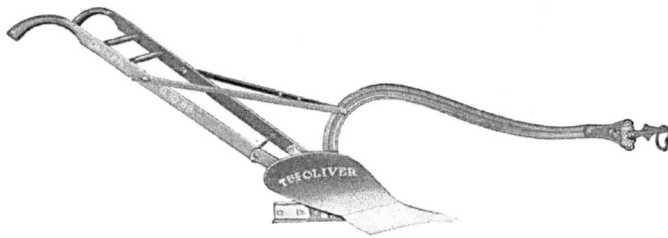

No. 310. Midway between a Breaker and a General Purpose plow. Suitable for new ground.

G-10 plow

The B is a wood beam chilled Vineyard plow with landside hinged in the center so it can be moved inward when working close to trees and vines.

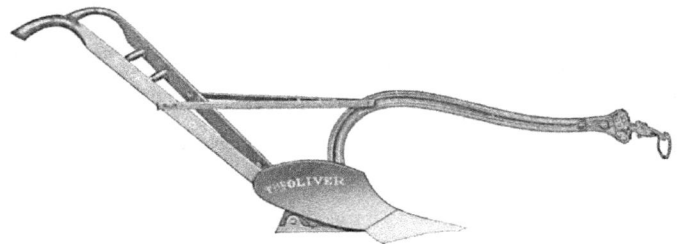

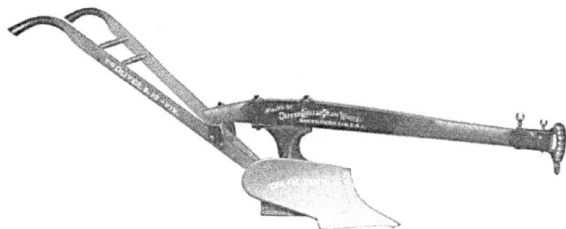

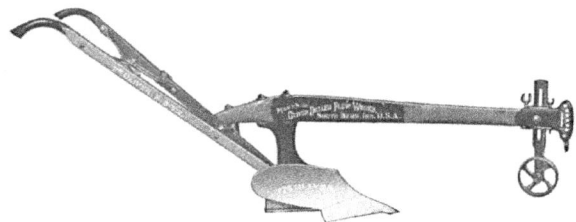

No. 13 wood beam chilled Vineyard plow

*No. 9 steel beam chilled Vineyard plow with dial hitch and cutter share. Handles are fully adjustable.*

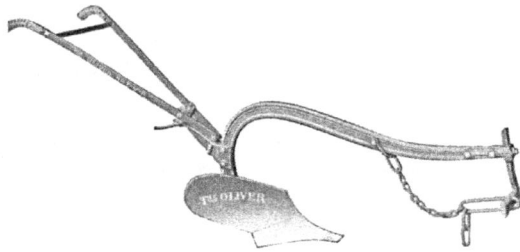

*7-VF is a steel beam Vineyard plow.*

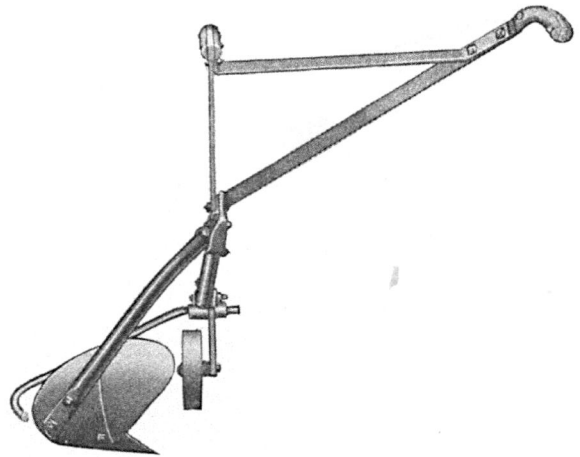

*The No. 2 is different in design but suited for its work.*

*Showing how it is possible to plow between the vines with No. 2. The offset bottom is easily guided around and between the vines while the horses and plowman walk between the rows.*

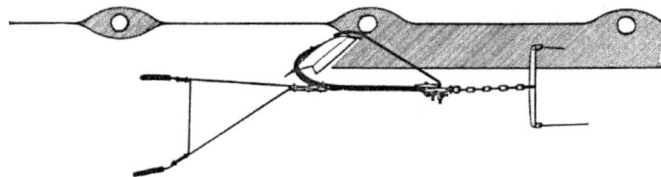

# Oliver

No. 53 ¹/², a full size Hillside plow

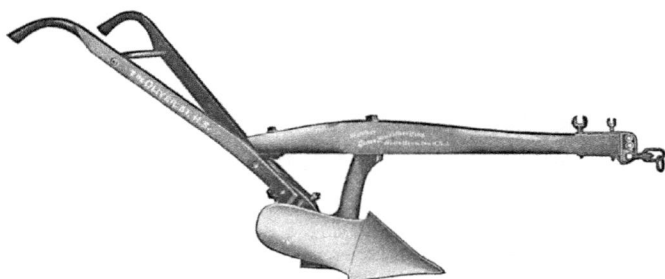

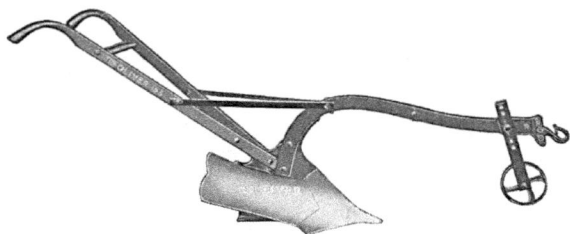

No. 51, the smallest of the Oliver Hillside
plows. A gardeners plow.

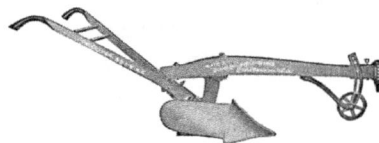

The 153 Trussed beam Hillside plow

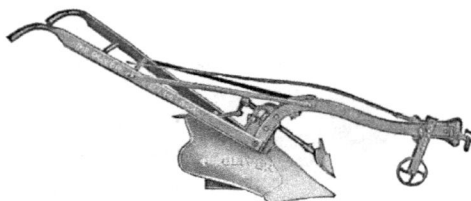

The No. 513 equipped with jointer,
depth wheel and shifting clevis.

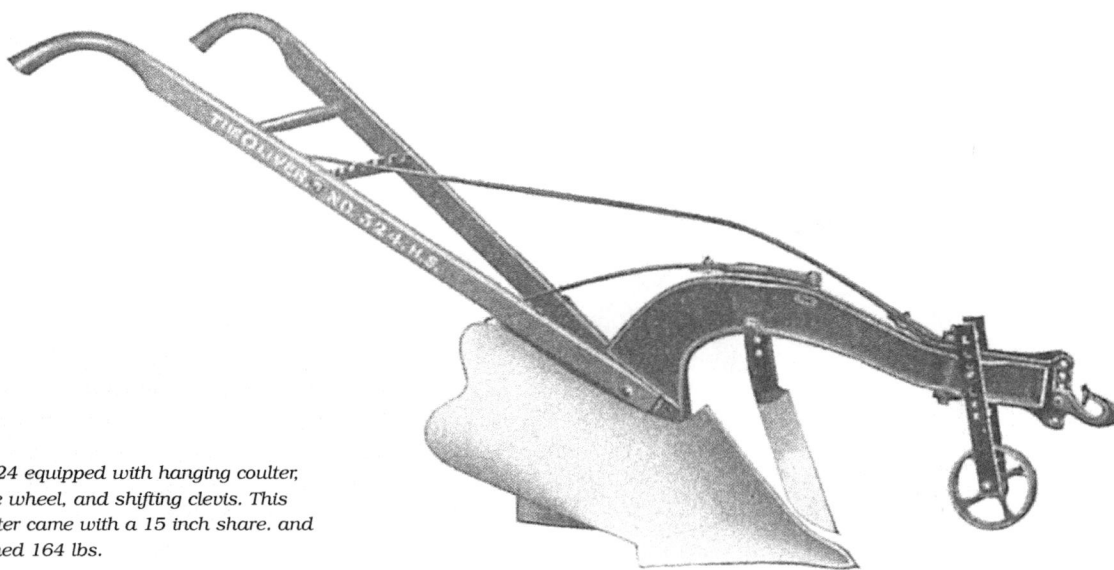

No. 524 equipped with hanging coulter,
gauge wheel, and shifting clevis. This
monster came with a 15 inch share. and
weighed 164 lbs.

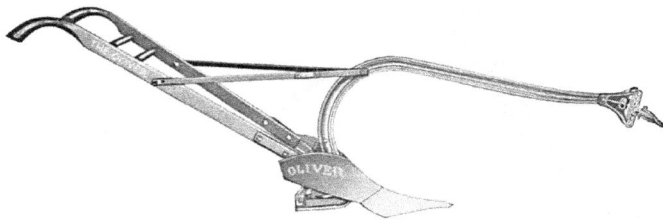

MB 12, a medium sized steel middlebreaker of 12 inch capacity.

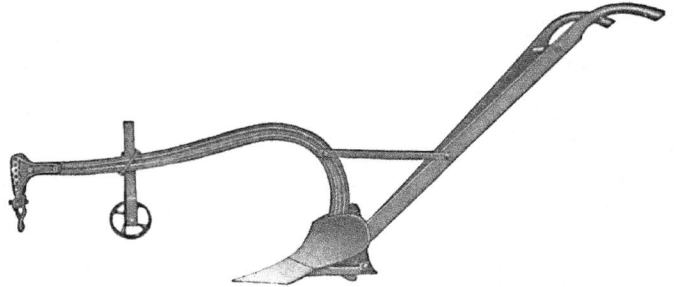

PB-10 chilled Middlebreaker for gritty soils.

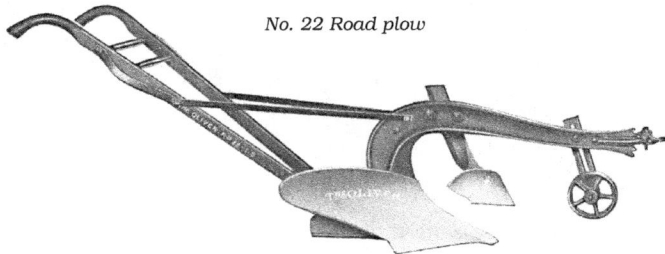

No. 22 Road plow

Oliver No. 25 Pavement plow built to do hardest jobs

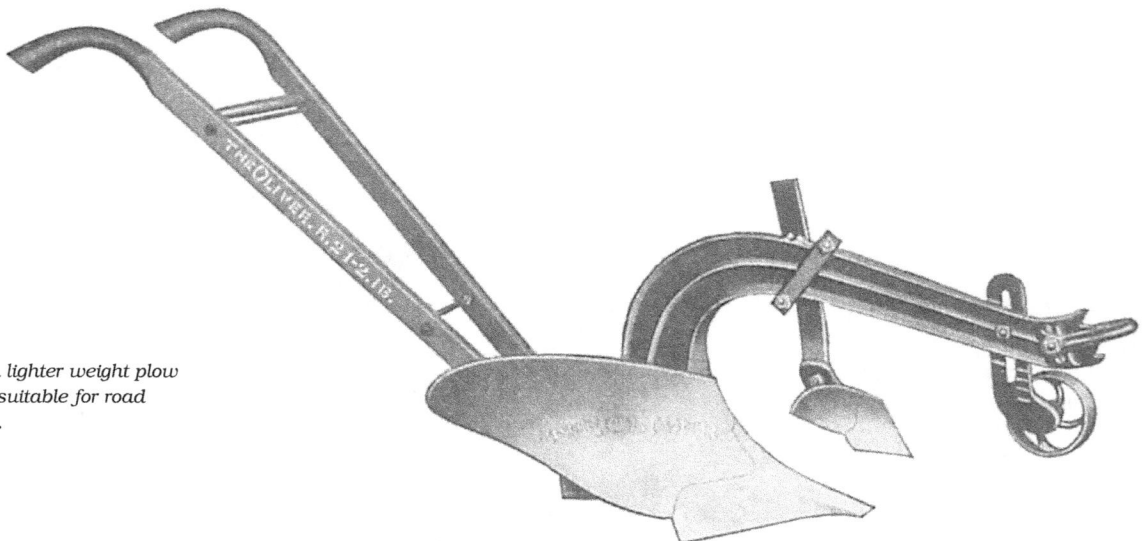

Oliver No. 2 1/2 is a lighter weight plow than the No. 22 and suitable for road work and hard soils.

# Oliver

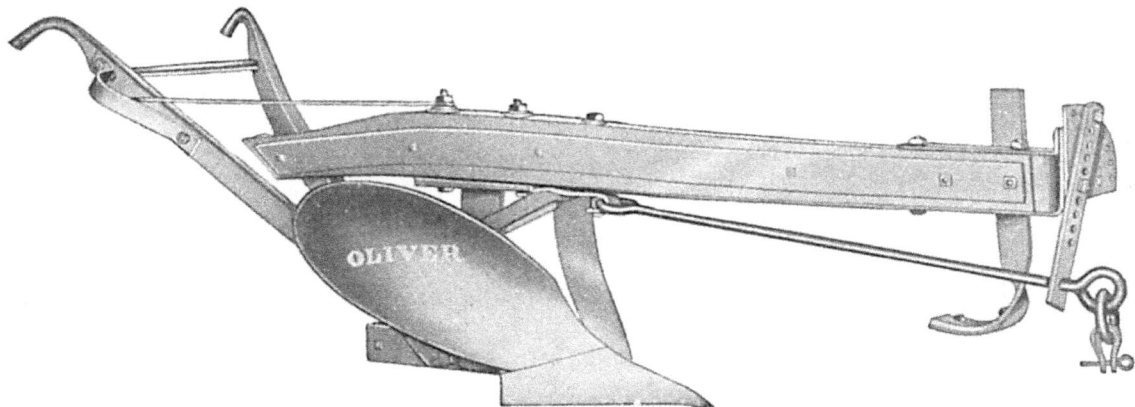

*The R and G 1-A wood beam and steel road and grading plow is built for hard work behind horses or tractors.*

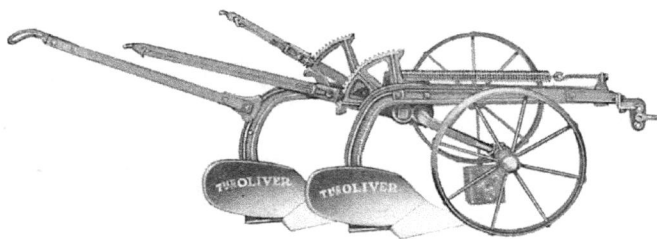

*Oliver 17A Improved walking gang*

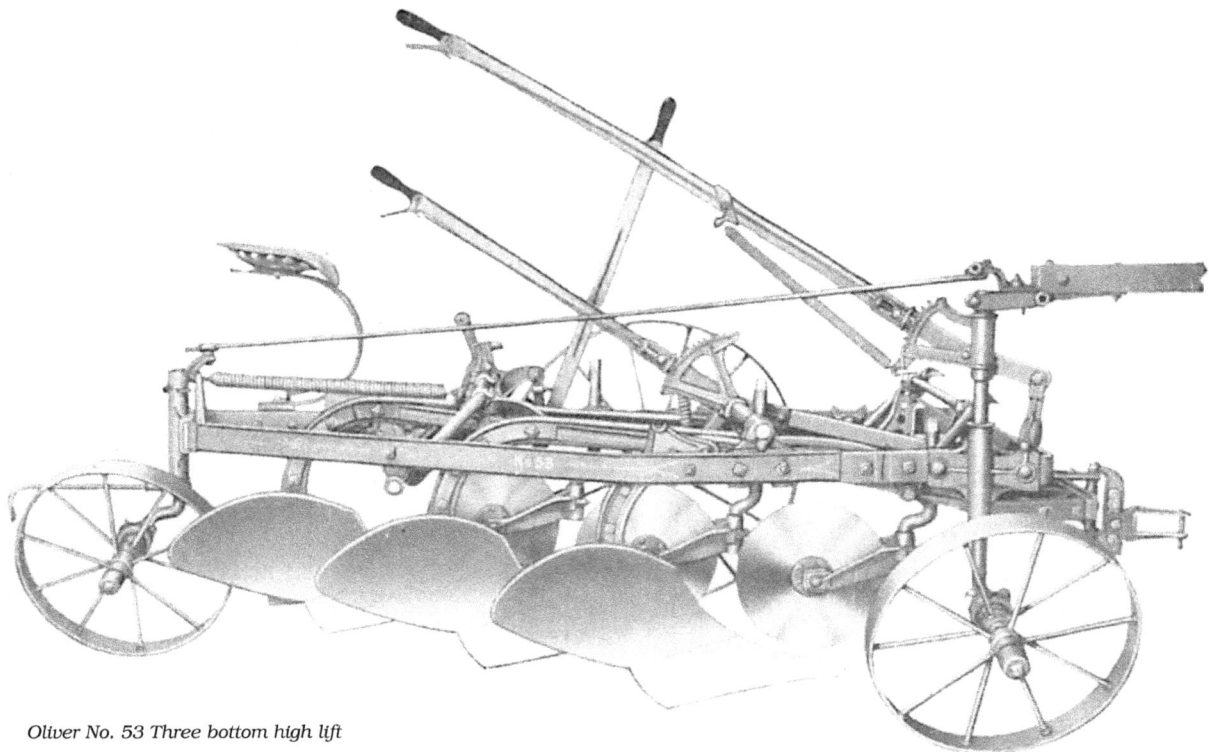

*Oliver No. 53 Three bottom high lift*

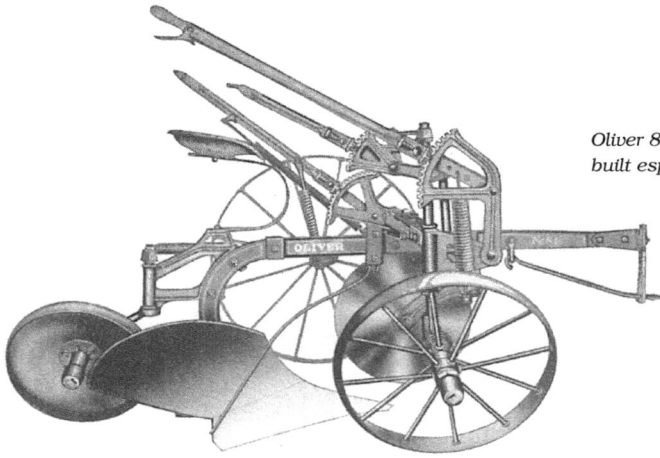

*Oliver 81 Sulky. The Oliver 81 and 82 models were built especially for the Argentine.*

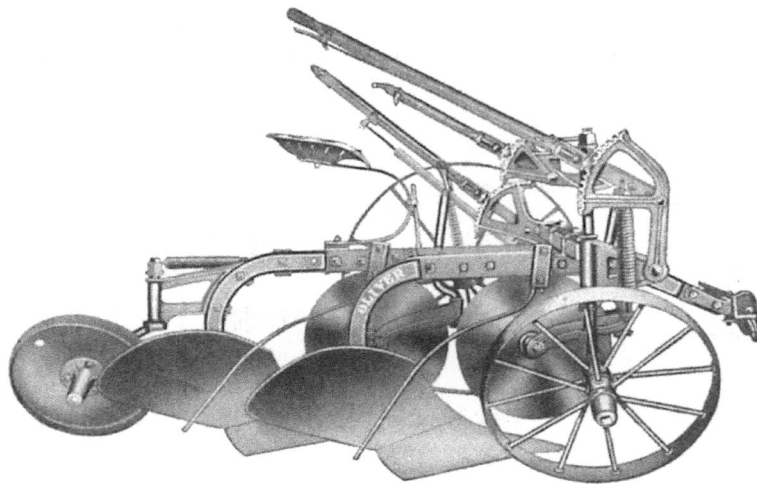

*Two bottom Oliver 82 gang*

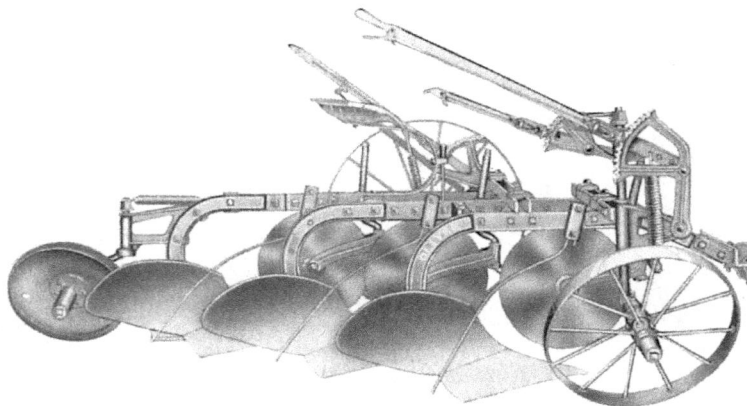

*Oliver No. 82 with Three bottoms*

# Parlin & Orendorff

1842—1915

Largest and Oldest
Permanently Established Plow Factory
on Earth

**P & O**

Trade Mark Registered

*CAPACITY:*
*Two Complete Implements*
*Every Minute*

SEVENTY-THREE YEARS OF "KNOWING HOW"
HAMMERED INTO EVERY ONE
OF THEM

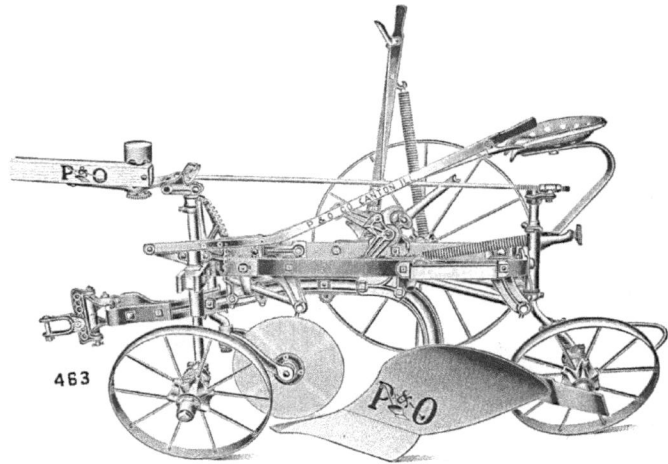

*They were the biggest plow making firm and they were the Mercedes Benz of plows. As the following pages will attest, their lineup of plows as of 1915 was incredible. After they were assimilated by McCormick Deering/ International something was indeed lost.*

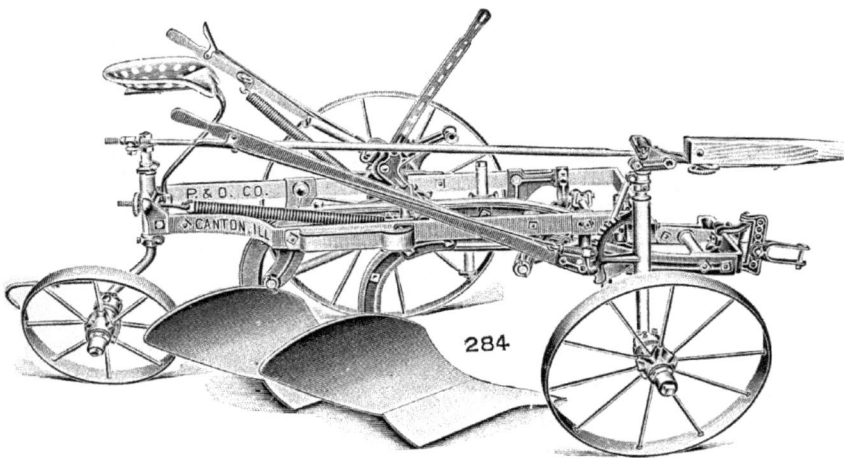

*No. 1 Diamond sulky, left hand, with mold extension.*

*No.1 Diamond gang.*

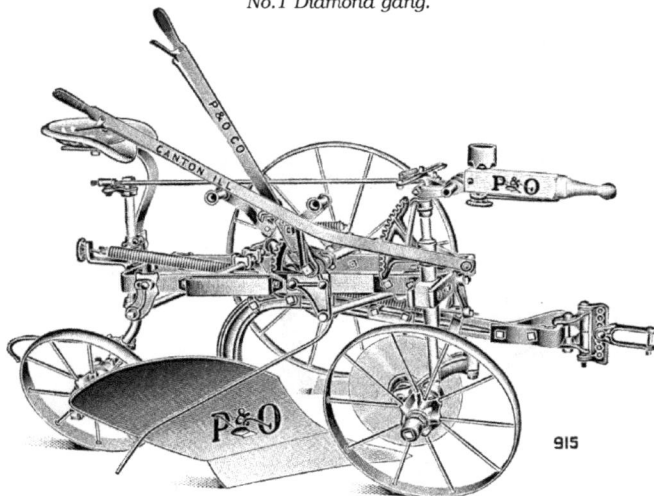

*No. 2 Diamond sulky.*

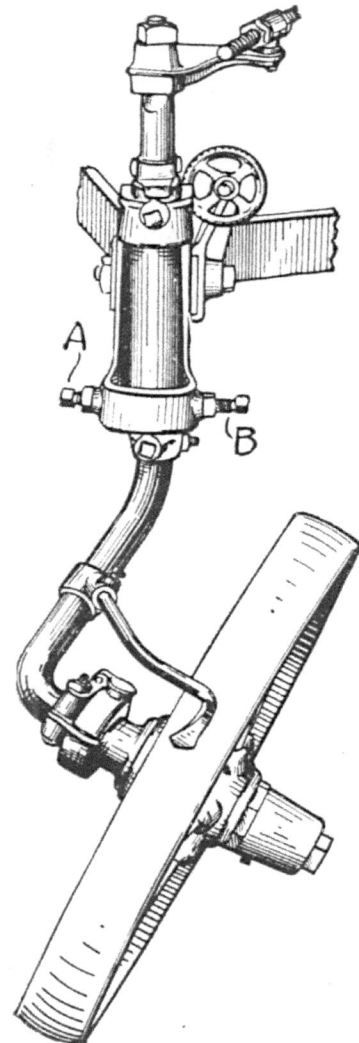

*Illustrating P & O rear axle adjustment features.*

*No. 2 Diamond gang*

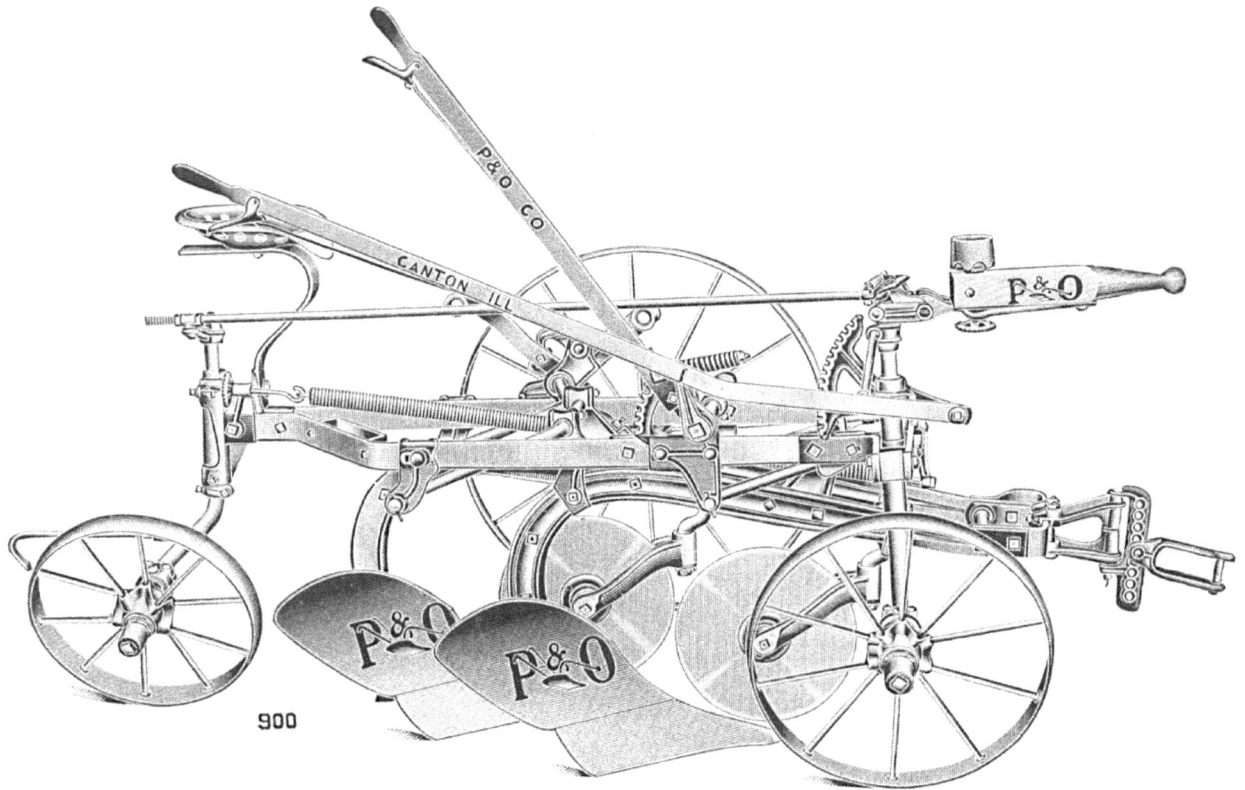

900

*No. 2 Three-furrow Diamond gang.*

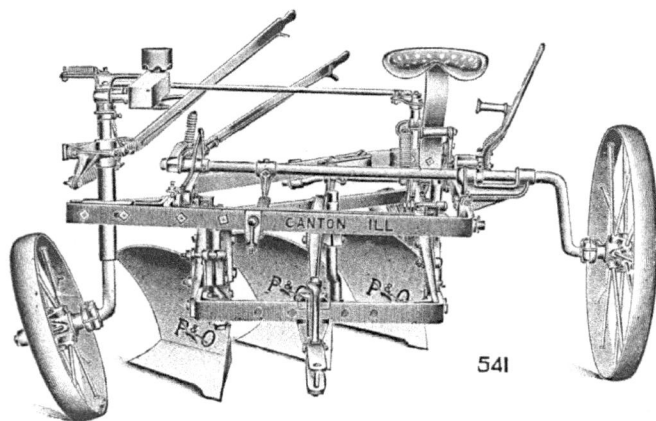

541

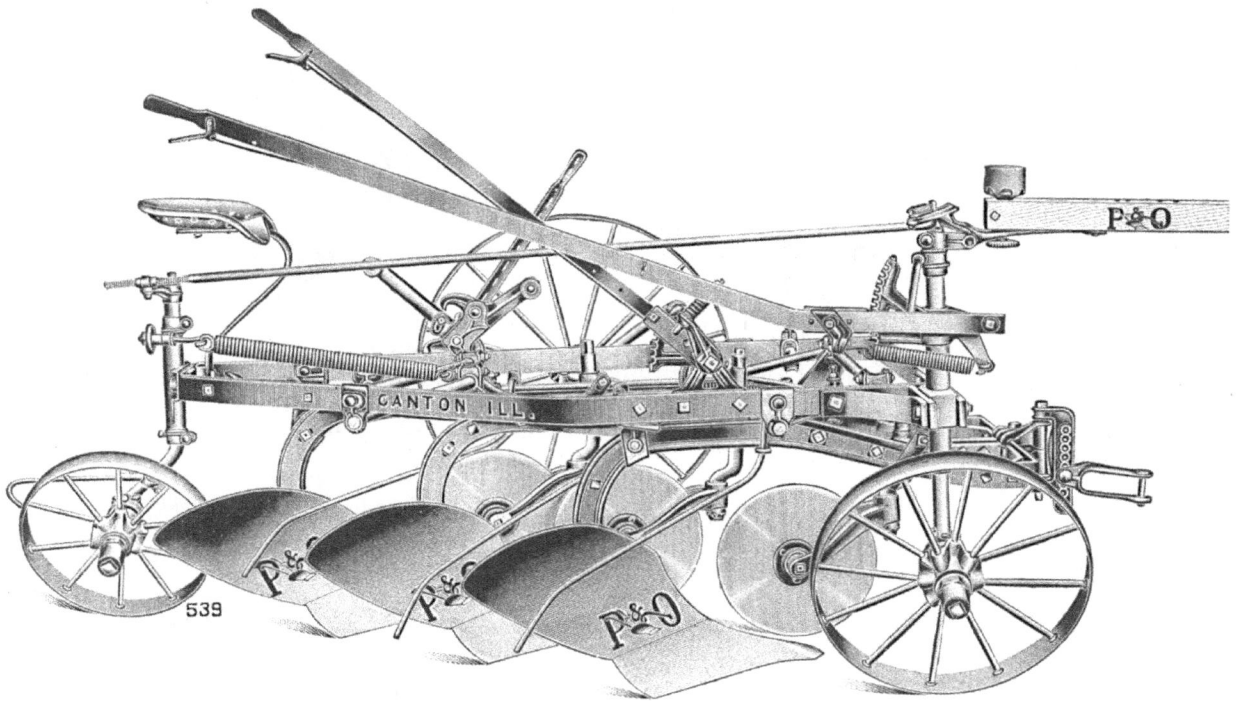

No. 2 Three-furrow Diamond gang.

No. 3 Three furrow Diamond, top view.

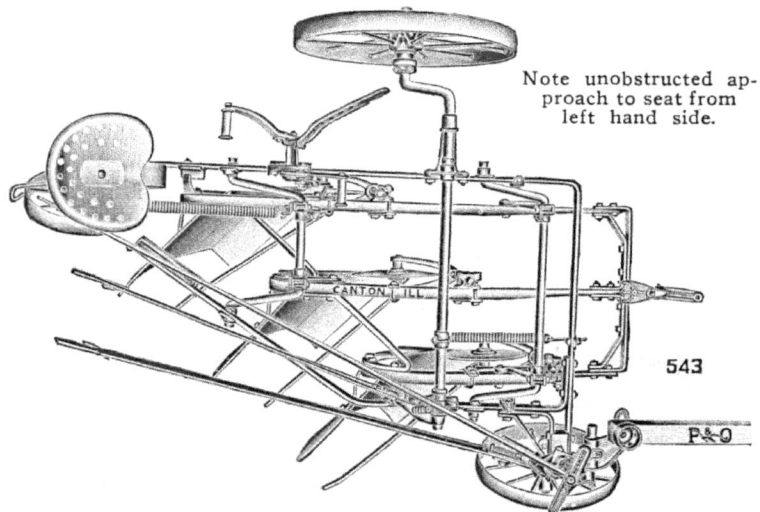

Note unobstructed ap-
proach to seat from
left hand side.

*No. 1 "Canton" sulky rear view.*

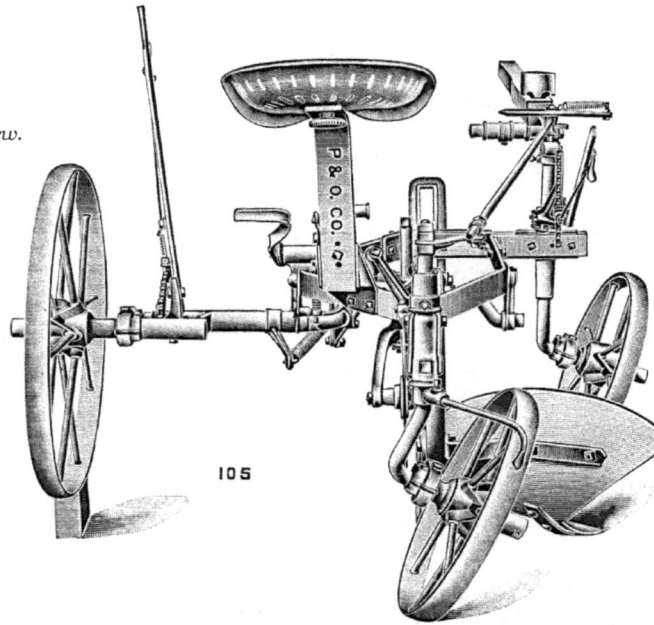

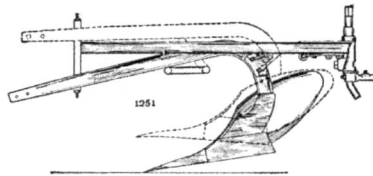

*P & O Canton sulky with single bail demonstrated this soil entering angle*

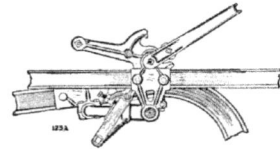

*Showing position of foot lift when the plow is lowered.*

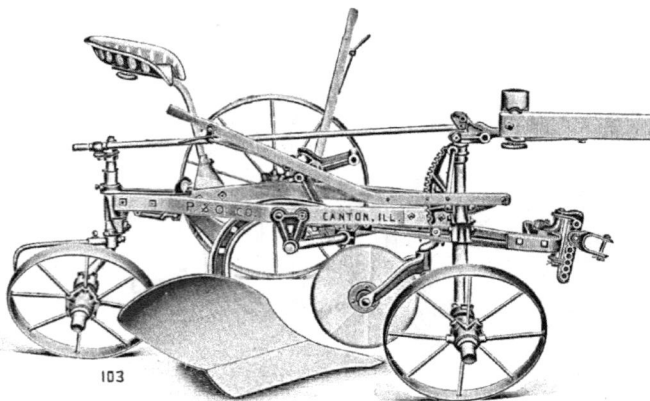

*No. 1 "Canton" sulky, moldboard view.*

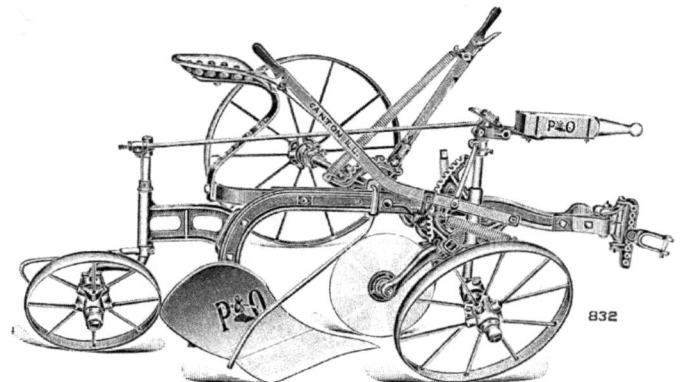

*No. 5 Success sulky, automatic control, moldboard view.*

# Parlin & Orendorff

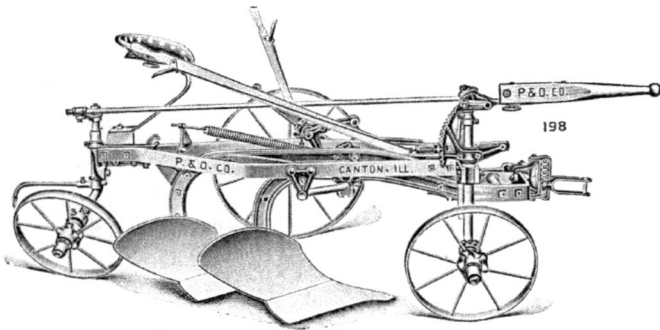

No. 1 "Canton" gang.

No. 5 Success sulky plow,
hand control, landside
view.

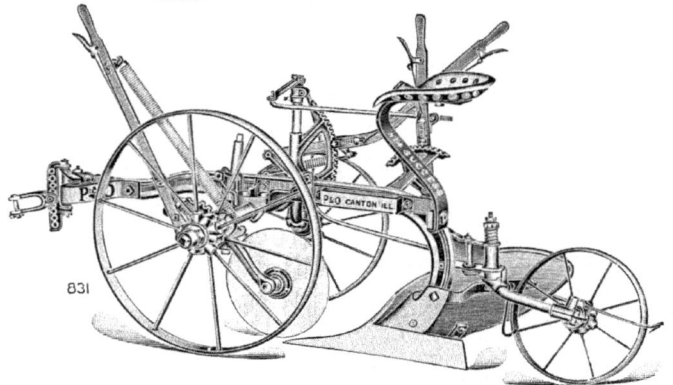

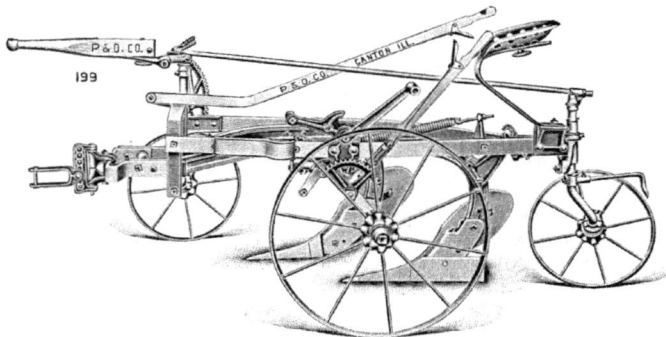

No. 1 "Canton" gang  landside view.

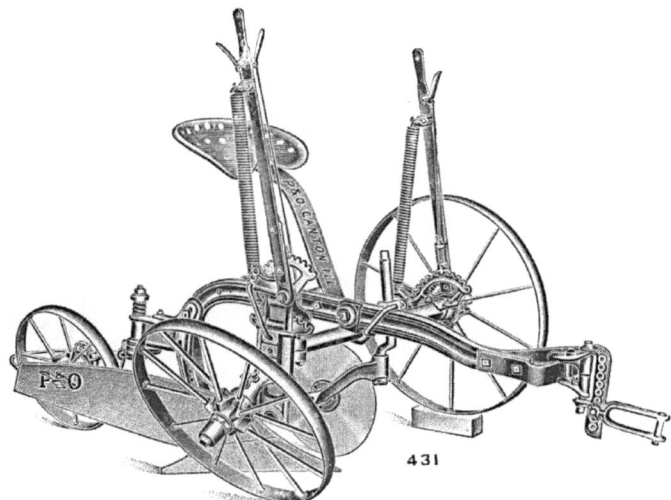

No. 6 Success sulky with breaker bottom.

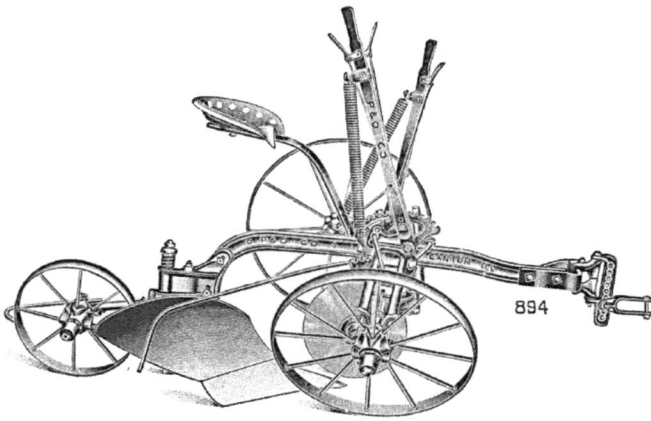

*No. 6 Success sulky.*

*Hitch shifting attachment.*

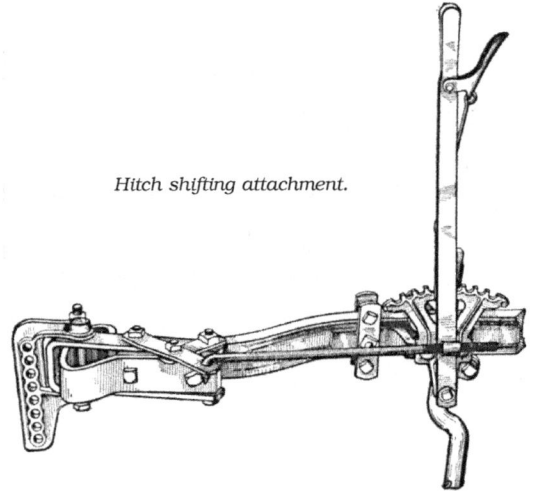

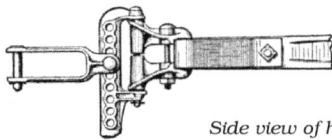

*Side view of hitch.*

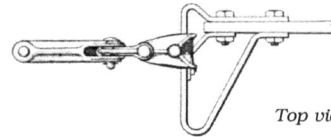

*Top view of hitch.*

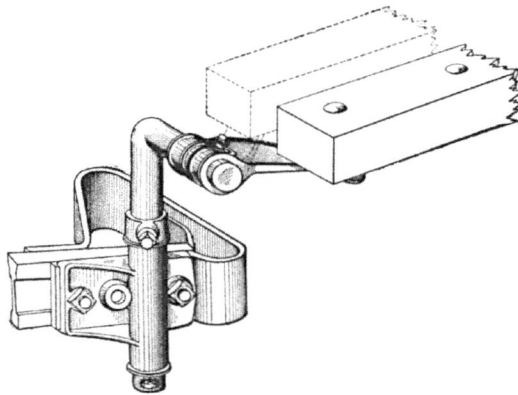

*Success sulky tongue attachment showing the tongue set for two horses; when three are being used the tongue plate is turned over and the tongue attached as indicated by dotted lines.*

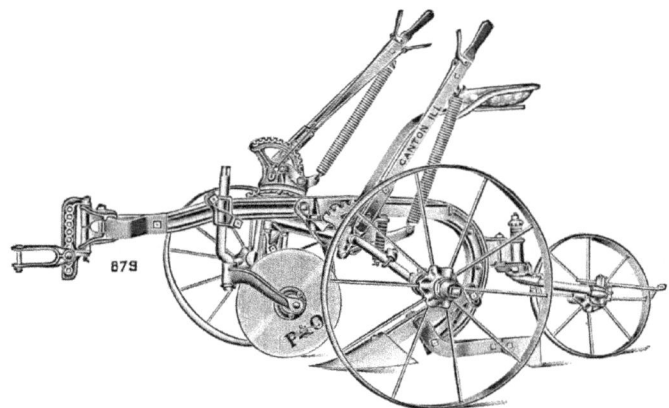

*No. 6 Success sulky with No. 5 steel sweep attachment.*

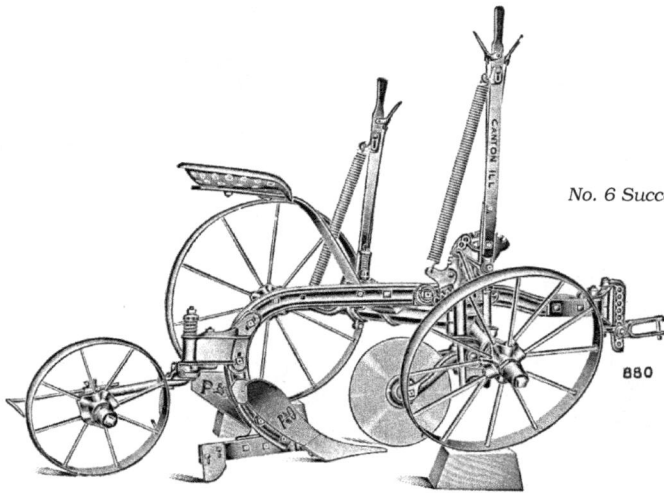

*No. 6 Success sulky with middle breaker attachment.*

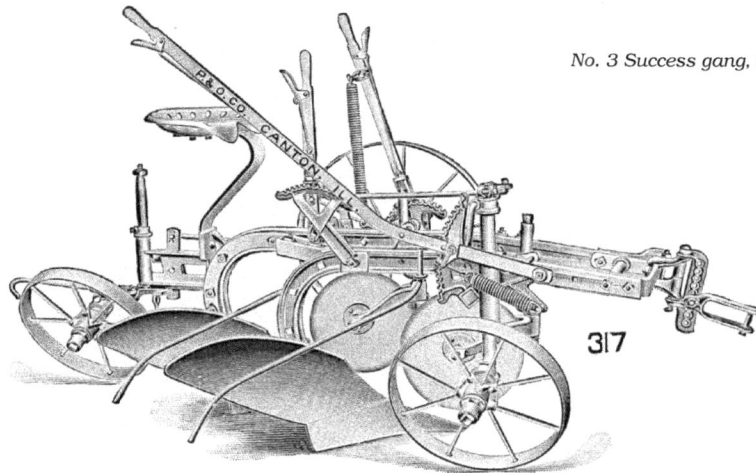

*No. 3 Success gang, rice pattern.*

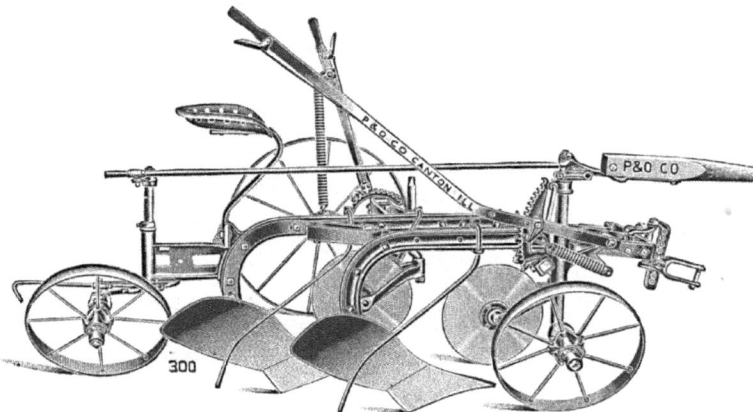

*No. 3 Success gang, automatic control.*

*Two way Success plow.*

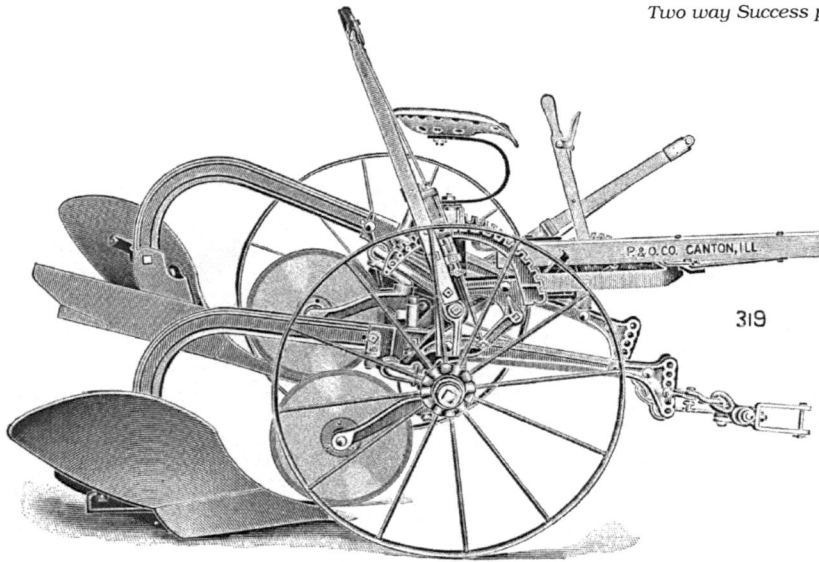

*Position of the tongue when set for three horses. No extra parts required.*

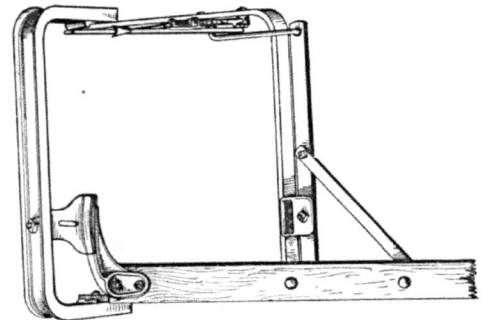

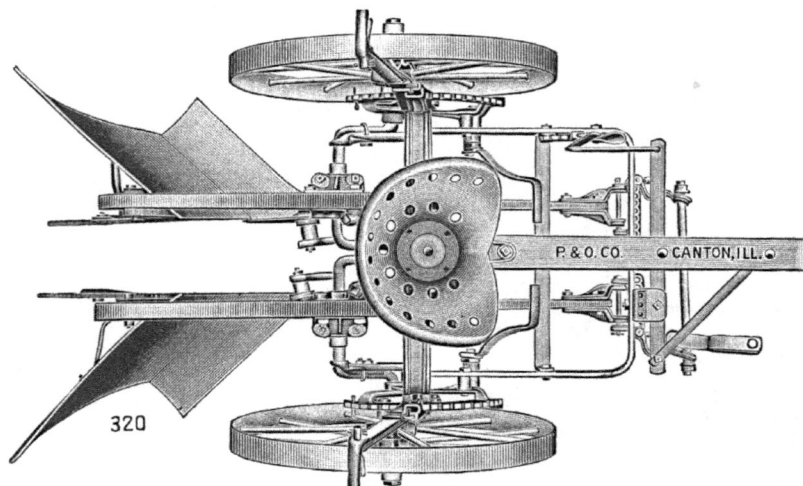

*No. 1 two way Success plow, top view.*

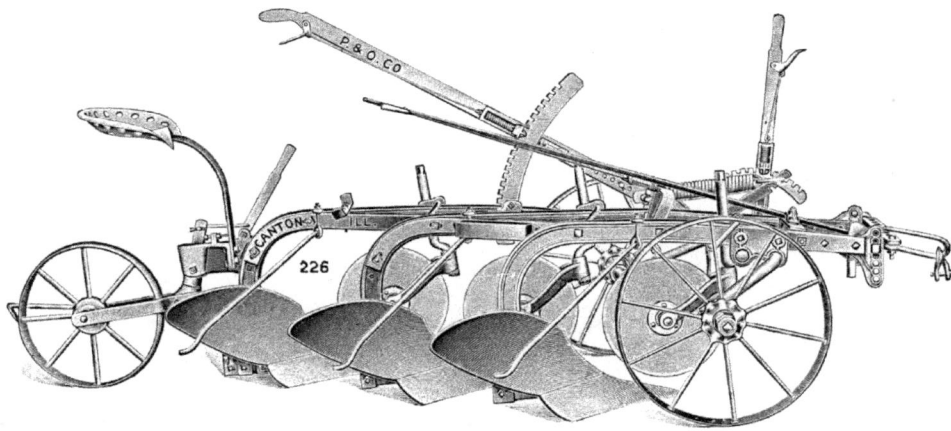

*Three furrow No. 1 Walking gang with riding attachment.*

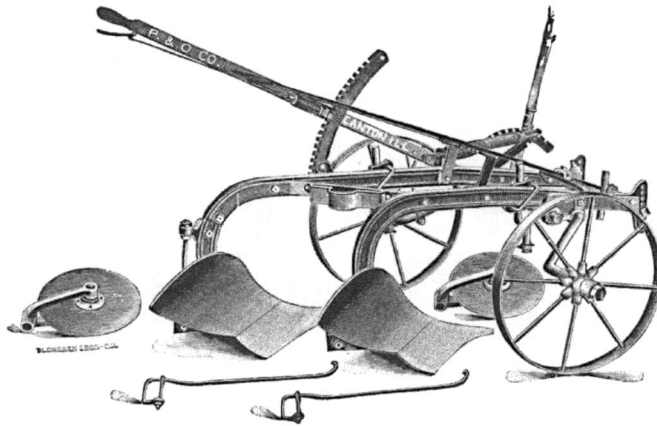

*No. 1 Two-furrow walking gang*

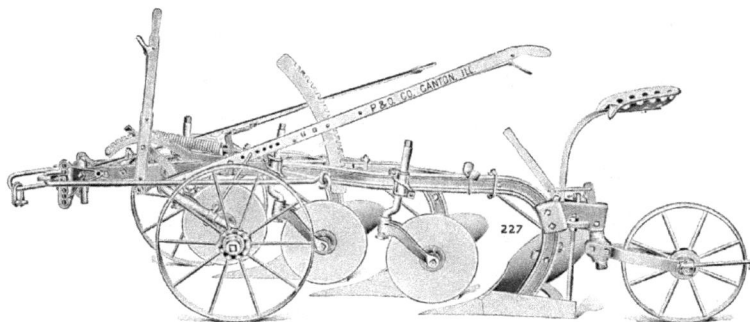

*Three -furrow walking gang with riding
attachment, landside view.*

No. 2 Walking gang with No. 2 riding attachment

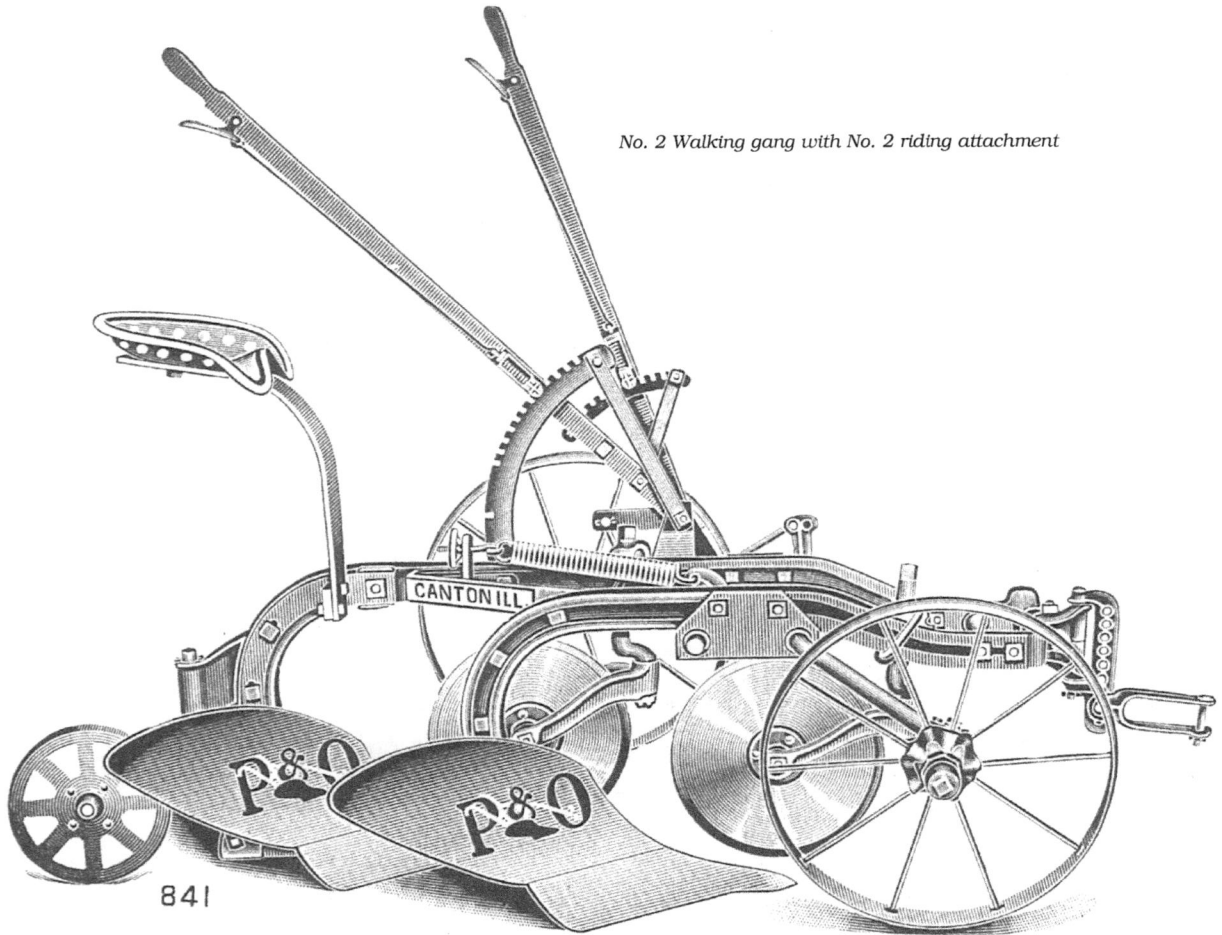

841

Two furrow No. 3 walking gang
with No. 3 riding attachment

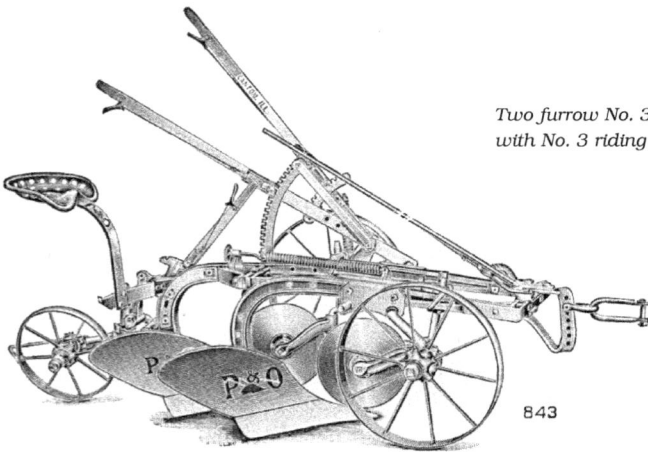

843

No. 2 walking gang, landside view.

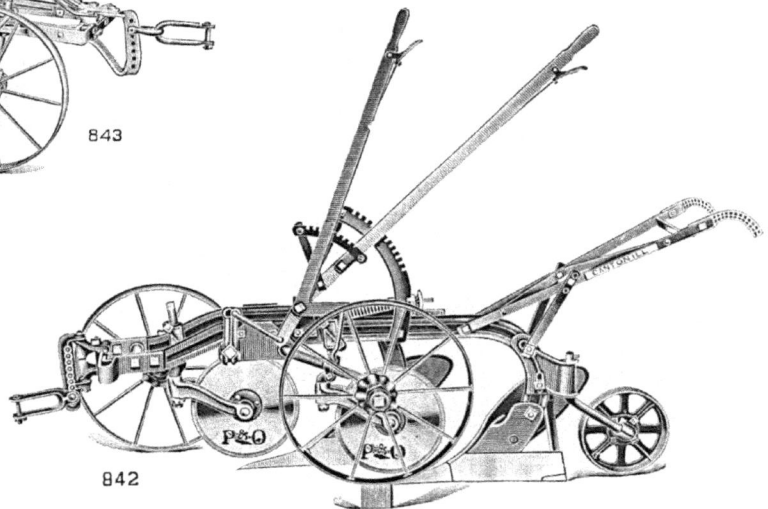

842

# Parlin & Orendorff

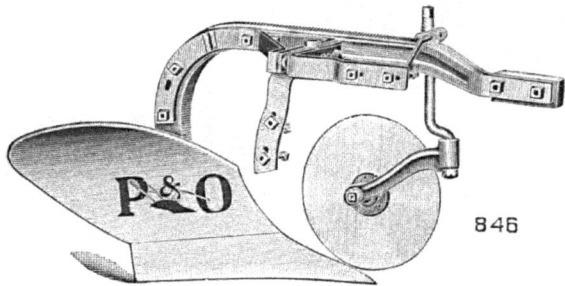

Third plow attachment for No. 3 walking gang, two-furrow.

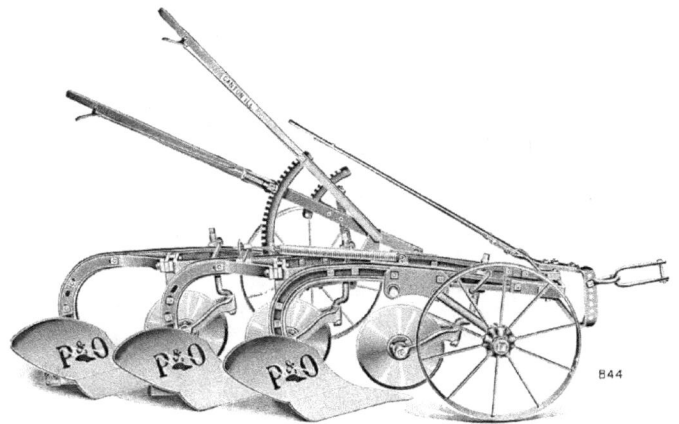

Three-furrow No. 3 walking gang

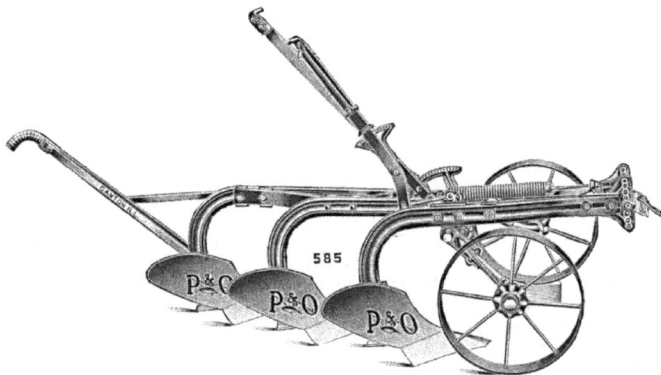

No. 4 Three-furrow Sunset Vineyard gang

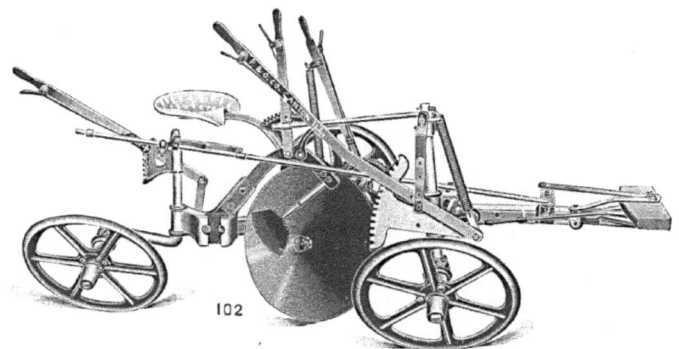

No. 3 Disc sulky

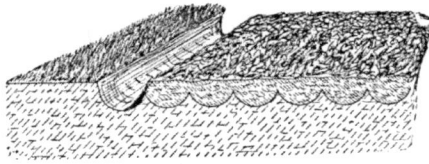

*Ground plowed without the cutter*

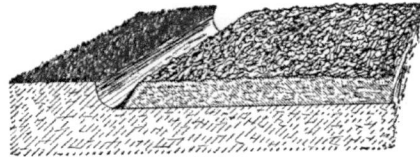

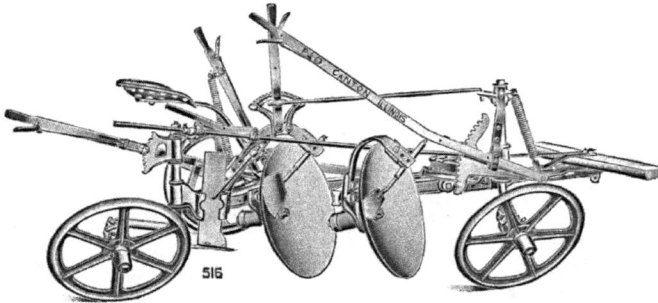

*Ground plowed with the cutter*

*No. 3 Disc gang with standing cutter*

*No. 4 Disc sulky with automatic controlling device.*

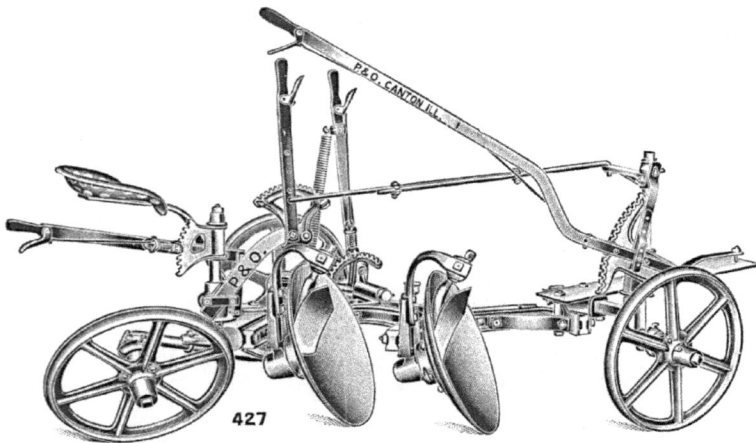

*No. 4 1/2 Two-furrow disc plow*

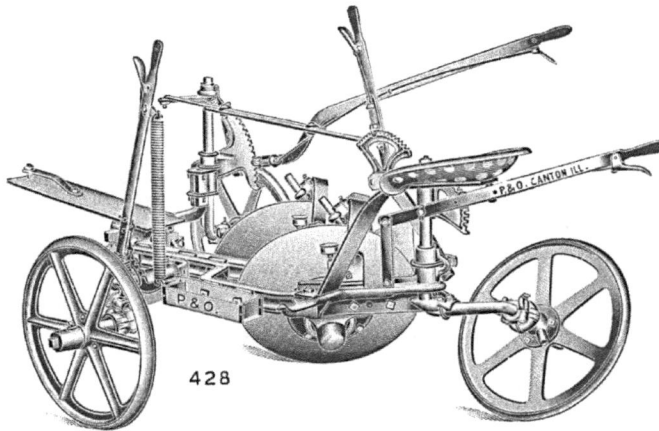

*No. 4 1/2 two-furrow rear view*

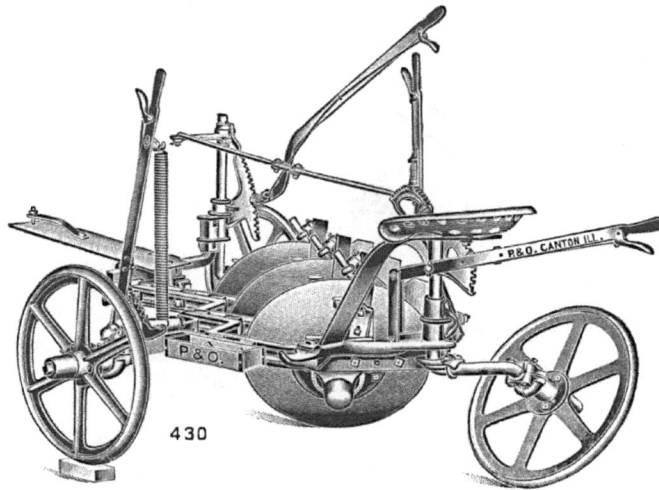

*No. 4 1/2 three-furrow rear view*

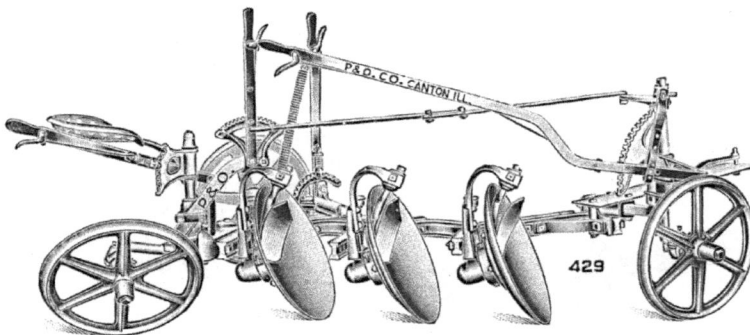

*No 4 1/2 Three-furrow disc gang plow*

*Wood beam stubble, "A" series*

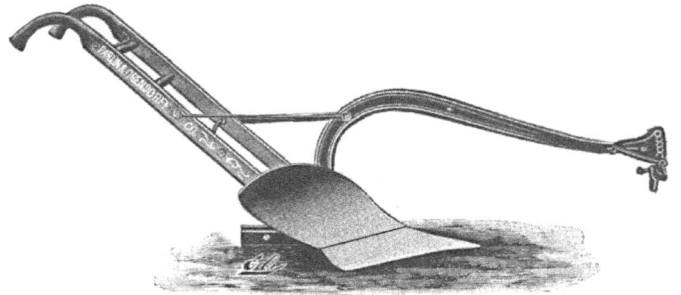

*Iowa stubble*

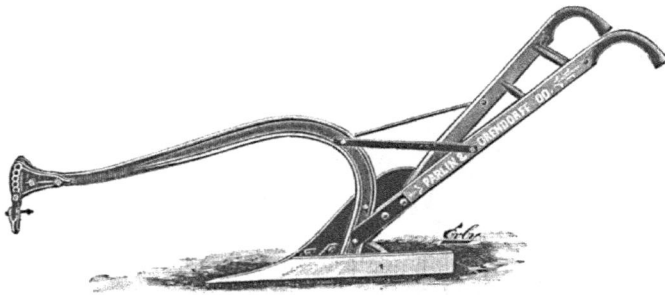

*Landside view of the Iowa stubble*

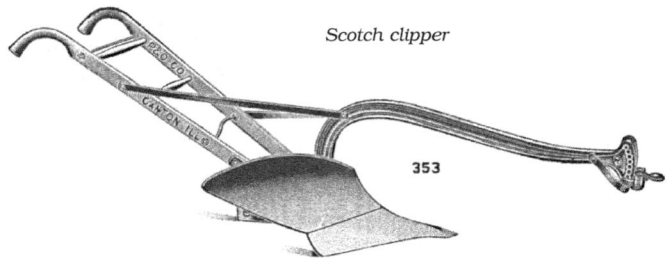

*Scotch clipper*

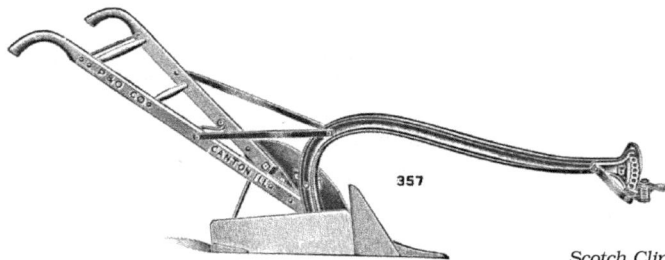

**353**

**357**

**356**

*Scotch Clipper High Landside, especially suited for deep plowing in loose sandy soil where the bank of the furrow crumbles, or in marshy swampy land where the fin cutter severs roots and vines.*

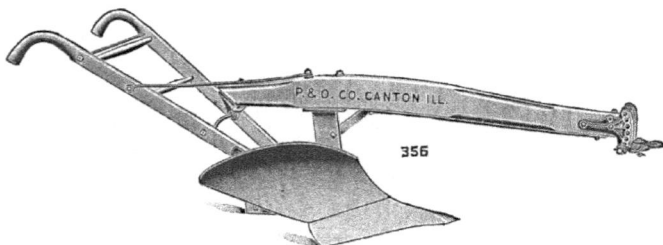

*Wood beam Scotch Clipper*

The Scotch Clipper High Landside featured a spacing block which when reversed landed the plow for three abreast

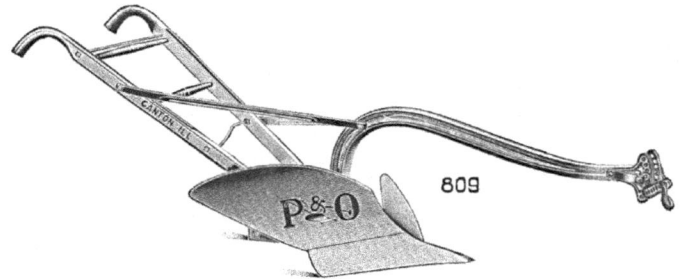

Alfalfa Scotch Clipper

Alfalfa Scotch Clipper showing the bottom side.

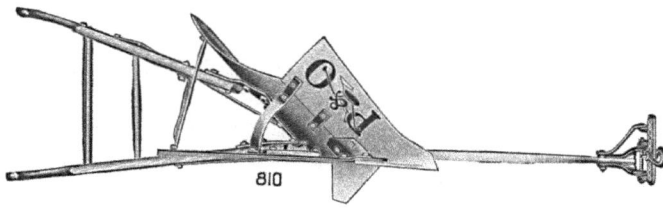

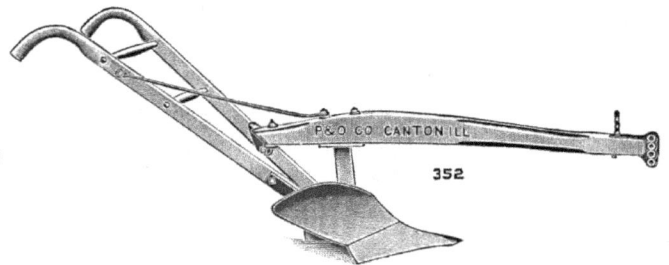

One-horse Corn plow

Scotch Clipper D Series was available 7" through 16"

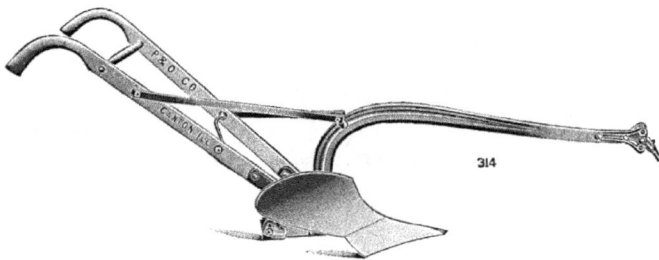

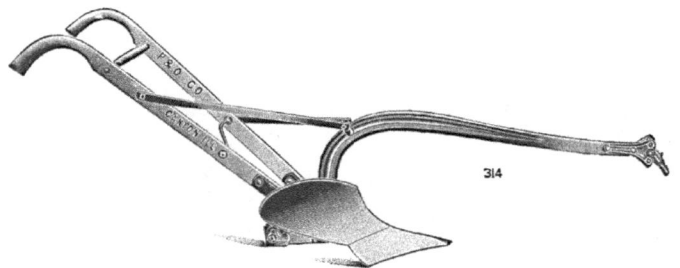

Blue Jay plows were the economy model

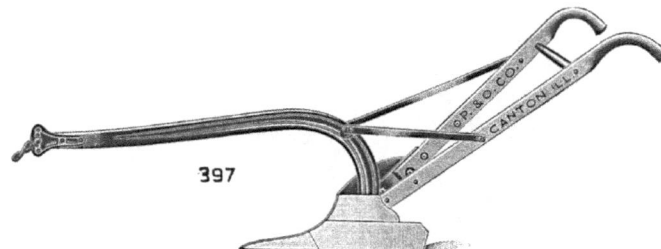

The backside of the Blue Jay plow

*Topside of the Sunset Vineyard from previous page. Handles are widely adjustable.*

*All steel Sunset Vineyard one horse plow, handles and hitch fully adjustable, came with 8 and 9 inch shares.*

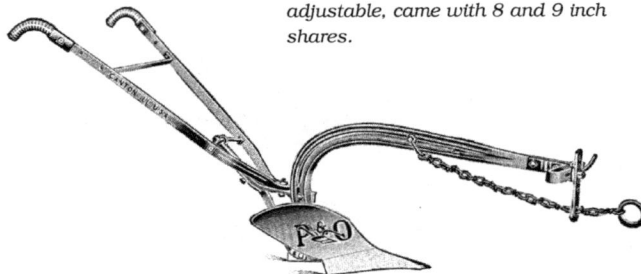

*Reversible Hillside plow*

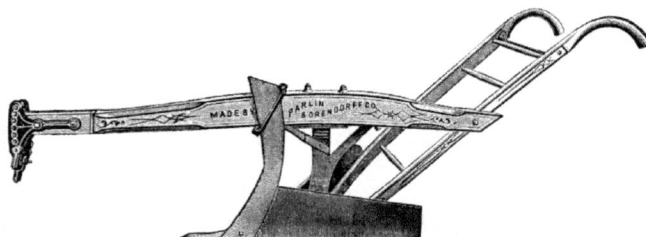

*Timber plow with Quincy cutter*

*Timber plow with knee cutter*

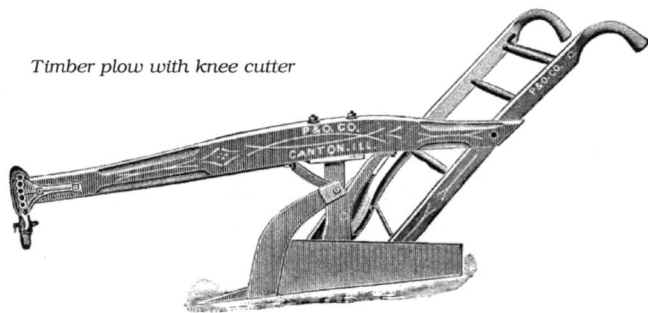

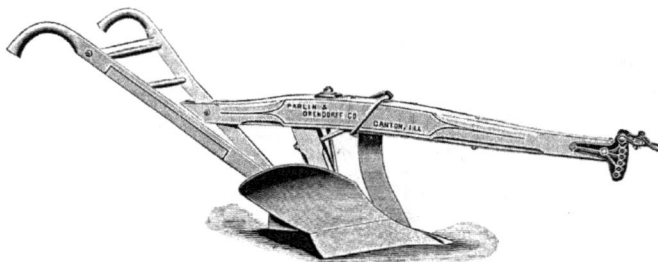

*Wood standard plow built for rough work in clay or underbrush.*

*Road plow with Quincy cutter*

*New Ground plow, 7 to 12 inch share and root knife.*

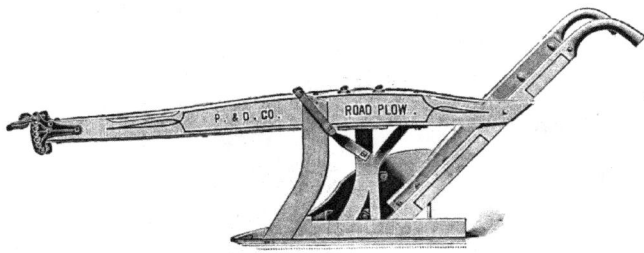

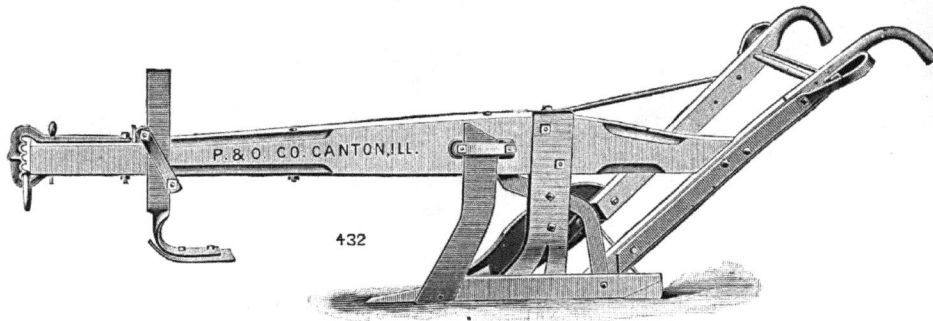

*Landside view of Road plow with Quincy cutter.*

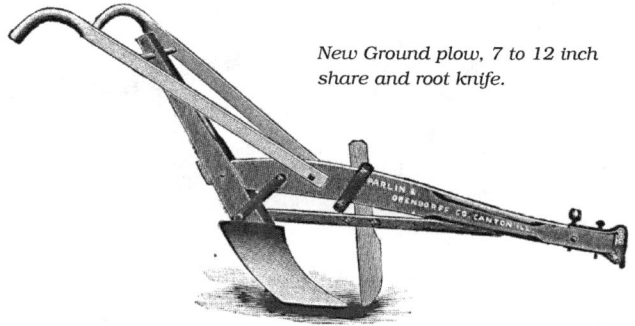

*Senior Grading plow with gauge shoe and Quincy cutter, built the heaviest and strongest for use in hard clay and shale.*

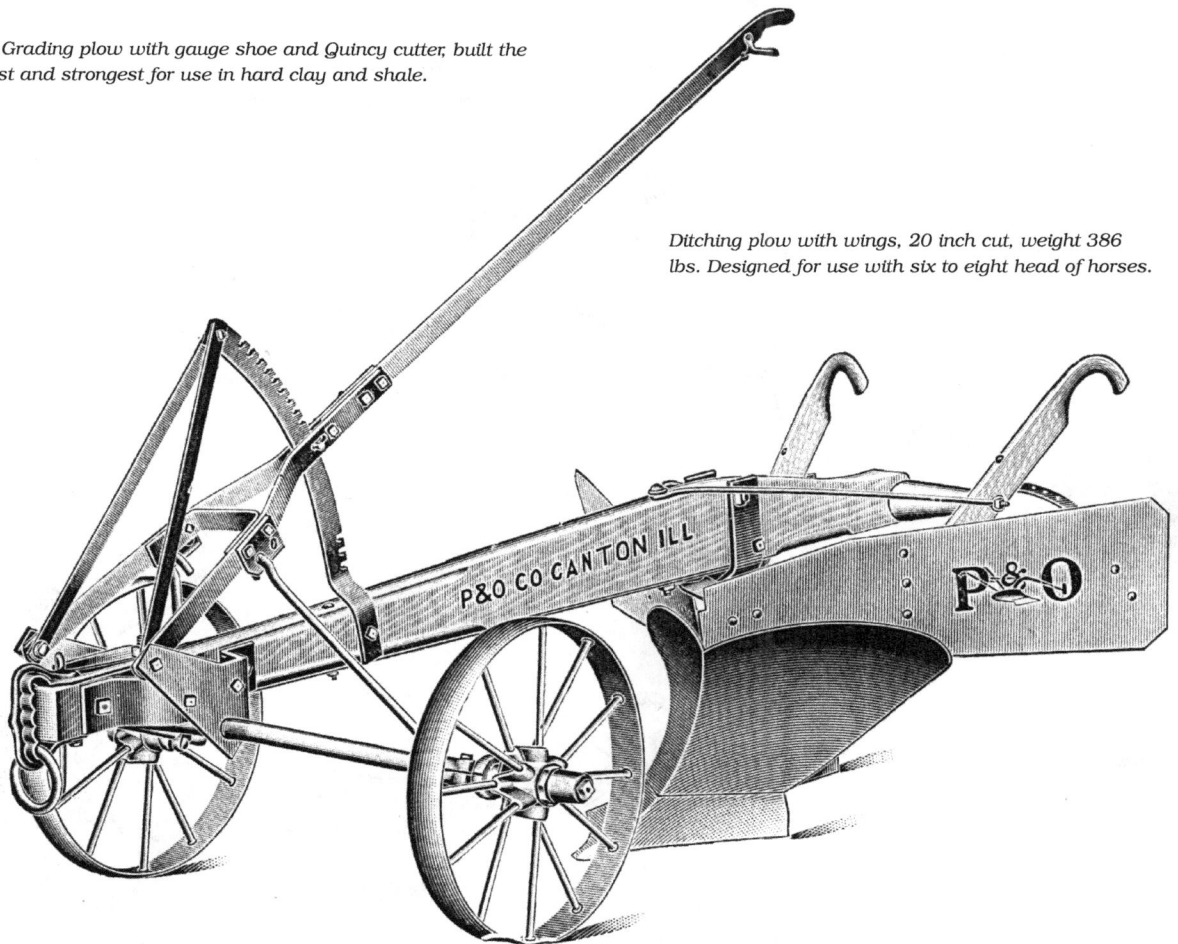

*Ditching plow with wings, 20 inch cut, weight 386 lbs. Designed for use with six to eight head of horses.*

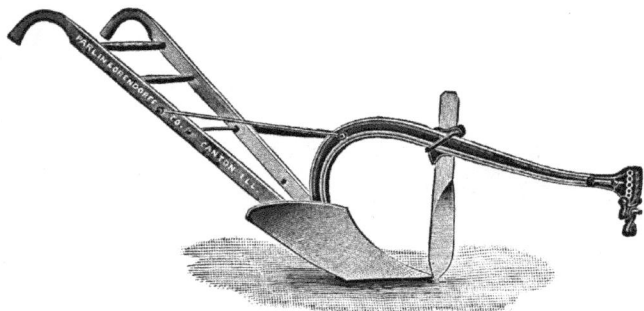

Root Ground plow with hanging cutter

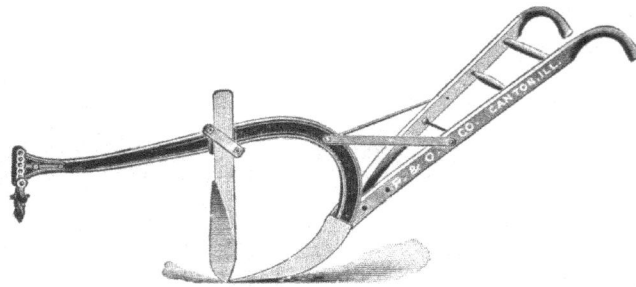

Landside view of Root Ground plow

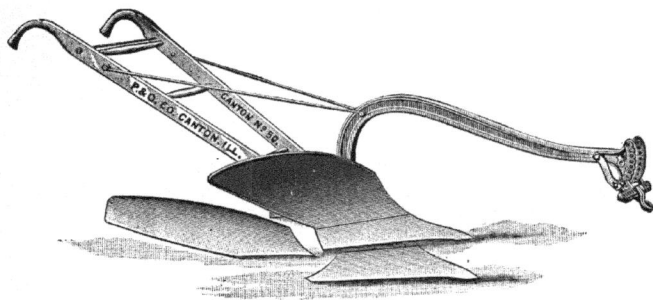

No. 50 Combination plow, convertible,
with attachments, to breaker.

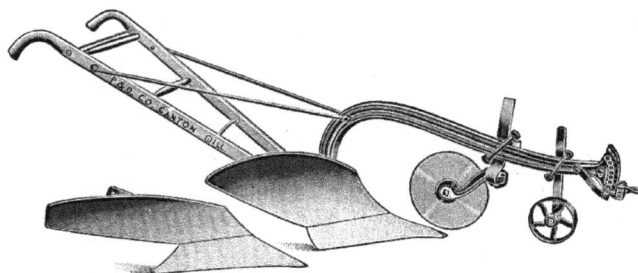

No. 50 Combination with rolling coulter,
depth wheel and breaker attachment.

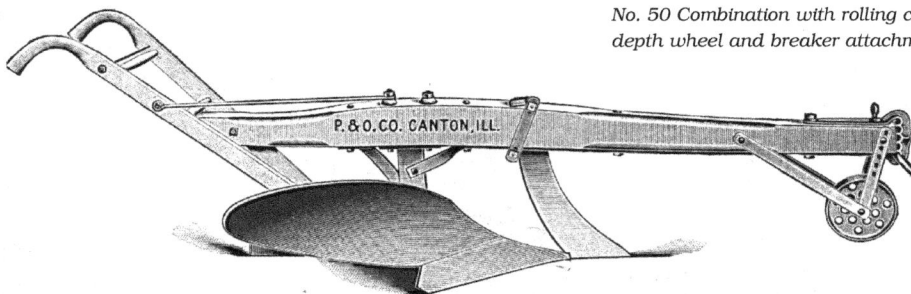

Brush breaker for plowing up hazel brush,
blackberry bushes and all underbrush. The standing
cutter is mortised at the heel to receive the rounded
point of the share preventing share from catching
under roots.

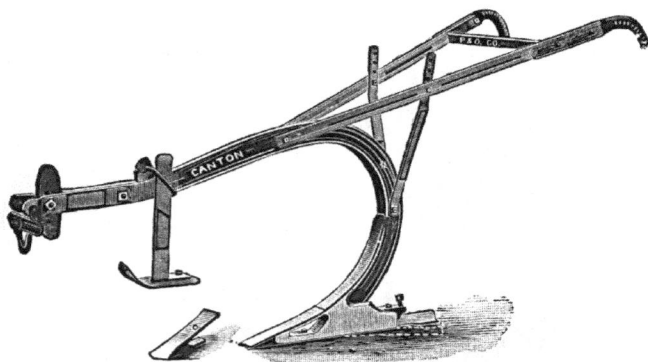

No. 4 Subsoil Plow

Brush Breaker plow with No. 5
truck wheel attachment.

Landside view of Wood standard plow with
Quincy cutter.

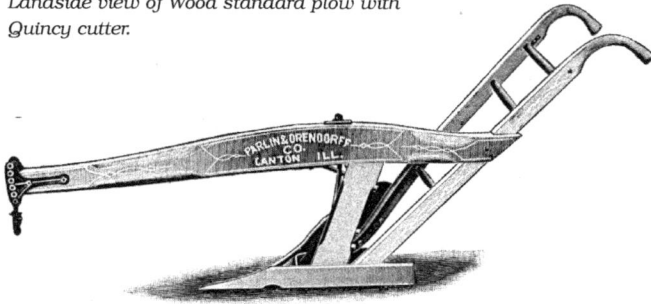

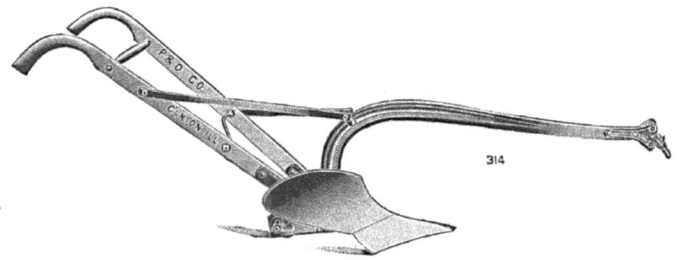

Sunset Special model made for
California and Pacific coast trade

Wood beam Sunset special

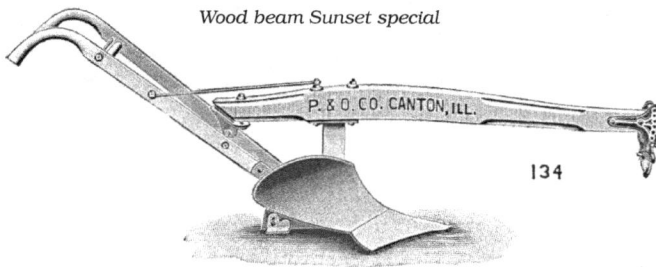

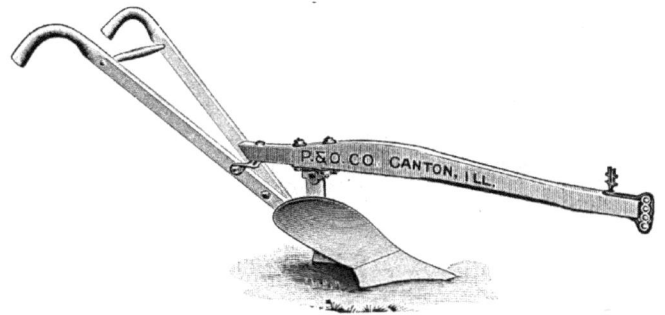

Sunset Vineyard plow intended for one horse.

Wood beam Prairie Breaker

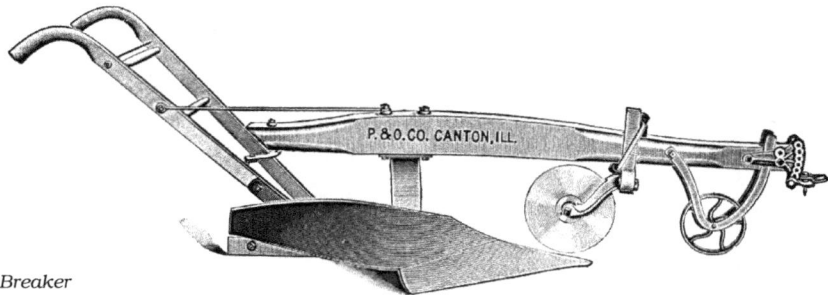

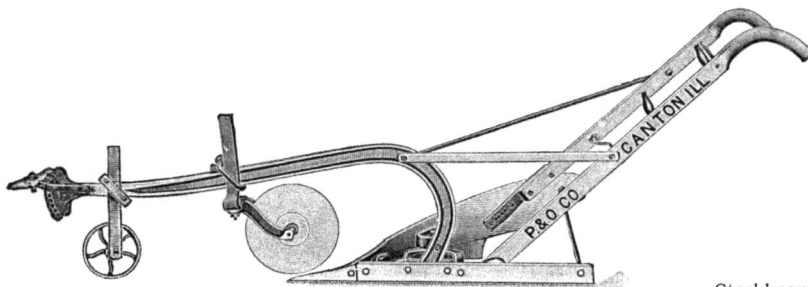

Steel beam Prairie Breaker

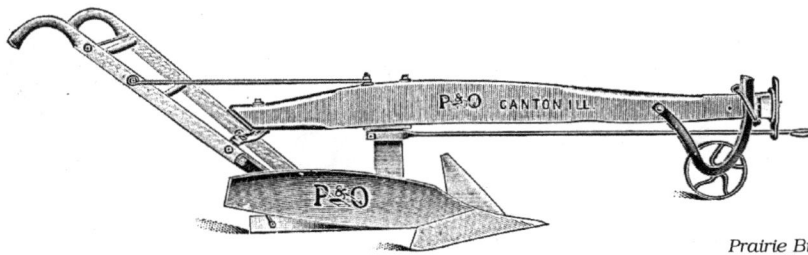

Prairie Breaker with draft bar

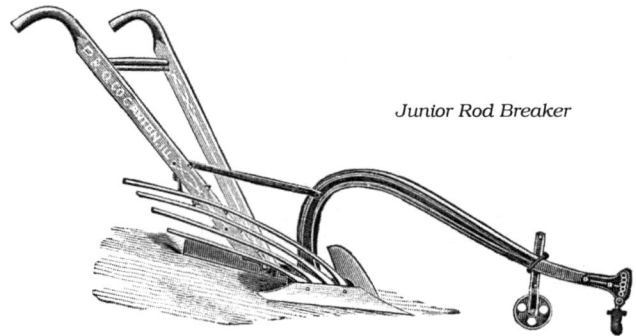

Junior Rod Breaker

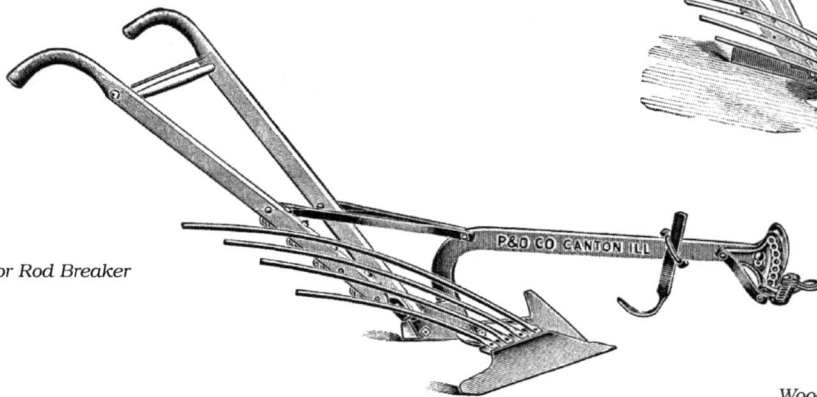

Senior Rod Breaker

Wood Beam Pioneer Rod Breaker

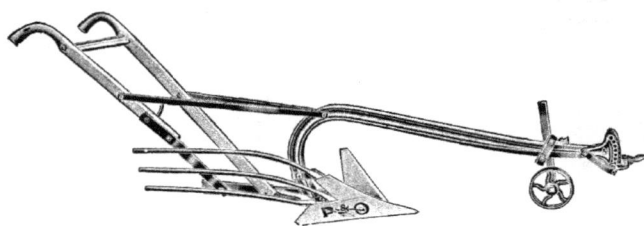

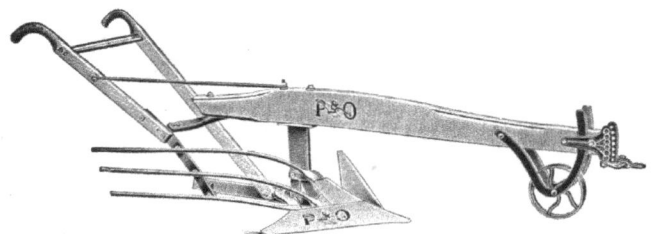

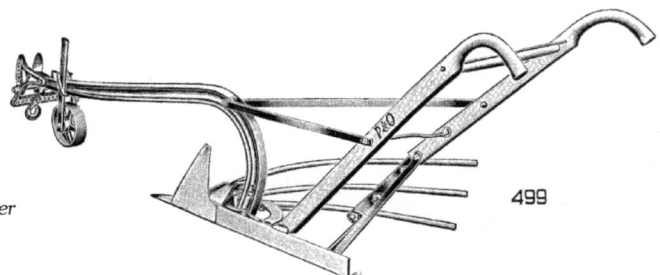

Pioneer Rod Breaker

Landside view of Pioneer Rod Breaker

499

# Parlin & Orendorff

Southern light one-horse or pony plow

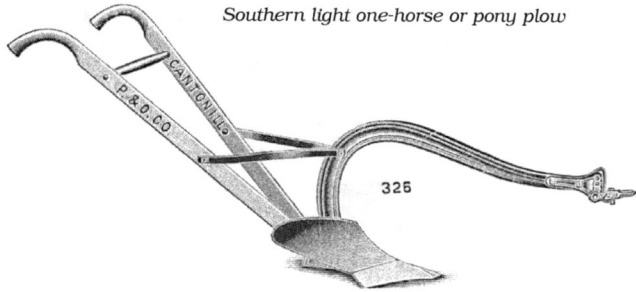

325

Landside view of Southern plow

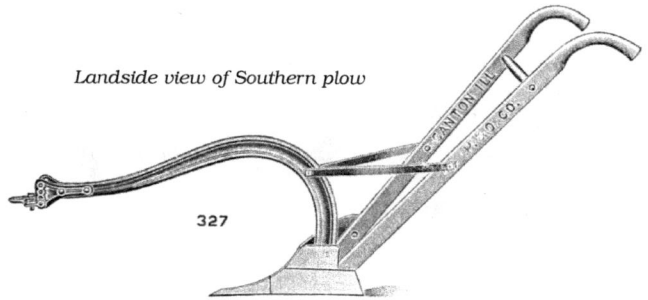

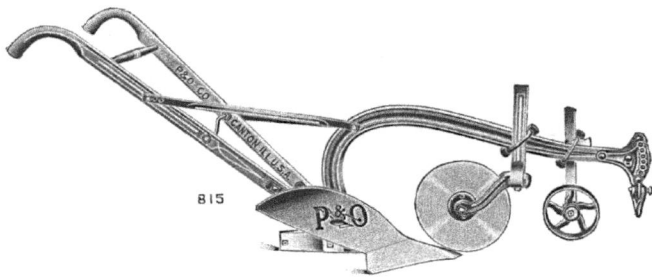

327

815

Southern Blackland plow

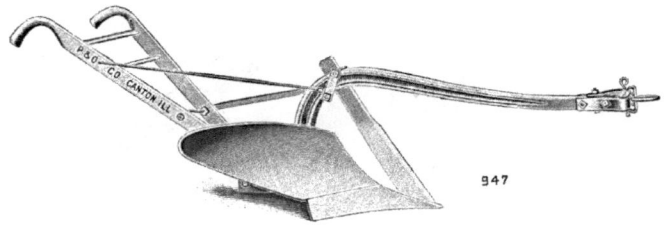

947

Sugar Land plow with hanging cutter

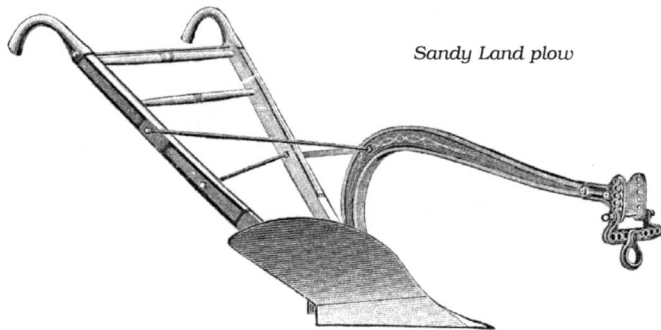

Sandy Land plow

Landside view of Sandy Land plow

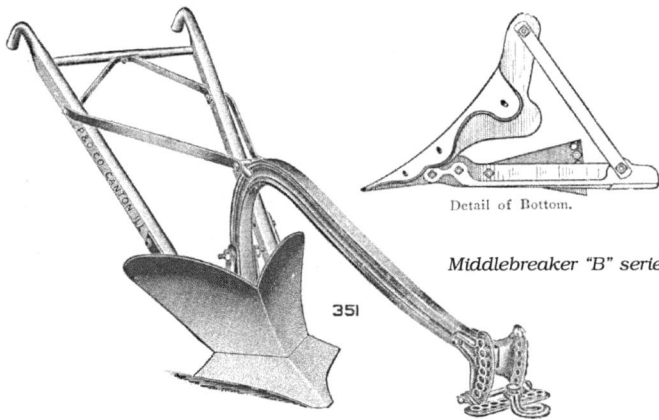

Detail of Bottom.

*Middlebreaker "B" series*

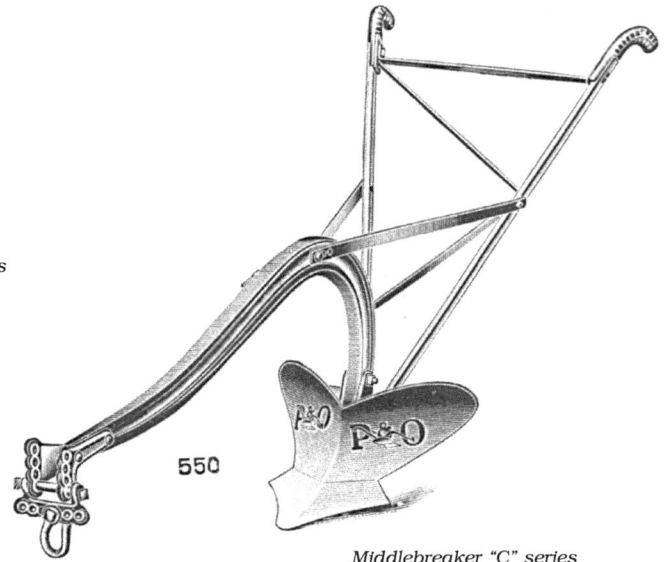

*Middlebreaker "C" series*

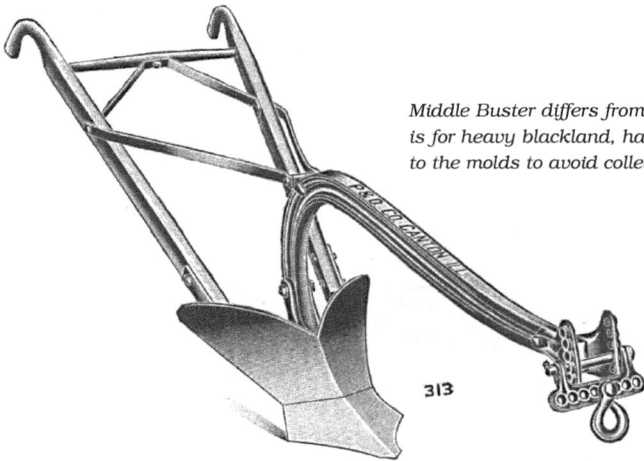

*Middle Buster differs from Middlebreaker as it is for heavy blackland, handles are attached to the molds to avoid collecting trash.*

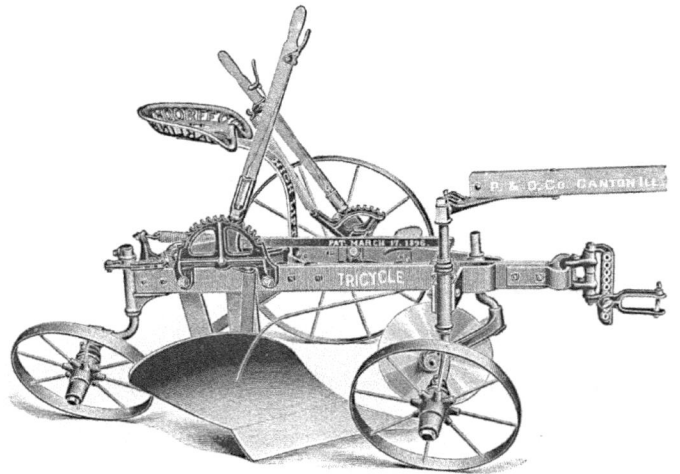

*Tricycle Sulky plow (extra heavy high-lift)*

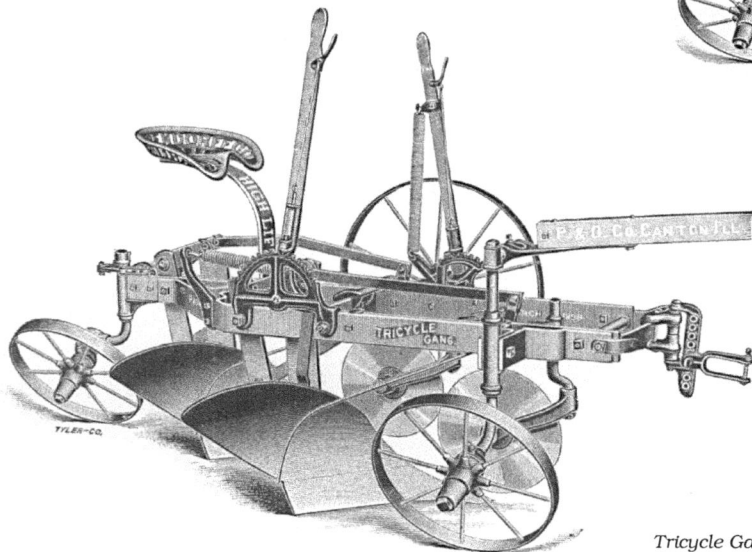

*Tricycle Gang plow*

Makes
& Models

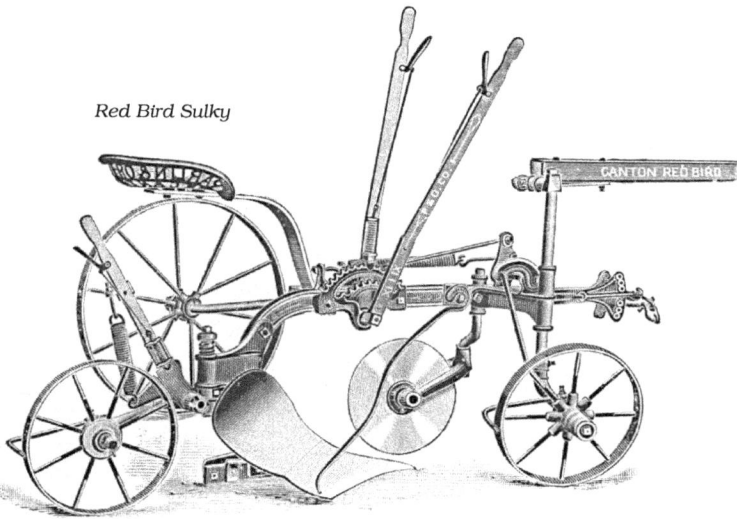

Red Bird Sulky

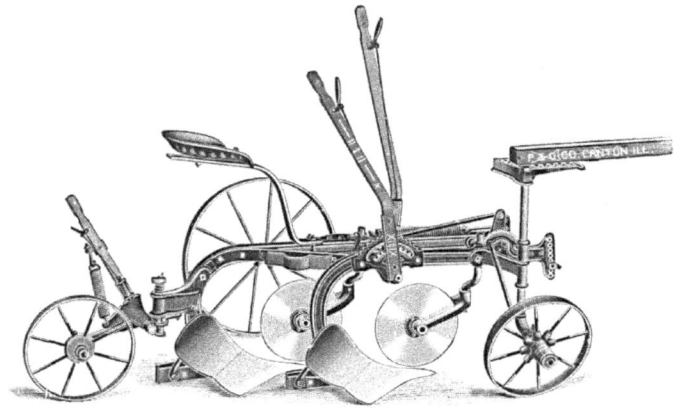

Orchard and Vineyard Gang. A small plow
built extra strong for use in ground with
shrubbery and roots.

Red Bird Gang

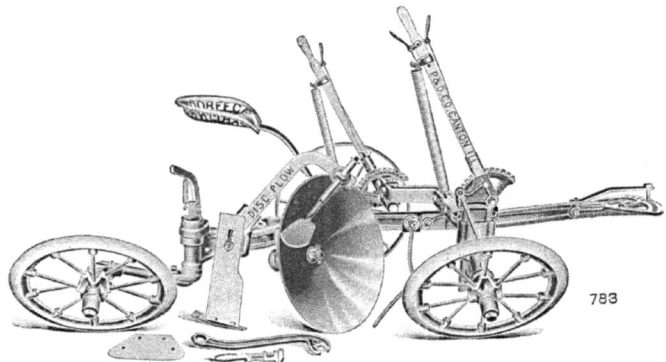

Canton Disc Plow with standing cutter

*Front view of Canton Disc plow*

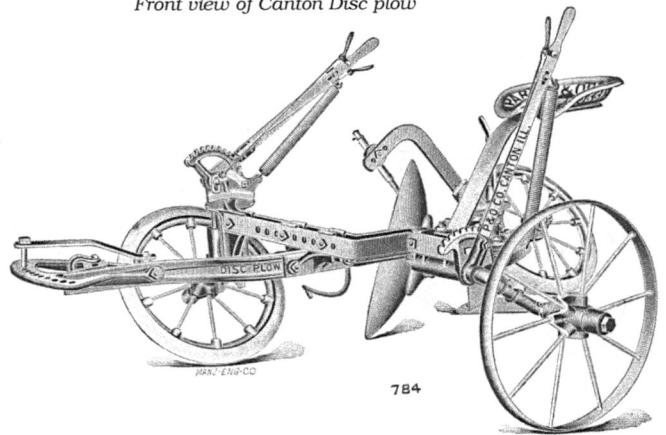

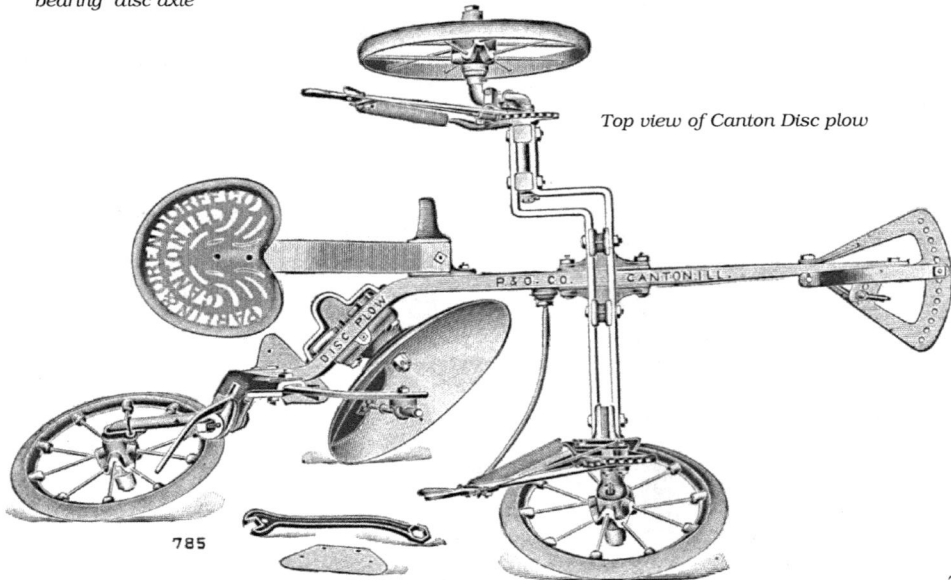

784

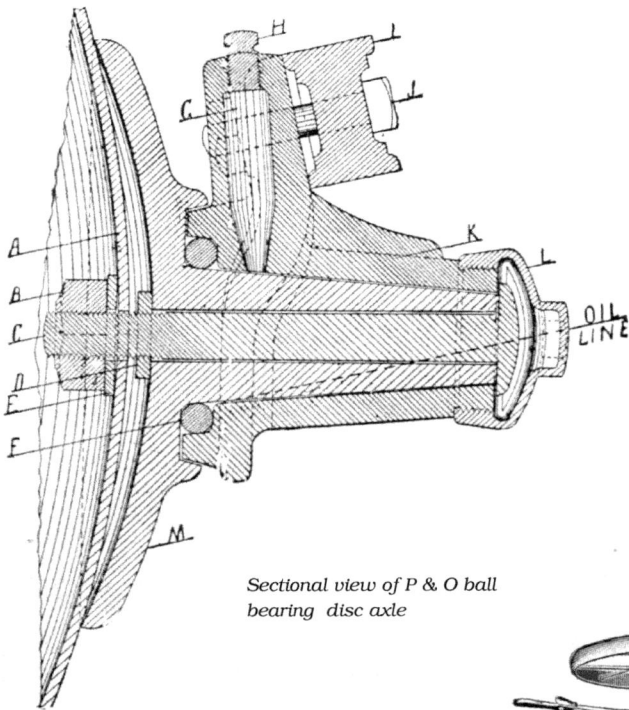

*Sectional view of P & O ball bearing disc axle*

OIL LINE

*Top view of Canton Disc plow*

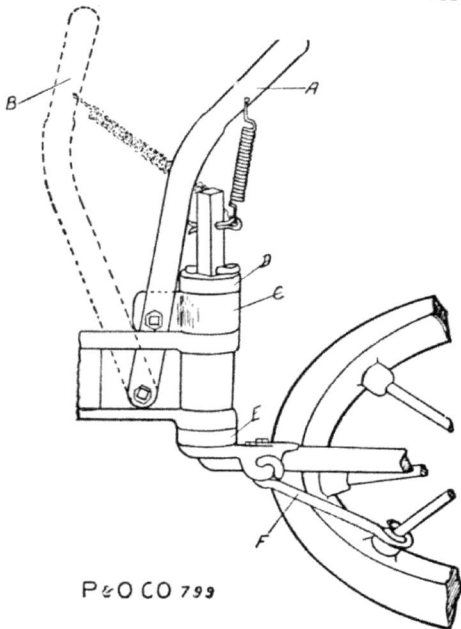

785

*The long scraper arm, standing cutter and automatic trip lever*

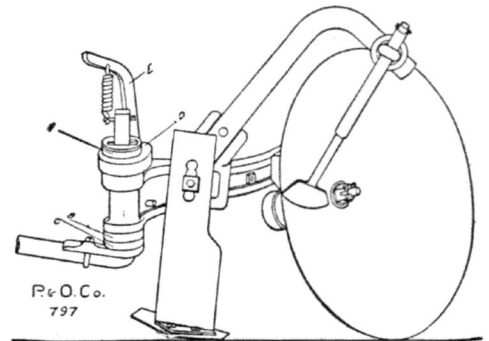

*(left) Automatic trip lever also showing brake hook holding rear wheel when necessary. D and E are the washers used in adjusting the frame for the three sizes of discs.*

P&O CO 799

P&O.Co. 797

515

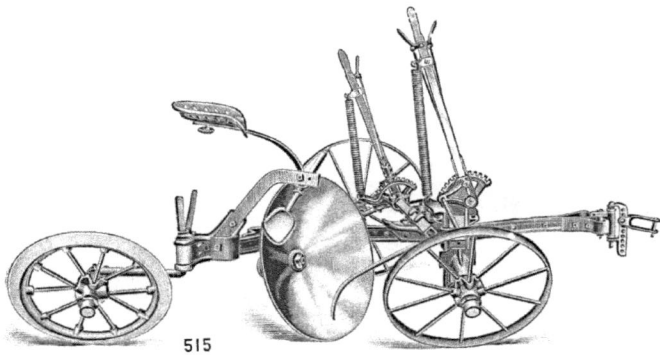

Canton Disc Plow without the standing share cutter

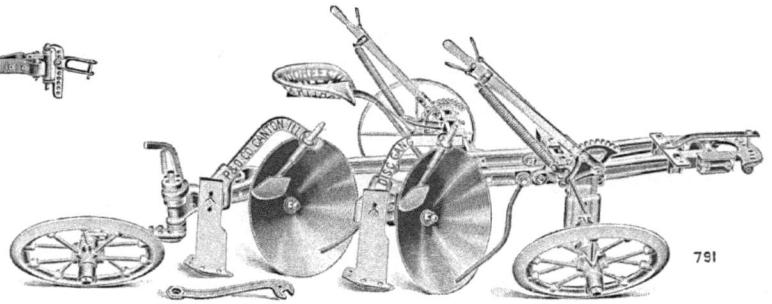

791

Canton Double Disc plow

912

Curlew gang plow with 8 inch shares and
reinforcements for hard baked ground

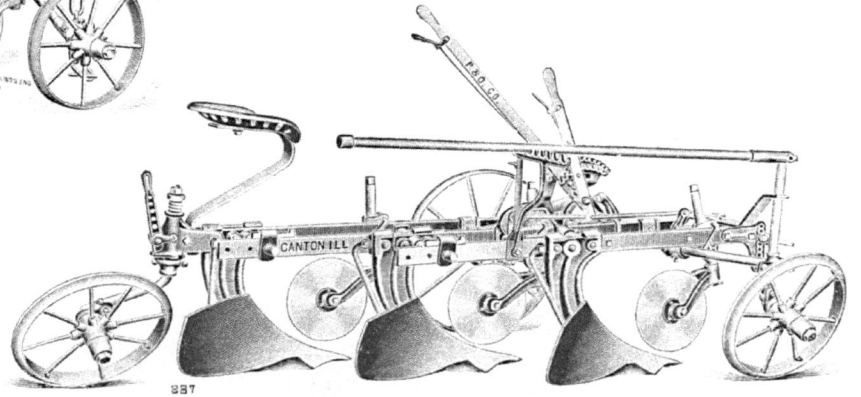

887

Curlew Steerage Gang with steering rod

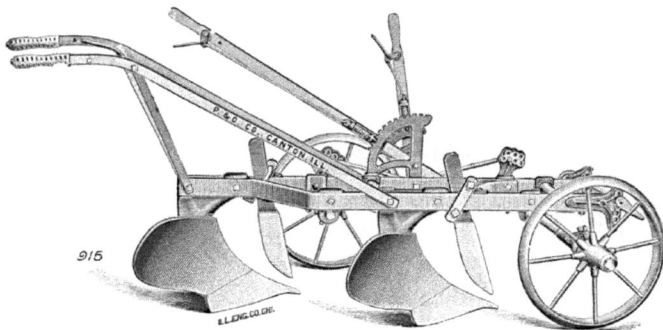

915

Wallaby Gang plow with hanging cutters and 8 inch stubble bottoms

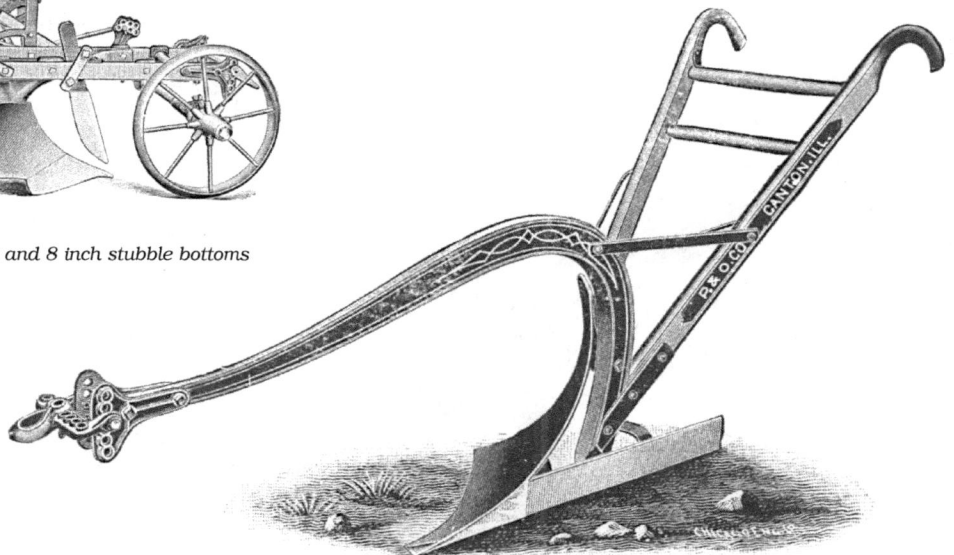

Stubble plow

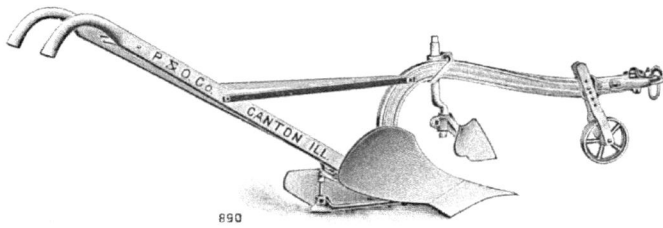

Canton Harlow plow with jointer and gauge wheel

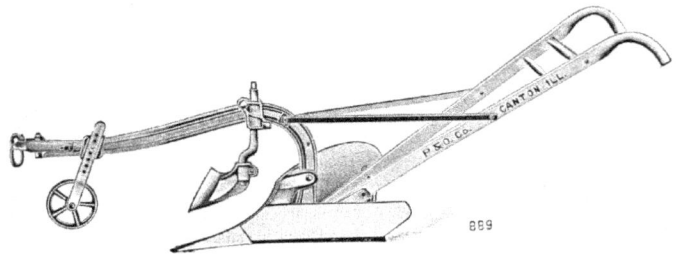

Landside view of Canton Harlow, dark line
represents removabvle cast shoe.

Timber Land plow for work amongst stumps and roots

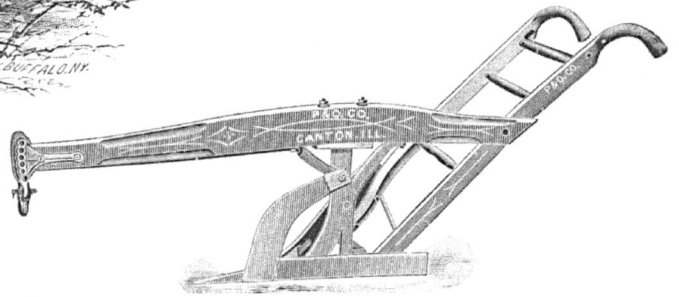

Timber Land equipped with Quincy cutter

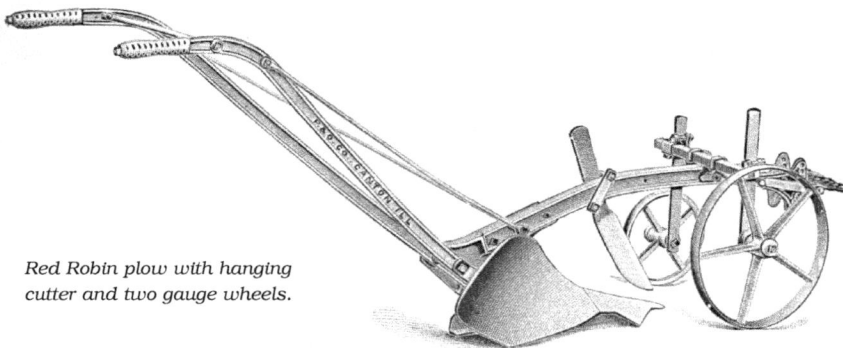

Red Robin plow with hanging
cutter and two gauge wheels.

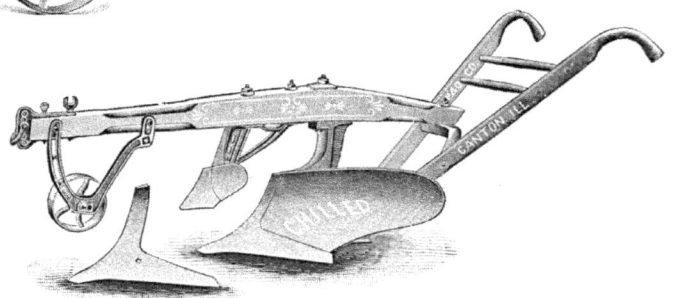

No. 40 Canton Chilled plow

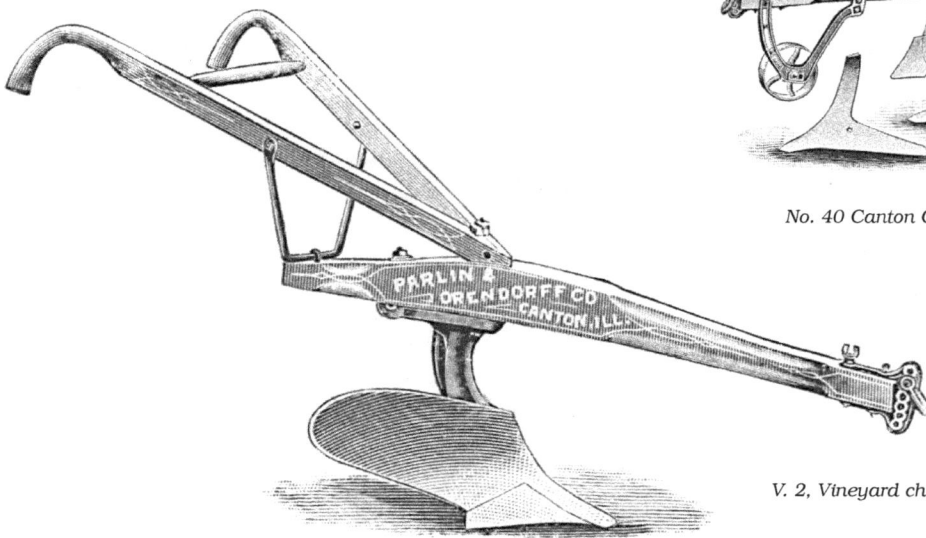

V. 2, Vineyard chilled plow

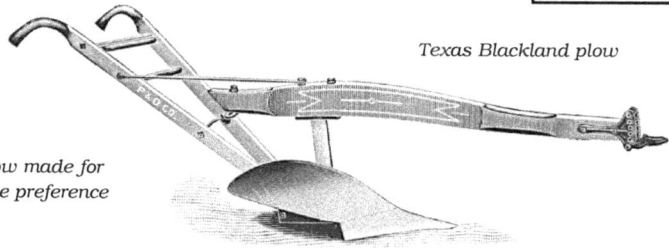

Texas Blackland plow

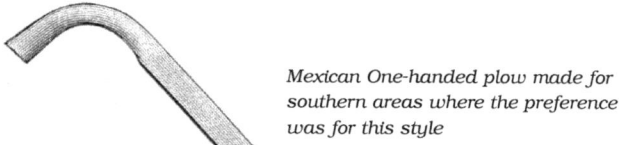

Mexican One-handed plow made for
southern areas where the preference
was for this style

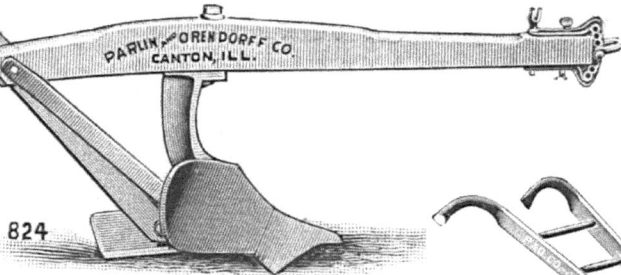

824

Dixie Medium blackland and sandy land plow

Dixie 7 inch plow

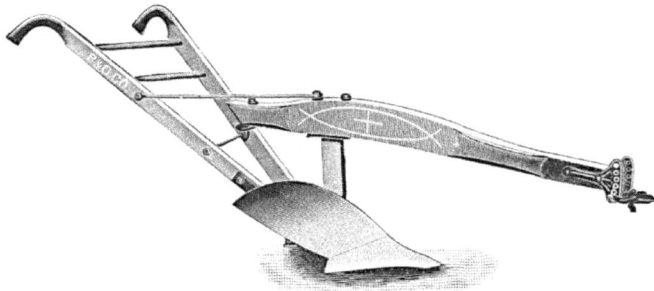

Dixie wood beam

Texas One Handle blackland

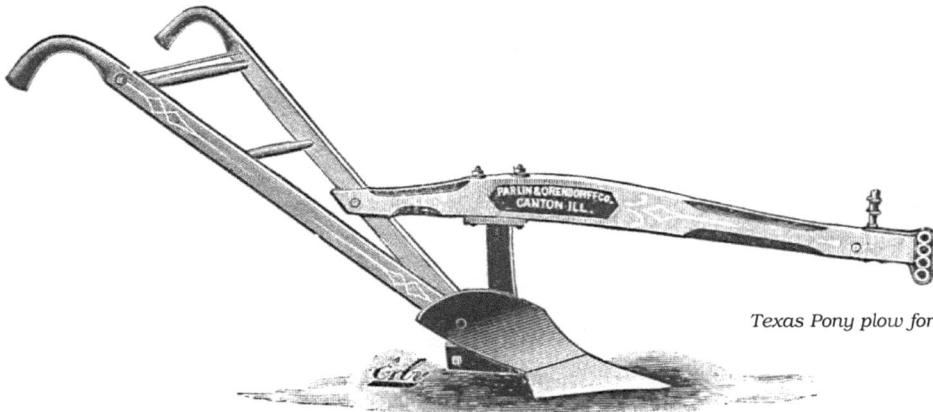

Texas Pony plow for blackland and sandy land.

Prairie Chief Breaker

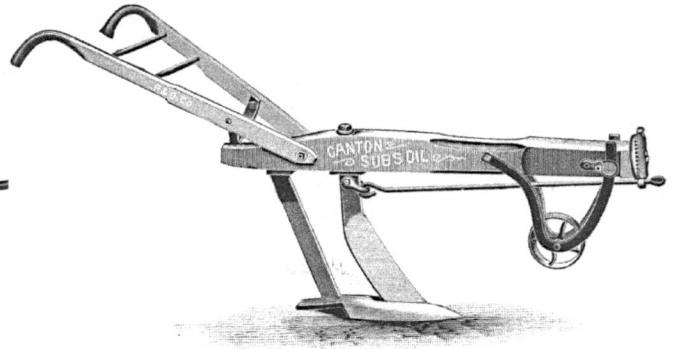

Subsoil plow with standing cutter

New Ground plow

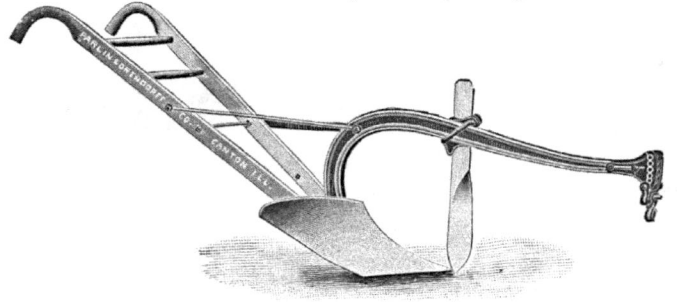

Root Ground plow

One Horse corn plow

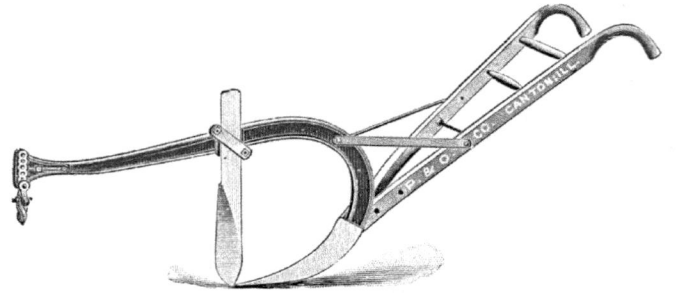

Landside Root Ground

Wood beam Lister with subsoiler

Hillside plow with double bottom,
beam and handles are mounted on
swivel. Very unusual

*Double Diamond Corn plow. A combination of right and left hand Corn plows coupled together. The right hand plow has two handles and the left hand plow but one, with an adjustable frame. Used for ridging up the ground where an ordinary cultivator fails.*

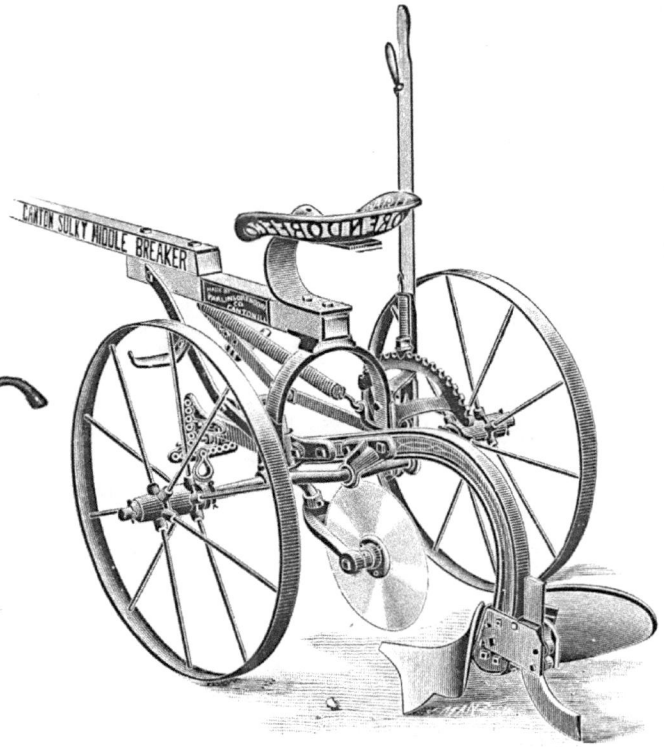

*The Sulky Middlebreaker with adjustable rudder.*

*Jack Eden with his six mules on gang plow in Corvallis, Montana, Photo by Helen Eden*

The Rock Island Plow Company was established in 1855
and built its implements in Rock Island, Illinois.

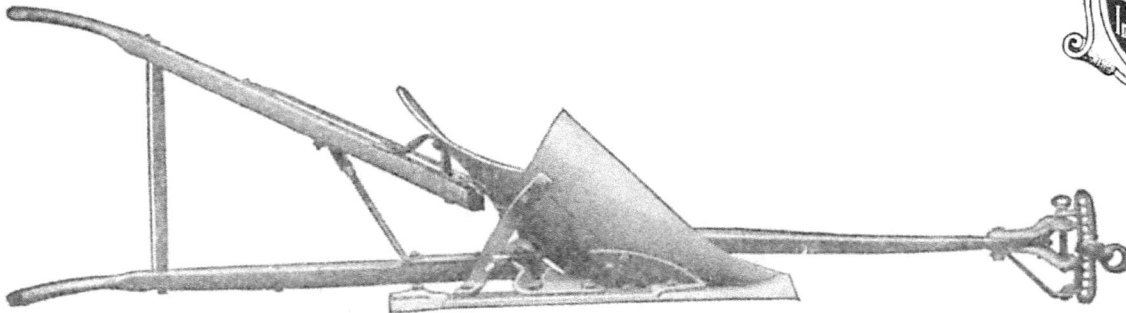

As this underside will attest, Rock Island put a premium on simplicity of design.

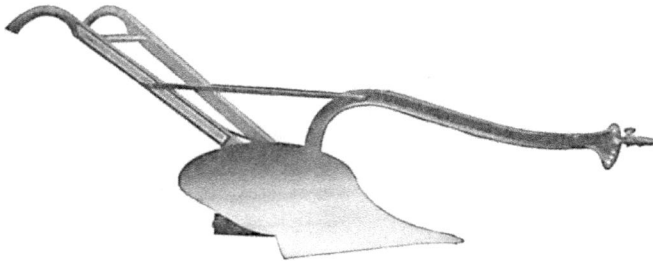

The Stubble series came with
12 to 18 inch bottoms

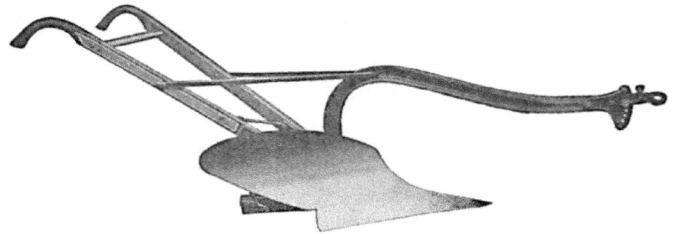

Turf and Stubble

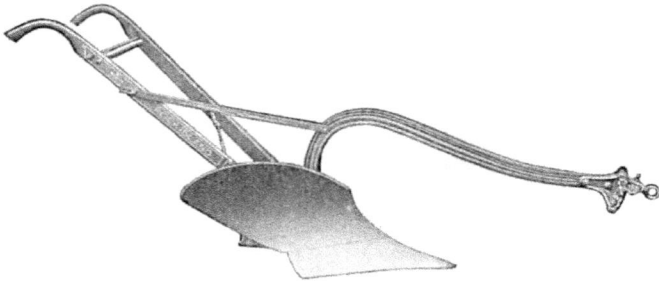

Scotch Clipper, general purpose

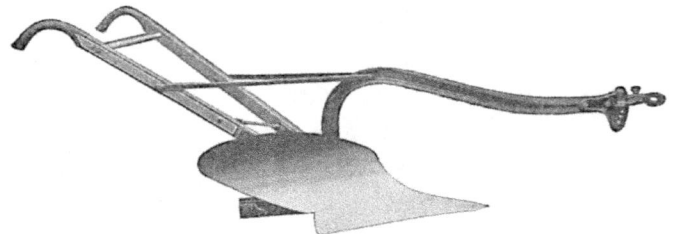

Challenge 10, heavy stiff work

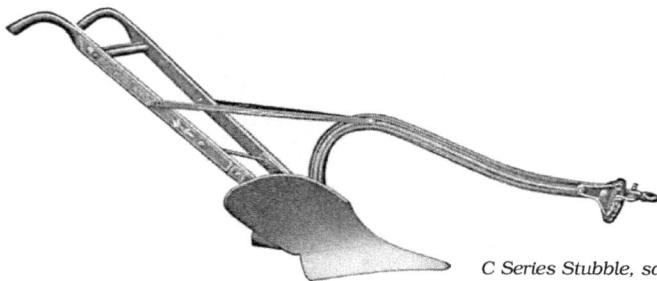

C Series Stubble, sandy land

# Rock Island

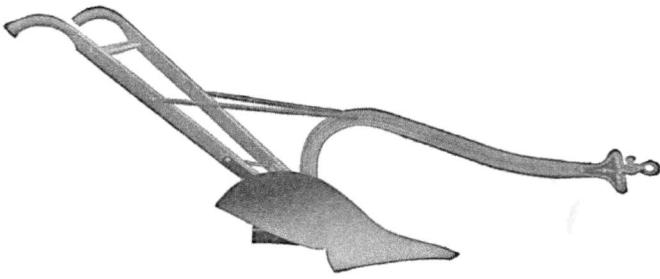

*Mixed Land series*

*TB series, black or waxy land*

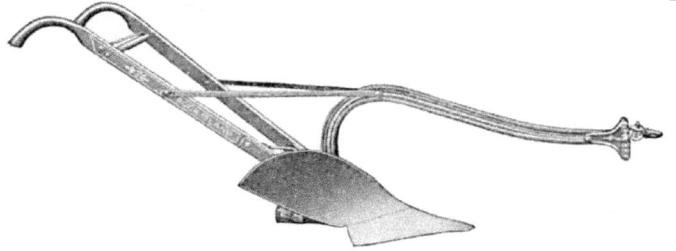

*T and W series, brush or hazel*

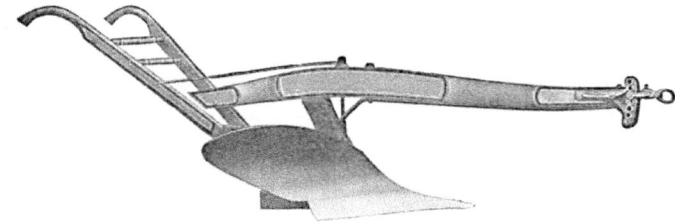

*Township plow*

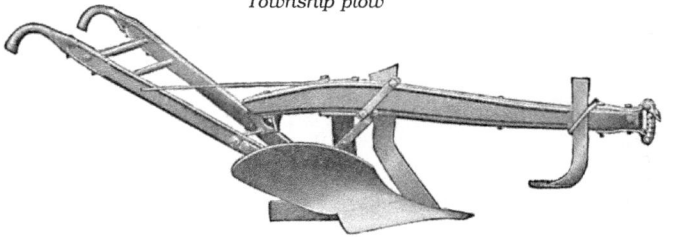

*Railroad plow*

*Vineyard plow*

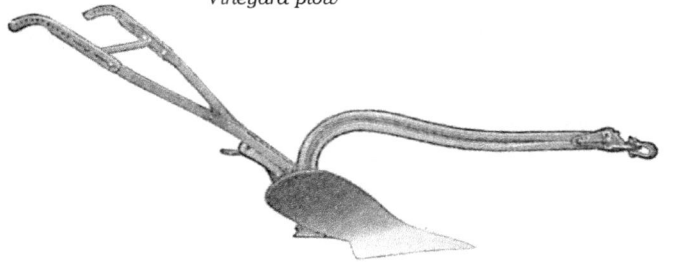

*Champion Prairie Breaker*

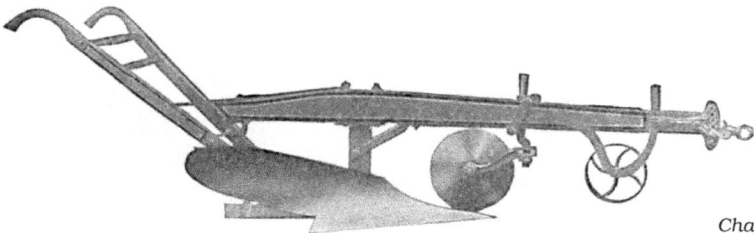

*Dakota Rod Breaker*

*Blackland Middlebreaker*

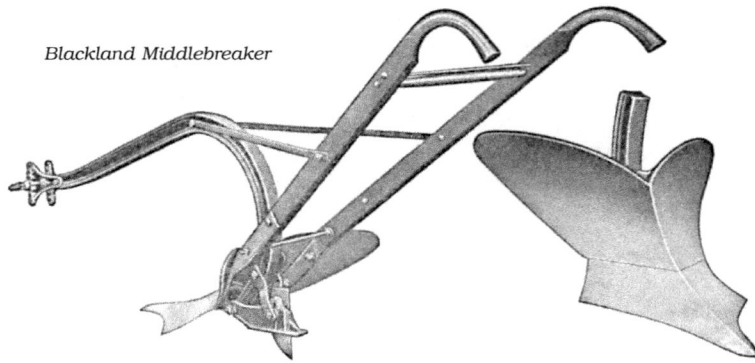

*Sandy Land Middlebreaker*

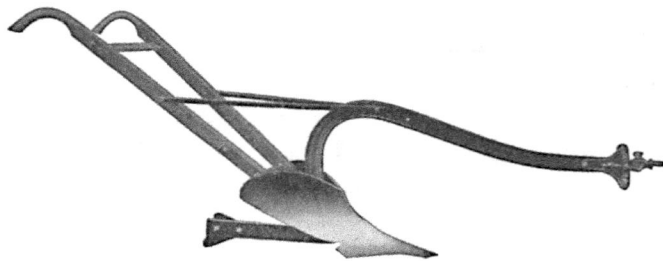

*Combination Jointer Coulter*

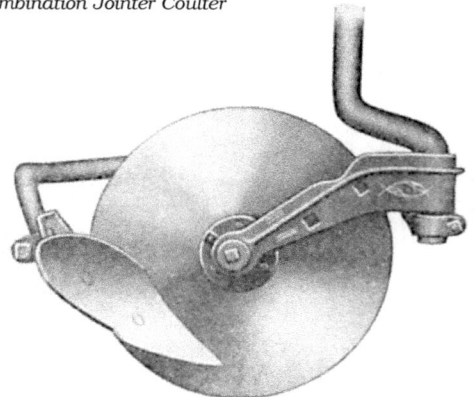

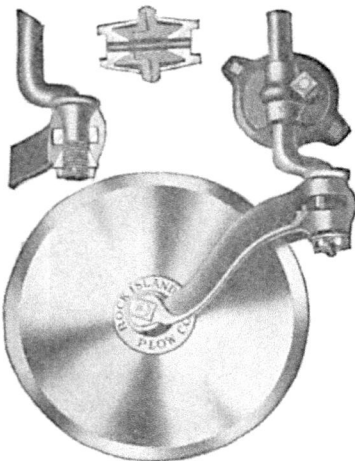

*Universal coulter*

# Rock Island

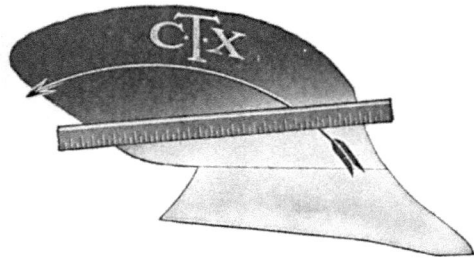

Rock Island came up with their own "CTX brand" variation on moldboard design claiming that its augur-like twist turns all the furrow over and allows no air pockets between the top and the subsoil.

Rock Island gang with CTX bottoms completely covering tall corn stubble.

In this picture the grain stubble ground is so thoroughly pulverized by the CTX bottoms that no harrowing is necessary.

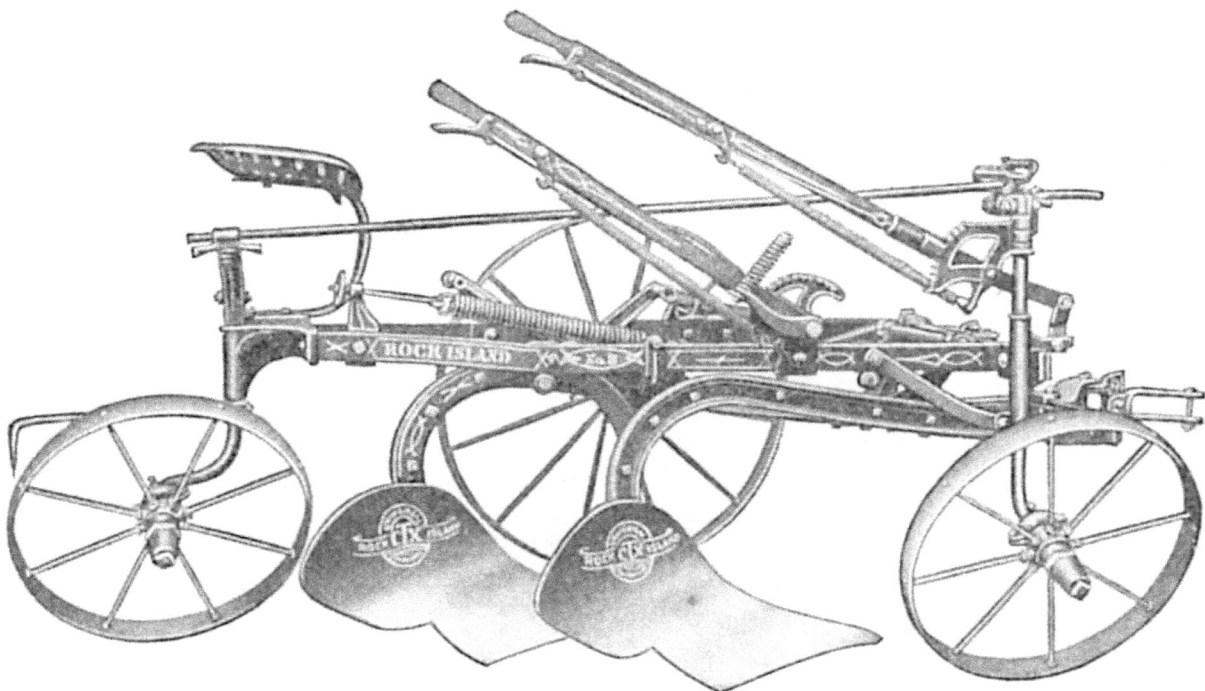

Rock Island No. 8 Gang

*Plowing with the Rock Island No. 8*

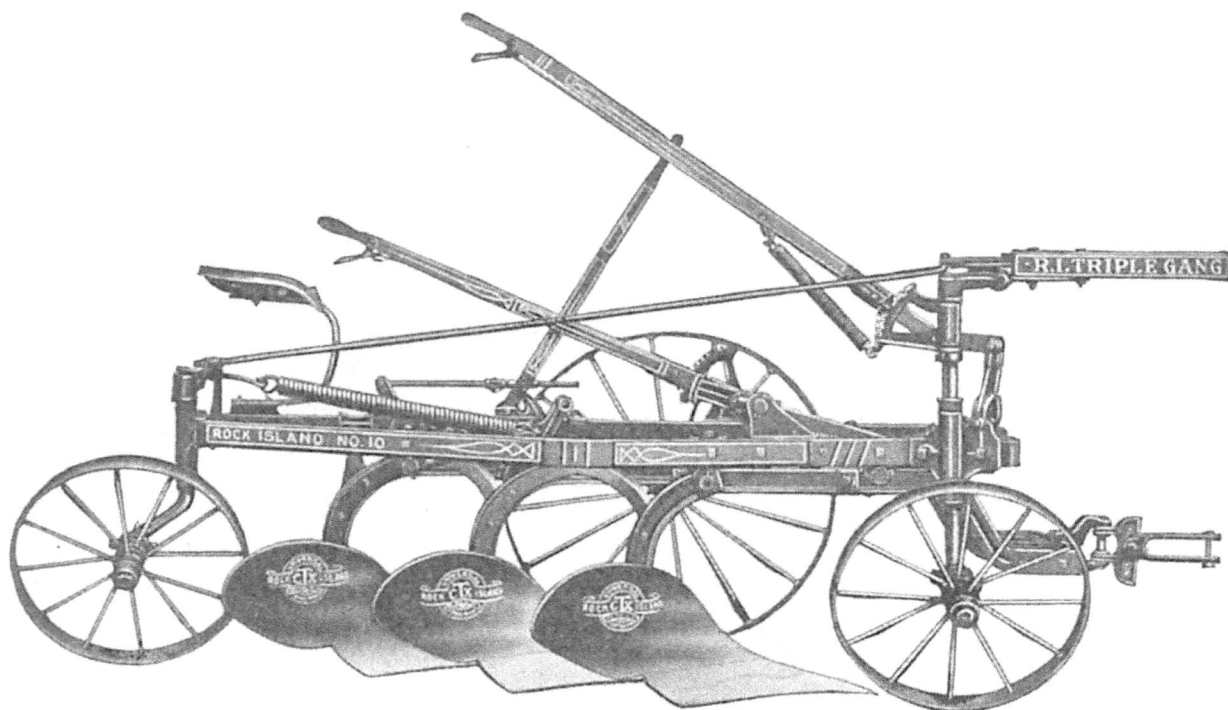

*Rock Island No. 10 Triple gang, both sides*

# Rock Island

*No. 6  Frameless Poleless Gang, two views*

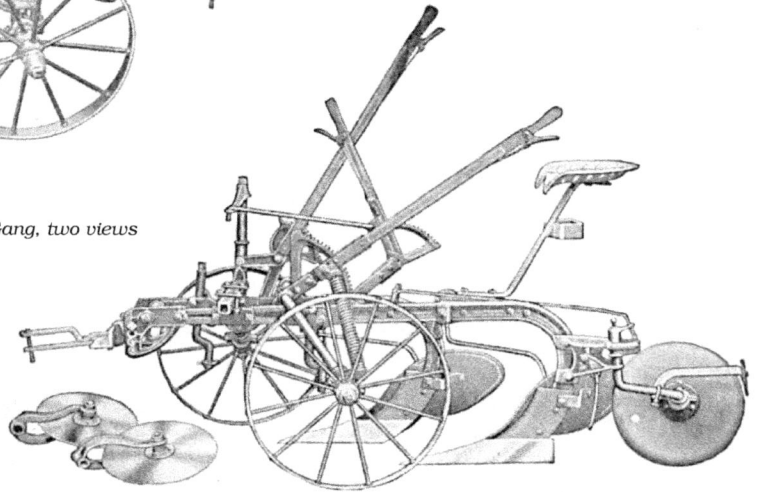

*No. 4. foot lift - plow raised*

*No. 4 foot lift - plow down*

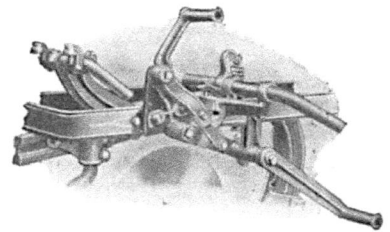

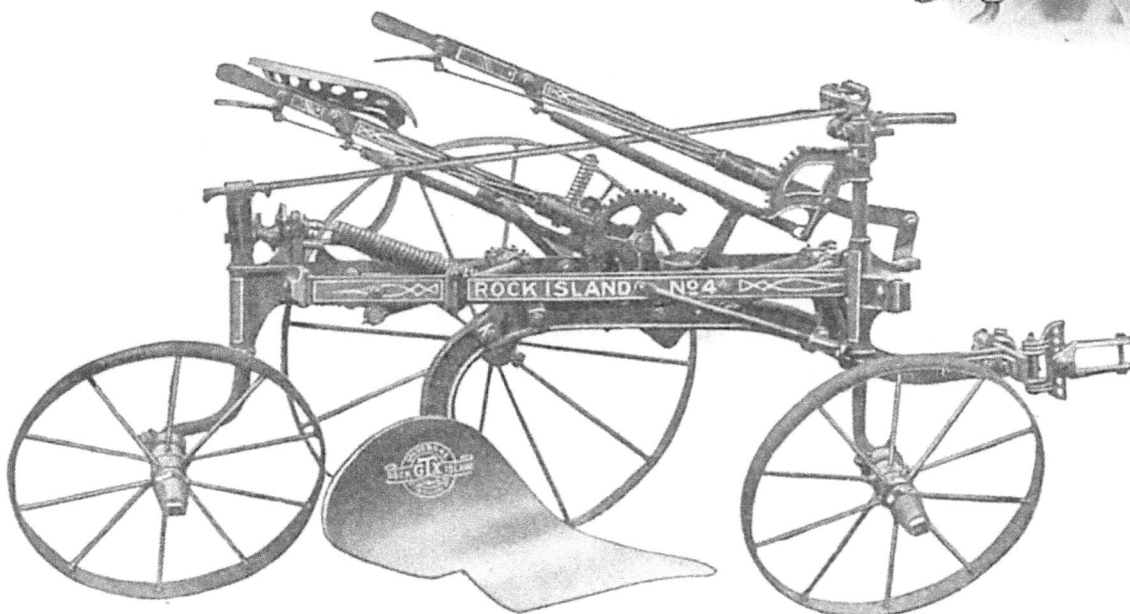

*Rock Island No. 4 Sulky*

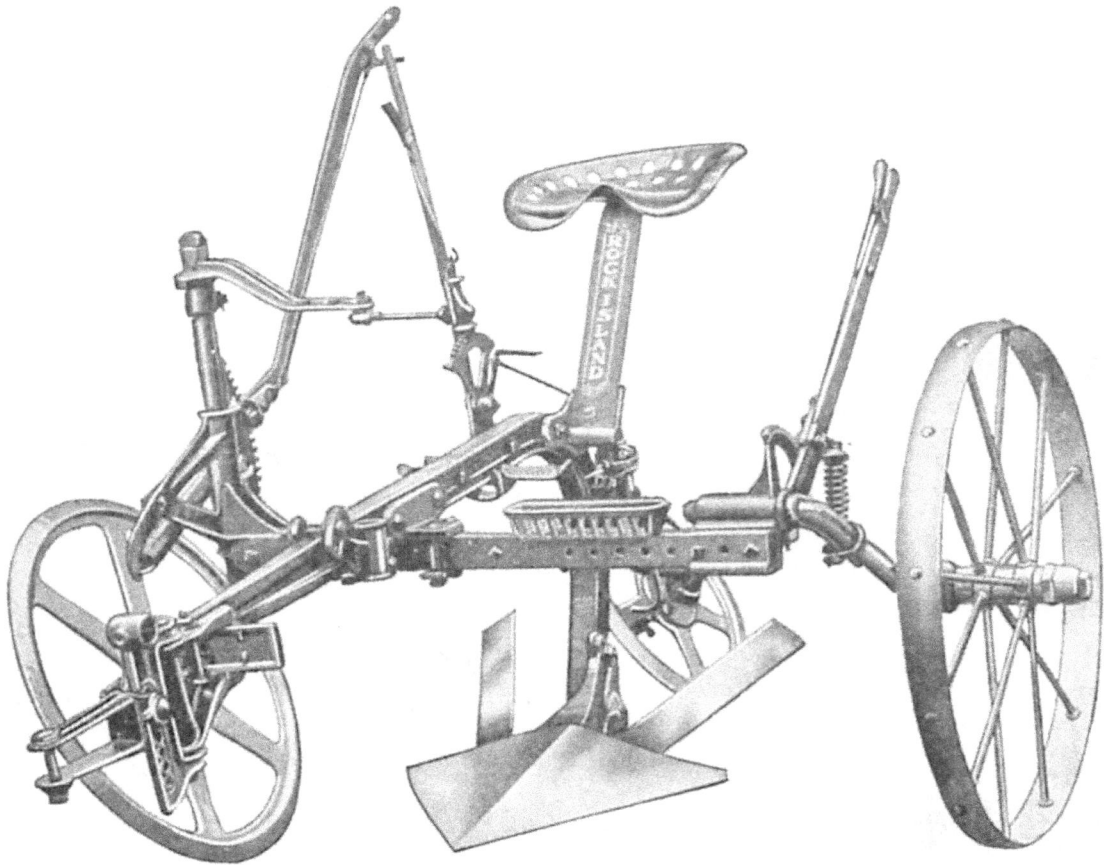

*Rock Island TBX Sulky with sweep attachment*

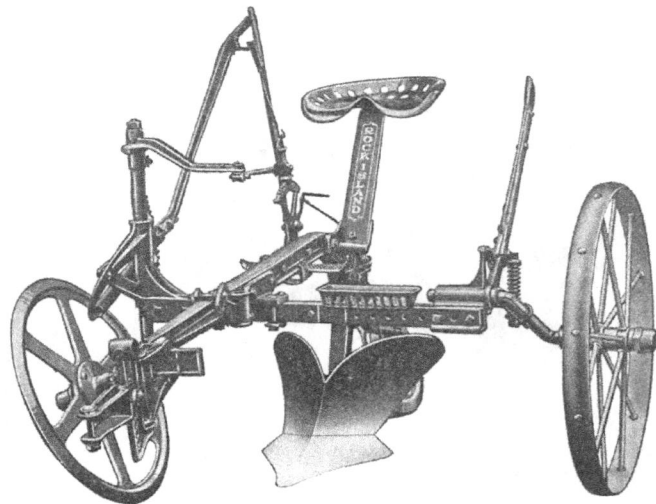

*TBX Sulky with Middle Burster Bottom*

*Rock Island TBX Sulky built for Southern lands with irregular shaped fields, stumps, rocks, and black and stiff lands.*

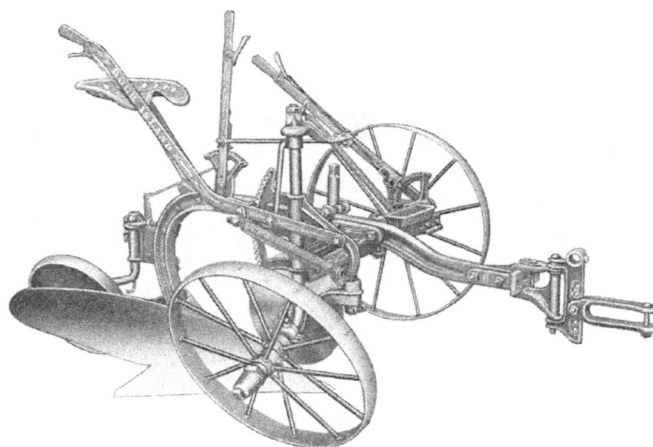

*No. 22 Northwest Special Sulky combination general purpose/ breaker.*

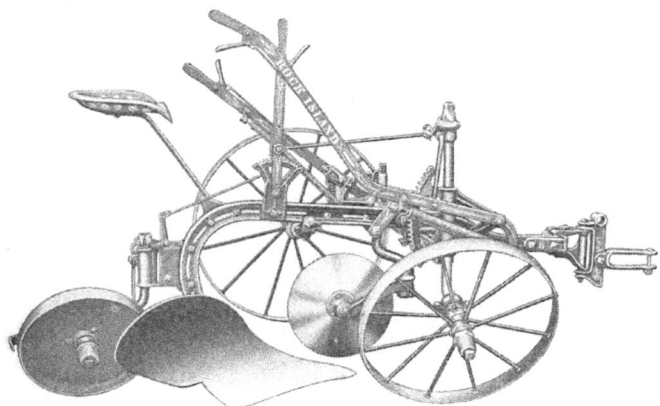

*No. 22 Northwest Special equippped with CTX bottom.*

*Rock Island No. 2*

No. 1 Litewate. Built of the lightest possible
construction for lighter soils and smaller teams

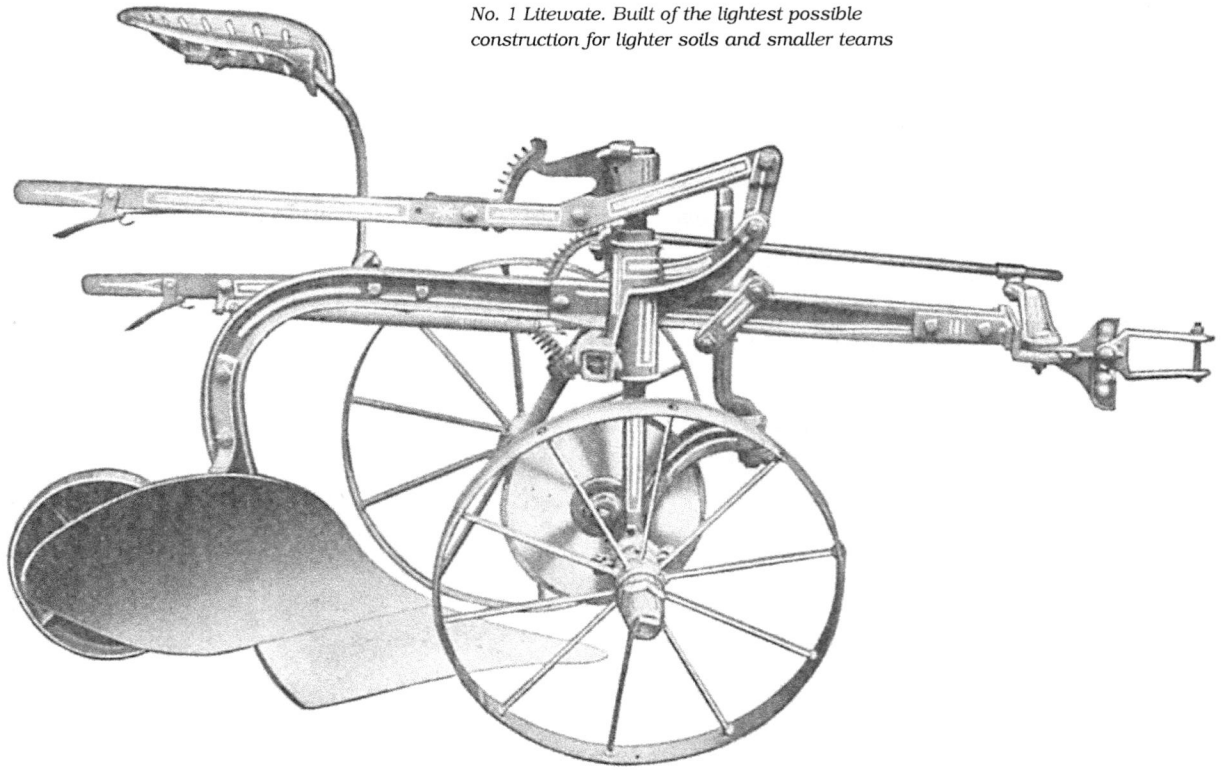

No. 1 Litewate landside view. Included a cushion spring to
make plow run level even across rough ground.

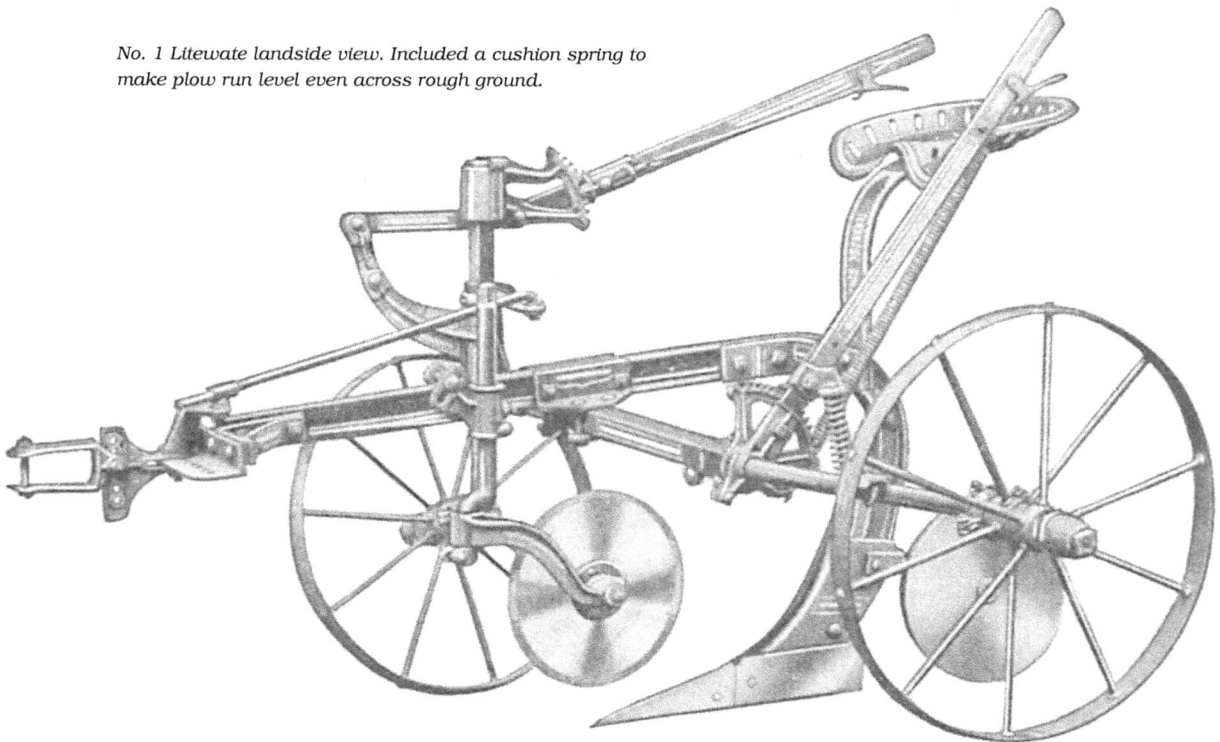

*Rock Island No. 3 Sulky (right or left hand),
The Square Corner Plow*

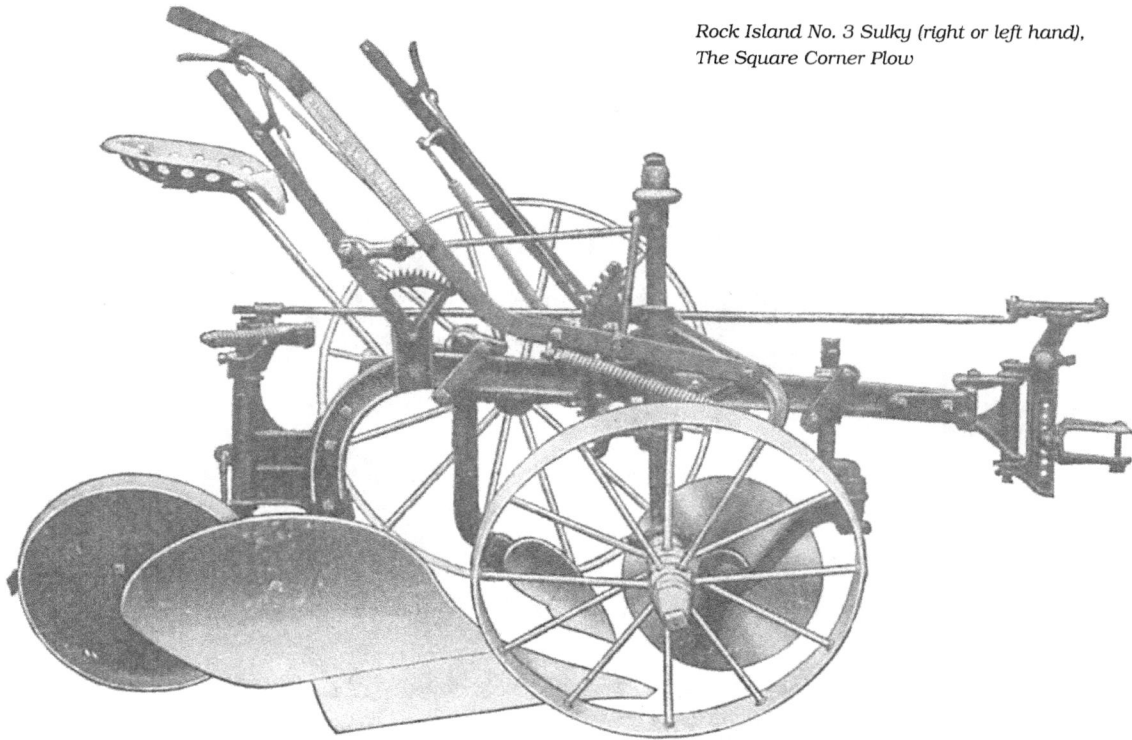

*Rock Island No. 5 Two Way Sulky*

*Rear view of the No. 5 Two Way Sulky*

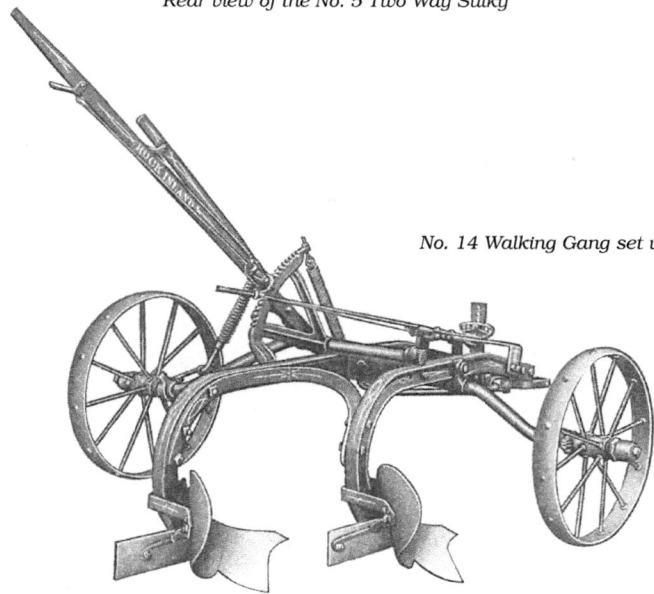

*No. 14 Walking Gang set up as Two Bottom.*

Rock Island

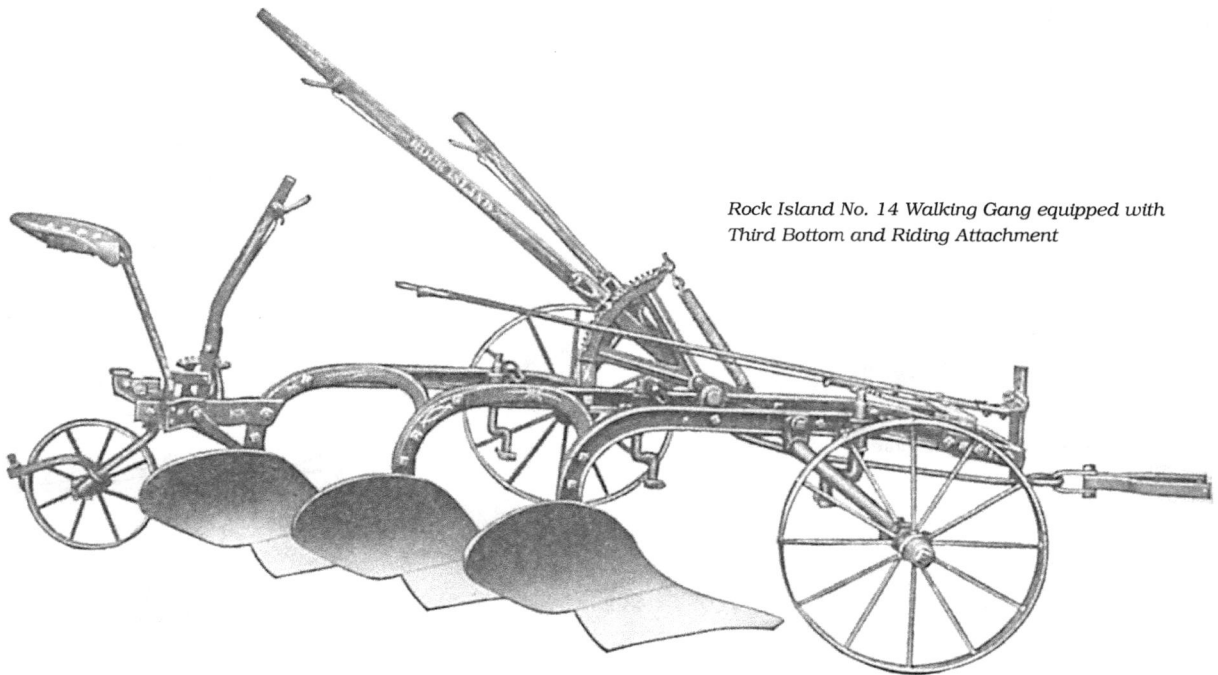

*Rock Island No. 14 Walking Gang equipped with
Third Bottom and Riding Attachment*

*No. 18 Vineyard Plow*

# The VULCAN PLOW COMPANY

The Vulcan Plow Co. was established in 1874 in Evansville, Indiana. The images we share with you came from their 1928 catalog.

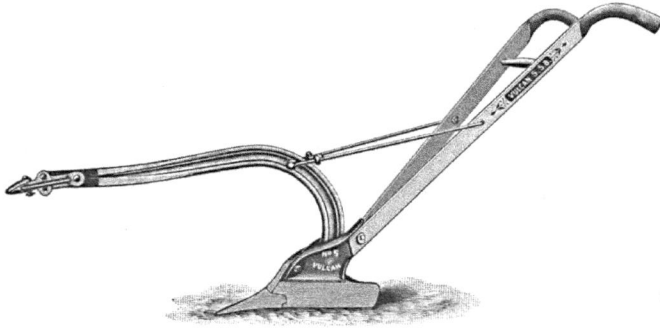

No. 5 Chilled plow with 6 inch share

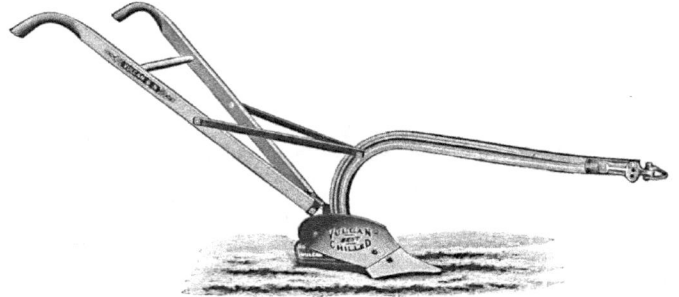

No. 6N Vulcan with 6 $^{1/2}$ inch share

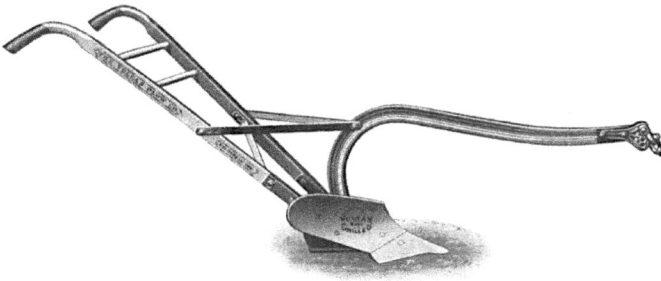

No. 7N Vulcan Chilled plow with 7 inch share

No. 11N  Vulcan rear view

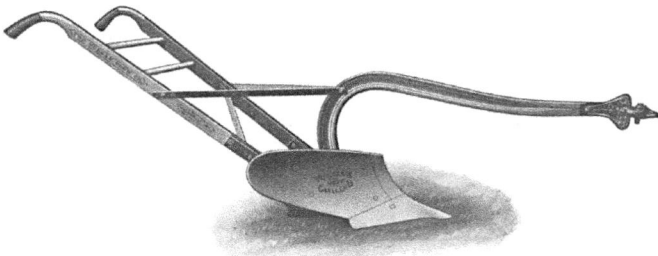

No. 12N general purpose two horse plow

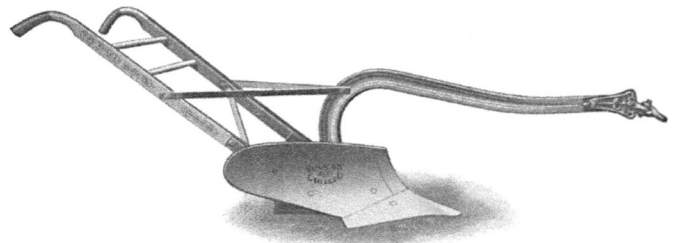

No. 14N with 14 inch share

# Vulcan

Makes &
Models

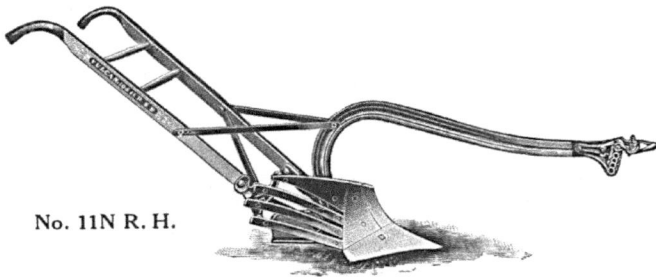

No. 11N R. H.

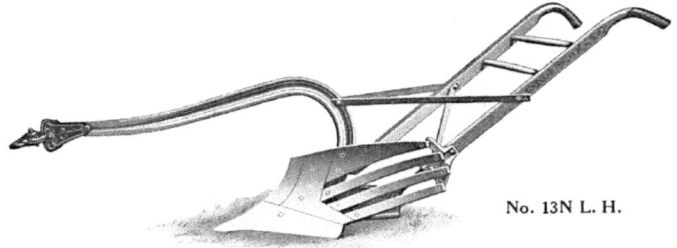

No. 13N L. H.

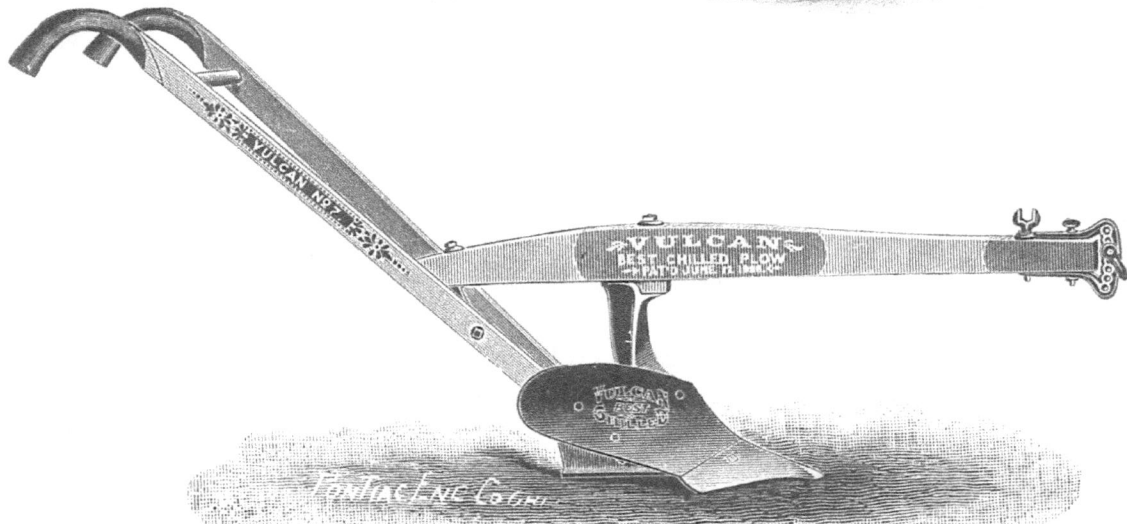

*One Horse Wood beam  came with shares from 7 inch through 16 inch.*

*Vulcan featured bottoms which had interlocking saddle, landside, and point. The marked points in these illustrations show where the overlaps and interlocks are.*

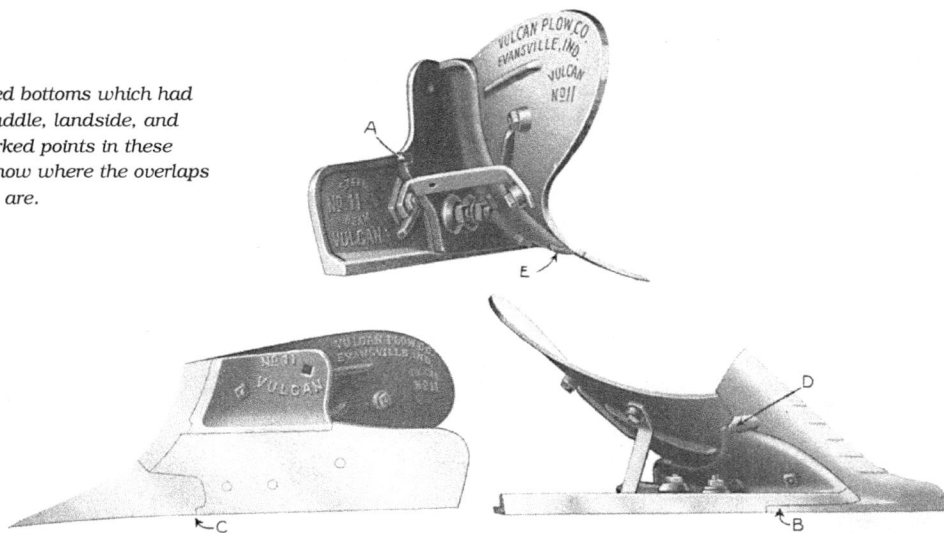

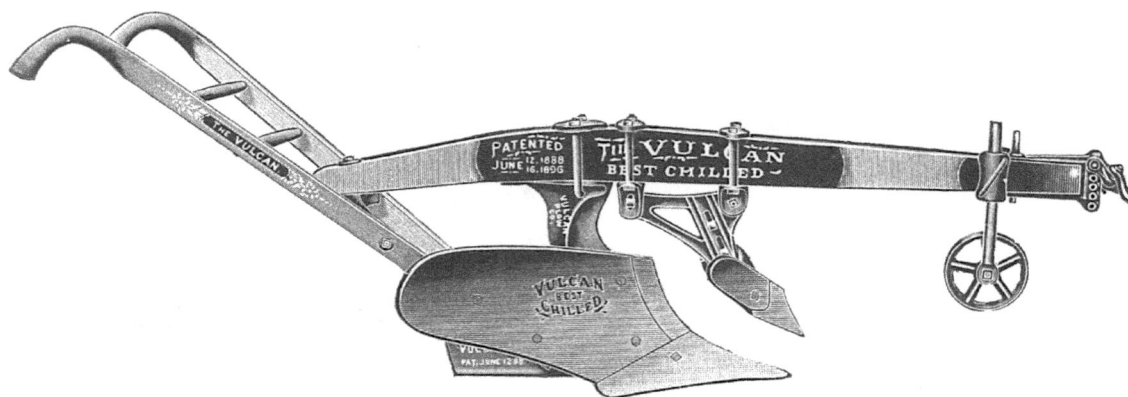

*No. 14 with special jointer and gauge wheel*

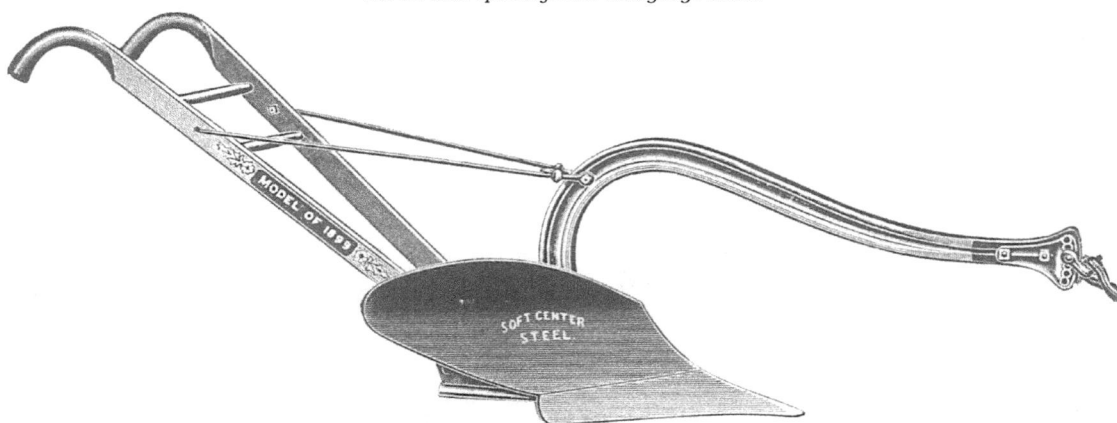

*Vulcan Scotch Clipper*

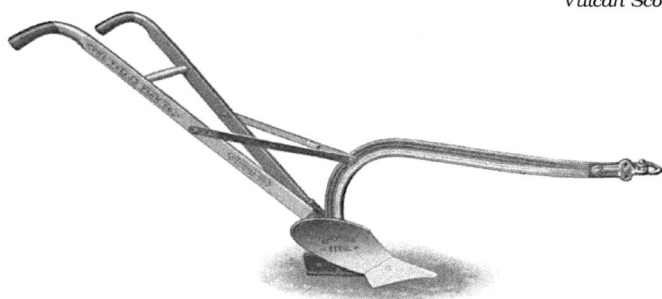

*No. 6 Rose Clipper*

*No. 8N Rose Clipper S.B.*

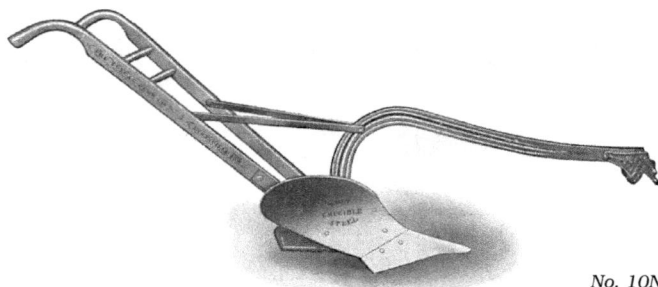

*No. 10N Rose Clipper*

# Vulcan

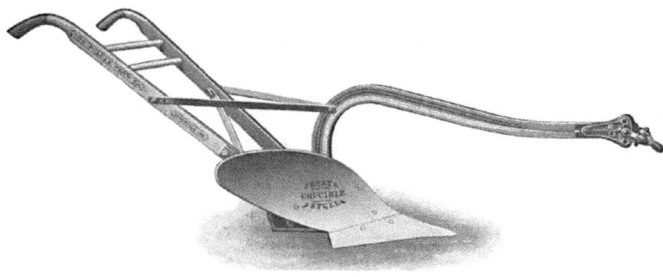

No. 12N Rose Clipper

Rear view of 12N

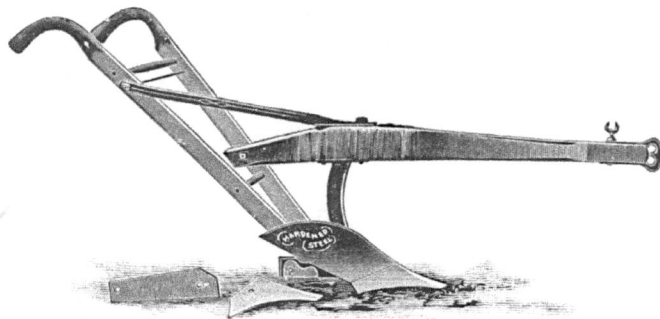

Pony Series with steel standard and cap, sloping
landside and adjustable heel.

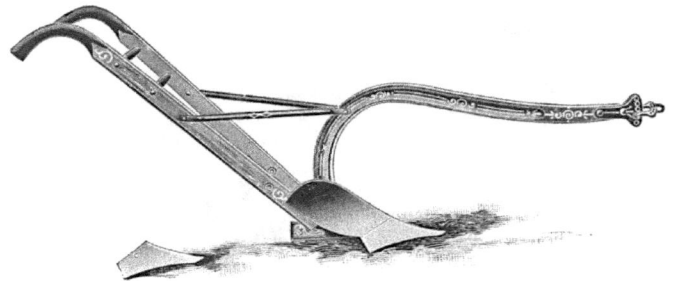

Steel beam Pony Series

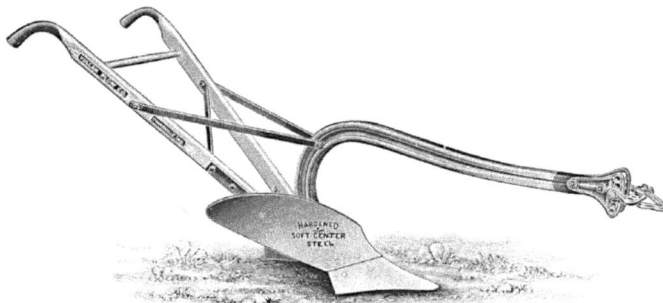

Vulcan Blackland plow

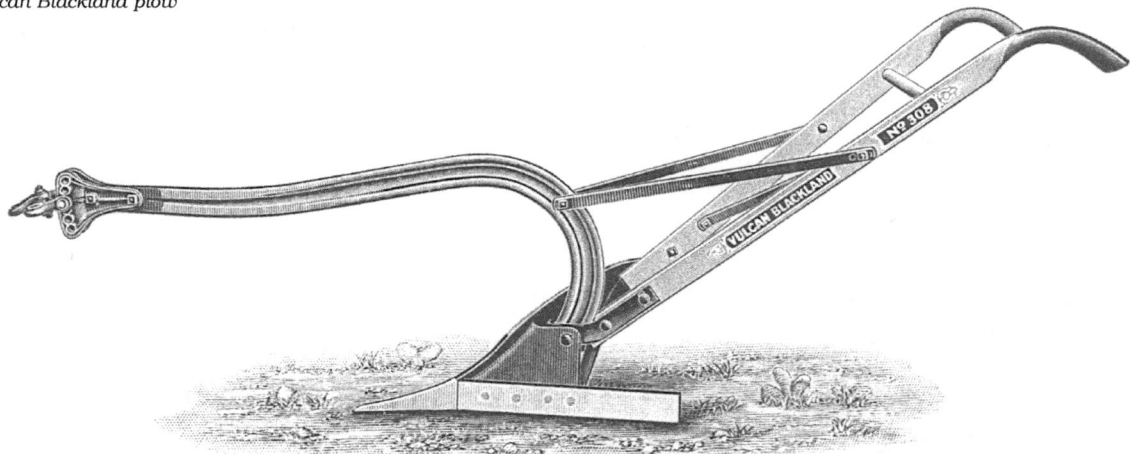

Vulcan Blackland No. 308, landside view

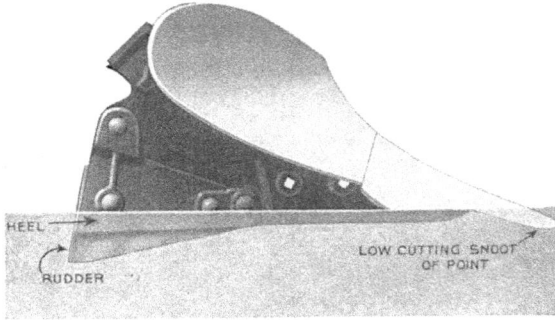

HEEL

RUDDER

LOW CUTTING SNOOT
OF POINT

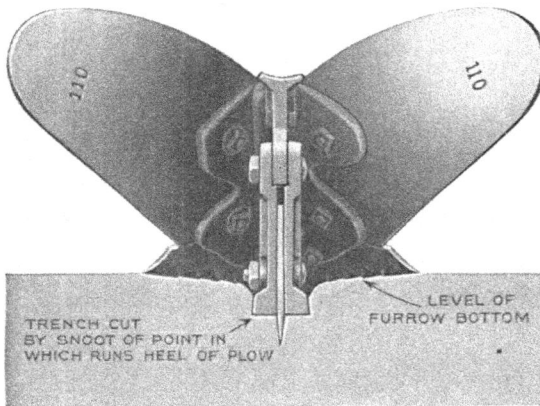

Mixed Land Middlebreaker

The shoe, on on all sizes of Vulcan Middlebreakers,
runs in a trench cut in furrow bottom by Duck Bill Snoot
of the Point. Notice how the low hanging snoot of Point
is on a level with shoe and is lower than balance of
furrow bottom

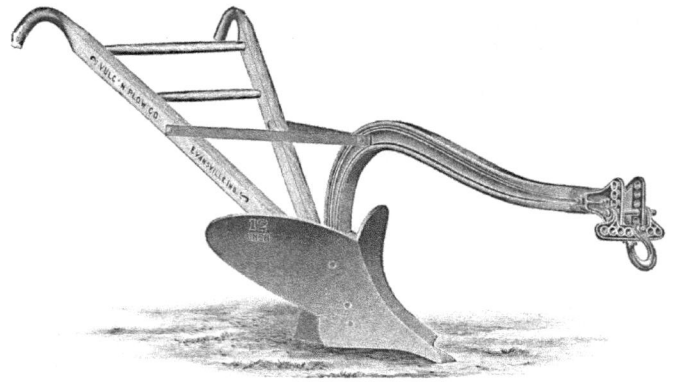

TRENCH CUT
BY SNOOT OF POINT IN
WHICH RUNS HEEL OF PLOW

LEVEL OF
FURROW BOTTOM

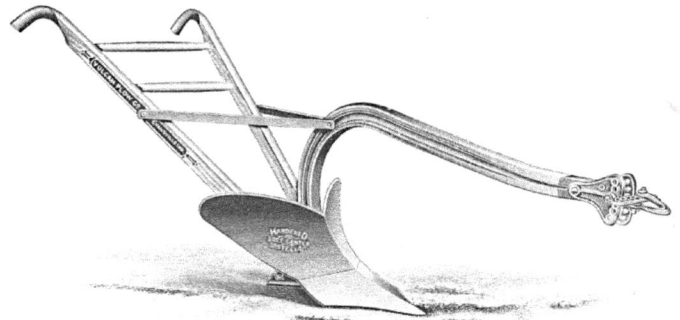

Blackland Middlebreaker

Rear view of Blackland Middlebreaker

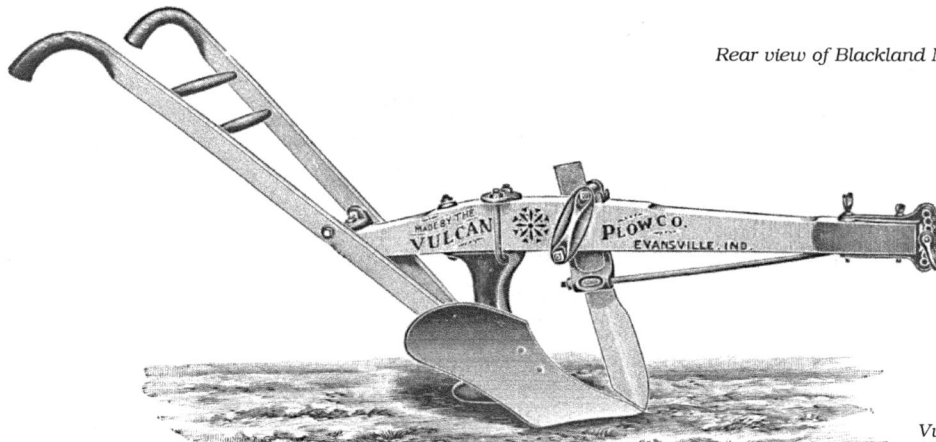

Vulcan No. 9 New Ground plow

No. 9 New Ground outfitted with 6 inch Single Shovel

No. 10 New Ground was claimed to be the first plow to ever use a rolling landside.

Landside view of the No. 10 New Ground showing the rolling landside

No. 10 Vulcan Hillside (reversible)

Landside view of No. 10 showing latch construction

No. 6 Steel beam Hillside

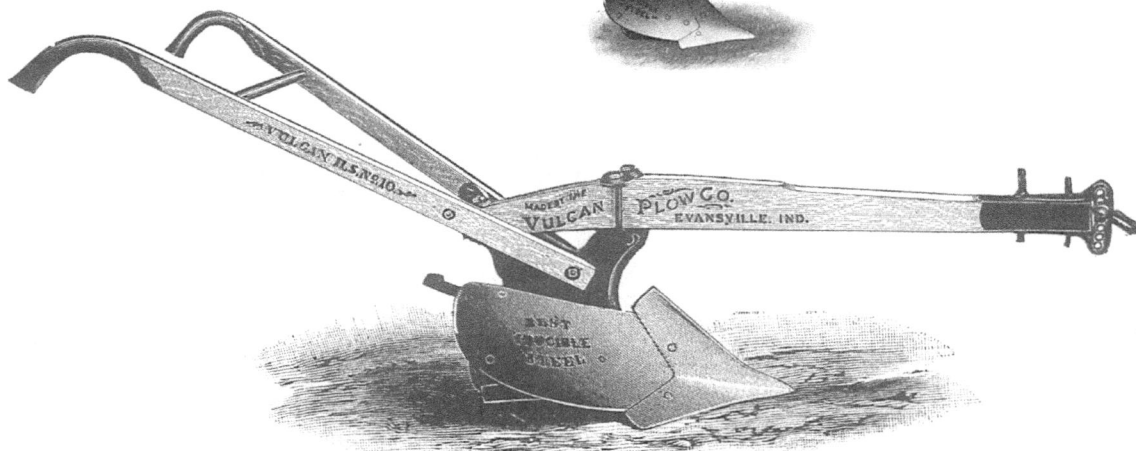

No. 10 Wood Beam Hillside

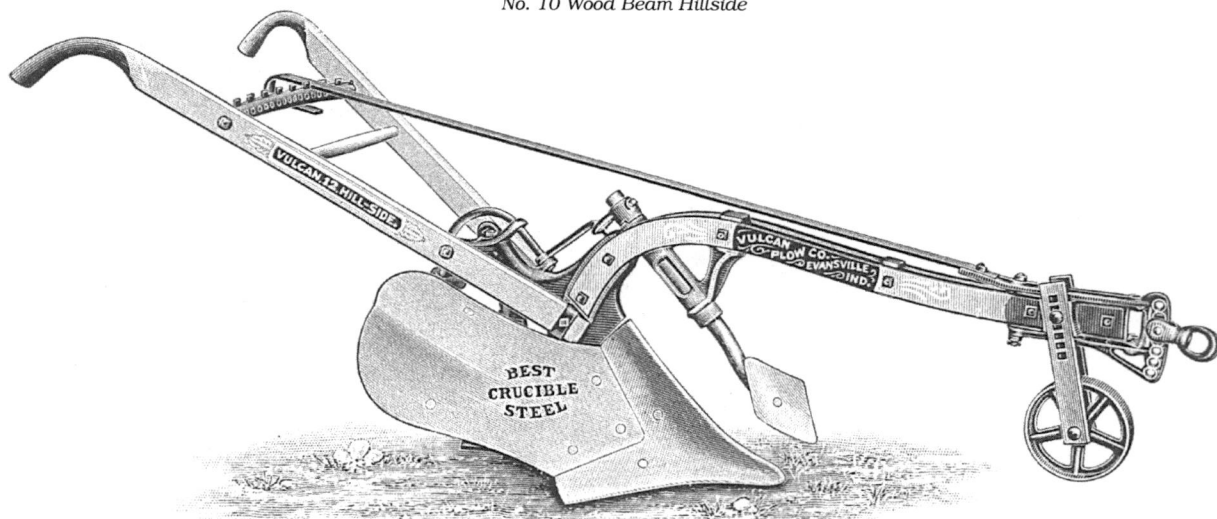

No. 12 Hillside with rollover jointer and adjustable hitch. This plow came with 12 to 15 inch share cut

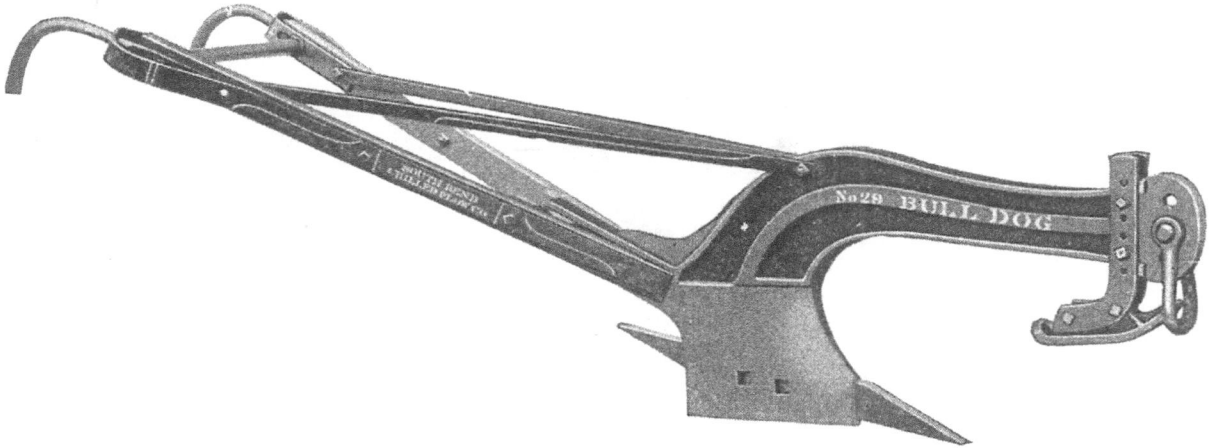

*No. 29 Bulldog Rooter plow weighed 330 lbs.*

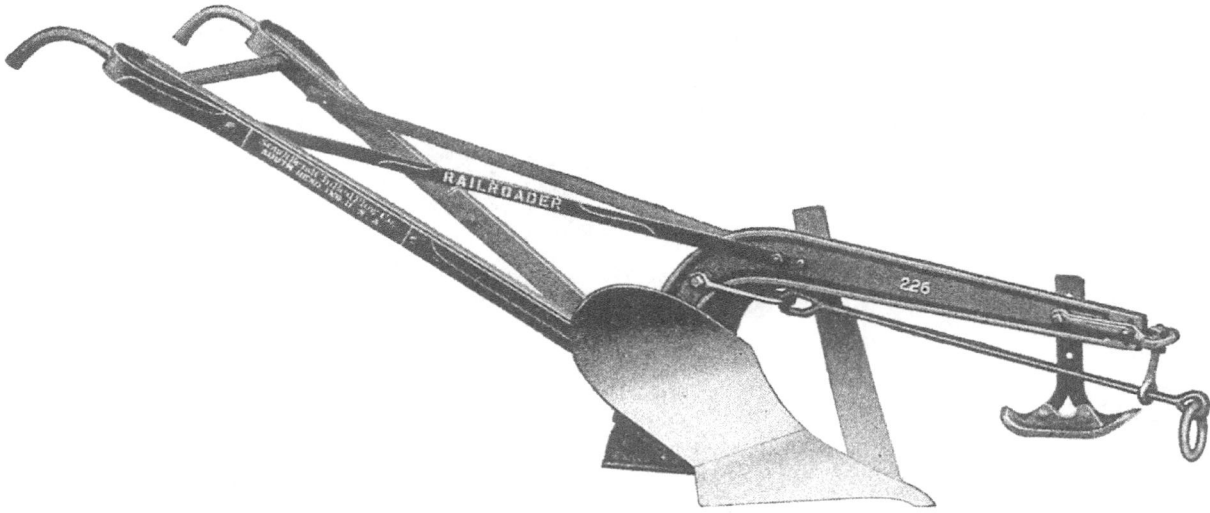

*No. 226 Railroader plow*

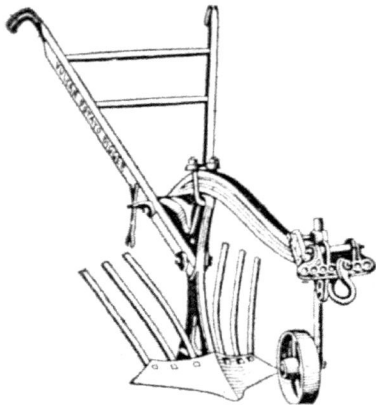

*Vulcan Subsoiler with Potato Digger Attachment*

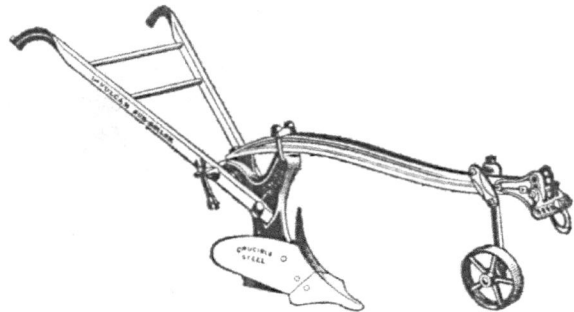

*Subsoiler with Middlebreaker attachment*

*Vulcan No. 12 Roadster*

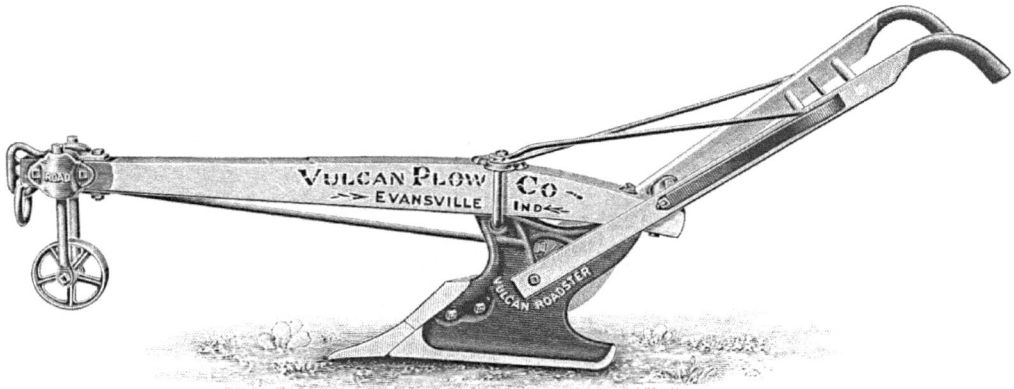

*Landside view of Roadster*

*A slightly older version of the Vulcan Roadster*

# Vulcan

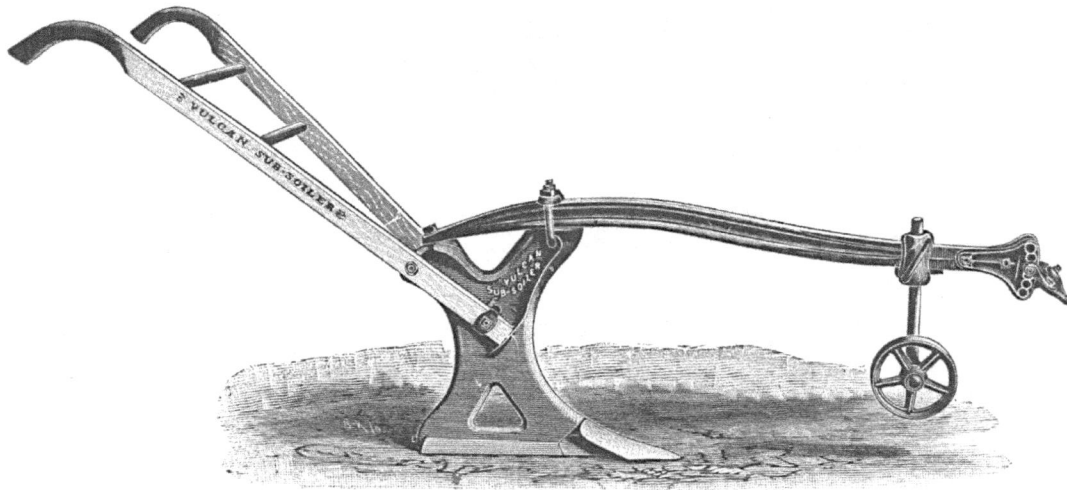

*Convertible Vulcan subsoiler (page 309 shows Potato Digger and Middlebreaker attachments for this plow)*

*No. 50 Vulcan Sulky with 9 through 14 inch bottoms, 15 inch rolling coulter, and cushioned land axle spring.*

*Victoria frameless sulky*

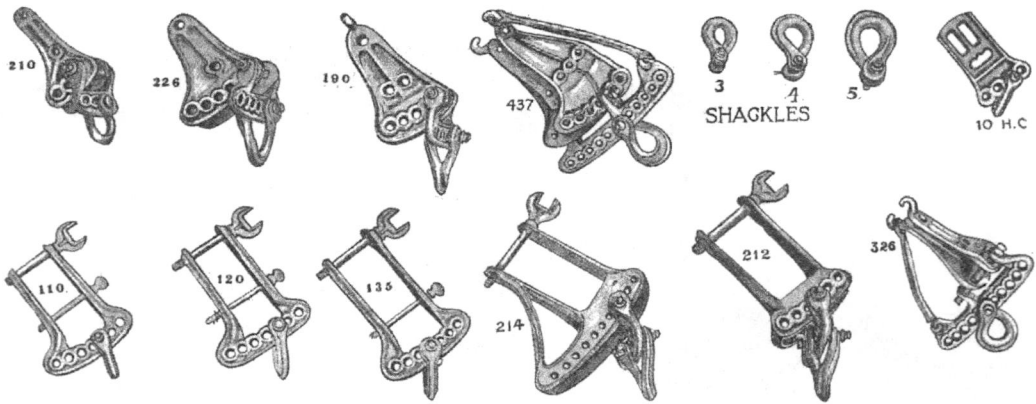

*Vulcan Plow Clevises*

# Le Roy Plow Company

The Le Roy Plow Company of Le Roy, New York advertised
in 1915 that they made 25,000 plows each year.
On the back color cover page we have run several Le Roy
plows, including their two way plow.

One-horse plow with shifting handles and beam.
Designed for general and orchard use. Highly unusual
design.

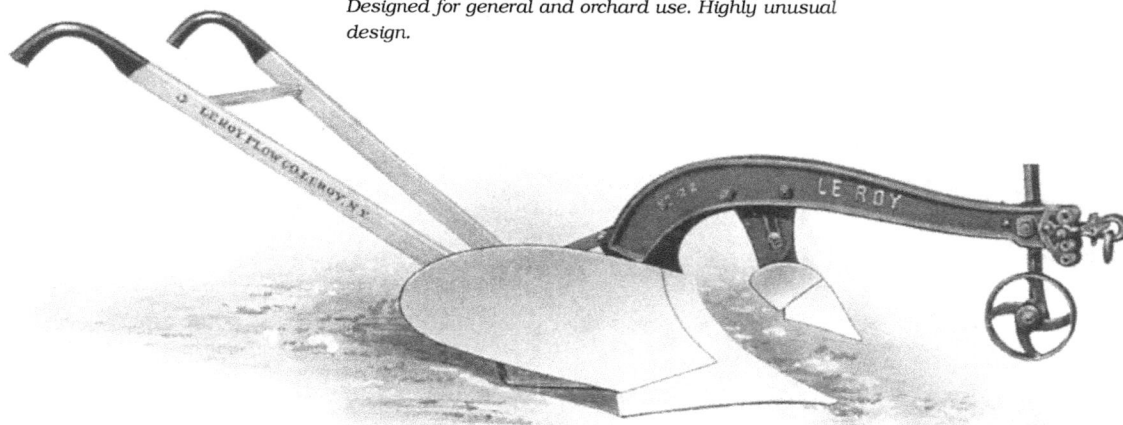

Le Roy Iron-Beam Chilled plow. For extra hard clay or cobble
stone land. A good clearing plow and excellent for turning sod
and sticky soils.

Le Roy Shovel Plow No. 10

# Empire Plows

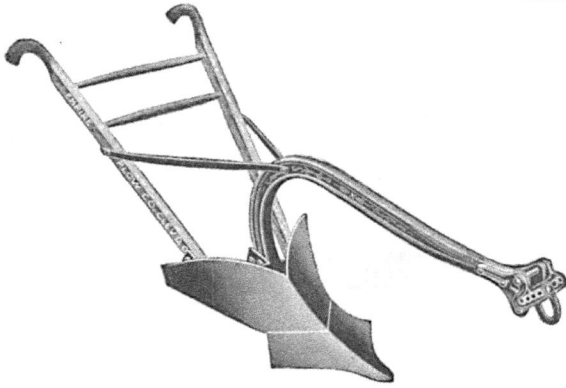

Empire Northern Double Moldboard
(otherwise known as a middlebreaker).

Empire wood beam garden plow

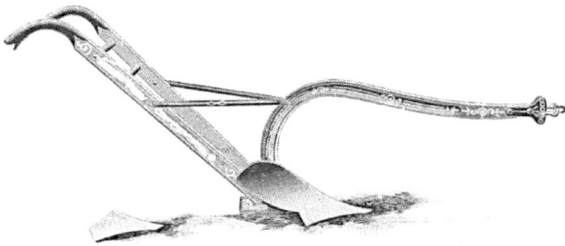

Empire steel beam garden plow

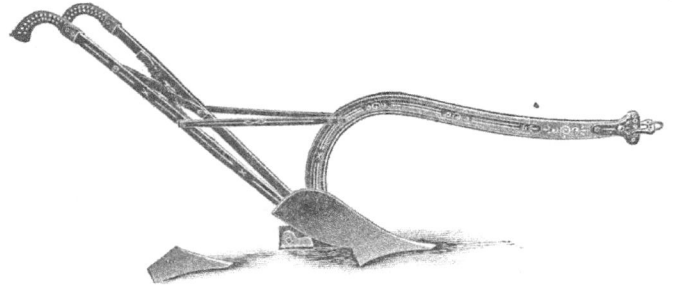

Empire steel beam with cast steel handles, sold in 1916 for $7.50

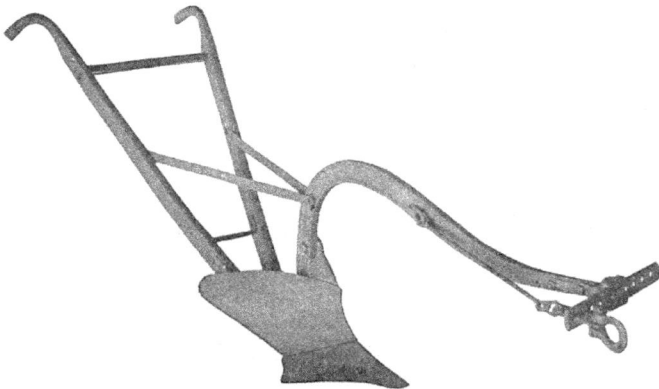

Empire Orchard plow

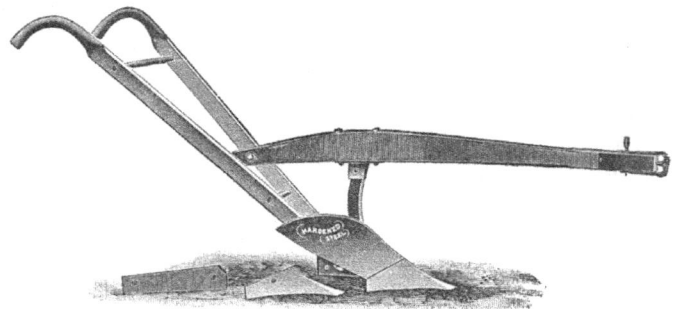

Empire A. O. Garden plow

Empire C.O. garden plow

# Sanders

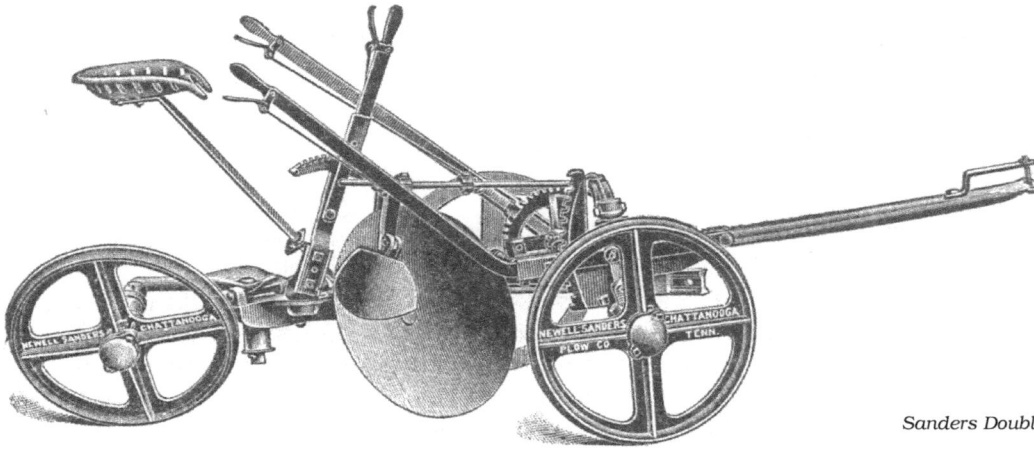

*Sanders Disc Sulky*

*Sanders Double Disc plow*

# South Bend Chilled Plow Company

*Railroader plow*

*Bull Dog Rooter*

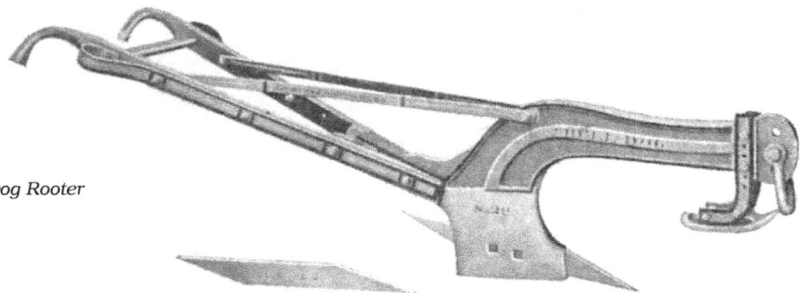

# Triumph

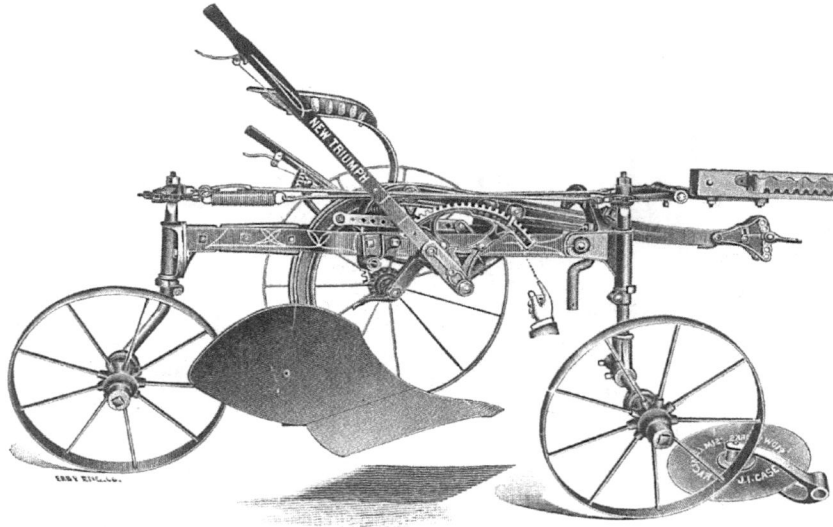

*New Triumph Sulky was touted as the one movement plow. All operations covered by one lever.*

*The one-lever movement New Triumph gang plow*

# Holbrook

*Pre 1900's hillside-rollover with hanging cutter.*

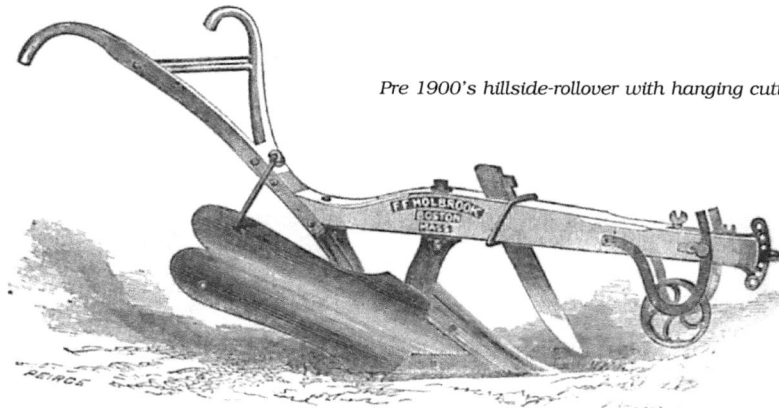

Makes & Models

*Square Deal gang plow*

*Square Deal single walking plow*

# Norwegian Plow Company

*of Dubuque, Iowa.*

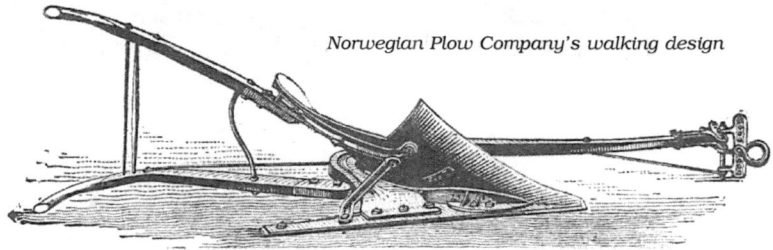

*Norwegian Plow Company's walking design*

*The Dutch Yankee frame sulky*

# Pekin Plow Company

*of Pekin, Illinois*

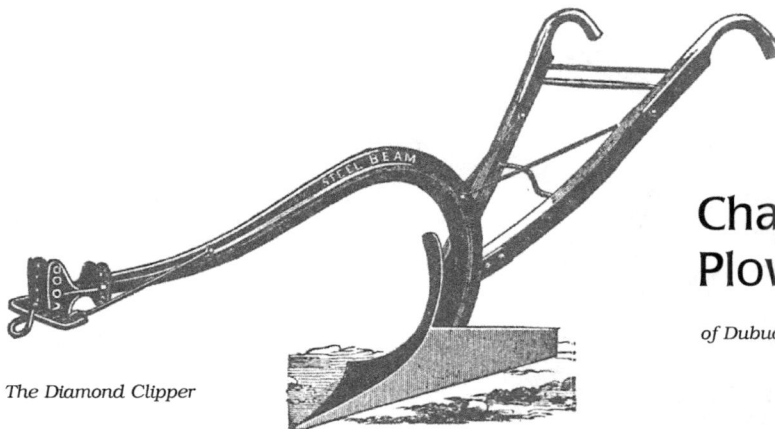

# Chamberlain Plows

*of Dubuque, Iowa*

*The Diamond Clipper*

## Smith Wagon & Implements

*The combined Headlight gang*

## J. Thompson & Sons Plow Company

*Ole Olson Sulky plow*

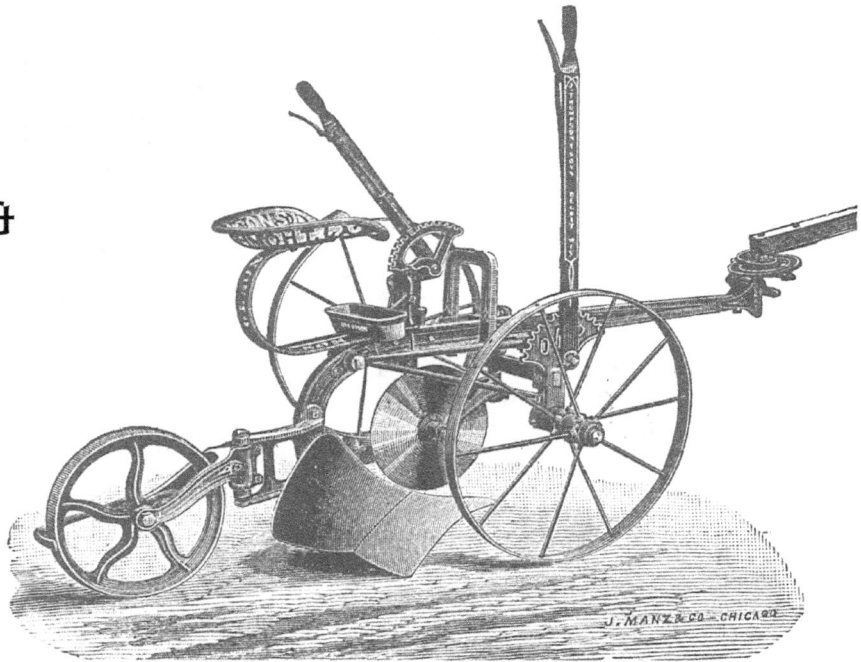

## Walton Plow Co.

*of Bloomington, Illinois*

*Wheel Walking plow*

# Roland Plow Works

## Bissel Plows

South Bend, Indiana

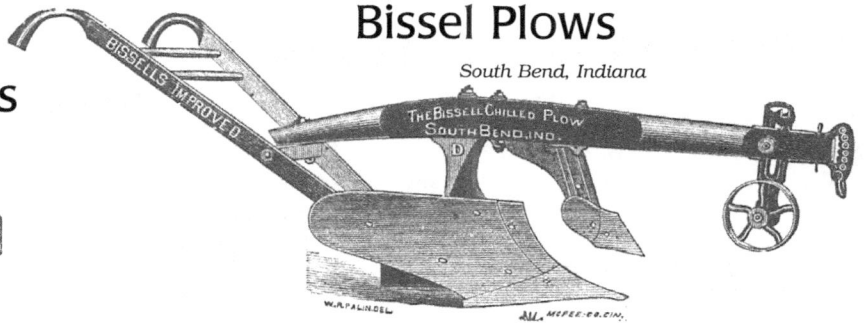

*Perfect Chilled plow*

# Gowanda Agricultural Works

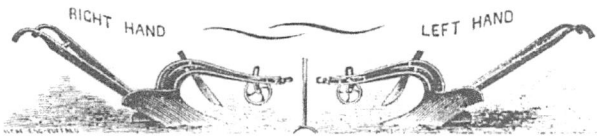

*Gowanda, N.Y.*

*Three Wheeled modified frame sulky*

## Hapgood Plow

*Three wheeled frameless sulky*

*Two wheeled frameless sulky*

# Cracker Jack Plow Company

# Weir Plow Company

*Monmouth, Illinois*

*Weir Riding Gang*

*Weir Rod Breaker*

*Weir sulky*

*Prairie Breaking bottoms used on Weir plows*

# The Bucher & Gibbs Plow Co.

*Hillside, Reversible, with 12 inch share*

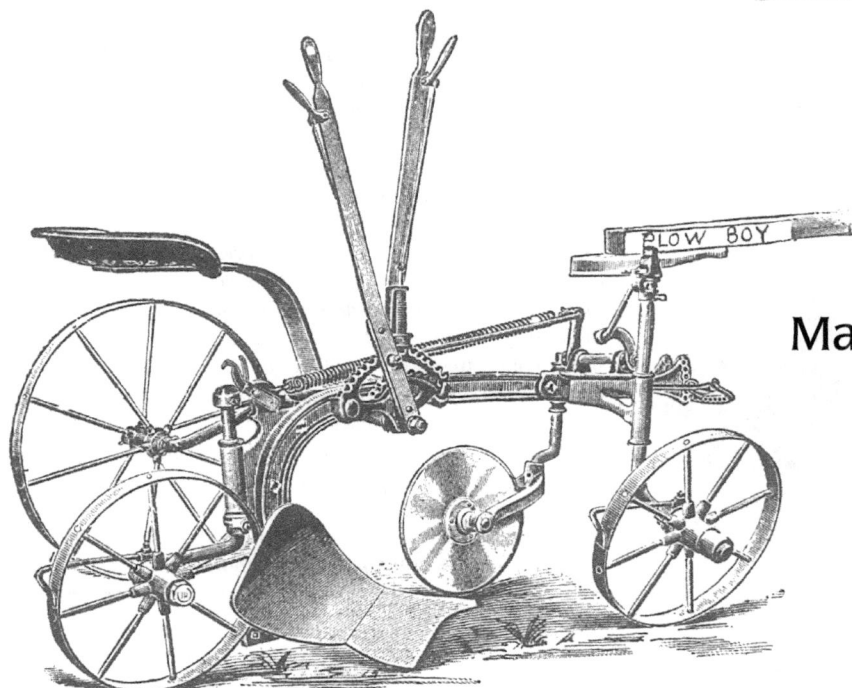

# Mayflower Plows

*Plowboy frameless sulky*

# OK Plows

*A Sulky plow attachment to fasten on to walking plows*

# Prairie Plows

*Two Wheeled sulky with 16 inch bottom*

*Our Award for the most lame-brained, or should we say 'unusual' plow goes to...*

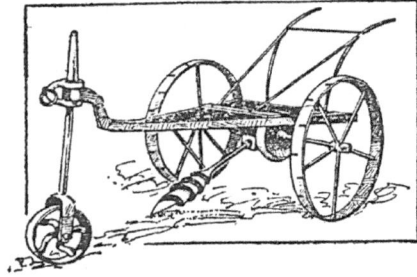

In the August 15 1914 issue of Farm and Home magazine this illustration and the copy below appeared.

*A New Type of Plow. A plow that works on the principle of the screw has been invented by S. H. Shuker of England. Instead of driving, in the usual way, as a nail or wedge is driven into a piece of wood, this works on the same principle as the screw by screwing into the soil, which will reduce the motive power very considerably. In fact, it is thought that instead of two horses drawing one plow and making one furrow only, two horses will be able to draw this plow with three scews and make three furrows instead of one. Two or three tests have been made in a small way in gardens of heavy soil, and the furrows left behind were said to be quite good. It is worked on the principle of cog-wheels driven by the rotation of its own wheels. The idea is somewhat similar to a breast-drill, one wheel being driven at right angles to another. The mechanism is quite simple.*

## A NEW GANG PLOW
### THE Gazelle

Dne Lever only used in operating

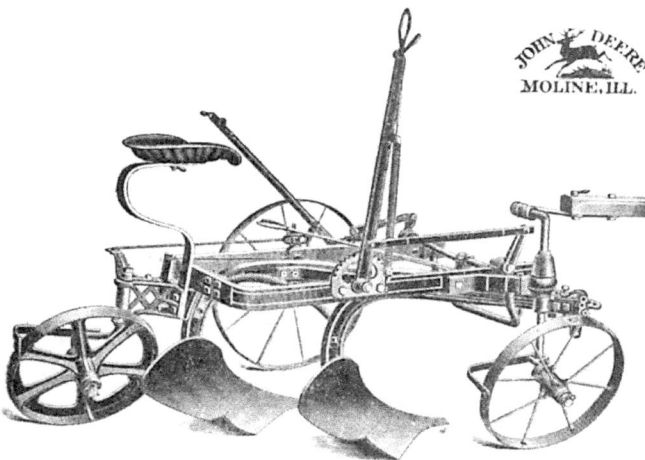

JOHN DEERE
MOLINE, ILL.

*John Deere's First Disc Plow*

JOHN DEERE
MOLINE, ILL.

# Pioneer Plows

As was shown several places throughout this text, the new models of **Pioneer Equipment**, from Dalton, Ohio are being used all across North America with tremendous success. Their current lineup includes walking plows, a frameless sulky plow design and the two bottom frame style gang.

*A closeup of the frame of the Pioneer Gang plow*

*The Pioneer Frameless Sulky with 14 inch Oliver Raydex bottom.*

*From the rear wheel of the Pioneer gang plow.*

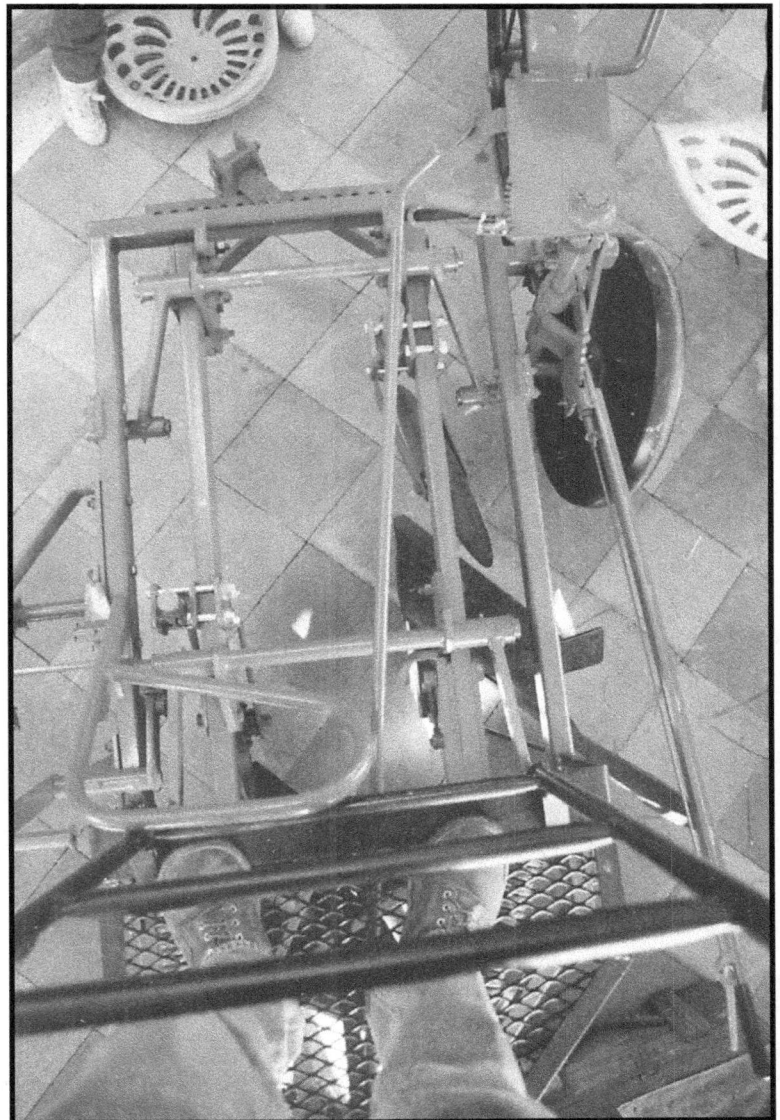

*The Pioneer sulky can be outfitted with optional pole and spring lift assist.*

*Looking straight down on the Pioneer gang framework.*

# White Horse Plows

In Gap, Pennsylvania the folks at White Horse Machine build a fine line of horsedrawn implements and eveners. Throughout this book are photos of their plows and rope and pulley systems at work. On page 90 is additional information on their offerings. On this page we offer two views of their hydraulically operated gang plow at work.

*For contact information for White Horse Machine and Pioneer Equipment, see the pages at the back of this book.*

*Fig. 449. Four abreast hooked to a John Deere gang plow.*

Chapter Twelve

# Maintenance, Repairs & Restoration

If the plowman wishes to work with an older make of plow, one which is not supported by an existing plow maker for parts and repairs, and if the plowman expects to do a considerable amount of plowing (i.e. from 10 to 80 acres per year for 6 years), this author strongly recommends having two or three junk plows of the same make and similar model to 'part out'. For example, it can be expensive to have a local machine shop 'build' an axle cap or handle. If the plow of choice is a John Deere Stag sulky, the plowman would be well served to have two similar plows, with same size bottoms, stashed behind the shed and waiting for 'high grade' exercises. Barring that, with the older makes it becomes critically important to know the available sources for shares, bolts, heels etc. and to do all repairs a full season ahead of time. If the plow of choice is one of the new models made by *Pioneer Equipment* or *White Horse Machine* there is NO worry about parts availability.

## Walking Plows

Maintenance and repair of walking plows is quite simple but does necessitate a basic understanding of their construction. You cannot be expected to repair a walking plow if you don't at least understand how things go together. Below is edited information which appeared within the Moline Plow Company catalogs. Additional information appears in the first two chapters of this book.

### Construction of Walking Plows
Walking plows can be divided into two general types, those having a footed beam and those having a steel frog. The

beams on plows of the first type usually attach directly to the landside, and on the latter are fastened to steel frogs.

### Bottom.
The bottom consists of five parts: Moldboard, share, landside bar and share pad, landside plate and brace.

The beam is footed, having been up-set and then flattened out, giving a wide support to the landside. It is fastened by two bolts with an extra bolt for the moldboard clip, making a very substantial construction.

### Steel Frog.
The steel frog construction shown below is one popular design. It holds the share, landside, moldboard and beam securely together, requires fewer bolts and clips, and will stand more severe strains and harder work. It also makes possible the use of a landing wedge

### Landing Wedge.
The beams on the 12, 13 and 14-inch plows when shipped from factory were usually landed for 2 horses. On these plows, when 3 horses are used, the front end of the beam should be on a straight line with the

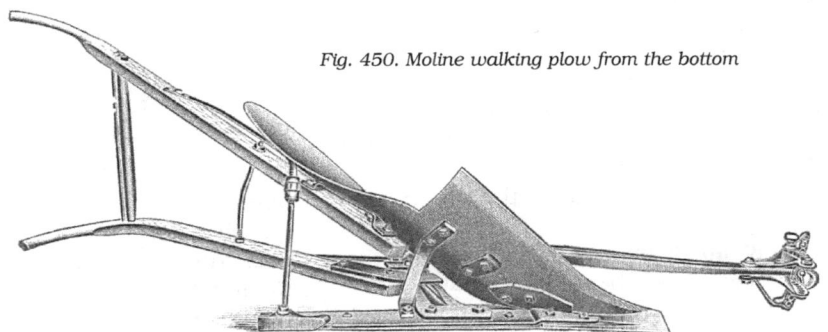

*Fig. 450. Moline walking plow from the bottom*

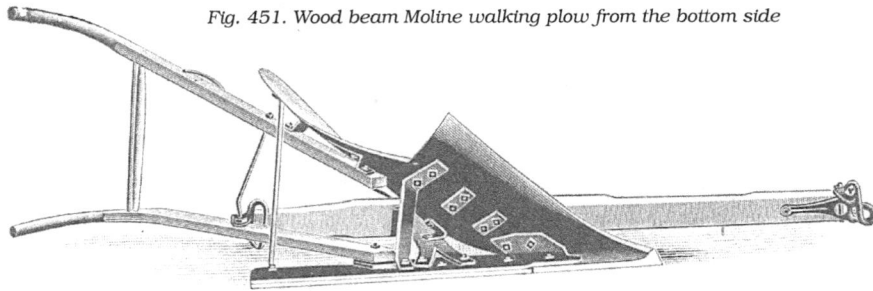

Fig. 451. Wood beam Moline walking plow from the bottom side

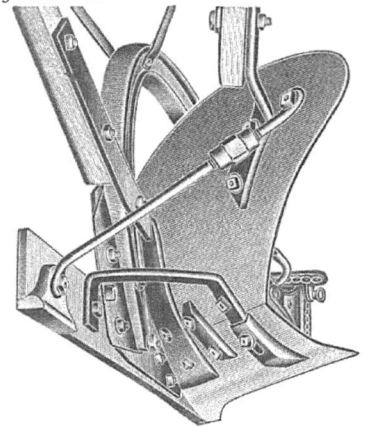

Fig. 452. A closeup of a walking plow framework

landside. This can be done by simply loosening the bolts which hold the frog and beam, sliding the landing wedge forward between them.

### Moldboard.

The moldboard has an extra heavy piece of steel welded to the shin, giving double thickness and wear at this point.

### Share.

Shares are made of either solid or soft center steel. The share point has an extra hard piece of steel welded on both the top and under side for wear. This is a very important feature as the point of the share must necessarily resist wear.

If the under side of a plow point is soft, as is usually the case when the landside of the plow extends to the front end of the point, the plow will soon loose its suction, which entails the expense and delay of resharpening.

It is impossible, when building a plow, to set the wing of a share to meet every condition of soil. A plow which would run perfectly in clay, black land or mixed land, might not work as well in sandy soil. The complaint is usually that the plow will lean to or from the land.

### Adjustable Wing Bearing.

Model CC and DD Moline plows featured an adjustable wing bearing, a feature few if any other plows had.

When the plow has a tendency to lean on the landside, it has too much wing bearing. By simply loosening the bolt holding the adjustable bearing, the share can be given more suction on the wing.

When the plow has a tendency to lean towards the plowed ground, the share has not enough wing bearing, and the adjustable wing bearing should be set down lower.

### Landside.

The landside of the share has a wide flange, forged before being welded, to give sufficient surface to secure a perfect weld. This also prevents the landside from cutting the share blade in the process of welding.

### Wood Beam.

Some wood beam plows have index castings at the rear end, to land for two or three horses.

Fig. 453. Place the plow on saw horses to check all the wheels for operational fit. It is important that the boxings be snug. Sloppy loose boxings will quickly convert to extra wear.

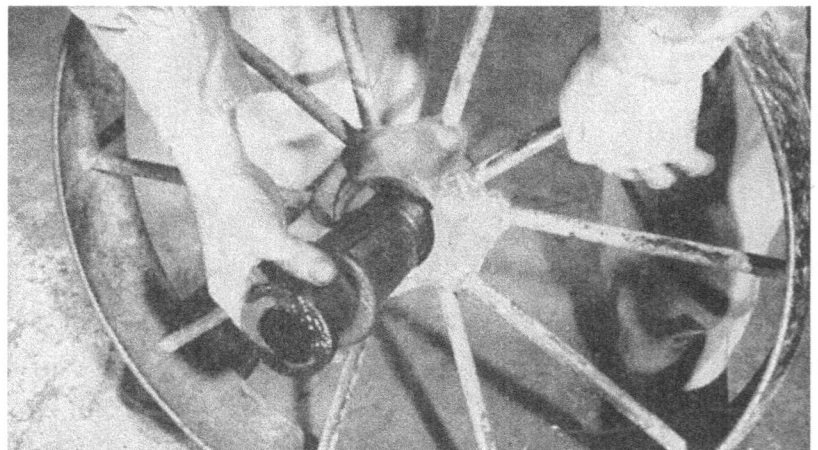

Fig. 454. Putting a new boxing in the furrow wheel. Before putting the wheel on, the axle should be cleaned with kerosene (or diesel). After the wheel is put on the axle, the hub cap should be packed with axle grease and screwed on, forcing grease through the wheel assembly. If the old sand band is badly worn it should be replaced with a new or better one.

# Riding Plows

### Overhauling Riding Plows.

These plows should be over-hauled and put in condition for the next season's work as soon as possible after finishing plowing. If new (or better) parts are needed before the plow can be used again, they should be ordered, attached, and the plow put in condition ready for the field, during slack seasons. Especially if you are using an older, possible rare make of plow. Don't expect to find a share or landside within a short distance of your farm. The following instructions are designed to aid in a thorough inspection of riding plows:

First - Examine the wheel boxings. If they are badly worn, they should be renewed; if not, they should be slipped off the axle and both axle and boxing washed clean with kerosene, and a fresh supply of grease applied. If the collar that holds the wheel to the axle is badly worn so as to allow excessive play of the wheel boxing on the axle, it should be replaced with a new one.

Second - Examine the shares. If they are not sharp, detach and have them sharpened and properly set. If they are excessively worn, get new shares that are made by the manufacturer of the plow.

Third - On the high lift, foot lift plows, suspended by one or two bails, examine the bail stops. These are located on the front frame bar and on the right frame bar. They should be so adjusted that when the plow is locked down in plowing position, the bails rest securely on the stops. The bail bearings should also be examined, and if they are worn loose, take the cap off and file or grind it until it fits snugly with all bolts tight. This will help greatly in keeping the plow running steadily and quietly.

Fourth - Examine rolling coulters and hub bearings. Coulters should be sharp and well polished. If the hub bearings are badly worn, replace them with new ones.

Fifth - Check the location of the rear axle collar. This collar should support rear end of the frame and transmit weight of rider to the rear wheel. If it has become loose and slipped down on the axle, weight of plow and rider will be carried on the bottom of the plow landside instead of on the rear wheel.

Sixth - The rear axle frame bearing carries the weight of the rider and plow and transmits the side pressure created by the moldboards to the rear furrow wheel. If the bearing becomes badly worn, the landside of the rear bottom will carry this pressure, resulting in heavy draft and excessive wear on the landside. On some makes of

plows, this bearing is provided with a take-up casting with bolts at both upper and lower ends. This makes the proper adjustment simple and easy.

Seventh - The front furrow wheel axle bearing should be reasonably snug, in order to keep the wheel running at the proper angle, and the front furrow to proper width.

Fig. 455. Examine the rolling landside or rear furrow wheel for boxing wear. This wheel should be set so that there will be sufficient clearing between the landside wheel and the furrow wall to place the fingers and to allow sufficient clearance beneath the landside wheel to pass the finger (see page 76).

Fig. 456. The land wheel should be checked to determine if there are worn parts. The area should be cleaned with kerosene (or diesel) and regreased.

Eighth - Look over the entire plow for loose nuts and worn bolts. A plow operates at great disadvantage when parts are loose due to bolts not fitting bolt holes and nuts being loose.

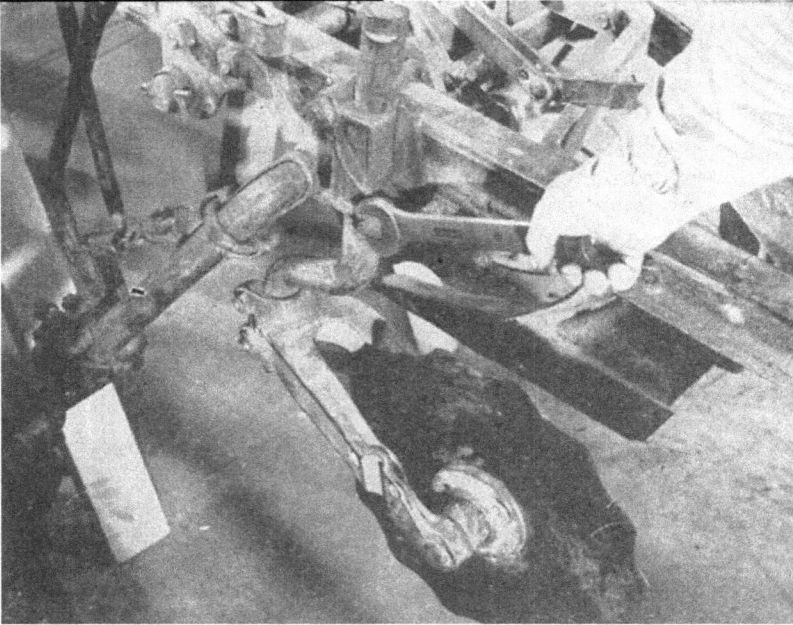

Inspect all parts of the plow. See that the eveners and plow are in good condition. Lever dog boxes should be oiled so they will work freely. And, above all, keep the polished parts free from rust.

*Fig. 457. (left) The rolling coulter should be properly adjusted. In most cases the center of the coulter should be directly above the point of the plow. The height of the coulter depends upon the depth of the plowing and the condition of the soil. When plowing about six inches deep, in ordinary soil, the coulter should be set about 1 $^{1/2}$ inches above the point of the share. The jointer should be set to cut 1 $^{1/2}$ to 2 inches deep.*

*Fig. 458. (center) Tag plow in position for determining the line of draft.*

*Fig. 459. Painting the plow. The entire surface has been cleaned of all dirt and grease and all rust has been removed or chemically neutralized. A primer coat was applied before the final coat of exterior grade implement paint.*

# INSTRUCTIONS FOR SHARPENING PLOW SHARES

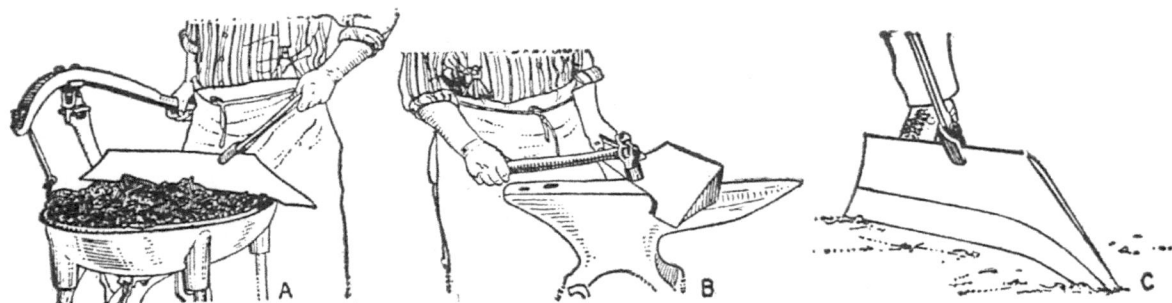

*Fig. 461. Steps in the sharpening of a plow share: A, only the part of the share to be pounded out should be heated; B, hammer on the top side; C, the right way to place a plow share for cooling.*

Many blacksmiths, after removing the share from the plow, plunge it into a big broad fire to heat preparatory to sharpening, often setting it on edge in the fire. This is wrong. It permits the heat to extend over the entire surface of the share, withdrawing the hardness that the manufacturers were so careful to conserve. It also causes it to warp and lose its original shape. The warped share is a real pain to put back on the plow bottom.

**To properly sharpen the plow share;**
Build a fire on the forge suitable for this particular work. This is done by banking the fire, allowing only a small opening in the side for the blaze and heat to escape. Start with the point of the share. Insert this into the fire just far enough to heat the part you wish to draw, never permitting the heat to extend farther back on the share than is absolutely necessary. Draw this down to the proper shape and thickness, which should be as near the original bevel as possible. After the point has been finished, work back toward the heel or wing of the share never heating more than 1 ½" from the edge and 2 ½" wide. It is important to keep hammering after the steel has changed from a red heat to a black as this makes the edge tough and hard, giving a wearing surface that will last much longer.

If once down the share is not enough, reheat; but restrict the heated part to the above measurements. In working along the cutting edge, keep it straight. Doing it this way will avoid having to go back and reset the edge.

It is impossible to give the exact amount of wing bearing on walking plow shares, but it should be from 1 to 1 3/4", according to the size of the plow. Shares used on wheel plows should have no wing bearing. These instructions refer to both hard and solid steel shares. (See chapter Two.)

# Restoring Plows

Before proceeding with any restoration of older plows the plowman should decide what purpose will be served by the work. Is it for resale? Is it for a lawn ornament? Is it to conceal some problem with the plow? Is it for pride of ownership? Is it to simply keep the plow in good using condition?

For the horsefarmer who is also handy in the farm shop an excellent side business may be built up by restoring older plows for resale. This author can attest to the fact that a strong and growing market exists for good horsedrawn implements. To this end we encourage you to restore the plow as if you would want to keep and use it yourself. If resale is your goal the quality of your work will either cause your market to grow or to shrink. Make it grow.

Decide whether or not you want to paint your plow. That may seem like a silly and obvious point but keep this in mind, the knowledgeable farmer-collector would prefer to purchase an implement bare of paint or with faded factory colors for the simple reason that it is easy to see what is being purchased. The rust, or lack of it, will be obvious. Any cracks, slight wear or sloppy repairs will be obvious. So a no-paint restoration is a logical option. Whether painted or not; disassemble the plow completely. Wash every part with diesel or kerosene using a wire brush. If only mild rust is present wire brush the surface till clean of rust. If heavy rust, power brush or grind and paint with a rust neutralizer (the author uses Ospho brand).

If the plow is to remain unpainted, brush or spray the entire dry surface (metal and wood) with an oil preservative mix (one part turpentine with three parts boiled linseed oil warmed to coffee drinking temperature). Do not use straight Linseed oil. After oiling reassemble.

If the plow is to be painted use a rust inhibiting implement or metal primer first, then apply at least two if not three coats of final colors. Paint the disassembled plow. Allow to dry completely then reassemble. Some minor touch ups might be necessary from wrenching marks.

Be ready for plenty of offers to buy your plow.

## Chapter Thirteen

# Plowing Contests

*Fig. 462. First place plowing in anybody's book.*

Although this chapter is indeed about plowing contests, matches, competitions, and demonstrations it also delivers bits of information on many miscellaneous plowing aspects and from unique and interesting perspectives.

This author was a regular plow match contestant in the 1970's in Oregon. The experiences were always exhilirating, educational and connecting.

Throughout this book there have been several suggestions that the beginner look to public "old-time" farming contests and demonstrations for learning options. They are also good places to meet potential teachers, mentors, and helpers.

They are in fact such gold mines of information

*Fig. 463. Jim Olson and his Percheron team, of Cashton, WI, are the picture of perfection as captured here by Bob Mischka. Jim was the winner of Men's Walking plow class at the Hritage Farm Plow match 1999 in Hudson, Iowa.*

and inspiration that agriculture in general and horsefarming in particular would be well served if many more plowing matches were started.

As an influence to that end, and to help round out the technical information in this book, we offer in this chapter the rules and regulations of 3 different plowing matches from New Jersey, Iowa, and Washington. Plus we offer a lengthy 1935 Canadian government paper from the Province of British Columbia's Department of Agriculture on *'ploughing and ploughing matches'* (lent to us by good friend John Erskine). This paper features terminologies and makes of plows (they call them ploughs) quite different from the U.S.

As will be seen from reading this material, there is variety from region to region but the goals are much the same. Those goals are:

- to showcase plowing with horses
- to challenge plowmen to be better
- to encourage the best horsemanship
- to educate the novice
- to attract new people

Look to the 'more information' page at the back of this book for contact numbers for the plowing matches.

## From Dick and Marie Brown's Heritage Farm Plowing in Iowa

We judge the plowing 50% on the plowing, 30% on the horse and the horse's appearance, and 20% on the teamster.

When plowing, the furrow is the first to judge. It must be done in a straight line, straight both horizontal and vertical, with no dirt clods and the furrow must be clean and flat. Second is the covering of the grass, straw, corn stalks or alfalfa you are plowing. The purpose of plowing is to get clean dirt for seed beds so the old crops must be turned under. The old rule for setting the depth of the furrow when horse plowing was to set your thumb on the land horizontally and your fingers to the bottom of the furrow, 5 to 6 inch was the practice.

Deeper plowing covers better and makes a nicer furrow, but you need more horsepower if the ground is hard or dry. The coulter must be set proper for a good furrow. The center of the coulter over the top of the point of the plow seems to work best in corn stalks. When plowing in alfalfa or grass, the coulter needs to be moved 1 ½ to 2 inch to the left, or outside when looking from the seat of the plow, to make a good furrow. The coulter needs to be deep enough to cut grass roots, but if too deep will lift the plow out of the ground especially evident on walking plows and sulky plows. The plow must not have any side draft. It seems to work best for us to use three horses on a 16" walking plow, 4-up on the 16" sulky, 6 on 2-14" gang and 9 on 3-14" gang plow.

When judging the horse, they must travel in a straight line, no heads turned, must walk in a slow pace, no running or prancing, so they can plow all day and not tire. The furrow horses must be willing to get in the furrow, and stay there on the bigger gangs. The teamster can not see the furrow from the plow seat so he has to trust that the furrow horses are in the furrow. You must also have a good lead team on the gangs that are willing to lead and work by voice and line command. The horses must be properly dressed for work. The collar and bit should fit so no sores develop. A show harness is not acceptable. The quarter straps and tugs should be in a straight draft line. The horses must not be over pulled. If a horse shakes in the shoulders or is winded, they are tired and must be excused. The horse must be well groomed and the harness in good repair. The horses need to be in good flesh and in shape to work. A good work horse must have a good trimmed foot to be able to work with ease. He can't have a long toe like the show horses are shod to make him look cow-hocked or sickle-hocked because it puts strain on his legs, tendons, and muscles and he cannot pull properly. A good judge can easily see if the horse is enjoying what he is doing.

The teamster must have a bond with his horses. He must be quiet, no yelling or hitting, and must handle his horses with ease. He must be able to adjust his plow easily. There are many adjustments on horse plows and if set right the plow will cut like a knife in hot butter. I have my plows set for our farm ground, but when we travel elsewhere to plow in different types of soil, I find my adjustments all wrong. That is when a teamsters knowledge of adjusting his plow is very helpful. Most of the adjustment is in the hind wheel, the hitch, and the tongue. The old timers used to wire the tugs to their single trees in rocky ground, then start their horses slowly and put the plow in the ground slowly. If a rock was hit, the wire would break before it threw the teamster off the plow.

In some plowing contests you are assigned a land with a stick on each end. Each plowman breaks open his own land called a back furrow. They make six rounds, then move to the left hand land and plow until land is all turned over, finishing out with a dead furrow. You are then judged for straightness of furrow, cover, neatness, even ends, and straightness of dead furrow in that order.

Other plowing matches plow in lands following one another. The plowers make one practice round and then 4 contest rounds. This is how we do it at our plowing match. This takes a GOOD judge. If the plowman in front of you messes up, it is hard for you to look good following a bad furrow. This is where the judge must take into consideration the teamster's ability to adjust his plow and horses to offset the problems he is faced with.

The real authorities of horse plowing are few as they are the "old timers" that did plowing to make a living on farms across our land. That is why we must keep this art of horse plowing alive. In the plowing events we have participated in, we have seen the plowers improve each year. We need to share our knowledge with others.

## From Howell Farm Plow Match in New Jersey

### How to judge fine plowing

**Straightness:** how straight are furrows? Are there hooks or skips in the plowing?

**Depth:** How uniform is the depth? Most plowing will be 7-8 inches.

**Width:** Width will vary with the size of the plow, so look for uniformity of the turned furrows.

**Coverage:** One of the functions of plowing is to turn under stubble and other organic matter lying on the surface. How well did the plower turn in or cover these materials?

**Evenness of ends:** When the plower comes to the end of the furrow, does he/she take the plow out of the ground at the same point as the previous furrows, or is the end of the plowed land zig-zagged or uneven?

**Draft:** Do the animals in the team pull evenly, or does one do more work than the other? A good team pulls together.

**Steadiness:** How steady is the pace of the team? Does the driver have to hold them back? Urge them on? A team that walks along at an even pace generally has better endurance, and makes it easier for the farmer to control the plow. Remember that plowers are not judged on time, and normally will pace themselves at the rate of about one acre plowed per day.

**Control:** How well do the animals listen to the commands and driving lines of the farmer?

**Adjustment of plow hitch:** Does the farmer have the plow adjusted and hooked so that basically, it works by itself or does he/she constantly fight the plow? Plowers who continually lean down on one handle or who have the plow in the ground by pushing down on the handles probably need to adjust their hitch.

**Appearance of hitch:** Do the horses appear to be sound, well muscled and fit? Does the equipment, including the harness, appear to be well maintained?

## Description of competition classes and awards

**Fine plowing**: participants can use one, two or three animals, including horses, mules, ponies, donkeys or oxen, to pull a right-hand, left hand or hillside walking plow of any width, weighing under 200 lbs. Plowing will be judged according to the following:
- back furrow
- evenness of furrow
- straightness of furrow
- penetration
- control of team
- condition of equipment
- depth of furrow
- trash and debris coverage
- evenness of ends
- condition of team

Awards:
1st place trophy
1-6th place ribbons

1-6th place premiums as follows: 1st: $130, 2nd: $115, 3rd: $100, 4th: $85, 5th: $70, 6th $55.

Ben Ellingson Award: the winner of the plowing match is named on a special trophy that recognizes the contribution that American farmers are making in international agriculture. The winner receives and keeps a special blue ribbon.

**Obstacle Course:** Participants use the team they used in the Fine plowing event to go through an obstacle course pulling a 12' log, 18" at the butt. Judging is based on appearance of the hitch, steadiness of the team, time, and number of faults thru a series of turns, gullies, a hillside and a creek bed.

Awards:
1st Place Trophy
1-6th place ribbons
1-6th place premiums as follows: 1st $100, 2nd $90, 3rd $80, 4th $70, 5th $60, 6th $50

**Old Timer Plowing:** Participants are farmers who know how to use a team and a walking plow, but who don't have their own horses and equipment. These contestants use a Howell Farm hitch to plow, and are judged on one round. Sign-up for this event is anytime prior to the event. Maximum participation is 8 teamsters. Plowing is judged according to the same characteristics as Fine Plowing, but with automatic points given for conditions of team and equipment.

Awards:
1st place trophy
1-6th place ribbons

**Novice Plowing:** Visitors from the crowd can sign up to try their hand at plowing. One or two quiet teams, driven by their owners, are used in this class. Visitors handle the plow only - not the horses. They are judged on either one pass, or one full round, depending on the number of people who want to try. Judging is on straightness and depth of furrow, and on coverage of trash and debris. Minimum age: 14 years old. Maximum entries: 10 if one team available; 20 if 2 teams available on adjoining lands.

Awards:
1st - 6th place ribbons.

### 1999 Rules and procedures
### Fine plowing:

**General description**: participants can use one, two or three animals - including horses, mules, or ponies - to pull a right-hand, lefthand or hillside walking plow of any width, weighing under --- lbs. Plowing will be judged according to the following: conditions of team, condition of equipment, evenness of furrow, depth of furrow, straightness of furrow, trash and debris coverage, penetration, evenness of ends, teamster's control of the team.

### Rules and Procedures:

1. Equipment needed to compete: Please remember to bring your own plow, eveners and wrenches. Arrive early enough to allow us to transport your plow and eveners to the field (see #6)

2. Practice plowing has been eliminated due to

*Fig. 464. Dick Brown with his six Percherons on a two bottom plow. Bob Mischka took this nice photo at the 1999 Heritage Farm Plow days in Hudson, Iowa.*

lack of extra lands. Anyone wanting to settle their horses before plowing can ground drive them. During ground driving, riding on the eveners is not permitted. Time: 8am - 9:30 in designated area.

3. Judging: You will be judged on your back furrow and a number of subsequent rounds determined by the judge. The judge will explain his judging criteria at 9:45, prior to starting the match. In the event of a tie, or very close call, the judge will have the option of asking those involved to "finish out" a land.

4. Plows must weigh under 200 lbs. Weights added to a plow are not permitted.

5. Once you have been judged, you may continue plowing on your land, but don't let visitors or friends plow, since it may result in poor work, making it hard for us to finish out later.

6. Transporting plow to and from the field: Between 8:30 am and 9:15 am we will transport your plow from your truck/trailer to the field, by wagon. When the match is over, please take your land number off your hame and put it into your plow handle; we will return the plow to your trailer, which will have your land number, in chalk, on the tire (we will mark your tire during check-in procedures).

## From the Chilliwack Washington Plow Match

Rules and Regulations
1. All classes will plow in sod.
2. Depth gauge on plow allowed in all classes.

3. Any type of jointer or cutter (coulter) permitted.

4. All classes (except div. III and IV) to plow not less than 14" and/or 16" wide and a furrow 6" deep.

5. Every plowman will be on the field by 10:00 am, at which hour lots will be drawn.

6. Every plowman will strike out his hand at the peg that corresponds with the number drawn by him.

7. Plowing will commence no later than 11:00 am

8. Every plowman or proxy will draw the number of the lot which he is to plow.

9. Competitors will furnish stakes.

10. One assistant allowed for the start. At the conclusion of two visible rounds, the assistant must leave the land. Except in Div. II, where a helper may be required to assist with the team in emergency. Also, Div. IV must have an adult helper accompany the contestant at all times.

11. After the first two visible rounds, no handling of the land will be allowed.

12. Every plowman will show six visible rounds at the back of his strike-out and then haw. Finished furrow must be thrown towards the contestant's own crown.

13. Every plowman must have six visible rounds finished by 1:00 pm or forfeit five points off his total points.

14. Every plowman must finish by 3:00 pm or he is disqualified.

15. Each plot will be plowed by each competitor.

16. The decision of the judges will be final.

17. Any plowman interfering with the judges or

infringing these rules will forfeit any chance of a prize.

18. Teams and equipment need not be the property of the contestant.

19. Contestants will be required to sign a responsibility waiver. Anyone under legal age must be signed for and accompanied by a parent or guardian.

20. 14 year olds may have the option of plowing in Div. II or Div. IV.

Note a ribbon for participation will be given to all participants in each division.

### Rules and Regulations for all Classes in
Chilliwack plowing match, except class 4

1. Entry fee for all classes $5.00, payable at time of making entry.

2. Wheel allowed in all classes.

3. All classes, high-cutting plows barred.

4. All classes, plowing shall be not less than 10 inches wide, 6 inches minimum depth. Measuring starts after 4 completed rounds.

5. Skimmers must be used on all horse drawn plows.

6. Every plowman, or proxy, will draw from a hat or box the number of the lot he is to plow.

7. All plowing starts at 11:00 am

8. Every plowman will strike out his land at the peg which corresponds with number drawn by him.

9. Competitors will furnish stakes.

10. One assistant allowed for setting and removing of stakes only in all classes. Further assistance will be allowed in all tractor classes, for two rounds, then helpers must leave plots. Further assistance on plots will be allowed in Class 7 and this class's exempt from Junior Championship.

11. Handling of the land for the first 2 visible rounds will be allowed for horse classes.

12. Horses - Every plowman will show twelve visible furrows for his strike-out, then haw. Finish furrow must be thrown towards contestants own crown.

13. Tractors - Every plowman will show eight or nine furrows on each side of his crown and then turn to the higher number and gather. Sole furrow must be thrown to own crown.

14. Every plowman must finish by 4:00 pm, or be disqualified. A competitor breaking any of these rules shall receive only one warning and automatically forfeit five points. Upon a second infringement, the competitor will be disqualified by the governing body.

15. Any plowman doing an unsatisfactory job of plowing or operating his equipment in a dangerous manner may be disqualified on the recommendation of the plowing supervisory committee.

16. Any protests must be made in writing and delivered to the secretary before the judges leave the field.

17. The decision of the judges and officials shall be final.

18. Any plowman interfering with the judges will be disqualified.

19. Competitors will be given a free lunch.

20. Competitors must provide feed for their horses.

21. Prize money for the best crown and finish will not be paid unless there are at least two entries in the class.

22. No person is allowed to compete in more than one class.

23. 5 point penalty for tractor plowmen not splitting

*Province of British Columbia*
*Department of Agriculture, Field Crop Branch*

# Ploughing and ploughing matches, 1932

### Introduction

Of the many operations connected with the production of a crop, good ploughing is one of the most important. It is observed that a good ploughman is a good farmer, that is to say the same careful attention which he gives to his ploughing is in evidence in all his farming operations.

It is in order to encourage good ploughing that ploughing matches have been in existence in some countries for years; whilst in this country, although of more recent introduction, ploughing matches are each year becoming more popular and consequently are increasing in number.

There are many young men on the farms in British Columbia who are good ploughmen and when asked the reason why they do not try their hand at ploughing matches, the answer is that they are not familiar with the requirements of, or methods adopted at ploughing matches. Those who have acquired the knowledge have had it passed on from older ploughmen.

The information contained in this circular may be of interest to many beginners whether they intend to take part in ploughing matches or wish to practice good ploughing at home. It is also intended that this circular shall be of value and assistance to those responsible for the organization of ploughing matches.

### The Strike Cut

The best way to make a start with the plain low cut or wide bottom plough, is by the use of scratches. These scratches are not furrows, but mere grooves about two inches deep. These grooves form a hinge for the first furrows to turn in on. The scratch furrows should be placed correctly in the center of the crown and it is generally permissible to handle them in order to straighten them out.

The scratches in the start should be from 20 to 24 inches apart according to the width of the plough used, but this can best be determined in practice. Having decided the width of the scratches, place the stakes half of that distance from the number peg at each end of the field; for 22 inch scratches this would be 11 inches. Set the stakes and make the first scratch; after this, set the stakes again on the opposite side of the number peg and draw the second scratch, then straighten the material out. Both scratches are of equal importance because the best one should be selected for the first crown furrow to rest on. If the second one is the best of the two, this one should be used even if it makes it necessary to slide empty to the other end of the field in order to make use of the straightest scratch. If the scratches are too far apart, an open crown will be the result, on the other hand if they are too close together, the first crown furrows will stand up too straight, which is perhaps the worst feature. The width and depth of the scratches will be decided upon by the contestant in his practice ploughing.

The first two crown furrows cover up the scratches and these furrows should be about four or five inches deep. They should close up neatly and must not crowd each other in the center. They should form a V shape. The next two furrows should be about five or five and one-half inches deep and should turn in neatly and cover all grass, and likewise with the rest. The third furrow should be about six inches and the fourth should be at the depth of the usual ploughing of the lot. The first six rounds constitute the crown and then the contestant starts to haw around, and as soon as he starts to do this he should make measurements concerning the width of land in relation to securing a good finish.

Some judges don't look at the cast away ridge at all, but it should all be judged except the first two furrows turned against your neighbor. If your neighbor left a very bad face and you were able to straighten it up, that would be so much to your credit or vice versa.

**Set of the Coulter**

In starting out, have the coulter well forward on the beam of the plough and down close to the point of the share, gradually raising it with each furrow until the fourth furrow, when it should be up two inches or more and half an inch or more to the left of the point of the share. In ploughing sod keep it nearly straight up and down over the point of the share. In that way you get the full turning capacity out of your share and mouldboard. The furrow starts to turn as soon as the point of the share cuts it if the coulter is there to cut it at the same time. The coulter should not be in a straight line from the beam of the plough to the point of the share. If so, take it to the blacksmith and get him to make a bend of about two and one-half inches to the right at the place where the cutting edge starts, and then move it back to the left so as it will cut the furrow at an angle. In this way an inch or so of grass will be left on the land side. By cutting an inch or so off the edge of the turning furrow, the grass can be turned in out of sight.

The coulter should be faced so that when a straight edge is laid on the land side of the coulter blade there will be about two inches of space between the heel of the plough and the inner side of the straight edge. The cutting edge of the coulter should be about three-sixteenths of an inch from the inner side of the straight edge, or half the thickness of the back of the coulter, this three-sixteenths to allow for the sharpening slope of the cutting face of the coulter. When standing exactly in between the handles of the plough, you should only see the back of the coulter, but by looking a little to the right or left you should see the coulter at the same angle. The coulter should look as if it would cut too much off land, or to the furrow side. A good practice to follow is to pass a half inch wide rule down the furrow side of the coulter, and if the rule

passes the point of the share, or has a clearance of half an inch to the left of the point of the share, this will indicate the coulter is properly placed. The coulter should be two inches up above the point of the share at the time of this trial. One may not get the right cut on the share or the right set on the coulter on the first trip to the blacksmith, but once in the right position you will notice how easy the plough is to hold. The coulter edge should be sharpened like a knife, not like a scissor blade, as it frequently is when it comes from he factory. The scissor blade style has too much pressure on one side of the coulter and is inclined to press the plough too much to the land side. The coulter should cut into the land clean and true, not sideways.

There should be some cut on the share also, that is the wing or feather should be raised higher than the point. In this way, the point and the first few inches of the share to the right of the point will bring out a firmer and thicker furrow, and when it comes up to the top of the mouldboard, there is more pressure for the top of the mouldboard to push it over.

After you get the plough set like this, you will be surprised to see how easy the plough will turn the furrow over and the furrow will have no inclination to fall back. Be sure not to lay the furrow over too flat because set this way you can hold an inch or so deeper with the same width, and that is where the body in the furrow and a big round comb is secured; furthermore, it makes the land easy to harrow and cultivate. There may be slight variations for the different styles of ploughs, but as a general rule this method works very well.

Wrong Method. If the coulter is set in a straight line from the beam to the share, back on the beam, and down close, and back on the share, it loses a lot of turning capacity and will not turn the grass edge in. By adopting this method the top of the furrow is skinny and hard to harrow and cultivate. Such a furrow has no body and no comb, and the grass is not turned in properly, and will look washed like and rooty after the rains of winter.

If the land is tough or full of grass roots, a cutting wheel is a good thing, that is a wheel with a cutting edge on it that will cut a seam 1-1/2 inches deep for the coulter to follow and make a clearance of couch grass roots that otherwise would collect on a straight coulter. The cutting wheel can be set at the same angle as the coulter.

No furrow should appear as if it was going to fall back. At the same time, it should not appear as if it was over too flat. Both sides of the comb (that is the top of the furrow) should be of about equal slope and shape. If it lies over too far it shows too much share face, and if it sets up too steep and is likely to fall back it is showing too much coulter face, and does not cover the grass well that way. All these things have to be taken into consideration in practice previous to the ploughing match.

The main evener or whipple tree should be from 40 to 44 inches long between the clevis bolt holes. There should be 12 inches of chain between the plough clevis and the clevis on the main evener to give the plough even balance and to give easier control of the plough, because the sidewise motion of the team is too easily transmitted to the implement.

The handles of the plough should not be any less than six feet, and some of them are eight feet in length. This gives an even balance and the plough can be easier handled.

### The Finish

It is advisable to use one horse for the last green furrow and the loose soil furrow in the finish. If two horses are used they tramp the land on either side of the finish, and this spoils the appearance of the finish.

In finishing up a land, the method to follow is thus: - plough around until coming within two widths of the plough furrows you have been cutting and then turn the split furrow towards the cast away side holding it about two inches shallower, then turn the last green furrow towards your own crown and then go round the edge of the field to the other end again and lift a sole furrow and also turn it towards your crown, keeping the land side of the plough four inches out from the base of the split furrow and turn the loose soil to within four inches from the top of the last green furrow. Some people advocate coming back with another sole furrow and placing it against the split furrow, and by doing this the finish is no wider than the width of the plough mouldboard. This method has a lot to recommend it as it covers up the base of the split furrow and the finish is much narrower, and is easier to harrow in.

### Score Card

The following score card is recommended by the Department of Agriculture for sod or stubble:

| | |
|---|---|
| Crown | 20 points |
| Straightness of furrows | 15 pts |
| Uniformity | 10 pts |
| Firmness and evenness | 10 pts |
| Covering of weeds, grass, stubble | 20 pts |
| In and out of ends | 10 pts |
| Finish | 15 pts |
| Total | 100 points |

## Explanation of Points

**Crown** - The crown consists of twelve furrows, that is, six rounds, and should be level along the top.

**Straightness** - Straight ploughing is not necessarily good ploughing, and crooked ploughing is not always bad, but the shortest distance between two points is a straight line and a man who ploughs straight is more likely to do good work.

**Uniformity** - A person should endeavor to retain the same width and depth of furrow throughout.

**Firmness and evenness** - An even top is required so that it will leave a level surface for seeding. Firmness is required so that there are no air spaces left underneath the furrow, as the land is more liable to become dried out if there are air spaces.

**Covering of weeds, grass or stubble** - A characteristic of good ploughing is the "complete" covering of weeds, grass or trash. No trash of any kind should appear between the furrow slices. By the complete covering of trash the insect population will be greatly decreased, especially cut-worms, since weeds and

other debris which lie on the surface offer the best possible protection for these pests. Proper set of the plough must be maintained to get the best results. Owing to the prevalence of couch or twitch grass, the covering of all grass and weeds is very important.

In the diamond cut or high cut and sod classes, the straight coulter is used with lots of side-cut, but in the novice and stubble classes, jointers, rolling coulters, wires or chains are a great help. The straight coulter does not mean that it should be in a straight line between the plough beam and the share, but it should have enough set to cut the furrow at an angle to allow the grass edge of the furrow to turn in.

**In and Out of Ends** - A light scratch furrow, such as has been outlined, makes it much easier to keep the ends of the furrows even. These light scratch furrows should always be thrown out towards the headland.

**Finish** - A straight, clean finish no deeper than the rest of the ploughing is desired, and as narrow as possible.

## Types of Ploughs

There are various types of ploughs on the market and the following is a description of some of them:

**No. 1 - High Cut**. Furrows are eight inches or less in width and from six to nine inches deep. One wheel or cutting wheel allowed, with straight or knee coulter. Coulter to be set with enough side cut and angle cut, to cover the grass.

There are several types of high cut ploughing, but the most useful type is where the grass is put flat in the bottom of the furrow, in contact with the subsoil. Burying the grass 3 inches deep and still leaving a round comb on the furrow, measuring six inches from the top of the comb to where the furrows join. This type of ploughing buries the grass absolutely, beyond chance of recovery, and it leaves a comb very easy to harrow and cultivate. There are no air spaces left underneath the furrow with this type.

**No. 2 - Sod plough**. This is a stepping stone to the high cut. This type is the most important, as it is the style of ploughing that is done every day in most parts of the province, where there is just one team of horses kept on the smaller farms. Furrows to be nine inches or over in width, and at least six inches deep. A sod plough is a plough of the following types; Judge, Punch, Judy, Wilkieson No. 3, Massey Harris 14P. No skimmers. One wheel or cutting wheel allowed with straight or knee coulter. Coulter to be set with enough side cut and angle to cover the grass properly, the same as in the high cut.

**No. 3 - Jointer Ploughs for novice and Junior classes.** Furrows to be ten inches or over in width and at least six inches deep. Jointer ploughs are ploughs equipped with skimmers and straight coulters, and have wide low-cut shares no less than eight inches in width, and with the breast measuring no less than six and one-half inches. (Measuring six and one-half inches from land side of the plough to out-side of bottom of mouldboard, from directly under the point where the top of the mouldboard touches the beam). Jointer ploughs are equipped with skimmers and straight coulters, and are the makes of ploughs like the Paris, Fleury 21, Dandy, Massey Harris 21 and 30, and

Fig. 465. Terry Beyer and Bob Champion with twelve mules on a 2 bottom plow. at Heritage Farm Plow Days

ploughs of these types.

A sod plough is sometimes used in the Jointer Class but as long as it has a share at least eight inches wide, cuts a furrow ten inches or over in width and uses a skimmer, there is usually no objection in BC but they were not allowed at the International Match at Stratford, Ontario, in October 1930, but were in competition in a few instances in the International Match of 1931.

Jointer ploughs are usually looked upon as utility or general purpose ploughs, and plough sod or stubble with the same equipment. A wheel is not often used with a jointer plough, the skimmer helps to take the place of the wheel. Wheels are not allowed in this class at the International Ploughing Match: there is not much room on the beam for a wheel, skimmer and coulter, and if a wheel is used the skimmer is usually crowded too close to the coulter.

**No. 4 - Stubble Ploughs** - Any of the makes of stubble ploughs work well if they are equipped with a skimmer, or skimmer and starlight coulter or swivel rolling coulter, but special mention might be made, for match work, of the 9x with a swivel coulter, a thirteen inch plough that will work either a twelve or fourteen inch furrow.

### Classes for smaller ploughing matches

The following classes are suggested for small ploughing matches:

No. 1 class. High cut. Open to all. Furrows to be eight inches or less in width, and not less than six inches deep.

No. 2 class. Sod class. Open to all who have not won a first prize in this class before. Furrows to be nine

to ten inches in width and not less than six inches deep. Sod ploughs only, no skimmers required.

No. 3 class. Novice. Open to all who have not won a first prize in this class before. Any style plough, skimmers optional.

No. 4 class. Juniors under eighteen years of age. Open to all who have not won two first prizes for ploughing in this class before. Any style plough, skimmers optional. In sod or stubble.

Suggested special prizes:
Best start in field.
Best finish in field.
Best working team in field.
Best groomed team in field.
Oldest ploughman.
Youngest ploughman.
Highest points in field.

### Classes for larger ploughing matches

The following classes are suggested for larger ploughing matches.

No 1 class. High cut. Open to all. Furrows to be eight inches or less in width, and not less than six inches deep. One cutting wheel allowed.

No. 2 class. Sod class. Open to all who have not won a first prize in this class before. Furrows to be from nine to ten inches in width. Sod ploughs only. No skimmers required, cutting wheel and coulter set the same as the high cut to cover the grass properly. This class should be of as much interest as the high cut and should be encouraged.

No. 3 class General purpose ploughs. Open to all who have not won a first prize in this class before.

Jointer ploughs only, with skimmers and coulter. Furrows to be from ten to twelve inches wide.

No. 4 class. Novice class. Open to all who have not won a first prize in this class before. Furrows to be from ten to twelve inches wide. Any style plough with skimmers.

No 5 class. Juniors under 21 years. Open to all who have not won a first prize in this class before. Any style ploughs with skimmers. Furrows to be from nine to twelve inches wide.

No. 6 class. Stubble. Boys under 17 years of age. General purpose of stubble ploughs with skimmers.

No. 7 class. Sulky.

No. 8 class. Gang.

No. 9 class. Tractors.

## Plan of Organization

1. A meeting in the winter where the whole question can be discussed and plans well laid ahead of time.

2. Live committees and the cooperation of everyone is required.

3. A good field, well located, is desirable. The field must be carefully staked out before the day of the match.

4. A reasonable amount of ploughing for each competitor.

High cut ploughs - 1/3 acre.

Walking ploughs - ½ acre

Sulky ploughs - 1 acre

Gang ploughs - 1-1/2 acres.

5. A carefully prepared list of rules and regulations. These should be explained to all competitors before starting their work.

6. Every director should be responsible for bringing at least one competitor (preferably a boy).

7. The competition should be open to all comers except men who have won a championship. A special class ought to be provided for them

8. The ploughing match should be held under the auspices of some properly recognized local association and all competitors should be members of that association, or their employees.

9. Uniformity in judging is necessary, by using the same score card, so that comparisons can be noted at different matches.

## Laying out a field.

It will be supposed that the field to be ploughed is 40 rods in length and the lands are made 33 feet or two rods in width.

A light furrow should be thrown out at both ends of the field. Starting at the side of the field, measure 22 feet and put in stake number one, and then place number two 33 ft. from number one, and all stakes along the end of the field are placed 33 ft. apart say, up to 20 lands or more. The stakes are also placed at the other end of the field. They should be painted white with black numbers and be well made so that they can be used year after year. For example let us suppose there are the following entries; - 6 walking ploughs; 3 sulky ploughs; 4 gang ploughs.

The walking ploughmen are got together; the numbers 1, 2, 3, 4, 5, 6 are put in a hat and each

draws. The men then get the land with the number corresponding to the one drawn; they set up their stakes (each ploughman furnishing his own), strike out and gather 6 visible rounds. After doing this they plough the land to the left of them, or haw, looking in the same direction as they did when they first made the start, and from the same end of the field.

On the other hand, since the sulky ploughmen are required to turn over one acre, they must have 2 lands. Therefore they are given the choice of lands 8, 10 and 12, and they carry on in the same manner as the walking ploughmen do.

The gang ploughmen have 1-1/2 acres to plough, therefore they must have 3 lands and so they have the choice of lands 15, 18, 21 and 24 and continue the same as the others.

It will be noticed that high cut gets the smallest lot to plough, because it is cutting a narrower furrow, but time is saved in the spring, as high cut ploughing properly done doesn't need any discing or seed drill. The seed is sown broadcast, and it comes up in rows the same as if a drill had been used. High cut ploughing, properly done, should put the grass flat on the bottom of the furrow, in close contact with the subsoil, where the grass is buried absolutely, with no chance of growing again among the grain. In high cut ploughing there should be no air space below the furrow, as is left with other ploughs. The hind part of the mouldboard should pack the furrow firmly up against the former furrow; the former furrow should be seen to move a little in the packing process. This packing process keeps the grain or peas from falling down between the furrows and getting lost or buried too deep. This is very attractive work when well done.

## Suggested rules for Ploughing Matches

The following rules have been compiled from those in force at numerous ploughing matches. They are intended as a guide to agricultural societies and may have to be changed to suit the particular ideas and conditions of any society:

1. Entry fee of $1.00 boys under 20 free. Anyone contributing $2.00 or over to the prize fund will be allowed an entry for himself free.

2. All entries must be made before ............... So that the ground may be staked out in time.

3. All ploughmen must be on the ground before 10:00 am when lots will be drawn for position. Ploughing will start by signal at 10:30 am and must be completed by 4:00 pm Any ploughman coming after lots are drawn will be obliged to take his lot in rotation with those who have drawn and finish not later than the time limit. Any ploughman not finished when time is called will be ruled out.

4. The land to be ploughed shall not be more than one-third of an acre for the following classes: high cut class, eight-inch wide class, and for nine inch wide sod class, and one-half of an acre for wider cutting walking ploughs.

5. Each ploughman will make one strike out and one finish. Scratch furrows to be thrown in, when break is made in sod. Six rounds shall constitute a crown.

6. No person will be allowed to interfere with or help the ploughmen, except in setting up and removing

poles, and no person will be allowed to accompany the ploughmen.

7. In all cases the crown must be opened out and finished with the same plough and team that is used throughout.

8. The depth of the ploughing is to be not less than six inches. The depth should be reached by the third round after opening up. The last green furrow in the finish to be turned towards the crown. The sole furrow is to be lifted, and also turned towards the crown. Ploughmen are not required to finish a furrow of full width.

9. Gauge wheels are optional in all cases, but skimmers must be used in the wider cutting ploughs, but not in the high cut, or in the nine inch wide sod class.

10. Chains or other devices for covering weeds or stubble to be allowed in the stubble classes only, but all weeds or stubble must be cut.

11. After first two visible rounds, no pulling or covering of weeds or stubble either by hand or foot, or tramping of land with feet is allowed. Anyone not conforming to this rule will be reduced one point for each offence. This refers to all classes.

12. In all cases where not more than one entry is made, the judge shall decide what prize, if any, shall be awarded. The judges have the right to withhold a prize if 50 points have not been made.

13. Any ploughman not conforming to the above rules will be disqualified.

14. No protests from decision of judges or committee will be allowed.

The Plow in Hand

J.I. CASE
PLOW-WORKS
RACINE, WIS.

The Sign of the
BEST Steel
Plows

*Fig. 466. The Case Plow Works L:ogo.*

## Judging Plowing.

*The information which follows is USDA textbook for judging the overall quality of plowing.*

Before going into the discussion of the design of the bottom for doing good plowing, one should consider first, what constitutes good plowing. Good plowing consists of turning and setting the soil into even, clean straight furrows of roundish confirmation. The main points to consider are the following:

1. The top of the furrow may be slightly ridged.

2. The soil must be pulverized thoroughly from the top to the bottom of the furrow.

3. Each furrow must be perfectly straight from end to end.

4. All back furrows must be slightly raised and all trash completely covered.

5. The outline of the furrows must be in a point without break or depression.

6. All trash must be buried completely in the lower right hand corner of the furrow.

7. Furrows must be thoroughly uniform with one another.

8. The depth of all the furrows must be the same, continuing in uniform depth.

9. The dead furrows must be free from all trash on the ground.

These rules are by which a plowing test may be judged. However, if these rules were followed in all sections of the country where different types of soil are found, the best seed bed would not always be made. The main things to consider in plowing are that the land be completely broken, that the soil is thoroughly pulverized, and inverted, with no air spaces left between the furrows. These are conditions that may be applied to any section. The whole bottom is essential for good plowing, the share cutting and slightly lifting the furrow slice, the landside controlling and steadying the plow, while the moldboard completes the lifting, pulverizing, and inverting of the furrow slice. It is upon the moldboard that the main part of successful plowing depends. The curvature and length of the moldboard determine the degree of pulverization the furrow slice will be given.

Fig 467. Good tractor plowing measured against good horse plowing. Some problems are solved by picking the better way of working.

## Chapter Fourteen

# Problem Solving

The efficiency of any plow will depend on the plowman's understanding of all the basic infor mation so far included in this text. The elimination of problems comes simply from the understanding of their cause, and through that understanding a correction. With horsedrawn plows that correction is usually either 1. a simple adjustment or 2. a broken or bent part or 3. intervention by the plow fairies. Numbers 1 and 2 we can help you with. Number 3 will take the patience of Job **and** the imagination of a child, neither of which this author fully understands. But please know this, should the plow fairies deem you worthy of their chicanery consider finding something else to do until they tire of waiting for you and move on to the neighbor.

Each class of plow, (moldboard and disc) as well as each type of walking and riding plow, has troubles and adjustments which are common to all. They also have troubles and adjustments that are applicable only to one certain type.

## Walking-plow troubles

**Running Too Deep.** Running too deep is a common problem. The plowman ends up working too hard trying to keep pressure on the handles. This tendency to plow too deep may be caused by:
1. too much suction;
2. the vertical hitch of the clevis may be too high;
3. or the beam may be bent upward.

It's not often that a plow will have too much down suction, but if this does happen the remedy is a very simple one: decrease the amount of suction. (See page

no. 28) If the hitch has been placed too high, this can be remedied by lowering the vertical hitch at the clevis; shortening the traces to bring the horses closer to the plow will also remedy this trouble. (See page no. 58) If the beam has been sprung or bent upward (See page no. 56, Fig. 127.), as it may with steel beams, it is very difficult to bend the beam back to the original shape. The best remedy is to get a new or different plow.

**Failure to Plow Deep Enough.** This trouble may be due to a number of causes, some of which are the reverse of deep plowing such as:
1. the hitch too low
2. and the beam bent down.

All these troubles result in the plowman having to lift up on the handles to make the plow penetrate the soil. If the hitch is too low this can be fixed by raising it at the clevis or lengthening the traces and moving the horses away from the plow. (See Chapter Three.)

If the beam is bent downward, the same remedy is used as in the case of the beam being bent upward.

If the share has become dull, it should be taken off and sharpened. It may also need repointing.

If the plow won't enter the ground because of its being too hard, the plow may be forced into the ground by adding weight. But it might be a better idea to wait for more favorable conditions (i.e. after a rainfall).

**Not Taking Enough Land.** When a plow won't take the required amount of land the following things may be wrong:
1. too little horizontal suction;
2. hitch too far to the left;
3. too much wing bearing;
4. beam sprung to the left;
5. improper landing of beam;
6. landed too much in the rear;
7. coulters may not be set properly.

The remedy for all these causes is simple and

*Fig. 468. The use of horses or mules does not necessarily guarantee good or better plowing. As this old photo will attest, crummy plowing can be accomplished if the plow is not set up properly.*

indicated by the cause of the trouble, and covered in Chapter Three.

**Taking Too Much Land.** This problem is just the reverse of the previous one, and the causes may also be just the opposite such as;
　1. hitch too far to the right;
　2. too much horizontal suction;
　3. not enough wing bearing;
　4. beam bent toward the open furrow;
　5. jointer or coulter set improperly, leading toward the land.

The remedies for these causes are obvious. and covered in Chapters One, Two, and Three.

**Failure to Scour.** Failure to scour is a very common problem in the Southwest, in the blackland section, and in the close grain soils of the river bottoms. The plow 'scours' when the soil comes clean from the moldboard. If your plow won't scour, the soil will stick to the plow and will not shed off. Non-scouring may result from a number of causes:
　1. The soil won't polish.
　2. Improper plow adjustment.
　3. Poor fitting of share and moldboard.
　4. Cutting edge of share not level
　5. No suction.
　6. Soil conditions not right.
　7. Soft spots or irregularities in the moldboard.
　8. Shape of bottom with relation to soil texture.
　9. Speed.
　10. Type of bottom not suited to soil.

The lack of polish may be caused by some covering over the moldboard (i.e. paint, varnish, crystalized oil or grease, etc.). There are some cheap plows, primarily circulating in third world trade, which have bottoms made of soft and inferior metals. These may work well only in certain easy soil conditions and only for a short time. There are also after-market shares, cast in quick and dirty fashion, which may look to fit well but which are fast to pit and difficult to scour. Any plow that is made of metal that is not hard enough to withstand the scratching of the soil and will not take a

good polish is always likely to give trouble. When plows were/are completed at the factory, the surface was/is finished by grinding. Grains left by the grinding process should run in the same direction that the furrow slice moves over the plow, or lengthwise of the moldboard. If this is done and the temper carefully made in the high carbon steel, very little trouble is likely to be encountered by non-scouring in the average soil.

Improper adjustments may cause non-scouring. For example if the new or replacement share doesn't fit snugly to the moldboard it could result in a depression or lip between share edge and moldboard edge. This uneven surface at the joint is likely to leave a place where there will not be enough pressure to prevent the soil from sticking. And a ribbon or spot of sticky soil will grow and cause non-scouring over the whole moldboard. So make sure the share and moldboard fit creates a smooth follow through on the curvature.

A dull cutting edge and no suction can be remedied by sharpening and placing more suction in the share to make it penetrate and give pressure on the moldboard. Non-scouring can be remedied, to some extent, by increasing the pressure of the furrow slice on the moldboard, as will be seen under the discussion of speed. The soil conditions may not always be just right for good scouring. The lack of a proper amount of moisture in the soil will sometimes cause non-scouring. Very little difficulty is ever encountered, however, in sandy and loamy soils. It is the clay and clinging soil that gives trouble. If the plow has not been hardened uniformly all over the surface, there by leaving soft spots, the surface will wear away faster and cause dents where soil will hang and cling. Non-scouring may also be due to the use of improper plow bottom shapes. (See page 29 Fig. 63.)

The lack of the proper speed will sometimes cause the plow not to scour. The plowman will often notice that if the speed is increased in the non-scouring part of the field, that the soil will shed much better. This is, of course, due to the pressure resulting from an increase of speed, practically forcing the soil off the surface of the moldboard. But the 'forcing action' will cost the plowman in quality as the speed will tend to 'throw' furrows with erratic coverage of sod, stubble and trash. And there are some conditions under which high speed may actually cause failure to scour.

**Methods used to aid scouring.** Our historical research has turned up many interesting approaches in the Southwest to provide a type of plow that will scour under almost all conditions. In a few places farmers reported making a practice of using plaster of Paris on their plow moldboards. It had to be replaced frequently, often once every day.

The best results were reportedly obtained from heating the moldboard. Those moldboards that were heated before plowing shed the soil and scoured well. Moldboards that were on the same plow that were not heated would not scour. This is directly in line with the observations made by other plowmen who have noted that after the plow has stood in the sunshine and become heated, it plows better than after it has become cooled.

There were experiments in England showing that when an electric current is passed through the soil having the moldboard as the negative electrode the soil would slip over the plow easier. It appears that the whole problem resolves into a soil problem and the effect of heat on soil colloids. Some complex research found that if the metal was hot, a decrease in the sliding friction was observed, but if the metal was cold and the soil was warm, the moisture films in contact with metal adhered to it.

**Excessive Draft.** Some of the things already discussed in this book with regard to excessive draft in ordinary walking plows are: the hitch, the condition of the share, side draft, position and type of moldboard, set of the plow, and the condition and set of the coulters.

## Sulky plow troubles and adjustments

Since the sulky plow is a riding plow, the troubles encountered will be somewhat different. The plow is mounted on three wheels (sometimes two) which support the weight of the plow bottom and at the same time influence the operation. Many of the troubles discussed under the walking plow are applicable to the sulky plow.

**Draft of Sulky Plows**. It is ironic that although the riding plow's claim to fame is reduced draft requirement, the principal trouble of this type plow is that of the heavy draft. This problem may be caused from a combination of several troubles: dull shares, too much suction, too much landside friction.

The remedy for dull shares is obvious. Too much vertical suction may result from the rear of the plow bottom being raised too high upon the rear furrow wheel shank or the rear of the beam elevated too much, as on the frame type of plow. (See page no. 76 Fig. 198.) Some term this heel clearance. The suction is given by elevating the rear of the plow. When new shares are placed on the plow, this may cause too much vertical suction, because the adjustment which has been made to suit the old worn share has not been changed.

**Adjustment of Wheels.** Too much landside friction may be caused by the rear furrow wheel not being set far enough to the left of the landside and not having enough lead away from the furrow wall. In the adjustment of the sulky plow, both frame and frameless, the rear furrow wheel is usually given a lead away from the furrow wall and gives landside clearance. This throws the landside away from the furrow face and prevents friction which would result from sliding along in contact with the furrow face, as in the case of the walking plow. The front furrow wheel is given a lead toward the furrow wall; it is also inclined. This is done to hold the plow in its proper place and to overcome the side draft caused by the pressure of the furrow slice upon the moldboard. It is essential that all sulky plows be run with the frame level.

**Scouring.** Scouring has been discussed under that of walking plows and will apply to the sulky plows.

## Gang-Plow Troubles

Since there is very little difference in the construction of the gang plow and that of the sulky, the same troubles will apply; however, there are one or two troubles that may be mentioned that are common only to gang plows.

**Uneven Furrow Crown**. In an old gang plow that has become badly worn, the furrow crown may be left uneven; that is, some of the furrows may be left high and others low. This may be due to the front bottom cutting deeper than its mate or the front bottom cutting wider than its mate. These troubles may be the result of the frame being loose at the joint, or to the bearing at various points allowing considerable play of the plow in operation.

The wheels of gang plows will influence the operation of the plow and are handled slightly different from those on sulky plows. The rear wheel is given a lead, usually away from the furrow wall as in the case of the sulky. The front furrow wheel is given a lead away from the furrow wall because of the additional amount of side draft in the gang plow. The rear wheel is also set to the left of the line of the landside to give landside clearance.

## Disk-Plow Troubles

The troubles and adjustments of the disk plow are different from those of moldboard plows. Yet there are many troubles that are common to both classes.

**Failure to Penetrate**. Failure to penetrate soil may be due to a lack of weight and the proper angle of the disk. The disk plow is weighted and partially forced into the ground. Therefore, if the frame is made rather light and the soil is hard, the plow may not penetrate easily. Provision is made on most disk plows for the placing of additional weight, especially on the rear wheel. Changing the angle of the disk to set nearer perpendicular will increase the tendency to penetrate. (See Chapter Seven)

**Width of Cut**. It is often necessary to change the adjustments on a disk plow as the hardness of the soil varies. With a soft soil a wide furrow is desirable, but with a hard soil a narrow furrow is essential. Moving the beam on the front furrow-wheel axle will influence the cut, as will changing the position of hitch and lead of the furrow wheels.

Under most conditions, the front furrow wheels should run straight forward, parallel to the line of the furrow. If the plowing is exceptionally hard, it may be given a slight lead to the furrow wall, Usually, the wheels are given a lead toward the plowed ground to counteract the side pressure of the furrow slices. Since disk plows do not have landsides, the wheels must hold the plow in position.

Should there be a tendency of the bottoms to trail, it may be due to the hitch being too far to the right. This arrangement of the hitch will have a tendency for each of the bottoms to cut a narrow furrow width due to their trailing behind one another.

Disk plows do not always cover trash as well as do moldboards. This is especially true when they are operated without scrapers. If the side is set rather flat from the vertical, it will not cover trash as well as when set more nearly straight up and down. When set straight up and down, the furrow slice will be thrown over more abruptly to the side. If the scraper is in use, the furrow slice will be taken from the disk and turned. The scraper having a curved surface will turn the furrow slice better than the straight type; however, the straight type will shed soil better and give less trouble when sticky soils are being plowed. As a general rule, the scraper should be set low and at an angle of about 35 degrees with the disk. It is also tilted to throw the soil toward the furrow.

## What causes a wheel plow to upset

Upsetting is most common on the three-wheel pole plows, although it occurs sometimes on the poleless plows. When hitching too close and too low on a three-wheel plow, the weight to the front furrow wheel and on the land wheel is lessened to a considerable extent. Instead of the weight being carried on these wheels in equal proportion with the rear wheel, additional weight is thrust upon the rear wheel and a considerable amount of it is carried through the hitch by the team. This suspends the plow between the horses' shoulders through the traces and evener, and the rear wheel. With the land wheel a considerable distance away from the plow, it is practically impossible to upset the plow toward the land or unplowed ground. The furrow wheel stands very close to the beam, allowing for just the width of furrow between the rim of the wheel and the beam, so it does not require a great amount of effort to upset the plow toward the plowed ground. This usually occurs when turning "gee" with a right-hand plow or on hillsides. The remedy for this condition is to lengthen the hitch or to raise it on the vertical clevis. In all cases, it is better to lengthen the hitch.

On practically all three-wheel plows, when equipped with pole, the pole is attached to the top of the furrow axle almost directly over the furrow wheel. The evener is attached to the point of the beam, or in case of a gang plow, to the cross clevis, some distance to land from the point to which the pole is attached. In the case of a fourteen-inch gang plow, this distance is considerable.

The neckyoke is attached to the pole by means of a sliding device, providing from 12 inches to 18 inches of room for the neckyoke to play on the pole forward or back. In turning "haw" with the right-hand plow, the neckyoke will slide forward on the pole, and the traces will slacken so the plow will turn freely, even though the hitch will be too short or too low. In turning "gee," however, the effect is exactly opposite. The neckyoke slides back on the pole to the stop, the traces tighten, and if the hitch is just a little too short, the plow will upset very easily. Again, the remedy is to lengthen the traces. In case of emergency, where traces do not permit sufficient adjustment, move the stop on the pole backward so the neckyoke can slip back farther on the pole when turning "gee". This will prevent the traces from tightening and pulling the plow over.

It is taken for granted that the plow is correctly assembled and all adjustments properly made; particularly important, the share must be in good condition. A share badly worn and rounded off like a sled-runner will cause the plow to ride out of the ground and upset easily.

## A Plow's Effect: good and bad.

There is a long standing debate on the pros and cons of the moldboard plow. This book would not be complete without some acknowledgement of those concerns. It is this author's contention that the diversity and attention to detail, which are earmarks of small scale low-tech agriculture, offer ample opportunities to nullify or ameliorate the plow's negative effects. That said what follows is the core debate.

*The goal for plowing and cultivation is to control the water supply by removing weeds and leaving the surface of the soil covered with a loose, dry mulch to retard evaporation.*

*The principle object of plowing is to loosen up the soil, for four purposes: (1) To enable the soil to absorb the rainfall more quickly and more freely than it would in its undisturbed condition; (2) to maintain more of the rainfall near the roots of plants; (3) to admit fresh air to the roots of plants; (4) to enable the roots of the young or quickly growing plants to penetrate the soil more easily.*

*A soil with a compact surface quickly dries out, and the water supply fluctuates rapidly and excessively, to the detriment of most crops during their growing period.*

*There is a serious defect in the principle of the common plow which, upon some soils and with certain kinds of plowing, is liable to have very serious effects. If a field is plowed for many successive years to a depth of 6 to 8 inches the tendency each time is to compact the soil immediately below the plow, thus rendering it more impervious to water; that is, the plow in being dragged along plasters the subsoil just as a mason with his trowel would smooth out a layer of cement to make it as close and impervious to water as possible. This is undoubtedly an advantage to some soils, but, on the other hand, it is very injurious to many.*

*The injurious effect of this compact layer formed by the plowing is twofold. It makes it more difficult for the rainfall to be absorbed as rapidly as it falls, and increases the danger of loss of water and injury to the soil by surface washing. Soils plowed at a depth of 3 to 4 inches, which is quite common in many parts of the country, would have a thin layer of loose material on the surface, with a compact subsoil below, into which water would descend rather slowly. With a rapid and excessive fall of rain, the light, loose topsoil is liable to be washed away by the excess of water, which can not descend into the subsoil as rapidly as it falls. This washing of the surface and erosion of fields into gullies occasion the abandonment of thousands of acres of land. The field will not wash so badly if it is not plowed, and, on the other hand, it will hardly wash at all if the cultivation is deeper and subsoil left in a loose and absorbent condition. The deeper the cultivation, the greater the proportion of rainfall stored away and the less danger of the erosion of the surface soil and the less serious the defect of our common method of plowing. While there is less danger from washing, however, with deep cultivation there is still a tendency toward the formation of a hardpan at whatever depth the land is plowed.*

Chapter Fifteen

# Deep Tillage Disc Plows

These plows are so unique, so obscure, and so specialized that we decided they deserved a little chapter all their own. The chances of seeing one of these, let alone owning one, are remote at best. We put this information near the back of the book because we deem it an oddity which may not see much use. That is unless some enterprising small firm decides to try to build them again. Something we would definitely encourage as this plow would be of tremendous value in some regions.

We discovered only three makers of the deep tillage disc plow but are certain others did exist. Far and away the literature and advertising campaigns of the Spalding out did the others. We offer here the illustrations we could find and some of the superlatives. Keep in mind that they were bent on selling these plows and may have been prone to stretch the truth only just a little. They did not yet have the successful model of the modern lying professional politician to pattern their advertising after.

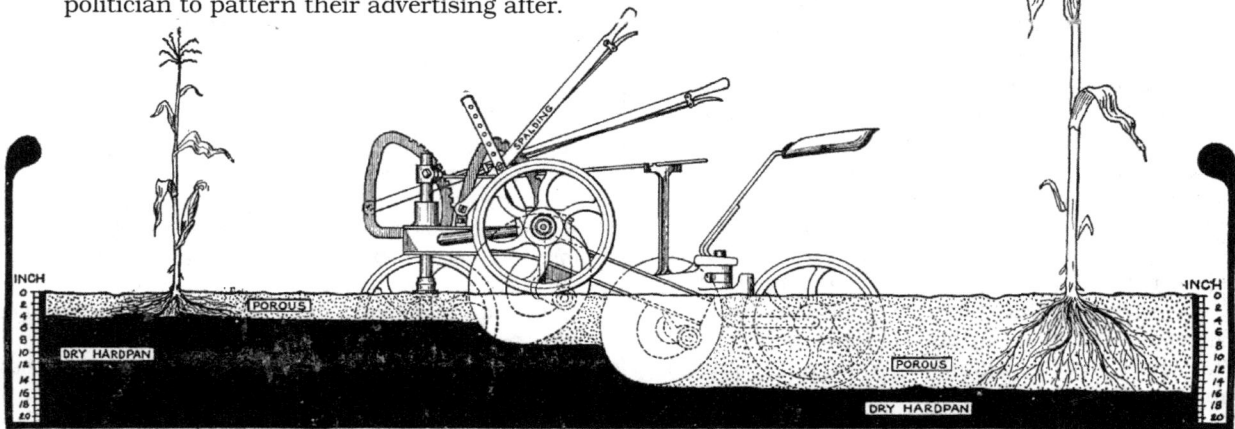

ORDINARY SEEDBED    Fig. 469 Spalding Deep Tilling Machine    SPALDING SEEDBED

Fig. 470. (left) Spalding plow at work

Fig. 471. California Deep Tilling Machine at work. (right).

Fir. 472. (below left) Crane's Combined Sulky, Gang and Deep Tilling Plow.

As a gang plow

# California Deep Tilling Machine
## and Combination Plow

As a deep tiller

Fig. 473. 'California' brand machine would plow 16 inches wide and 16 inches deep. The plow was quickly convertible to gang or sulky allowing for more general use.

"This machine was specially designed for plowing the soil to a greater depth than heretofore accomplished by any mechanical means, or by any implement (other than the spade) and to preapre a thoroughly good and well drained seed bed at a single operation."

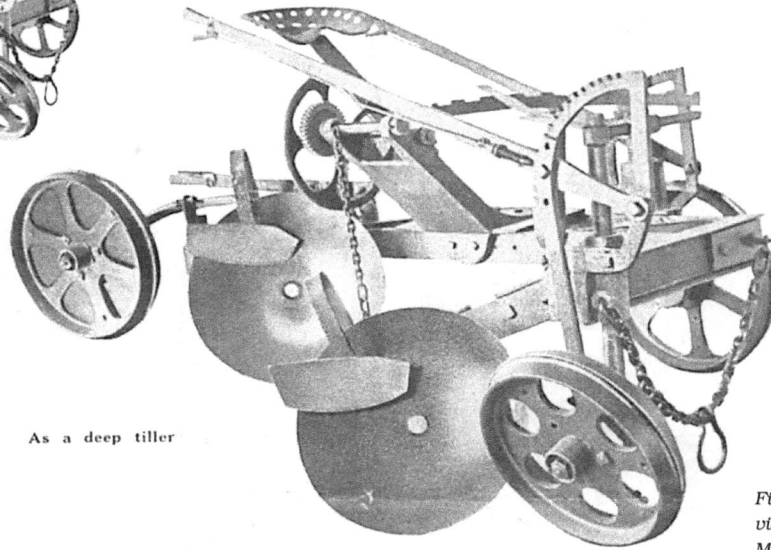

Fig. 475. (below) A front view of the Spalding Machine

Fig. 474. The Spalding Machine working up a garden site.

Fig. 476 &  477 Two view os the Spalding Plow in different field conditions .

# For Your Land's Sake

## USE THE SPALDING DEEP TILLING MACHINE

Fig. 478

Fig. 479. The Spalding seen from behind while in a turning position.

## The Operation of the Spalding

"The front disk cuts from four to eight inches and turns the top trash over into the bottom of the furrow. The rear disk follows in the furrow of the front disk and cuts from six to nine inches deeper. The greater portion of the earth turned by the front disk comes within the path of the movement of the earth cut and lifted by the rear disk, and as the earth passes across the face of this revolving rear disk it is disintegrated, pulled apart and thoroughly mixed with the earth plowed by the front disk, leaving the top trash turned under by the front disk completely covered in the bottom of the furrow..... In reality it is a verticle gang, plowing two furrows deep at a single operation."

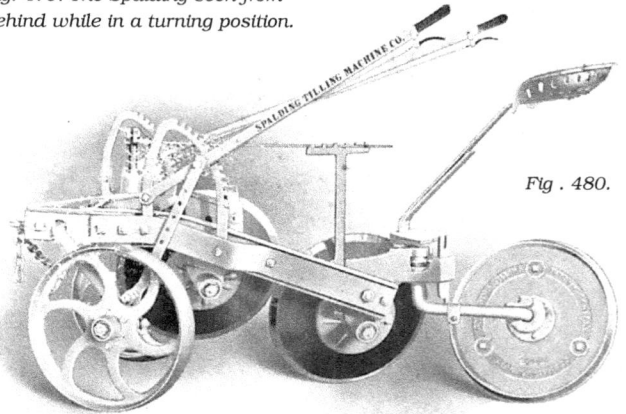

Fig . 480.

"...goes straight to the downright enjoyment and true delight that may be had out of converting the whole farm into a garden..."

SPALDING DEPARTMENT

# GALE MANUFACTURING CO.

## ALBION, MICHIGAN

### MAKERS OF AGRICULTURAL IMPLEMENTS

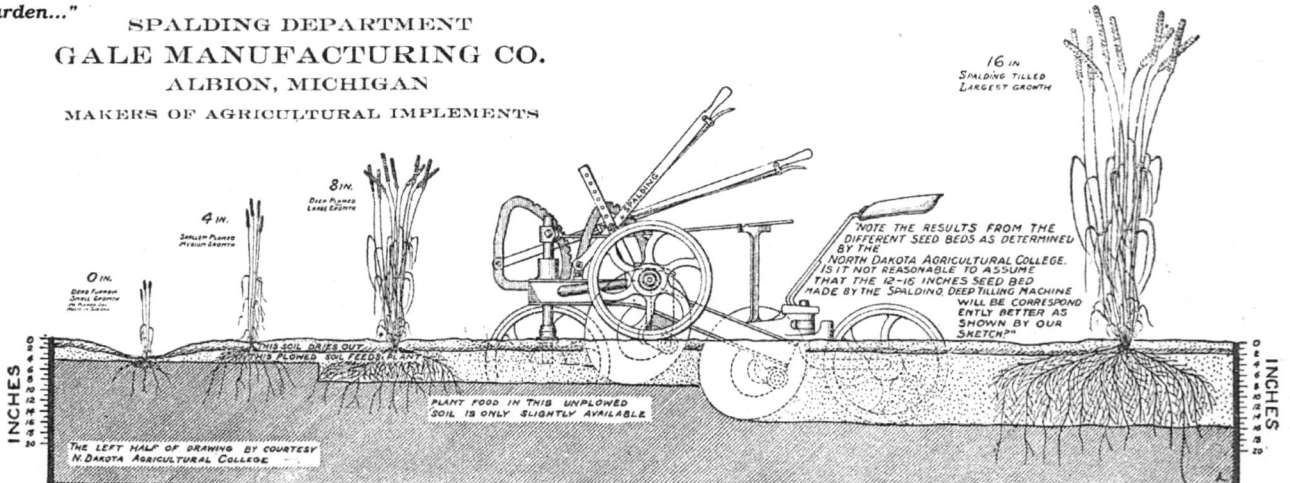

Fig. 481. The Spalding literature had many nifty illustrations to dramatically portray the net effect of its operations.

Chapter Sixteen

# How They Sold Their Plows

Some reprints of the ads and literature the plow companies
employed to entice the discriminating farmer to buy their goods.

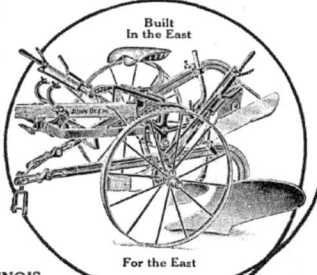

NO company went to such heights in its efforts to woo the farmer with common sense information than did the John Deere Company. For every one of their pieces of equipment they produced attractive, informative brochures with a great deal of the type of information that farmers will always appreciate. We are reproducing the entire contents of one of those brochures here for your information and as a final salute to what was truly an amazing chapter in North American history, the golden era of agriculture.

## John Deere-Syracuse No. 210 Sulky

### A Light-Draft, All-Wheel-Carried Plow for Every Type of Soil

The John Deere-Syracuse No. 210 Sulky Plow is acknowledged to be the lightest-draft plow of its type.

It does an extra good job of plowing in any kind of soil and under all conditions.

It runs level and plows at uniform depth, always--even when turning square corners.

It's the all-wheel-carried plow that has established its superiority wherever the use of this type of plow is practical.

The advantages of the No. 210 over the ordinary sulky are many. The special design of rolling landside, and the fact that the plow can be used with either the Syracuse or John Deere clean-shedding bottoms are features responsible for the extremely light-draft and good working qualities of the No. 210 in a variety of conditions.

# JOHN DEERE
## SYRACUSE
### NO. 210
## SULKY PLOW

*The* **LIGHT-DRAFT PLOW THAT MAKES GOOD WORK EASY**

When you buy John Deere Implements you are sure of prompt repair service during their long

*Showing the John Deere-Syracuse No. 210 Sulky equipped with 16-inch bottom with wing extension. This equipment assures clean covering and is used effectively in sections infested by the European corn borer.*

# The Rolling Landside Carries the Rear Weight

### Rolling Landside Assures Light Draft

The rear weight of the No. 210 is supported by the rolling landside--an original Syracuse feature--which serves as the third leg of a triangular rolling support. All of the body and frame weight is properly balanced and rolls on three bearings that run in oil. Thus, friction between the plow and the bottom of the furrow is eliminated, and side pressure against the furrow wall is minimized.

### Rolling Landside Means Share Economy

Shares stay sharp and last longer on the No. 210 Sulky than on other riding plows, because the rolling landside prevents the plow from running on its nose--a position ruinous to shares. Neither is there any dragging friction, which is equally destructive to share life and light draft.

### Requires No Adjustment In the Field

The relation between the rolling landside and the share is permanent--it is "built in" at the factory. No adjustment in the field is ever required.

### Front Furrow Wheel Governs Plow

Aside from the rolling landside and the interchange of bottoms suitable for all sections, the No. 210 has one big, outstanding feature that is found only on this sulky--it's the front furrow caster wheel that actually governs the plow, the controlling device that automatically keeps the plow cutting furrows at uniform width and depth at all times.

*Solid furrow wheels do not gather trash, weeds or grass.*

*This view of the No. 210 shows the rolling landside to good advantage. Notice the solid, smooth surface and cleaning device which prevent clogging.*

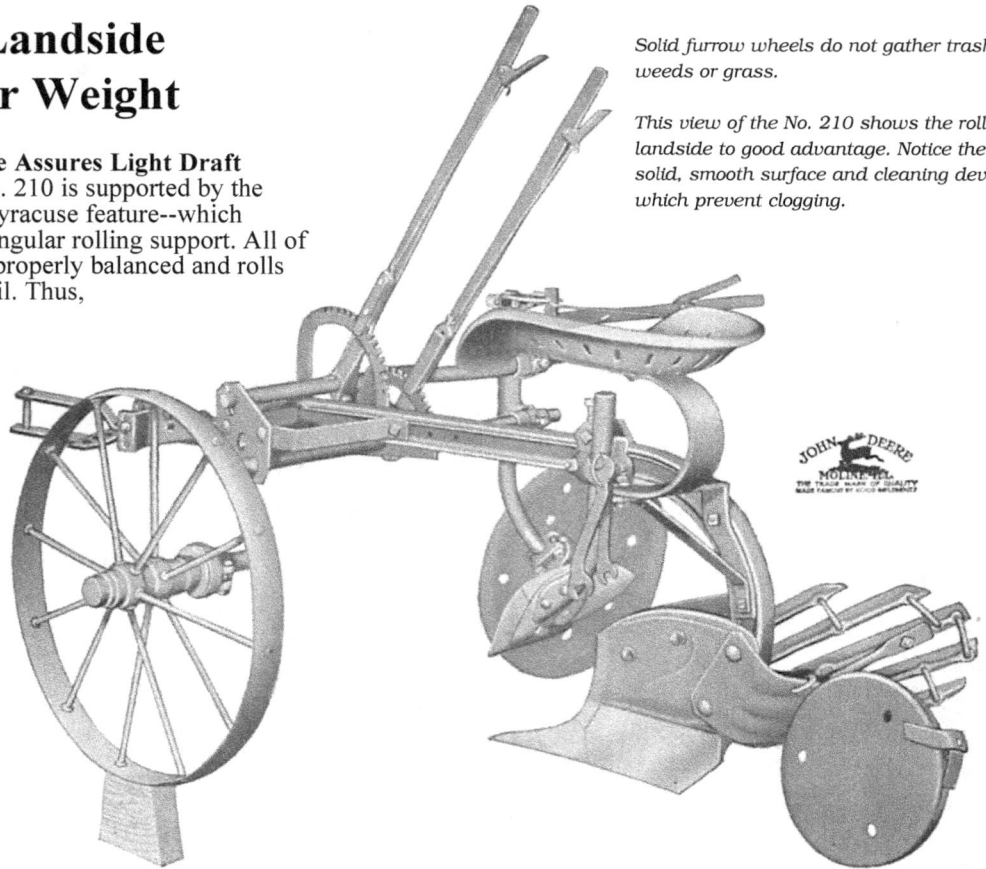

### How It Works

With the sulky settled to its work in the straight furrow, the controlling lever is in correct position when the plunger is resting in the center notch. The sulky is then being governed automatically. Just before reaching a corner, the latch is released by a simple hand movement.

As the team and plow again settle to work on the straight-away, the lever seeks its proper position and the plunger engages automatically. Except for the movement required when releasing the latch at the turns, the operator has both hands free at all times for managing his team. The whole operation is unusually simple; the control of the sulky at the turns is a revelation to plowmen.

### Special Front Furrow Wheel Support

The front furrow wheel support on the John Deere-Syracuse No. 210 Sulky is the first to be really successful. It does not cramp or bind--faults which are frequently found with a slide bearing.

The support consists of two round steel axle bails, set one above the other, and having duplicate offset. A single lever operates both axle bails in unison. The inner and outer ends of the bails revolve in oiled bearings, making the vertical motion smooth and practically frictionless--there is no binding or cramping.

### Adjustment for Furrow Width Is Easily Made and Is Positive

In order to adjust the front furrow wheel in or out for a narrower or wider furrow, it is only necessary to loosen one set-screw on each bail, set the wheel where desired, and tighten the set-screws.

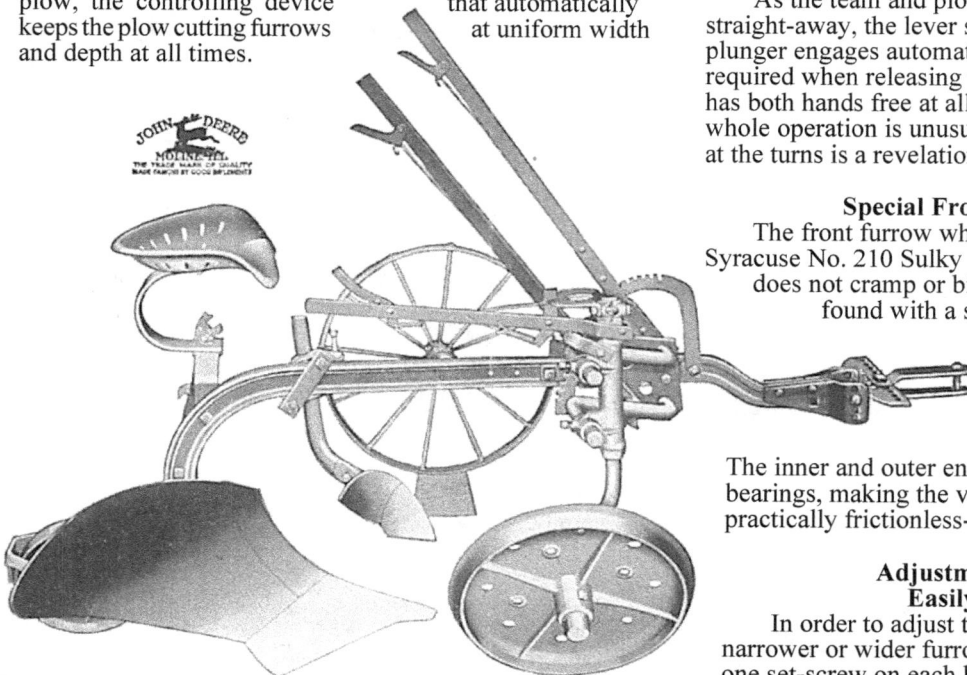

*This view shows the John Deere-Syracuse Right-Hand No. 210 Sulky with Jointer and Syracuse chilled moldboard with detachable shin piece.*

# John Deere-Syracuse No. 210 Sulky Plow

The John Deere-Syracuse No. 210 Sulky is furnished in both right-hand and left-hand styles.

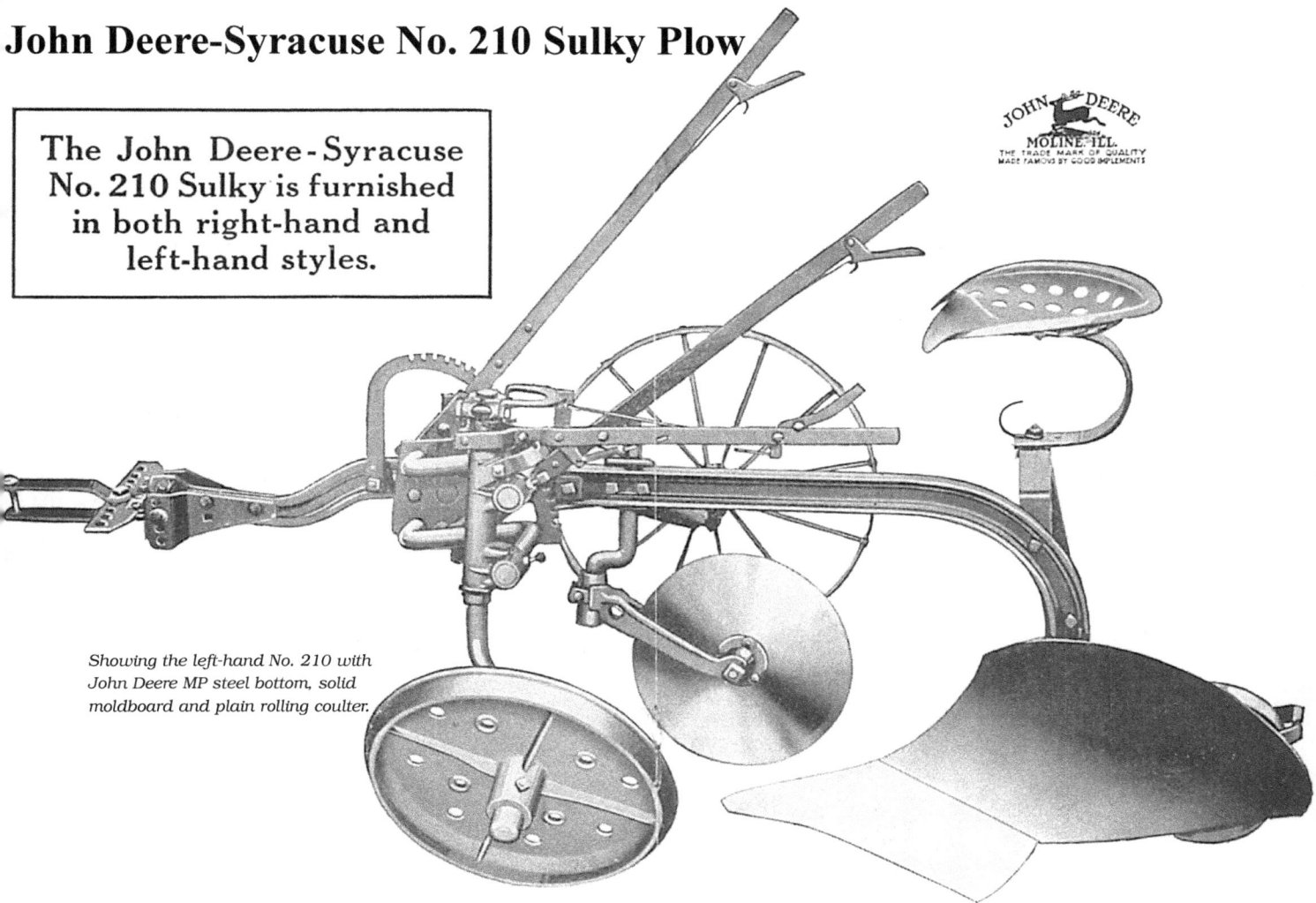

*Showing the left-hand No. 210 with John Deere MP steel bottom, solid moldboard and plain rolling coulter.*

### Sand-Proof Bearings--Trash-Proof Furrow Wheels

The front furrow wheel and land wheel are provided with the most up-to-date type of magazine axle box obtainable; also sand caps, which completely cover the hubs and prevent sand from working into the bearings. Both furrow wheels are of the web, or solid construction, spindle type, and will not gather trash, grass or weeds that otherwise might become wound around the axle. The rear furrow wheel, or rolling landside, revolves on a chilled bearing.

### Strong Clevis Has Wide Range of Adjustment

The evener clevis is of the rigid type. With this clevis, the evener is held up when the team slackens, as in turning, making it difficult for the horses to step over the traces. A wide range of adjustment is provided, both lateral and vertical, to take care of any size of two- or three-horse team.

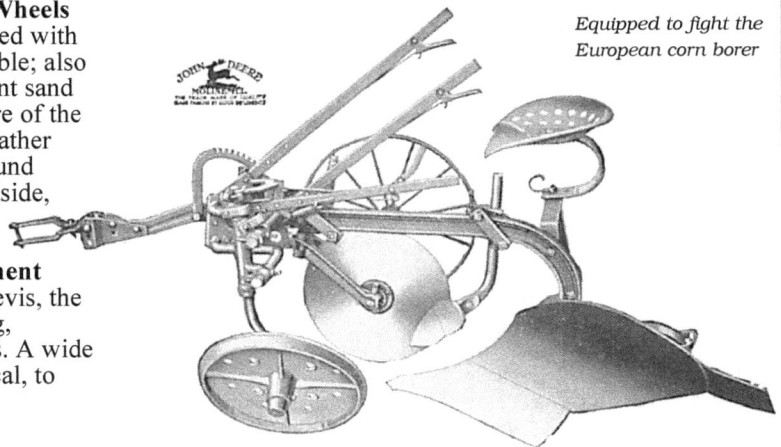

*Equipped to fight the European corn borer*

*John Deere-Syracuse No. 210 equipped with 16-inch bottom, moldboard extension, 18-inch rolling coulter and independent jointer; 10-foot trash wire also included.*

EQUIPMENT FOR JOHN DEERE-SYRACUSE CORN-BORER PLOWS
*This view shows the equipment furnished on each bottom of these plows: Moldboard wing extension, 18-inch rolling coulter, independent jointer and trash wire (10-ft.).*

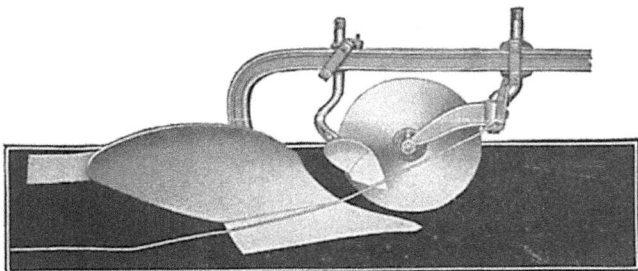

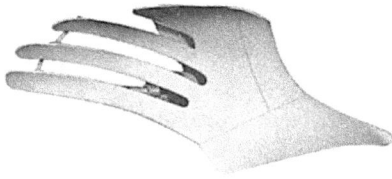

*Syracuse chilled Slat Moldboard*

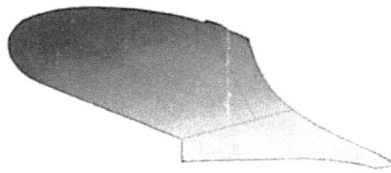

*Syracuse Chilled General-Purpose Bottom*

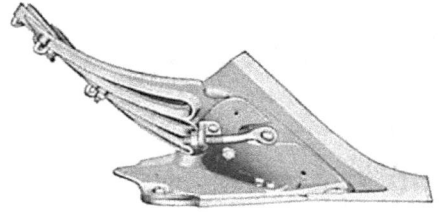

*Syracuse Quick-Detachable Share*

### Convenient Location of Seat and Levers

For convenience of operation, also, the No. 210 Sulky has no equal. The seat is placed to one side, well to the rear and low down. The foot rest is convenient and substantial. The man on the seat is made comfortable; he has all operating levers within easy reach.

# John Deere-Syracuse
# Plow Bottoms

John Deere-Syracuse plow bottoms are superior to other chilled bottoms. They are the result of many years' experience in plow manufacture as well as constant study of the various soil conditions that require their use.

Each bottom is carefully inspected, and passed only when it conforms to the John Deere high standard of quality. The materials used are of the highest grade obtainable.

Syracuse chilled iron is the hardest metal used in plow construction. All Syracuse chilled moldboards, shares, landsides and shin pieces are poured against the chill with the wearing surface down. This fact is important, because when iron is poured, the impurities rise. The result is that every Syracuse part, by reason of the Syracuse method of chilling, will show a smooth, regular and even wearing surface without holes or pits.

A secret annealing process is employed to prevent castings from warping when exposed to the air. This process assures uniformity and an entire absence of shrinkage cracks.

In addition to the wearing qualities that are insured by the material and skill employed in their manufacture, Syracuse plow bottoms are designed with an intimate knowledge of the soil requirements in every locality where they are sold. They scour perfectly and create the most productive seed beds.

Syracuse bottoms equipped with the Syracuse slat moldboard have no equal for cleaning, shedding and pulverizing in sticky land, and they are noted for their extremely light draft.

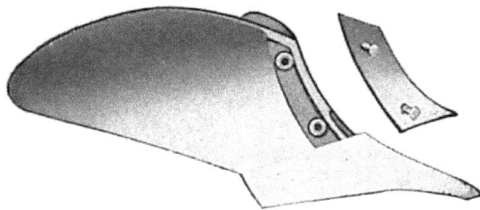

*Syracuse Detachable Shin Piece.*

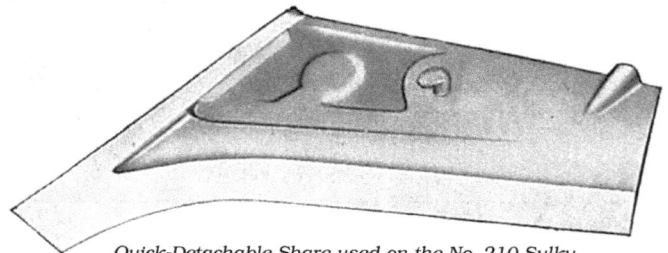

*Quick-Detachable Share used on the No. 210 Sulky.*

### Quick-Detachable Shares

All bottoms for the No. 210 are furnished regularly with quick-detachable shares.

The quick-detachable share is easy to put on and take off.

The share is held firmly in place by an eyebolt hooked over a stud.

Simply loosen the nut, lift the eyebolt from the stud, and the share can be easily slipped off.

Slip the share back in place, hook the eyebolt, tighten the nut, and the share is on tight.

And here is an important point to consider: When the nut is tightened, the share fits closely to the moldboard at every point of contact. It stays tight.

The nut is within easy reach from the rear of the plow. It isn't necessary to turn the plow over to change shares.

### Detachable Shin Piece

Chilled bottoms for the No. 210 Sulky are equipped with a chilled iron detachable shin piece, which can, when worn, be replaced with a new shin piece, enabling the operator to maintain a keen-cutting edge at small expense.

### Equipment

The Syracuse chilled jointer and John Deere three-horse steel evener are regular equipment.

The plain rolling coulter or John Deere combination rolling coulter and jointer can be furnished in place of the jointer when desired.

Both the Syracuse chilled jointer and plain rolling coulter can be used at the same time by attaching the rolling coulter just ahead of, and with blade slightly overlapping, the jointer point.

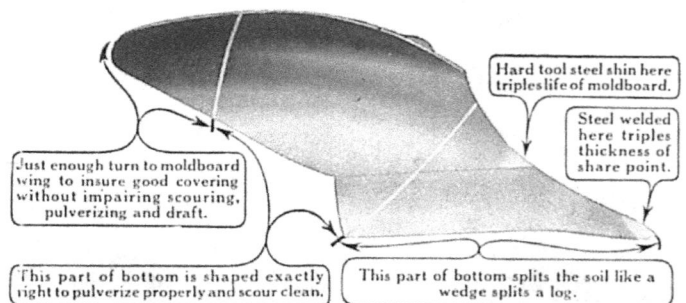

Hard tool steel shin here triples life of moldboard.

Steel welded here triples thickness of share point.

Just enough turn to moldboard wing to insure good covering without impairing scouring, pulverizing and draft.

This part of bottom is shaped exactly right to pulverize properly and scour clean.

This part of bottom splits the soil like a wedge splits a log.

*John Deere General-Purpose Steel Bottom.*

Chapter Seventeen

# In Closing

Back with the first edition I said this was a book not so much by me as by everyone, over a hundred year history, who had cared enough about the subject to leave a mark by way of innovation, design, recorded practice, evidenced style, advocated approach. We, dozens of us, all did this book together.

I worked to keep my quirky ego under control. I slipped a couple of places (i.e. plow fairies and the prevarications of politcos come to mind). I ask that I be forgiven. I was tired, too many late nights with this lump of a book. It's a big book. It could have been much bigger. There's stuff missing that some of you might think should have been in here. I'm okay with that. What this book represents, in the very best sense, is an extension of the life of this technology - these methods. That was our goal. And I believe we succeeded.

This book has been a true labor of love. In the beginning I had a clear picture of what this book needed to look, feel and act like. The finished volume you hold in your hand is quite different from those first mental pictures. Necessity and the sheer scale of the project have dictated an unpredictable form. The book has a life and identity of its own. And that is as it should be.

And this book belongs to each of you readers. More than just the physical ownership of the bound volume, you, by looking through and reading this material, now 'own' something of this craft.

And then there are all those people whose long hours of worry, work and sacrifice actually made this book. The first edition of this book thanks Kathy Blann, Suzanna Clarke, Amy Evers, Kristi Gilman-Miller, Betty Gilman, Marilyn Warren, and Lisa Booher. Plus this book thanks many individuals who contributed, loaned, and found materials for us. They include but are not limited to John & Heather Erskine, Helen Eden, Justin Miller, Bob Mishka, Gene Westberg, Frank Cina, The Denton Family, Judith Hoffman, and the forty some odd thousand readers of Small Farmer's Journal and our other publications and books.

And the Second Edition was made possible in large part by Eric Grutzmacher.

It perhaps does not need to be pointed out that, aside from an interesting report included in the chapter on Plowing Contests, there is nothing in this book specific to Canadian and British 'ploughing' & 'ploughs'. Please believe that this was not done to suggest relative merit. I know only slightly more than nothing about plows in those places and have very little material in my library. When the materials for this book started measuring in stacks feet high, decisions had to be made about what to include and what to trim. The Canadian and British plows, along with the extensive illustrated history of plows before 1890, were editing casualties. Perhaps therein lay two separate and important books for the future?

And as a final passing note on what some few of you find laughable - our notion that the use of animal power for small scale farming is viable for the twenty-first century - please give this some thought;

What do we measure at the end of a life worth living? Do we measure worldly possessions or how fast acres were covered? Or do we measure and value those times well spent in the presence of exhilaration, love, grace, peace, a satisfied spirit?

For thirty years I have chosen work horses as motive power for my farming. Some of the most rewarding moments in my life have been whole days in communion with my working partners and the sky and land we use as our little church. Those times spent resulted also in some produce that went for sale, to sustain animals or to sustain our family, but that was not the higher good, the higher value. That came from the communion.

If its a race you want I'll give you a race. You pick your way of working and I'll pick mine and we'll see who gets to heaven smiling. It's my wish we all do.

Here's wishing you a wonderful day's plowing.

*February 2000 and February 2018 - Lynn Ralph Miller*

*L. R. Miller plowing 1996 by Kristi Gilman-Miller*

## Bibliography

**Allis-Chalmers Manufacturing Co., Farm Practices,** *Plowing, Harrowing, Planting, Cultivating, Milwaukee, Wisconsin, Allis-Chalmers Manufacturing Co.*

**Avery, B. F. & Sons, Tillage Implements & Harvesting Machinery,** *Catalog 104, Louisville, Kentucky.*

**Bailey, L. H., Cyclopedia of American Agriculture,** *A Popular Survey of Agricultural Conditions, Practices and Ideals in the United States and Canada, New York, The MacMillan Company, 1907*

**Bricker, Garland A., Illustrated Lessons in Agriculture,** *Columbus, Ohio, The Kauffman-Lattimer Company, 1919.*

**Burkett, Charles William and Stevens, Frank Lincoln and Hill, Daniel Harvey, Agriculture for Beginners,** *Boston, Ginn & Company, 1903.*

**Carruthers, Ian and Rodriguez, Marc, Tools for Agriculture,** *A guide to appropriate equipment for smallholder farmers. London, Intermediate Technology Publications, Ltd., 1992*

**Cook, G. C, (The Late), Scranton, L. L. and McColly, H. F., Farm Mechanics Text and Handbook,** *Danville, Illinois, The Interstate Printers & Publishers, 1951.*

**Courtney, Wilshire S., The farmer's and mechanic's manual,** *New York, E. B. Treat & Co., 1869.*

**Crozier, William and Henderson, Peter, How the Farm Pays,** *The Experiences of Forty Years of Successful Farming and Gardening, New York, Peter Henderson & Co., 1902.*

**Curwen, E. Cecil and Hatt, Gudmund, Plough and Pasture,** *The Early History of Farming, New York, Henry Schuman, 1953.*

**Davis, Kary Cadmus, Ph.D., Productive Farming,** *Cornell University, Philadelphia and London, J. B. Lippincott Co., 1911.*

**Draft Horse & Mule Association of America, Inc., Horse & Mule Power** *in American Agriculture, Advantages of Farm Grown Power, St. Louis, Missouri, Draft Horse & Mule Association of America, 9700 Music Ave., St. Louis, MO 63123.*

**Evans, George Ewart, The Horse in the Furrow,** *London, Faber and Faber, 1960.*

**Farm Implement News, The, A Monthly Illustrated Newspaper Devoted to the Manufacture, Sale and Use of Agricultural Implements and their Kindred Interests, Vol. X, No. 1,** *Chicago Illinois, January, 1889.*

**Farm Tools, Inc., Farms Tools, Inc.,** *Mansfield, Ohio.*

**Faulkner, Edward H., Plowman's Folly,** *New York, Grosset & Dunlap, 1943.*

**Gustafson, A. F., Drainage, Plowing, Seedbed Preparation, and Cultivation in New York,** *Ithaca, New York, Cornell Extension Bulletin, New York State College of Agriculture at Cornell University, 1929*

**Horse Association of America, Horse-Mules, Power-Profit,** *Chicago, Illinois, Union Stock Yards, 1928.*

**International Correspondence Schools, Scranton, PA, Green Manure,** *Great Britain, International Textbook Company, 1936.*

**International Correspondence Schools, Scranton, PA, Soil Drainage,** *Great Britain, International Textbook Company, 1933.*

**International Correspondence Schools, Scranton, PA, Soils,** *Great Britain, International Textbook Company, 1935.*

**International Correspondence Schools, Scranton, PA, Tillage,** *London, International Textbook Company, 1937.*

**International Harvester Company, Instructions for Setting Up and Operating the P & O No. 2 Diamond Gang Plow,** *Chicago, Illinois, Harvester Press.*

**Jennings, B. A., Plow Adjustment,** *Cornell Extension Bulletin 381, Ithaca, NY, New York State College of Agriculture at Cornell University, 1937.*

**John Deere, John Deere Syracuse No. 210 Sulky Plow,** *The Light-Draft Plow that makes good work easy.*

**John Deere, The Operation, Care and Repair of Farm Machinery, Eighth Edition,** *Moline, Illinois, John Deere.*

**John Deere, The Care and Operation of Plows and Cultivators,** *Moline, Illinois, John Deere.*

**Johnson, Paul C., Farm Inventions in the Making of America,** *Des Moines, Iowa, Wallace-Homestead Book Company, PO Box BI, Des Moines, IA 50304, 1976.*

**Jones, Mack M., M.S., Shopwork on the Farm,** *New York and London, McGraw-Hill Book Company, Inc., 1945.*

**Lipscomb & Company, H. G., Wholesale Hardware and Associate Lines,** *Nashville, TN, H. G. Lipscomb & Company, 1913.*

**Malden, Walter J., Workman's Technical Instructor,** *Morton's Handbooks of the Farm, No. X., London, Vinton & Co., Ltd., 9, New Bridge Street, E.C., 1896.*

**McLennan, John, Ph.M., A Manual of Practical Farming,** *New York, The Macmillan Company, 1913.*

**Miller, L. R., Training Workhorses, Training Teamsters,** *Sisters, Oregon, Small Farmer's Journal, PO Box 1627, Sisters, OR 97759, 1-800-876-2893.*

**Miller, Lynn R., Work Horse Handbook,** *Sisters, Oregon, Small Farmer's Journal, PO Box 1627, Sisters, OR 97759, 1-800-876-2893.*

**Mitchell, Lewis & Staver Company, Catalog No. 16,** *Agricultural Implements, Wagons, Pumps Engines, Pipe, Wind Mills, etc., etc., Portland, Oregon, 1916*

**Moline Plow Co., Catalog No. 47,** *Moline Plows and other Flying Dutchman Farm Tools, Tillage, Seeding and Haying Machinery, Moline, Illinois, Moline Plow Co. Printing Department.*

**Moreland, Wallace S., A Practical Guide to Successful Farming,** *Garden City, New York, Halcyon House, 1943.*

**Oliver, James, Oliver Sulky and Gang Plows,** *South Bend, Indiana, Oliver Chilled Plow Works.*

**Oliver Farm Equipment, The Book of Oliver, Volume 1,** *Chicago, Illinois, Oliver Farm Equipment Sales Company, 1931.*

**Parker, Edward C., Field Management and Crop Rotation,** *Planning and Organizing Farms; Crop Rotation Systems; Soil Amendment with Fertilizers; Relation of Animal Husbandry to Soil Productivity; and Other Important Features of Farm Management, St. Paul, Minnesota, Webb Publishing Co., 1915.*

**Parlin & Orendorff Plow Co., Agricultural Implements,** *Catalog No. 59, Chicago, Illinois, Inland Press.*

**Parlin & Orendorff Plow Co., P & O Light Draft Plows,** *Catalog No. 73, Chicago, Illinois, Hammond Press, W. B. Conkey Company.*

**Ramsower, Harry C., Equipment for the Farm and the Farmstead,** *Boston, Ginn and Company, 1917.*

**Rock Island Plow Company, Rock Island Implements, Catalog No. 41,** *Rock Island, Illinois, Rock Island Plow Company.*

**Rock Island Plow Company, Making Farm Life Easier,** *Rock Island, Illinois, Rock Island Plow Company.*

**Sampson, H. O., B.Sc., B.S.A., Effective Farming, A** *Text-Book for American Schools, New York, The MacMillan Company, 1918.*

**Seymour, E. L. D., B. S. A., Farm Knowledge, A** *Complete Manual of Successful Farming Written by Recognized Authorities in All Parts of the Country; Based on Sound Principles and the Actual Experience of Real Farmers -- "The Farmer's Own Cyclopedia". Garden City, New York, Prepared Exclusively for Sears, Roebuck and Co. by Doubleday, Page & Company, 1918.*

**Shearer, Herbert A., Farm Mechanics,** *Machinery and its use to save hand labor on the farm, Chicago, Frederick J. Drake & Co., 1918.*

**Small Farmer's Journal,** *Lynn Miller, Editor, Sisters, Oregon, PO Box 1627, Sisters, OR 97759, 1-800-876-2893.*

**Smith, Harris Pearson, A.E., Farm Machinery and Equipment,** *New York and London, McGraw-Hill Book Company, Inc., 1937.*

**Taylor, Dr. W. E., Soil Culture and Modern Farm Methods,** *Minneapolis, Minnesota, Deere & Webber Company.*

**Thomas, John J., Farm Implements and Farm Machinery,** *and the Principles of their Construction and Use: with Explanations of the Laws of Motion and Force as Applied on the Farm, New York, Orange - Judd Company, 1883.*

**Thompson & Hoague Co., Farm Implements and Equipment,** *Concord, New Hampshire, Thompson & Hoague Co.*

**Todd, S. Edwards, The Young Farmer's Manual: or, How to Make Farming Pay,** *Giving Plain and*

*Bibliography
Continued*

*Practical Details of General Farm Management*, New York, F. W. Woodward, 37 Park Row, 1867.

**U. S. Department of Agriculture**, **Plowing Terraced Land**, *Leaflet No. 214, Washington, D.C., 1942.*

**U. S. Department of Agriculture**, **Laying Out Fields for Tractor Plowing**, *Farmers' Bulletin No. 1045, Washington, D.C.*

**U. S. Department of Agriculture**, **Plowing with**

**Moldboard Plows**, *Farmers' Bulletin No. 1690, Washington, D.C.*

**U. S. Department of Agriculture**, **Disk Plows and their Operation**, *Farmers' Bulletin No. 1992, Washington, D.C., 1948.*

**U. S. Department of Agriculture**, **Disk Plows**, *Farmers' Bulletin No. 2121, Washington, D.C. Revised, 1963.*

**U. S. Department of Agriculture**, **Moldboard Plows**, *Farmers' Bulletin No. 2172, Washington, D.C.*

**Vulcan Plow Co., The,** **Catalog No. 55**, *Evansville, Indiana, Keller-Crescent Co., 1928.*

**Wallace, Corcoran & Co.**, **Agricultural Implements, Engines, Wagons, Etc. Catalog No. 6,** *Illustrated Catalogue of Farm Implements and Machines, Barn*

Equipment, Wind Mills, Pumps, Grading Machinery, Sheep Shearing Machines, Wire Fencing, Wagons, Gas Engines, Etc., Portland, Oregon.

**Warren, G. F., Elements of Agriculture**, *New York, The MacMillan Company, 1914.*

**Weir, Wilbert Walter, BS(A), MS, Ph.D., Productive Soils,** *Chicago, Illinois, J.B. Lippincott Company, 1946.*

**Widtsoe, J. A. and Stewart, George, Western Agriculture**, *St. Paul, Minnesota, Webb Publishing Company, 1928.*

**Wilson, A. D., Minnesota Farmers' Institute Annual Number 21** - *1908, Minneapolis, Minnesota, Kimball-Storer Company, 1908.*

**Wilson, A. D. and Speer, Ray P., Minnesota Farmers' Institute Annual Number 25** - *1912, Minneapolis, Minnesota, Kimball-Storer Company, 1912.*

**Wright, Philip A., Old Farm Implements**, *North Pomfret, Vermont, David & Charles, 1961.*

# Photo Credits

# Index
## Horsedrawn Plows & Plowing

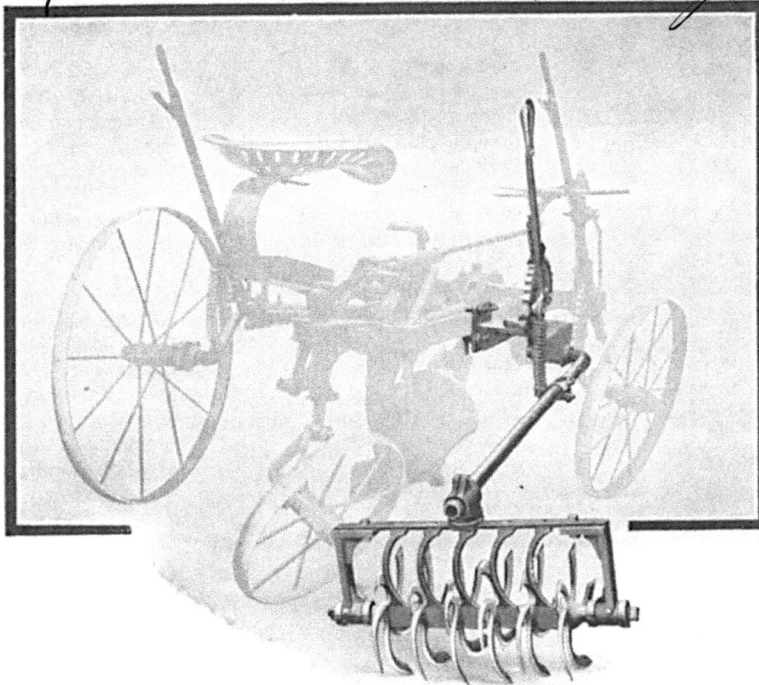

Index

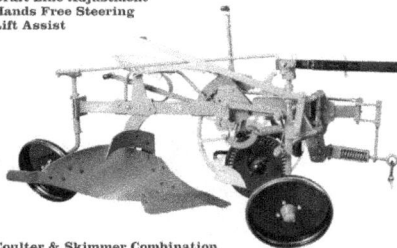

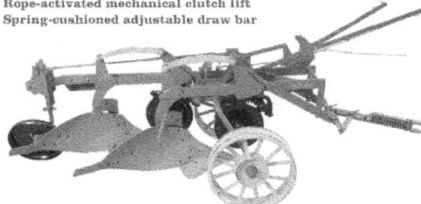

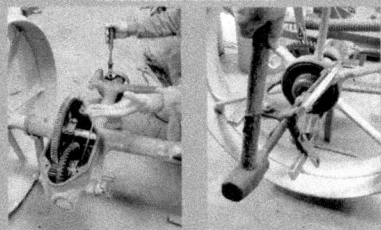